Sci

10X 3/02
7/02

APR 22 1997

HANDBOOK of AIR TOXICS

Sampling, Analysis, and Properties

Lawrence H. Keith
Mary M. Walker

3 1336 04203 9199

LEWIS PUBLISHERS
Boca Raton New York London Tokyo

Library of Congress Cataloging-in-Publication Data

Catalog record available from Library of Congress.

This book contains information obtained from authentic and highly regarded sources. Reprinted material is quoted with permission, and sources are indicated. A wide variety of references are listed. Reasonable efforts have been made to publish reliable data and information, but the author and the publisher cannot assume responsibility for the validity of all materials or for the consequences of their use.

Neither this book nor any part may be reproduced or transmitted in any form or by any means, electronic or mechanical, including photocopying, microfilming, and recording, or by any information storage or retrieval system, without prior permission in writing from the publisher.

CRC Press, Inc.'s consent does not extend to copying for general distribution, for promotion, for creating new works, or for resale. Specific permission must be obtained in writing from CRC Press for such copying.

Direct all inquiries to CRC Press, Inc., 2000 Corporate Blvd., N.W., Boca Raton, Florida 33431.

© 1995 by CRC Press, Inc.
Lewis Publishers is an imprint of CRC Press

No claim to original U.S. Government works
International Standard Book Number 1-56670-114-7
Printed in the United States of America 1 2 3 4 5 6 7 8 9 0
Printed on acid-free paper

NOTICE: Computerized Version and Copyright Permission

This printed book, ***Handbook of Air Toxics: Sampling, Analysis, and Properties***, is a derivative of a computerized (electronic) version which is produced in a fully searchable hypertext format using Microsoft® Windows™ under the title ***Instant EPA's Air Toxics***. The computerized version was produced in order to provide complete searching flexibility sometimes needed with a complex technical reference book such as this because of the limitations of a printed index. ***Instant EPA's Air Toxics*** is available from the publisher, Instant Reference Sources, Inc. (Telephone: 800-301-0359; FAX: 512-345-2386) and from other distributors. Instant Reference Sources, Inc. has given permission for the copyright of ***Handbook of Air Toxics: Sampling, Analysis, and Properties*** to Lewis Publishers.

ACKNOWLEDGMENTS

The editors thank Ms. Christine Baker of Mary Walker and Associates, Inc. for technical research and editorial assistance.

Dr. Lawrence H. Keith

Dr. Lawrence H. Keith is a Corporate Fellow at Radian Corporation in Austin, Texas. A pioneer in the development of methods for environmental analyses, Dr. Keith started his career at the EPA Water Research Laboratory in Athens, GA in the mid 1960's and then joined Radian Corporation in 1977. At Radian he is involved in helping to develop and market new environmental reference materials for use in trace level analyses and he also directs projects involving environmental sampling and analysis and analytical method development.

He is also past chairman of the American Chemical Society (ASC) Division of Environmental Chemistry and a current Executive Committee Member of that Division and the Editor of the Division Newsletter, *EnvirofACS*. He is also Chairman of the ACS Committee on Environmental Improvement Subcommittee on Monitoring and Analysis. Dr. Keith has lectured widely in the U.S. and other countries and also teaches short courses on Practical Environmental Sampling and Analysis.

He became interested in computerized publications in 1985 and founded Instant Reference Sources with his wife, Virginia Keith, in order to publish some of the earliest electronic books.

Mary M. Walker

Mary M. Walker is owner and president of Mary Walker & Associates, Inc., an environmental consulting an publications firm located in Lookout Mountain, TN. Consulting specialties of the firm include mediation of environmental conflict, environmental risk management, regulatory liaison, environmental policy development, and facilitation of public participation for businesses, industry, and government. Current publications include electronic databases in collaboration with Dr. Keith, and a regional environmental magazine called *EnviroLink*.

Ms. Walker is an environmental chemist and senior level Certified Hazardous Materials Manager with more than twenty-five years of varied experience. She has been an environmental chemist and computer systems designer for the Cancer Research Center of the Bowman Gray School of Medicine, the U.S. Environmental Protection Agency, and the Tennessee Valley Authority. As an environmental consultant over the past six years, she has counseled businesses and industries on environmental risk management and public interaction, and managed environmental site assessments.

Ms. Walker is currently Program Chair for the Division of Environmental Chemistry of the American Chemical Society, a member of the Executive Committee of the Environmental Division, and a member of the National Committee for Environmental Improvement of the American Chemical Society.

Preface

This book is a printed version of a four volume electronic database series for DOS, and a combined single volume hypertext version using Microsoft Windows. The DOS version was produced first (in 1991), before Windows was in such frequent use, and it is also published by Lewis Publishers. The Microsoft Windows version was produced in the fall of 1994 as a key member of the **Professional PC References / Windows** series in order to provide this important information in the new hyperlinked format. The Windows version is published by Instant Reference Sources, Inc. under the title ***Instant EPA's Air Toxics*** and is hyperlinked to other members of this series by chemical names and also by analytical methods.

These three publications thus provide a comprehensive reference work in any of the three formats from which someone may wish to access it (DOS, Windows, or as a printed and bound book). The publications are intended to help regulators, and the regulated community, meet the challenges of sampling and analysis, emissions reductions, and health and safety issues related to human exposure. Although this book contains the same information as the two electronic versions, it cannot be searched like them and thus lacks the power to find any and every occurrence of keywords singly or in combination with each other. However, it is a helpful compilation of data in its own way and we hope it will be a useful addition to your library of references.

Much of this data was obtained from the National Toxicology Program's Chemical Database and other referenced literature sources. For substances indicated in the Clean Air Act Amendments as compound groups, many hundreds of chemicals belong to each group. The Editors have selected examples of members of each group and provided data available from our reference sources for these examples. These representative compounds and elements are listed in Chapter 3. We have also provided the Clean Air Act's definitions or notes for the groups as appropriate. Many of the air toxics have several synonyms in common usage and Chapter 3 includes lists of common synonyms with each compound. The order of air toxics in this publication (and the Appendix of Air Toxics Names) varies slightly from the order given in the Clean Air Act Amendments, due to differences in conventions used to alphabetize the lists.

Reference numbers for the literature sources for most items of chemical and physical data are given in brackets following the data entry in this book. Where no reference number is given, the information comes from the National Toxicology Program's Chemical Database but was not further referenced in that source.

Larry Keith
Mary Walker

January, 1995

Table of Contents

Chapter 1

Sampling EPA's Air Toxics

Introduction

The Clean Air Act Amendments of 1990 renew and intensify national efforts to reduce air pollution at a level that surpasses all previous efforts in environmental regulation. The Amendments list 189 toxic air pollutants, including 172 individual compounds and 17 compound groups and require the Environmental Protection Agency to promulgate new control standards for the principal sources of such emissions. The designated chemicals were not previously regulated under the National Ambient Air Quality Standards, which applied to a small number of the most common pollutants. The Clean Air Act Amendments require immediate sampling and analysis to obtain data for the determination of emission factors which will be used to determine control measures. Chapter 2 provides summaries of the analytical methods which have been successfully tested or which are most likely to be successful in analyzing the air toxics.

It is important to emphasize that some of the methods referenced in this book are not yet validated for the air toxics, may not have been used with some of the air toxics yet, and must be modified and then verified before they can be used with confidence. However, all the methods in this book represent EPA's best estimates at this time for methods that can be used for sampling and analysis of the air toxics.

Because of the concern over the presence of low concentrations of these 189 hazardous air pollutants in air, existing sampling and analytical methods were reviewed in order to determine the applicability of these methods to the 189 hazardous air pollutants. For some of these hazardous air pollutants, the existing sampling and analytical methodologies are directly pertinent to the analyte of interest in emissions from stationary sources, and the performance parameters for the compounds have been completely defined. For some portion of the 189 hazardous air pollutants, data were available to substantiate, at least in part, performance of an analytical methodology, but no information was available in the literature to establish whether the compound could be sampled quantitatively. For a significant percentage of the list of 189 hazardous air pollutants, assignment to a sampling and analytical methodology could be made tentatively only on the basis of chemical or physical correspondence to other compounds which have been sampled and analyzed with the methodology. For most of the 189 hazardous air pollutants, sampling and analytical methodology available at the present time can serve only as screening methods to establish presence or absence under a given set of conditions. [1]

The ultimate goal of both the regulatory and the regulated acommunities is to have validated test methods available for any analyte which may require testing. However, the need for information is immediate since the Clean Air Act Amendments have been passed, and completely validated test methodologies are presently available for only a small number of analytes. The regulatory requirements will not allow the gathering of information to be deferred until validated test methodologies are available for each of the 189 hazardous air pollutants, nor are resources available for the Environmental Protection Agency to provide validated methodology for every possible analyte. It will therefore be necessary to use methodology which presently exists to gather screening information for the broadest possible number of analytes, until broad-based methods are validated for large numbers of analytes. The purpose of the catalog of methods presented here is to identify methods with the broadest possible applicability to the 189 hazardous air pollutants listed in the Clean Air Act Amendments of 1990. [1] Many single-analyte methods are already validated for specific source categories, but the focus of these method summaries is a compilation of methods that contain broad coverage.

Under an EPA program, the literature has been surveyed to determine the applicability of existing methodology. The primary goal was to use methodology applicable to the largest number of analytes listed in the Clean Air Act Amendments, with the full realization that a broad coverage by a methodology may require some sacrifice of sensitivity and accuracy. Specialized sampling and analytical methodologies may be available to apply to a single analyte, for example, or to one particular family of analytes. If a survey method is used instead of a specialized methodology, detection limits will be higher and there is a risk that trace quantities of the analyte in question will not be observed. On the other hand, if the survey methodology indicates the presence of significant quantities of an analyte for which a specialized methodology is available, an informed decision can be made on whether to use the specialized methodology in subsequent testing. The need for validation of proposed methodologies has been recognized. Assignments of analytes to a specific methodology have been made on the basis of previous validation studies and/or physical properties. [1]

Many of the sampling and analytical procedures need additional development and validation efforts to improve accuracy and precision. It should be stressed that a method which requires validation is not an inferior method; the method simply requires additional experimentation to define precision and bias. The performance of the method for a given analyte and source may be entirely acceptable, but until validation data are available, the user cannot know that the performance of the method will be acceptable prior to use. [1]

Nevertheless, many of the methods cited in this book are presently in use, regardless of validation status. Remote sensing techniques, such as Fourier Transform Infrared (FT-IR) and Fourier Transform Ultraviolet (FT-UV), have potential as emission inventory and air toxics factor measurement methods. Air toxics emission rate analysis using remote sensing measurement methods has not been validated at this time, and extensive field testing is required to establish accuracy and precision for these methodologies. Newer surface analytical techniques to measure semivolatile and condensable toxic air pollutants that may be associated with particulate materials (especially PM-10) are also being investigated. Present techniques for sampling and analysis require laborious, time-consuming, and costly extraction procedures to concentrate and analyze toxic organics on particles. Some of the newer instrumental techniques such as laser-induced mass analysis or time-of-flight techniques may prove feasible for future air toxic analysis. [1]

Note that some of the chemical names selected for the compounds listed in the Clean Air Act are not the best or commonly selected synonyms. Nevertheless, the names listed in the Appendix are the primary names used here because this is the way they are officially listed. Chapter 3 contains a field with synonyms that can be used to verify the more commonly used names of some of the air toxics.

Problems Unique to Sampling Air

Due to its variability, air is a unique medium when compared to all other matrices. Results from air samples collected at the same location but at different times can differ by orders of magnitude due to changes in predominant wind direction and in on-site activities. Even results from samples collected directly downwind of an emission source at different times within a day can vary by an order of magnitude or more due to changes in meteorological parameters and site emissions. With many measurements all but the most severe analytical errors will usually be overwhelmed by errors in extrapolating the data from a limited time period to a much longer time period. Therefore many of the statistics that are used to assess data from other matrices are not applicable to many air data.

In addition, many analytes of interest in air are rather reactive compounds, present in very low concentrations. They may be distributed between two or more phases, e.g., gases and solids or gases and liquids. Proper sampling requires a single phase if inferences are to be made from the chemical composition of samples. However, since most sampling devices immediately perturb the equilibrium of distribution of analytes between solid and gaseous phases, conclusions about phase distributions from most air samples is tenuous. [2]

Air can contain a large number of organic compounds. As a general rule, organic compounds are present in ambient air in very low concentrations. Sometimes there are higher concentrations inside and around certain industries. Low concentrations of analytes always complicate the process of obtaining samples for analysis. Analyzing low levels of organic compounds in the air generally requires large sample volumes using solid sorbents for vapor phase compounds and filters for solid (particulate) phase compounds. [2]

Large variations in analyte concentrations over short periods of time are common with air samples. Consequently, unrepresentativeness and variations from sampling may be large orders of magnitude from those involving laboratory analysis. In addition, air exposure concerns include short term acute and chronic risks in addition to the long term exposure risks characteristic of water, soil, and biota matrices. Thus, pollutant action levels at specified concentrations may be in minutes rather than days or years. [2]

Obtaining Representative Samples

Air is an extremely variable matrix in which concentrations can vary naturally by orders of magnitude due to changes in the on-site meteorology. Thus, for air, representative sampling is the degree to which a sample or group of samples characterizes the site conditions, both spatially and temporally.

Vapor pressure and polarity are two of the most important physical properties affecting compound sampling from air. Important contributors to obtaining representative air samples include the efficiency of the collection apparatus, integrity of the sample entering and being removed from the apparatus, location of the sampler, and timing of the collection. [2]

For stationary emission sources such as stacks, the sampling site and number of traverse points used for collection affect the quality of the data. Emission tests are based on the assumption that a sample obtained at a given point is representative; if this assumption is wrong it can cause serious problems. Since analyte concentration gradients can be very large, screening analyses are usually needed to measure their values prior to the main sampling effort. One method of assessing the magnitude of analyte concentration variations due to sampling techniques is to place two or more identical samplers next to one another. Assuming the analytical precision is known, the remaining variation (imprecision) in the data may usually be the result of sampling efforts. [2]

If samples are to retain representativeness, gaseous analytes must not react irreversibly with filter or sample container surfaces or with collected aerosol particles. Furthermore, collected aerosols must be handled in way that will minimize analyte release or retention; however, it is impossible to prevent losses of volatile analytes collected on filters. [3] Also, filters must not be contaminated with gaseous materials that can convert to aerosols on the filter media because this will cause false positives.

Sorbent sampling is fraught with contamination, interferences, analyte capture efficiency problems, and recovery efficiency problems. It is also a very important and useful technique when used with appropriate controls. Volatile compound breakthrough before sampling completion will cause errors in their measured concentrations; therefore, two sorbent cartridges are used in series. When no detectable concentration of the analytes of interest (above background) is measured in the second cartridge, no measurable breakthrough has occurred. Also, unknown reactions may occur on a sorbent during solvent extraction or during thermal desorption. Using two different sorbents to compare analytical results may provide documentation that unknown reactions did not occur. [2]

From the above it is clear that valid sampling methods, sample stability, appropriate blanks, and control samples are extremely important in verifying that representative samples of air have been taken. Never overlook these factors!

Selecting Sampling Devices

Solid sorbents are commonly used to collect volatile and semivolatile organic compounds (VOCs and SVOCs). Solid sorbent media may be divided into three categories: organic polymeric sorbents, inorganic sorbents, and carbon sorbents. Both the capture properties and the recovery process must be considered (and verified if they are not documented) when sampling with solid sorbents.

In the vapor phase nonvolatile organic compounds have negligible concentrations in the atmosphere. They are usually bound to solid particles and can be collected with filtration devices. However, high molecular weight hydrocarbons (above C_{10}) can still be in sufficient concentrations in a gaseous state to be relevant for ozone studies. [2]

Devices for Collecting Volatile Compounds

Volatile compounds are generally considered to be below the molecular weight of C_{10} hydrocarbons. They may also consist of oxygenated, sulfur-containing, or nitrogen-containing species. When ambient air samples for these types of compounds are taken, eliminate ozone and nitrogen oxides so that reactions with these oxidants will not take place in the sampling device. Alternatively, experiments in which the sampling conditions are duplicated can demonstrate that the pollutants of interest are inert to those oxidants over the course of sample collection and analysis. [2]

Steel canisters, with interior walls specially electropolished to prevent decomposition of the collected organic compounds, are commonly used for collecting ambient air samples. Canisters may be either evacuated in advance and then filled with air or pressurized using pumps with inert interior surfaces. [2] The most common sizes of canisters are 6 and 15 liters.

Tedlar/Teflon bags are another type of container used to collect whole air samples. These must always be checked for leaks before use. They are filled using pumps with inert interior surfaces or indirect pumping. With the latter technique, the deflated bag is placed in an airtight container which is then evacuated; the bag fills as it expands in the container. [2]

Samples of soil atmospheres are obtained from boring small diameter holes to a depth of three or more feet. Sampling ports, such as plastic tubes with perforations at their base, are installed in these holes. Air samples are extracted with peristaltic pumps and pumped into Tedlar/Teflon bags. Common problems include bore hole preparations and sample handling. [2]

Condensing volatile organic compounds from air into a cryogenic trap is an attractive alternative to sorbent sampling, particularly when it is combined with ambient air sampling in appropriate containers. The advantages of this technique are that it collects and measures a wide range of organic compounds, greatly reduces contamination problems, and obtains consistent recoveries. [2]

Impingers allow an air sample to bubble through a solution which collects a specific contaminant by either chemical reaction or absorption. For long sampling periods, the impinger may need to be kept in an ice bath to prevent the solution from evaporating during sampling. The sample is drawn through the impinger using a sampling pump; more elaborate sampling trains with multiple impingers may also be used. However, caution must be used to avoid spillage of impinger solutions during sample collection, storage, and shipping.

Passive monitors use a sorbent or reactive medium contained in a protected environment. The principle of diffusion of the analytes, from ambient air into the interior of the monitoring device, is used to calculate their concentrations in the ambient atmosphere from their sampling volumes. [2]

The most widely used procedure for sampling ambient air for volatile (VOC) and semivolatile (SVOC) compounds is to pass measured volumes of air (typically 2 to 100 L for most VOCs and 2 to 500 cubic meters for SVOCs) through a solid material which sorbs the component of interest. Unfortunately, solid sorbents, like most concentration techniques, are not compound specific, and unwanted compounds must be separated from the target compounds with which they are co-collected. In addition, breakthrough of the compounds of interest can occur, resulting in lower than actual measured concentrations. A second solid sorbent cartridge in series placed behind the first will verify that breakthrough did not occur when the analyte concentrations from the second cartridge do not statistically exceed those from the corresponding blanks. However, the sum of the quantities of analytes collected in the two cartridges should not be assumed to represent the total quantity of analytes in the air sample. Since most SVOCs are phase distributed, a particle filter is usually placed in front of the sorbent cartridge. However, the distribution of SVOCs between the filter and the sorbent will not accurately reflect atmospheric phase distributions, so the filter and sorbent are combined for analysis. [3]

High volume PS-1 samplers draw a sample through polyurethane foam (PUF), or a combination foam and XAD-2 resin plug, and a glass quartz filter at a rate of 5 to 10 cfm (cubic feet per minute) (or 2.4 to 4.7 liters per second). This system is often used for sampling low concentrations of semivolatiles, PCBs, pesticides, or chlorinated dioxins in ambient air.

Personal sampling pumps are reliable, portable sampling devices that draw air samples through a number of sampling media including resin tubes, impingers, and filters. Flow rates are usually adjustable from 0.1 to 4 liters per minute and can remain constant for up to 8 hours on one battery charge or continuously with an alternating current charger/converter.

After collecting the compounds of interest, solvent extraction or thermal desorption is used to recover organic compounds from the sorbent. Contaminants from the sorbents in these devices commonly interfere with subsequent analyses, thus they must be thoroughly precleaned before use, and blank samples must be analyzed to document background levels of potential interferants. [2]

The advantage of solvent extraction is that only an aliquot of the extract is analyzed. This makes replicates possible, and the concentration of analytes can be optimized. The principal disadvantage of solvent extraction is that only small aliquots (typically 0.1 to 1%) of the extract can be analyzed, resulting in higher levels of detection and quantitation than could be achieved if all the sample were used. Collecting large volumes of sample may counteract this disadvantage. [2]

The advantage of thermal desorption is that the entire sample is used in the analysis, so smaller air volumes suffice. This can be a disadvantage if replicates are needed. Another disadvantage is that the analytes may be decomposed by pyrolysis. Also, thermal desorption is not possible or practical with most semivolatile compounds. [2]

Sorbent Sampling Materials

Inorganic sorbents include silica gel, alumina, magnesium aluminum silicate (Florisil), and molecular sieves. These sorbents are more polar than the organic polymeric sorbents. They more efficiently collect polar organic compounds; unfortunately, water is also efficiently captured, and this causes rapid sorbent deactivation. [4] In addition, isomerization of organic compounds may be catalyzed by some of these sorbents. Therefore, inorganic sorbents are not often used for sampling volatile organic compounds.

Carbon cartridges consist of primary and backup sections. Ambient air is drawn through them so that the backup section verifies that breakthrough the analytes on the first section did not occur, showing that sample collection was quantitative. The adsorbed compounds must then be eluted, usually by solvent extraction (but sometimes using thermal desorption) and analyzed.

Activated carbon sorbents are relatively nonpolar compared to the inorganic sorbents, and water is less of a problem. However, water may still prevent analysis in some cases, especially where the relative humidity is high. The irreversible sorption of some organics to activated carbon and the potential reactions promoted by the high surface area of this sorbent also may cause problems. [4]

Organic polymeric sorbents include materials such as a porous polymeric resin of 2,4-diphenyl-p-phenylene oxide (Tenax-GC), styrenedivinylbenzene copolymer (XAD) resins, and polyurethane foam (PUF). These materials have the important feature of collecting minimal amounts of water in the sampling process.

Tenax sorbents are widely used in sampling ambient air for organic compounds (Tenax-GC, a polymeric solid sorbent originally developed as a support for gas chromatographic columns, is the most widely used solid sorbent). Tenax-GC has a low affinity for water and a high thermal stability, which permits the thermal desorption of collected volatile materials. [4] However, it is not very good for collection of some highly volatile organics.

Polyurethane foam (PUF) is often used for sampling semivolatile organic compounds such as pesticides and polynuclear aromatic hydrocarbons. One advantage is its low resistance to air flow; large air volumes can be passed through it. It also provides good desorption recoveries with common solvents. [2]

Devices for Collecting Nonvolatile Compounds

Nonvolatile compounds, i.e., elemental carbon and compounds with high molecular weights, have negligible concentrations in the vapor phase but these compounds are often bound to solid particles. They are sampled by the usual methods for atmospheric particles, e.g., filtration. High-volume samplers are commonly used as well. [4]

Filter packs are sometimes used as prefilters for sampling air. Teflon or similar prefilters that remove particles from the airstream may be followed by a nylon filter to remove nitric acid. A cellulose final filter treated with potassium or sodium carbonate or bicarbonate (and sometimes also with glycerol) can ensure a moist surface. [5]

PM-10 samplers collect particulates with a diameter of 10 microns or less from ambient air. Particulates of this size represent the respirable fraction, and thus are of special significance. PM-10 samplers can be high volume or low volume. The high volume sampler operates in the same manner as the TSP sampler at a constant flow rate of 40 cfm drawing the sample through a special impactor head which collects particulates 10 microns or less. The particulate is collected on an 8 by 10 inch filter. The low volume sampler or low volume PM-10 sampler operates at a rate of approximately 17 L/min (liters per minute). The flow must remain constant through the impactor head to maintain the 10 micron cut-off point. The low volume PM-10 collects its sample on 37-mm Teflon filters.

Influence of Meterology on Sampling Air

The primary meteorological effects that must be considered when obtaining air samples are wind direction, wind speed, temperature, atmospheric stability, atmospheric pressure, and precipitation. Wind direction is the most important factor, and it must be constantly documented during air sampling activities. A change in wind direction can provide large variations in the amounts of air pollutants collected in just a short period of time. [2]

Wind speed may affect the volatilization rate of contaminants from liquid sources and also the concentrations of pollutants downwind. As the wind speed increases, it may increase the volatilization rate of some liquids; however, increased wind speed also usually dilutes the concentration of vaporized pollutants downwind of a source. On the other hand, greater wind speed may increase the concentration of nonvolatile contaminants sorbed to particulates such as soil and dust. This is because, under windy conditions, large particulates and greater quantities of them can be transported from contaminated sites. [2]

Another factor that increases volatilization is higher temperature. Both solar radiation and air temperatures must be considered, but in general, solar radiation usually has the larger influence causing the volatilization of liquids with high vapor pressure. [2]

Atmospheric instability relates to vertical motions of the air. In unstable atmospheric conditions, dispersion of contaminants in the air increase throughout various vertical levels. Downwind contaminant concentrations are usually higher when stable atmospheric conditions exist. [2]

Atmospheric pressure affects the migration of gases through and out of landfills. Volatile contaminants from landfills may be released at higher rates during periods of low atmospheric pressure changes. The reverse may also occur, so slower volatilization rates may be observed when high atmospheric pressure are in effect. However, if significant lag times are associated with these pressure changes they may not be very noticeable, and each landfill must be considered separately. [2]

Precipitation decreases overall airborne contaminants, although contaminant concentrations will usually be highest in the initial precipitation samples. During precipitation events, contaminants are removed from the air when particulate matter is physically carried down with the precipitate and gaseous contaminants are dissolved in it. Particulate matter transported by the wind is also usually insignificant when the soil or other particulates are wet. Also, volatile contaminants emitted from wet soils may be at much lower rates than from dry soils; however, the humidity of the air at less than 100% generally has little effect on either volatilization or transport of contaminants in air. [2]

Influence of Topography on Sampling Air

Mountains, hills, valleys, lakes, and seas can significantly affect the wind direction and the amount of mixing or dispersion of contaminants in the air. Within valleys, the air generally flows along the axis of the valley with a bi-directional distribution. During evenings and nights, colder air from high slopes may drain down into a valley causing a thermal inversion of the air. This increases concentrations of contaminants from valley sources because of decreased vertical dispersion. If the valley itself has a slope, the colder air will tend to flow down that slope and carry the contaminants with it. The reverse situation can also occur in the morning and daylight hours, but this phenomenon is not as frequent; the up-valley and up-slope winds are not as strong as the cooler down-valley and down-slope winds. [2]

Near the sea coast and large lakes, temperature differences resulting from daylight and nighttime effects also cause bi-directional wind changes. Generally, during the day, the wind will move from the cooler body of water toward the land and then exhibit an upward motion over the land. The reverse situation usually occurs at night when rapid cooling from the land causes air over the land to become cooler than air over the water. These "land breezes" usually are not as strong as the sea or lake breezes. [2]

Chapter 2

Analysis of EPA's Air Toxics

Introduction

The method summaries provided here can help you decide which method may be best for your specific needs. You will often find more than one method that may be applicable for any one of the 189 Clean Air Act analytes; then the question becomes which of them is best for your needs. Criteria for making these decisions include:

- instrumentation available,
- specificity of the detector needed,
- sensitivity desired, and
- potential interferences.

One of the first considerations must be availability of both sampling and analytical instrumentation. If, for example, the method selected requires a mass spectrometer for analysis and one is not available, then clearly either another method or another laboratory must be selected. As another example, specific gas chromatographic (GC) columns may be required and, if they are not available, then the choice becomes one of delaying analyses until the required column can be obtained or using another column on hand and verifying that the analytes of interest separate from each other and any sample interferences.

Some methods have better specificity than others and this will affect the degree of confidence in the identification of specific analytes as well as the possibility of false positive detections. Note that there is an important difference between detection and identification. Detection involves determining whether a signal produced by using a specific method is from the sample instead of being an artifact from instrument noise, background contamination or other types of interferences. A signal that meets detection criteria and that has the characteristics of the analyte of interest (for example, a peak in a gas chromatogram at the correct retention time for that analyte) is often assumed to also identify that analyte. This is not necessarily true; identification usually requires multiple identification characteristics for an assumed identification to be valid. In the illustration above, repeating the analyses using a different GC column so that a second and different chromatographic retention time of the analyte can be compared to a standard or using a mass spectrometer as a selective detector are examples of acceptable methods for confirming an analyte's identity. The exception to this rule may occur when pollutants are simply being monitored. In monitoring situations the identity of the pollutants of interest have been previously established and it has also been shown that their signals using the method of choice are free from all interferences (including contamination).

Sensitivity can be an important consideration when concentration levels of the analytes of interest are likely to be very low. Sensitivity varies among the methods in this database for most of the analytes. Therefore, detector selection is especially important for organic compounds. Analytical sensitivity and selectivity characteristics must be weighed versus one another in making these decisions. An expert system called the GC Advisor has been written based on rules deduced from answers to questions about the user's needs and it also summarizes major advantages and disadvantages of each of the candidate detectors. The GC Advisor is accompanied by a second expert system called The QC Advisor; it provides advice on what types of QC samples to collect based on your needs for estimating bias and/or precision from laboratory and/or sampling sources. [6]

References for Chapters 1 and 2

1. "Screening Methods for the Development of Air Toxics Emission Factors", EPA-450/4-91-021, Inventory Guidance and Evaluation Section, Emission Inventory Branch, TSD, U.S. EPA, Research Triangle Park, NC, 27711, September, 1991.

2. Keith, L.H., Chapter 4, "Sampling Air Matrices", in Environmental Sampling and Analysis - A Practical Guide, pp 41-50, Lewis Publishers, Inc., 121 South Main St., Chelsea, MI, 48118, 1991.

3. Lewis, R.G., "Problems Associated With Sampling for Semivolatile Organic Chemicals in Air", Proceedings, 1986 EPA/APCA Symposium on Measurement of Toxic Air Pollutants," APCA Special Publication VIP-7 Air and Waste Management Association, Pittsburgh, PA, pp. 134-145, 1986.

4. Clements, J.B. and R.G. Lewis, "Sampling for Organic Compounds", in Principles of Environmental Sampling, L.H. Keith, Ed., American Chemical Society, 1155 16th St., N.W., Washington, DC., 20460, p. 287, 1988.

5. Hicks, B.B., T.P. Meyers, and D.D. Baldocchi, "Aerometric Measurement Requirements for Quantifying Dry Deposition", in Principles of Environmental Sampling, L.H. Keith, Ed., American Chemical Society, 1155 16th St., N.W., Washington, DC., 20460, p. 297, 1988.

6. Keith, L. H., "The GC Advisor and the QC Advisor," Lewis Publishers, Inc., 121 South Main St., Chelsea, MI, 48118, 1991.

Analytical Method Summaries

The method summaries below all include methods that have been reviewed by EPA and which are considered to be potentially useable for analysis of EPA's Clean Air Act Air Toxics. Some are valid methods for these substances and others have not yet been validated for use with them. Most of these are EPA methods but some are from NIOSH, California Air Resources Board (CARB) and one is from OSHA.

EPA METHOD 0010 - Sampling For Semivolatile Organic Compounds

Modified Method 5 Sampling Train

REFERENCE:

Test Methods for Evaluating Solid Waste, Third Edition. Report No. SW-846. U.S. Environmental Protection Agency, Office of Solid Waste and Emergency Response. Washington, DC: 1986.

1.0 SCOPE AND APPLICATION

EPA Method 0010 is used to determine the Destruction and Removal Efficiency (DRE) of semivolatile principal organic hazardous constituents (POHCs) from incineration systems. This method may also be used to determine particulate emission rates from stationary sources, as per EPA Method 5 (see the Reference Table included in the description of Method 0010 in SW-846). Method 0010 has been applied to semivolatile compounds, including polychlorinated biphenyls (PCBs), chlorinated dibenzodioxins and dibenzofurans, polycyclic organic matter, and other semivolatile organic compounds.

2.0 SUMMARY OF METHOD

Gaseous and particulate pollutants are withdrawn isokinetically from an emission source and collected in a multicomponent sampling train. Principal components of the train include a high-efficiency glass- or quartz-fiber filter and a packed bed of porous polymeric adsorbent resin (typically XAD-2 ® or polyurethane foam (PUF) for PCBs). The filter is used to collect organic-laden particulate materials and the porous polymeric resin to adsorb semivolatile organic species (compounds with a boiling point above 100°C). Comprehensive chemical analyses, using a variety of applicable analytical methodologies, are conducted to determine the identity and concentration of the organic materials.

3.0 INTERFERENCES

Oxides of nitrogen (NOx) are possible interferents in the determination of certain water-soluble compounds such as 1,4-dioxane, phenol, and urethane. Reaction of these compounds with NOx in the presence of moisture will reduce their concentration. Other chemical reactions are possible, since SO_2, O_3, and other reactive species may be present in the emissions. Other problems that could produce a positive or a negative bias, depending upon the compounds of interest, are:

- Stability of the compound of interest in methylene chloride, since at least one of the common analytical methods relies upon methylene chloride extraction of the sorbent media from the sampling train;

- Formation of water-soluble organic salts on the resin in the presence of moisture (compounds that form water-soluble salts can be recovered by appropriate control of pH during the extraction process); and

- Solvent-extraction efficiency of water-soluble compounds from aqueous media.

When gas chromatography/mass spectrometry is used as the analytical technique, compounds that coelute chromatographically can frequently be deconvoluted if their mass spectra are different. Using two or more ions per compound in quantitative analysis can overcome interference at one mass; however, if the concentration of the compound of interest is sufficient to saturate the detector at a given mass, an alternative mass MAY NOT be selected. In this case, the extract must be diluted to bring the concentration of the compound of interest into the calibration range in order to obtain accurate quantitative analysis.

4.0 METHOD TARGET COMPOUNDS

Method 0010 is an extremely powerful and versatile methodology. A single analytical methodology cannot simultaneously address all semivolatile compounds for which Method 0010 might serve as a sampling methodology. Method detection limits are a function of volume sampled, and the volume that is sampled will also vary according to the analyte.

5.0 APPLICABLE CAA AMENDMENTS POLLUTANTS *

* The appropriate analytical methodology for each pollutant is shown in parenthesis.

acetamide (Method 8270)
acetophenone (Method 8270; 8270 target)
2-acetylaminofluorene (Method 8270)
acrylamide (Method 8270)

acrylic acid (Method 8270 with derivatization)
4-aminobiphenyl (Method 8270; 8270 target)
aniline (Method 8270; 8270 target)
o-anisidine (Method 8270)
benzidine (Method 8270; 8270 target)
benzotrichloride (Method 8270)
benzyl chloride (Method 8270)
biphenyl (Method 8270; also Method 8310)
bis (2-ethylhexyl) phthalate (Method 8270; 8270 target)
bromoform (Method 8270)
caprolactam (Method 632)
captan (Method 8270)
carbaryl (Method 632)
catechol (Method 8270)
chloramben (Method 515/615)
chlordane (Method 8270; 8270 target)
chloroacetic acid (Method 8270 with derivatization)
2-chloroacetophenone (Method 8270)
chlorobenzene (Method 8270)
chlorobenzilate (Method 8270)
o-cresol (Method 8270; 8270 target)
m-cresol (Method 8270)
o-cresol (Method 8270; 8270 target)
cresylic acid (mixture of cresol isomers; Method 8270)
cumene (Method 8270)
2,4-D salts and esters (2,4-D, Method 8270; 8270 target; salts and esters, Method 515/615)
DDE (Method 8270; 8270 target)
dibenzofurans (Method 8280, Method 8290)
1,2-dibromo-3-chloropropane (Method 8270)
dibutyl phthalate (Method 8270; 8270 target)
1,4-dichlorobenzene (Method 8270; 8270 target)
3,3'-dichlorobenzidine (Method 8270; 8270 target)
dichloroethyl ether (Method 8270)
1,3-dichloropropene (Method 8270)
dichlorvos (Method 8270)
diethanolamine (Method 8270, possibly Method 632)
N,N-diethylaniline (Method 8270)
diethyl sulfate (Method 8270)
3,3'-dimethoxybenzidine (Method 8270)
dimethylaminoazobenzene (Method 8270; 8270 target)
3,3'-dimethylbenzidine (Method 8270)
dimethyl carbamoyl chloride (Method 531)
dimethyl formamide (Method 8270)
dimethyl phthalate (Method 8270; 8270 target)

dimethyl sulfate (Method 8270)
4,6-dinitro-o-cresol and salts (Method 8270 for
4,6-dinitro-o-cresol; Method 515/615 for
4,6-dinitro-o-cresol and salts)
2,4-dinitrophenol (Method 8270; 8270 target)
2,4-dinitrotoluene (Method 8270; Method 8270 target)
1,4-dioxane (Method 8270)
1,2-diphenylhydrazine (Method 8270; 8270 target)
epichlorohydrin (Method 8270)
ethylbenzene (Method 8270)
ethyl carbamate (Method 632)
ethylene dibromide (Method 8270)
ethylene glycol (Method 8270)
ethylene thiourea (Method 632)
glycol ethers (Method 8270)
heptachlor (Method 8270; 8270 target)
hexachlorobenzene (Method 8270; 8270 target)
hexachlorobutadiene (Method 8270; 8270 target)
hexachlorocyclopentadiene (Method 8270; 8270 target)
hexachloroethane (Method 8270; 8270 target)
hexamethylene-1,6-diisocyanate (Method 8270)
hexamethylphosphoramide (Method 632)
hydroquinone (Method 8270)
isophorone (Method 8270; 8270 target)
lindane (Method 8270)
maleic anhydride (Method 8270)
methoxychlor (Method 8270; 8270 target)
methyl isobutyl ketone (Method 8270)
4,4'-methylene bis(2-chloroaniline) (Method 8270)
methylene diphenyl isocyanate (Method 8270 or Method 632)
4,4'-methylenedianiline (Method 8270)
naphthalene (Method 8270; 8270 target; also Method 8310)
nitrobenzene (Method 8270; 8270 target)
4-nitrobiphenyl (Method 8270; also Method 8310)
4-nitrophenol (Method 8270; 8270 target)
2-nitropropane (Method 8270)
N-nitroso-N-methylurea (Method 8270, possibly Method 632)
N-nitrosodimethylamine (Method 8270; 8270 target)
N-nitrosomorpholine (Method 8270, possibly Method 632)
parathion (Method 8270)
pentachloronitrobenzene (Method 8270; 8270 target)
pentachlorophenol (Method 8270; 8270 target)
phenol (Method 8270; 8270 target)
p-phenylenediamine (Method 8270)
phthalic anhydride (Method 8270)

polychlorinated biphenyls (Method 8270 with very high detection
limits; Method 680)
1,3-propane sultone (Method 8270)
beta-propiolactone (Method 8270)
propoxur (Method 632)
quinoline (Method 8270)
quinone (Method 8270)
styrene (Method 8270)
styrene oxide (Method 8270)
2,3,7,8-tetrachlorodibenzodioxin (Method 8280, Method 8290)
1,1,2,2-tetrachloroethane (Method 8270)
tetrachloroethylene (Method 8270)
toluene (Method 8270)
2,4-toluenediamine (Method 8270, Method 632)
2,4-toluenediisocyanate (Method 8270, Method 632)
o-toluidine (Method 8270)
toxaphene (Method 8270; 8270 target)
1,2,4-trichlorobenzene (Method 8270)
1,1,2-trichloroethane (Method 8270)
2,4,5-trichlorophenol (Method 8270; 8270 target)
2,4,6-trichlorophenol (Method 8270; 8270 target)
trifluralin (Method 8270)
xylenes: o-xylene, m-xylene, p-xylene (Method 8270)
polycyclic organic matter (Method 8270; Method 8310)

EPA METHOD 0011 - Sampling & Analysis For Aldehydes & Ketones

DRAFT METHOD Sampling for Formaldehyde Emissions from Stationary Sources

REFERENCE:

Test Methods for Evaluating Solid Waste, Third Edition. Report No. SW-846. U.S. Environmental Protection Agency, Office of Solid Waste and Emergency Response. Washington, DC: 1986.

1.0 SCOPE AND APPLICATION

EPA Draft Method 0011 is used to determine the Destruction and Removal Efficiency (DRE) of formaldehyde. Although this methodology has been applied specifically to formaldehyde, many laboratories have extended the application to other aldehydes and ketones. Compounds derivatized with 2,4-dinitrophenylhydrazine (DNPH) can be detected in concentrations as low as 6.4 x 10-8 lbs/cu ft (1.8 ppbv) in 40 cu ft of stack gas sampled over a 1-hour period.

2.0 SUMMARY OF METHOD

Gaseous and particulate pollutants are withdrawn isokinetically from an emission source and are collected in aqueous acidic DNPH solution. Formaldehyde (and other aldehydes and ketones) present in the emissions reacts with DNPH to form the dinitrophenylhydrazone derivative. The dinitrophenylhydrazone derivative is extracted, solvent-exchanged, concentrated, and then analyzed by high-performance liquid chromatography.

3.0 INTERFERENCES

A chromatographic method is subject to interference from coeluting components of the matrix. A decomposition product of DNPH, 2,4-dinitroaniline, can be an analytical interferent if concentrations are high. High concentrations of oxygenated compounds, especially acetone, that have the same retention time or nearly the same retention time as the dinitrophenylhydrazone of formaldehyde and that also absorb at 360 nm will interfere with the analysis. Contamination of the aqueous acidic DNPH reagent with formaldehyde and 2,4-dinitroaniline is frequently encountered. The reagent must be prepared within five days of field use and must be stored in an uncontaminated environment both before and after sampling in order to minimize blank problems. Because acetone is ubiquitous in laboratory and field operations, some level of acetone contamination is unavoidable; however, it must be minimized to the extent possible.

4.0 METHOD TARGET COMPOUNDS

Aldehydes
Ketones

5.0 APPLICABLE CLEAN AIR ACT LIST COMPOUNDS

acetophenone
acrolein
2-chloroacetophenone
formaldehyde
isophorone
methyl ethyl ketone
methyl isobutyl ketone
propionaldehyde
quinone

EPA METHOD 0012 - Sampling Metals Using A Sampling Train

Multi-Metal Train

REFERENCE:

Test Methods for Evaluating Solid Waste, Third Edition. Report No. SW-846. U.S. Environmental Protection Agency, Office of Solid Waste and Emergency Response. Washington, DC: 1986.

1.0 SCOPE AND APPLICATION

EPA Method 0012 is used to determine metals from municipal waste incinerators and similar combustion processes. In addition, the method may be modified to determine particulate emissions.

2.0 SUMMARY OF METHOD

Method 0012 consists of a stack sampling train and a number of analysis techniques for the recovered samples. The source sample is withdrawn isokinetically from the stack through a heated probe. The majority of the particulate emissions are collected on a filter in a heated filter holder that is located after the probe outside the stack. The gaseous emissions are collected in a series of chilled impingers located after the filter. Two of the impingers contain dilute nitric acid in hydrogen peroxide, and one or two contain acidic potassium permanganate solution.

Sample train components are recovered and digested in acid as separate fractions. The impingers and digested probe and filter solutions are analyzed for mercury by cold vapor atomic absorption spectroscopy. All the sampling train components, except the permanganate solution, can be analyzed by inductively coupled argon plasma emission spectroscopy (ICAP) or atomic absorption spectroscopy (AAS). Graphite furnace atomic spectroscopy (GFAAS) is used for analysis of arsenic, cadmium, lead, antimony, selemium, and thallium if greater analytic sensitivity is required. For convenience, AAS may be used to analyze for all metals and/or a combined sample in a single analytical determination, provided the resulting detection limits meet the data quality objectives of the testing program. Detection limits for the target compounds range from 0.11 to 27 $\mu g/m^3$ for all methods except GFAAS. For selected metals, detection limits range from 0.03 to 0.8 $\mu g/m^3$ with GFAAS.

3.0 INTERFERENCES

Organic interferences will be removed with complete digestion. Use of the train to quantify particulate emissions may impact on mercury determinations, although data acquired to date show that less than 2 percent of the mercury is lost in the particulate catch.

Iron can be a spectral interference during analysis of arsenic, chromium, and cadmium by ICAP. Aluminum can be a spectral interference during the analysis of arsenic and lead by ICAP. Matrix modifiers should be used to eliminate interferences in all GFAAS analyses. Refer to EPA Method 6010 (SW-846) for details on potential interferences.

4.0 METHOD TARGET COMPOUNDS

arsenic*
barium**
beryllium*
cadmium*
chromium*
copper**
lead*
manganese**
mercury*
nickel*
phosphorus**
selenium**
silver**
thallium**
zinc*

*Primary target compounds

**Secondary target compounds

5.0 APPLICABLE CAA AMENDMENTS POLLUTANTS

NOTE: Draft Method 0012 analytical methods do not speciate inorganic compounds.

antimony compounds
arsenic compounds
beryllium compounds
cadmium compounds
chromium compounds
cobalt compounds
lead compounds
manganese compounds
mercury compounds
nickel compounds
selenium compounds
titanium tetrachloride

EPA METHOD 0030 - Sampling Volatile Organics Using A Sampling Train (VOST)

Volatile Organic Sampling Train

REFERENCE:

Test Methods for Evaluating Solid Waste, Third Edition. Report No. SW-846. U.S. Environmental Protection Agency, Office of Solid Waste and Emergency Response. Washington, DC: 1986.

1.0 SCOPE AND APPLICATION

EPA Method 0030 is used to calculate destruction and removal efficiency (DRE) of volatile principal organic hazardous constituents (POHCs) from stack gas effluents of hazardous waste incinerators, and enable a determination that DRE values are equal to or greater than 99.99 percent. For the purpose of definition, volatile POHCs are those POHCs with boiling points less than 100°C. If the boiling point of a POHC of interest is less than 30°C, the POHC may break through the sorbent under the conditions of the sample collection procedure.

2.0 SUMMARY OF METHOD

A 20-L sample of effluent gas is withdrawn from an emission source at a flow rate of 1 L/min, using a glass-lined probe and a volatile organic sampling train (VOST). The gas stream is cooled to 20°C by passage through a water-cooled condenser and volatile POHCs are collected on a pair of sorbent resin traps. Liquid condensate is collected in an impinger placed between the two resin traps. The first resin trap (front trap) contains approximately 1.6 g Tenax ® and the second trap (back trap) contains approximately 1 g each of Tenax ® and petroleum-based charcoal (SKC lot 104 or equivalent), 3:1 by volume.

An alternative set of conditions for sample collection has been used. This method involves collecting a sample volume of 20 L or less at a reduced flow rate. (Operation of the VOST under these conditions has been referred to as SLO-VOST.) This method has been used to collect 5 L of sample (250 mL/min for 20 min) or 20 L of sample (500 mL/min for 40 min) on each pair of sorbent cartridges. Smaller sample volumes collected at lower flow rates should be considered when the boiling points of the POHCs of interest are below 35°C.

3.0 INTERFERENCES

The sensitivity of this method depends on the level of interferences in the sample and the presence of detectable levels of volatile POHCs in the blanks. The target detection limit of this method is 0.1 μg/m^3 (ng/L) of flue gas, to permit calculation of a DRE equal to or greater than 99.99 percent for volatile POHCs that may be present in the waste stream at 100 ppm. The upper end of the range of applicability of this method is limited by breakthrough of the volatile POHCs on the sorbent traps used to collect the sample. Laboratory development data have demonstrated a range of 0.1 to 100 μg/m^3 (ng/L) for selected volatile POHCs collected on a pair of sorbent traps using a total sample volume of 20 L or less. Interferences arise primarily from background contamination of sorbent traps prior to or after use in sample collection. Many interferences can be due to exposure of the sorbent materials to solvent vapors prior to assembly and exposure to significant concentrations of volatile POHCs in the ambient air at hazardous waste incinerator sites. A sufficiently high background level in the source (for example, a THC level above 100 ppm) can make trace analysis in the samples impossible.

4.0 METHOD TARGET COMPOUNDS

None

5.0 APPLICABLE CLEAN AIR ACT LIST OF CHEMICALS

acetaldehyde++
acetonitrile++
acrolein++
acrylonitrile++
allyl chloride*
benzene*
bromoform***
carbon disulfide*
carbon tetrachloride*
chlorobenzene (validated)***
chloroform*
chloromethyl methyl ether++
chloroprene*
cumene***
1,4-dichlorobenzene***
1,3-dichloropropene***
1,1-dimethylhydrazine++
1,4-dioxane++
epichlorohydrin++
1,2-epoxybutane++
ethyl acrylate++
ethylbenzene***
ethyl chloride**
ethylene dibromide***
ethylene dichloride*
ethylidene dichloride*
hexane*
methyl bromide**
methyl chloride**
methyl chloroform*
methyl ethyl ketone++
methyl hydrazine++
methyl iodide*
methyl isobutyl ketone++
methyl isocyanate++
methyl methacrylate++
methyl tert-butyl ether++
methylene chloride*
propionaldehyde++
propylene dichloride*
propylene oxide++
1,2-propylene imine++
styrene***
1,1,2,2-tetrachloroethane***
tetrachloroethylene (validated)***
toluene (validated)***

1,1,2-trichloroethane***
trichloroethylene*
2,2,4-trimethylpentane*
vinyl acetate++
vinyl bromide**
vinyl chloride (validated)**
vinylidene chloride*
xylenes***
o-xylene***
m-xylene***
p-xylene***

* Boiling point between 35°C and 100°C. Should work well with this methodology.

** Special precautions must be taken to avoid breakthrough when these compounds are analytes.

*** Boiling point above 100°C. Can be observed using VOST methodology but VOST should not be used to provide quantitative data unless specific validation is performed.

++ Polar and/or water soluble. In general, perform well in the VOST methodology. Recovery can be improved by modifying the standard purge and trap analytical method by elevating the purge temperature to 60°C and adding 1 g of sodium chloride or sodium sulfate to condensate before purging (see EPA-600/8-87-008). With modified analytical methodology, the VOST methodology may be applicable to these compounds (validation required before use of the methodology to generate regulatory data).

EPA METHOD 0050 - Sampling HCl & Chlorine By Isokinetic Sampling Train

DRAFT METHOD Isokinetic HCl/Cl_2 Emission Sampling Train

REFERENCE:

Test Methods for Evaluating Solid Waste, Third Edition. Report No. SW-846. U. S. Environmental Protection Agency, Office of Solid Waste and Emergency Response. Washington, DC: 1986.

1.0 SCOPE AND APPLICATION

EPA Draft Method 0050 is used to collect hydrogen chloride (HCl) and chlorine (Cl_2) in stack gas emission samples from hazardous waste incinerators, municipal waste combustors, and boilers and industrial furnaces. Collected samples are analyzed using EPA Method 9057. Method 0050 collects the emission sample isokinetically and is therefore, particularly suited for sampling at sources, such as those controlled by wet scrubbers, that emit acid particulate matter (e.g., HCl dissolved in water droplets).

Method 0050 is not acceptable for demonstrating compliance with HCl emission standards less than 20 ppm. This method may also be used to collect samples for subsequent determination of particulate emissions (EPA Method 5).

2.0 SUMMARY OF METHOD

Gaseous and particulate pollutants are withdrawn from an emission source and are collected in an optional cyclone, on a filter, and in absorbing solutions. The cyclone collects any liquid droplets. The cyclone is not required if the source emissions do not contain liquid droplets (See EPA Method 0051). The Teflon ® mat or quartz-fiber filter collects other particulate matter, including chloride salts. Acidic and alkaline absorbing solutions collect gaseous HCl and Cl_2, respectively. Following sampling of emissions containing liquid droplets, any HCl/Cl_2 dissolved in the liquid in the cyclone and/or on the filter is vaporized and ultimately collected in the impingers by pulling Ascarite IIþ-conditioned ambient air through the sampling train. In the acidified water absorbing solution, the HCl gas is solubilized and forms chloride ions. The Cl_2 gas present in the emissions has a very low solubility in acidified water and passes through to the alkaline absorbing solution where it undergoes hydrolysis to form a proton (H+), chloride ions, and hypochlorous acid (HClO). Chloride ions in the separate solutions are measured by ion chromatography (EPA Method 9057). If desired, the particulate matter recovered from the filter and the probe can be analyzed (EPA Method 5).

3.0 INTERFERENCES

Volatile materials that produce chloride ions upon dissolution during sampling are interferences in the measurement of HCl. Cl_2 disproportionates to HCl and HClO upon dissolution in water, and will interfere with the HCl analysis. Cl_2 exhibits a low solubility in water, however, and the use of acidic rather than neutral or basic solutions for collection of HCl greatly reduces the dissolution of any Cl_2 present.

4.0 METHOD TARGET COMPOUNDS

chlorine
hydrogen chloride

5.0 APPLICABLE CAA AMENDMENTS POLLUTANTS

chlorine
hydrogen chloride

EPA METHOD 0051 - Sampling HCl & Chlorine By Midget Impingers

DRAFT METHOD Midget Impinger HCl/Cl_2 Emission Sampling Train

REFERENCE:

Test Methods for Evaluating Solid Waste, Third Edition. Report No. SW-846. U. S. Environmental Protection Agency, Office of Solid Waste and Emergency Response. Washington, DC: 1986.

1.0 SCOPE AND APPLICATION

EPA Draft Method 0051 is used to collect hydrogen chloride (HCl)and chlorine (Cl_2) in stack gas emission samples from hazardous waste incinerators, municipal waste combustors, and boilers and industrial furnaces. The collected samples are analyzed using EPA Method 9057. Method 0051 is designed to collect HCl and Cl_2 in their gaseous forms; sources such as those controlled by wet scrubbers that emit acid particulate matter (e.g., HCl dissolved in water droplets) must be sampled using an isokinetic HCl/Cl_2 sampling train (EPA Method 0050).

2.0 SUMMARY OF METHOD

An integrated gas sample is extracted from the stack and passes through a particulate filter, acidified water, and finally through an alkaline solution. The filter removes particulate matter, such as chloride salts, that could potentially react and form analyte in the absorbing solutions. In the acidified water absorbing solution, the HCl gas is solubilized and forms chloride ions. The Cl_2 gas present in the emissions has a very low solubility in acidified water and passes through to the alkaline absorbing solution where it undergoes hydrolysis to form a proton (H+), chloride ion, and hypochlorous acid (HClO). The chloride ions in the separate solutions are measured by ion chromatography (EPA Method 9057).

3.0 INTERFERENCES

Volatile materials that produce chloride ions upon dissolution during sampling are obvious interferences in the measurement of HCl. One interferent for HCl is Cl_2, which disproportionates to HCl and HClO upon dissolution in water. The Cl_2 gas exhibits a very low solubility in water, however, and the use of acidic rather than neutral or basic solutions for collection of HCl greatly reduces the dissolution of any Cl_2 present. Sampling a 400 ppm HCl gas stream containing 50 ppm Cl_2 with this method does not cause a significant bias. Sampling a 220 ppm HCl gas stream containing 180 ppm Cl_2 results in a positive bias of 3.4 percent in the HCl measurement.

Reducing agents such as SO_2 may cause a positive bias in the Cl_2 measurement by the following reaction:

$HClO + HSO_3^- = H_2SO_4 + Cl^-$

4.0 METHOD TARGET COMPOUNDS

chlorine
hydrogen chloride

5.0 APPLICABLE CAA AMENDMENTS POLLUTANTS

chlorine
hydrogen chloride

EPA METHOD 5040 - Thermal Desorption & GC/MS Analysis Of Volatile Organics

Analysis of Sorbent Cartridges from VOC Sampling Train

REFERENCE:

Test Methods for Evaluating Solid Waste, Third Edition. Report No. SW-846. U.S. Environmental Protection Agency, Office of Solid Waste and Emergency Response. Washington, DC: 1986.

1.0 SCOPE AND APPLICATION

EPA Method 5040 is used to analyze Tenax ® and Tenax ®/charcoal cartridges containing volatile principal organic hazardous constituents (POHCs) from wet stack gas effluents from hazardous waste incinerators. Volatile POHCs are defined as those POHCs with boiling points less than 100°C. Method 5040 is based on the quantitative thermal desorption of volatile POHCs from the Tenax ® and Tenax® / charcoal traps and analysis by purge-and-trap gas chromatography/mass spectrometry (GC/MS).

2.0 SUMMARY OF METHOD

POHCs are collected on Tenax ® and Tenax ® /charcoal sorbent cartridges using a volatile organic sampling train (VOST), EPA Method 0030. The contents of the sorbent cartridges are spiked with an internal standard and thermally desorbed for 10 minutes at 180°C with organic-free nitrogen or helium gas (at a flow rate of 40 mL/min), bubbled through 5 mL of organic-free water, and trapped on an analytical adsorbent trap. After the 10 minute desorption, the analytical adsorbent trap is rapidly heated to 180°C, with the carrier gas flow reversed so that the effluent flow from the analytical trap is directed into the GC/MS. The volatile POHCs are separated by temperature-programmed gas chromatography and detected by low resolution mass spectrometry. The concentrations of the volatile POHCs are calculated using the internal standard technique.

EPA Method 5030 and 8240 may be referenced for specific requirements for the thermal desorption unit, purge-and-trap unit, and GC/MS system.

3.0 INTERFERENCES

Interferences from hydrocarbons may be a problem if a source with significant hydrocarbon content is sampled. The mass of low molecular weight POHCs (such as acetonitrile, with a mass of 41) would coincide with the mass of ions common to all hydrocarbons, and acetonitrile would thus not be identifiable by low resolution mass spectrometric techniques.

4.0 METHOD TARGET COMPOUNDS

Volatile POHCs.

5.0 APPLICABLE CLEAN AIR ACT LIST COMPOUNDS

acetonitrile*
acrylonitrile*
allyl chloride
benzene
bis(chloromethyl) ether*
carbon disulfide
carbon tetrachloride
chlorobenzene**
chloroform
chloromethyl methyl ether*
chloroprene
1,1-dimethylhydrazine*
1,4-dioxane*
1,2-epoxybutane*
ethyl acrylate*
ethylbenzene**
ethyl chloride***
ethylene dibromide

ethylene dichloride
ethylene imine*
ethylidene dichloride
hexane
methyl bromide***
methyl chloride***
methyl chloroform
methyl ethyl ketone*
methyl hydrazine*
methyl iodide
methyl isobutyl ketone*,**
methyl isocyanate*
methyl methacrylate*
methyl tert-butyl ether*
methylene chloride
propylene oxide*
1,2-propylene imine*
tetrachloroethylene**
toluene**
1,1,2-trichloroethane**
trichloroethylene
triethylamine*
2,2,4-trimethylpentane
vinyl acetate*
vinyl bromide***
vinyl chloride***
vinylidene chloride

* A protocol has been developed for sample analysis for water-soluble volatile POHCs and PICs (EPA-600/8-87-008). Application of this protocol may improve recoveries for polar water-soluble compounds.

** Boiling point above 100°C.

*** Boiling point below 30°C. Special care must be taken to avoid breakthrough on sorbent.

EPA METHOD 5041 - Thermal Desorption & HRGC/MS Analysis Of Volatile Organics

DRAFT METHOD Protocol for the Analysis of Sorbent Cartridges from Volatile Organic Sampling Train (VOST): Wide-Bore Capillary Column Technique

REFERENCE:

Test Methods for Evaluating Solid Waste, Third Edition. Report No. SW-846. U.S. Environmental Protection Agency, Office of Solid Waste and Emergency Response. Washington, DC: 1986.

1.0 SCOPE AND APPLICATION

EPA Draft Method 5041 is used to the analyze Volatile Principal Organic Hazardous Constituents (POHCs) collected from the stack gas effluents of hazardous waste incinerators using the VOST methodology which uses EPA Method 0030 for sampling. Draft Method 5041 is most successfully applied to the analysis of non-polar organic compounds that vaporize between 30°C and 100°C. Data are applied to the calculation of destruction and removal efficiency (DRE).

The VOST methodology may also be used to collect and analyze many compounds that boil above 100°C. However, sampling method collection efficiency, sampling tube desorption efficiency, and analytical method precision and bias should be determined for these compounds. An organic compound with boiling point below 30°C may break through the sorbent under the conditions used for sample collection. The quantitative value obtained for such a compound must be qualified, since the value represents a minimum value for the compound if breakthrough has occurred.

The VOST analytical methodology can be used to quantify volatile organic compounds (VOCs) that are insoluble or slightly soluble in water. When volatile water-soluble compounds are included in the VOST organic compound analyte list, quantitation limits can be expected to be approximately ten times higher than quantitation limits for water-insoluble compounds (if the compounds can be recovered at all) because the purging efficiency from water (and possibly also from Tenax ® is so poor. Purging efficiency from water can be improved by modifying the VOST analytical methodology to include addition of salt (sodium chloride or sodium sulfate) to the purge water and heating the purge water (see EPA-600/8-87-008).

Overall sensitivity for the method depends on the level of analytical interferences encountered in the sample and the presence of detectable levels of volatile POHCs in the blanks. The target detection limit for the method is 0.1 $\mu g/m^3$ (ng/L) of flue gas, to permit calculation of a DRE equal to or greater than 99.99 percent for volatile POHCs that may be present in the waste stream at 100 ppm. The upper end of the range of applicability of this method is limited by the dynamic range of the analytical instrumentation, the overall loading of organic compounds on the exposed tubes, and breakthrough of the volatile POHCs on the sorbent traps used to collect the sample.

2.0 SUMMARY OF METHOD

Sorbent tubes are thermally desorbed by heating and purging with organic-free helium. The gaseous effluent from the tubes is bubbled through pre-purged organic-free water and trapped on an analytical sorbent trap in a purge-and-trap unit. After desorption, the analytical sorbent trap is heated rapidly and the gas flow from the analytical trap is directed to the head of a wide-bore column under subambient conditions. The VOCs desorbed from the analytical trap are separated by temperature-programmed high resolution gas chromatography and detected by continuously-scanning low resolution mass spectrometry. Concentrations of VOCs are calculated from a multi-point calibration curve, using the method of response factors.

3.0 INTERFERENCES

Sorbent tubes that are to be analyzed for VOCs can be contaminated by diffusion of VOCs (particularly Freon ® refrigerants and common organic solvents) through the external container (even through a Teflon ®- lined screw cap on a glass container) and the Swagelok ® sorbent tube caps during shipment and storage. The sorbent tubes can also be contaminated if organic solvents are present in the analytical laboratory. The use of blanks is essential to assess the extent of any contamination. Field blanks must be prepared and taken to the field. The end caps of the tubes are removed for the period of time required to exchange two pairs of traps on the VOST sampling apparatus. The tubes are recapped and shipped and handled exactly as the actual field samples are shipped and handled. At least one pair of field blanks is included with each six pairs of sample cartridges collected.

At least one pair of blank cartridges (one Tenax ®, one Tenax ® -charcoal) must be included with shipment of cartridges to a hazardous waste incinerator site as trip blanks. These trip blanks will be treated like field blanks except that the end caps will not be removed during storage at the site. This pair of traps will be analyzed to monitor potential contamination that may occur during storage and shipment.

Analytical system blanks are required to demonstrate that contamination of the purge-and-trap unit and the gas chromatograph/mass spectrometer has not occurred or that, in the event of analysis of sorbent tubes with very high levels of organic compounds, no compound carryover is occurring. Tenax ® from the same preparation batch as the Tenax ® used for field sampling should be used in the preparation of the method (laboratory) blanks. A sufficient number of cleaned Tenax ® tubes from the same batch as the field samples should be reserved in the laboratory for use as blanks.

Cross-contamination can occur whenever low-level samples are analyzed after high-level samples, or when several high-level samples are analyzed sequentially. When an unusually concentrated sample is analyzed, this analysis should be followed by a method blank to establish that the analytical system is free of contamination. If analysis of a blank demonstrates that the system is contaminated, an additional bake cycle should be used. If the analytical system is still contaminated after additional baking, routine system maintenance should be performed: the analytical trap should be changed and conditioned, routine column maintenance should be performed (or replacement of the column and conditioning of the new column, if necessary), and bakeout of the ion source (or cleaning of the ion source and rods, if required). After system maintenance has been performed, analysis of a blank is required to demonstrate that the cleanliness of the system is acceptable.

If the emission source has a high level of non-target organic compounds (for example, hydrocarbons at levels of hundreds of ppm), the presence of these non-target compounds will interfere with the performance of the VOST analytical methodology. If one or more of the compounds of interest saturates the chromatographic and mass spectrometric instrumentation, no quantitative calculations can be made and the tubes that have been sampled under the same conditions will yield no valid data for any of the saturated compounds and may not yield valid data for other compounds. In the presence of a very high organic loading, even if the compounds of interest are not saturated, the instrumentation is so saturated that the linear range has been surpassed. When instrument saturation occurs, it is possible that compounds of interest cannot even be identified correctly because a saturated mass spectrometer may mis- assign masses. Even if compounds of interest can be identified, accurate quantitative calculations are impossible at detector saturation. No determination can be made at or beyond detector saturation, even if the target compound itself is not saturated. At detector saturation, a negative bias will be encountered in analytical measurements and no accurate calculation can be made for the DRE if analytical values may be biased negatively.

Compounds that coelute chromatographically with the compounds of interest may interfere with the analysis if the coeluting compounds have ions at the same masses as the compound of interest. An alternate ion can be selected for the compound of interest (provided that detector saturation has not occurred) to avoid the interference at the primary quantitation mass.

4.0 METHOD TARGET COMPOUNDS

acetone
acrylonitrile
benzene
bromodichloromethane
bromoform
bromomethane
carbon disulfide
carbon tetrachloride
chlorobenzene
chlorodibromomethane
chloroethane
chloroform
chloromethane
dibromomethane
1,1-dichloroethane
1,2-dichloroethane
1,1-dichloroethene
trans-1,2-dichloroethene
1,2-dichloropropane
cis-1,3-dichloropropene
trans-1,3-dichloropropene
ethylbenzene
iodomethane
methylene chloride
styrene
1,1,2,2-tetrachloroethane
tetrachloroethene
toluene
1,1,1-trichloroethane
1,1,2-trichloroethane
trichloroethene
trichlorofluoromethane
1,2,3-trichloropropane
vinyl chloride
xylenes

5.0 APPLICABLE CLEAN AIR ACT LIST COMPOUNDS

acetonitrile (requires modification of the methodology for water-soluble compounds, but may require high resolution mass spectrometry)
acrylonitrile
allyl chloride
benzene
bis(chloromethyl) ether (will require modification of the methodology for water-soluble compounds)
bromoform
carbon disulfide
carbon tetrachloride
chlorobenzene (above 100°C boiling point)
chloroform
chloromethyl methyl ether (water-soluble)
chloroprene
1,1-dimethylhydrazine (water-soluble)
1,4-dioxane (water-soluble)
ethyl acrylate (water-soluble)
ethylbenzene (above 100°C boiling point)
ethyl chloride (low boiling point; requires special care)
ethylene dibromide (above 100°C boiling point)
ethylene dichloride
ethylene imine (water-soluble)
ethylidene dichloride
hexane
methyl bromide (low boiling point; requires special care)
methyl chloride (low boiling point; requires special care)
methyl chloroform
methyl ethyl ketone (water-soluble)
methyl hydrazine (water-soluble)
methyl iodide
methyl isobutyl ketone (above 100°C boiling point; water-soluble)
methyl isocyanate (water-soluble)
methyl methacrylate (water-soluble)
methyl tert-butyl ether (water-soluble)
methylene chloride
propylene dichloride
propylene oxide (water-soluble)
1,2-propylene imine (water-soluble)
styrene (above 100°C boiling point)
1,1,2,2-tetrachloroethane (above 100°C boiling point)
tetrachloroethylene (above 100°C boiling point)
toluene (above 100°C boiling point)

1,1,2-trichloroethane (above 100°C boiling point)
trichloroethylene
triethylamine (water-soluble)
2,2,4-trimethylpentane
vinyl acetate (water-soluble)
vinyl bromide (low boiling point; requires special care)
vinyl chloride (low boiling point; requires special care)
vinylidene chloride
xylenes (above 100°C boiling point)

EPA METHOD 8080 - Chlorinated Pesticides & PCBs By GC/ECD Or GC/ELCD

Organochlorine Pesticides and PCBs

REFERENCE:

Test Methods for Evaluating Solid Waste, Third Edition. Report No. SW-846. U.S. Environmental Protection Agency, Office of Solid Waste and Emergency Response. Washington, DC: 1986.

1.0 SCOPE AND APPLICATION

EPA Method 8080 is used to determine the concentration of various organochlorine pesticides and polychlorinated biphenyls (PCBs) in soil at levels of parts-per-billion. The method can be modified in order to analyze air samples. Method 8080 should be used only for non-stack stationary sources.

2.0 SUMMARY OF METHOD

Prior to analysis, appropriate sample extraction techniques must be used. Water samples are extracted with methylene chloride using either EPA Method 3510 or EPA Method 3520. Solid samples are extracted using EPA Method 3540 or EPA Method 3550. The extraction solvent must be exchanged to hexane. A modification to accommodate air samples may be necessary. Once sample preparation is completed, the sample is injected, either manually or by an autosampler, into a gas chromatograph (GC). The compounds in the GC effluent are detected by an electron capture detector (ECD) or a halogen-specific detector (such as an electroconductivity detector (ELCD) in the halogen mode). The identification and concentration of the target compounds are based on a comparison of the retention times and peak areas of calibration standards with the sample peaks. A minimum of five concentration levels is required for calibration.

Method 8080 recognizes an Aroclor profile rather than individual PCBs. Since Aroclors are made up of numerous PCBs, their chromatograms are multi-peak and identification of an Aroclor is by pattern recognition. Method 8080 should not be applied to stack samples because Aroclors that have been combusted no longer exhibit the pattern of an Aroclor and consequently will not be identified. Spiking each sample, standard, and blank with the surrogates dibutyl chlorendate and/or 2,4,5,6-tetrachloro-meta-xylene is recommended in order to monitor the performance of the extraction, cleanup (when used), and analytical system.

3.0 INTERFERENCES

DDT and endrin degrade easily in the injection port of the GC as a result of buildup of high-boiling residue from sample injection. The degradation products of these compounds can interfere with peak detection and identification of certain compounds. Therefore, proper GC maintenance procedures are recommended in order to avoid these problems.

4.0 METHOD TARGET COMPOUNDS

aldrin
alpha-BHC
beta-BHC
delta-BHC
lindane
chlordane
4,4'-DDD
4,4'-DDE
4,4'-DDT
dieldrin
endosulfan I
endosulfan II
endosulfan sulfate
endrin
endrin aldehyde
heptachlor
heptachlor epoxide
methoxychlor
toxaphene
Aroclor-1016
Aroclor-1221
Aroclor-1232
Aroclor-1242
Aroclor-1248
Aroclor-1254
Aroclor-1260

5.0 APPLICABLE CLEAN AIR ACT LIST COMPOUNDS

polychlorinated biphenyls
4,4'-DDE
heptachlor
lindane
methoxychlor
toxaphene
chlordane

EPA METHOD 8270 - Semivolatile Organic Compounds By HRGC/MS

GC/MS for Semivolatile Organics: Capillary Column Technique

REFERENCE:

Test Methods for Evaluating Solid Waste, Third Edition. Report No. SW-846. U.S. Environmental Protection Agency, Office of Solid Waste and Emergency Response. Washington, DC: 1986.

1.0 SCOPE AND APPLICATION

EPA Method 8270 is used to determine the concentration of semivolatile organic compounds in extracts prepared from all types of solid waste matrices, soils, and ground water. It is also applicable to an extract from sorbent media in conjunction with Method 0010. Direct injection of a sample may be used in limited applications.

Method 8270 can be used to quantify most neutral, acidic, and basic organic compounds that are soluble in methylene chloride and capable of being eluted without derivatization as sharp peaks from a gas chromatographic fused-silica capillary column coated with a slightly polar silicone. Such compounds include polynuclear aromatic hydrocarbons, chlorinated hydrocarbons and pesticides, phthalate esters, organophosphate esters, nitrosamines, haloethers, aldehydes, ethers, ketones, anilines, pyridines, quinolines, aromatic nitro compounds, and phenols, including nitrophenols.

The following compounds may require special treatment when being determined by this method:

- Benzidine can be subject to oxidative losses during solvent concentration. Also, chromatography is poor.

- Alpha-BHC, gamma-BHC, endosulfan I and II, and endrin are subject to decomposition under the alkaline conditions of the extraction step. Neutral extraction should be performed if these compounds are expected.

- Hexachlorocyclopentadiene is subject to thermal decomposition in the inlet of the gas chromatograph, chemical reaction in acetone solution, and photochemical decomposition.

- N-Nitrosodimethylamine is difficult to separate from the solvent under the chromatographic conditions described.

- N-Nitrosodiphenylamine decomposes in the gas chromatographic inlet and cannot be separated from diphenylamine.

- Pentachlorophenol, 2,4-dinitrophenol, 4-nitrophenol, 4,6-dinitro-2-methylphenol, 4-chloro-3-methylphenol, benzoic acid, 2-nitroaniline, 3-nitroaniline, 4-chloroaniline, and benzyl alcohol are subject to erratic chromatographic behavior, especially if the gas chromatographic system is contaminated with high-boiling material. Derivatization can help to stabilize the chromatographic performance of these compounds.

- Acrylic acid will require derivatization to apply the analytical methodology.

- The phthalate esters exhibit better recoveries under acid extraction conditions.

- Bromoform, chlorobenzene, cumene, 1,3-dichloropropene, 1,4-dioxane, epichlorohydrin, ethylbenzene, ethylene dibromide, methyl isobutyl ketone, 2-nitropropane, styrene, toluene, 1,1,2,2-tetrachloroethane, 1,1,2-trichloroethane, o-, m-, and p-xylene and tetrachloroethylene, are within the boiling point range for Method 8270, but they are sufficiently volatile that care must be taken in sample concentration to avoid loss. An adjustment in chromatographic conditions will be required to resolve the relatively volatile compounds from the solvent.

- 4-Aminobiphenyl, aniline, o-anisidine, benzidine, 3,3'-dichlorobenzidine, N,N-diethylaniline, 3,3'-dimethoxybenzidine, dimethylaminoazobenzene, 3,3'-dimethylbenzidine, 4,4'-methylene bis (2-chloroaniline), 4,4'-methylenedianiline, p-phenylenediamine, quinoline, and 2,4-toluenediamine are basic nitrogen-containing compounds and careful control of pH during extraction is required to optimize recovery.

The practical quantitation limit for Method 8270 is approximately 50 μg/mL of extract. The entire sorbent module with filter is typically extracted and concentrated to 1 mL (a final volume of 5 mL is used to avoid loss of volatile compounds), and this final extract volume represents the entire volume of gas sampled.

2.0 SUMMARY OF METHOD

Method 8270 describes conditions for gas chromatography/mass spectrometry (GC/MS) to allow for the separation of the semivolatile compounds on the method target list. Sample extraction, purification, and concentration techniques are addressed in other methods. EPA Methods 3510, 3520, 3540, 3550, 3580 may be applicable to the sample preparation. The following purification methods may be used prior to GC/MS analysis:

phenols, Methods 3630, 3640, 8040
phthalate esters, Method 3610, 3620, 3640
nitrosamines, Method 3610, 3620, 3640
organochlorine pesticides and PCBs, Method 3620, 3640, 3660
polynuclear aromatic hydrocarbons, Method 3611, 3630, 3640
nitroaromatics and cyclic ketones, Method 3620, 3640
haloethers, Method 3620, 3640
chlorinated hydrocarbons, Method 3620, 3640
organophosphorus pesticides, Method 3620, 3640
petroleum waste, Method 3611, 3650
all priority pollutant base, neutral, and acids, Method 3640.

3.0 INTERFERENCES

Raw GC/MS data from all blanks, samples, and spikes must be evaluated for interferences. If an interference results from the preparation and/or cleanup of samples, corrective action can be taken to eliminate the problem. If the problem is a very high sample background of alkyl or aromatic hydrocarbons, very little can be done to resolve the problem other than dilution of the samples, which raises the detection limit. If chromatographic coelution occurs, deconvolution of the coeluting components by mass spectrometric techniques will be effective if the compounds are not chemically related and their mass spectra can be resolved. If isomers coelute and their mass spectra are similar, the coelution cannot be resolved.

Contamination by carryover can occur whenever high level and low level samples are analyzed sequentially. To reduce carryover, the sample syringe must be rinsed carefully with solvent between sample injections. The chromatographic column should be allowed to remain at a high temperature until all late-eluting components have eluted from the column in order to avoid chromatographic carryover problems. Whenever an unusually concentrated sample is encountered, it should be followed by the analysis of clean solvent to check for cross-contamination. If contamination is observed, the injections of solvent should be repeated until the contamination is no longer observed before another sample injection is performed.

4.0 METHOD TARGET COMPOUNDS

acenaphthene
acenaphthylene
acetophenone
aldrin
aniline
anthracene
4-aminobiphenyl
Aroclor-1016
Aroclor-1221
Aroclor-1232
Aroclor-1242
Aroclor-1248
Aroclor-1254
Aroclor-1260
benzidine
benzoic acid
benzo(a)anthracene
benzo(b)fluoranthene

benzo(k)fluoranthene
benzo(g,h,i)perylene
benzo(a)pyrene
benzyl alcohol
alpha-BHC
beta-BHC
delta-BHC
gamma-BHC (lindane)
bis(2-chloroethoxy)methane
bis(2-chloroethyl)ether
bis(2-chloroisopropyl)ether
bis(2-ethylhexyl) phthalate
4-bromophenyl phenyl ether
butyl benzyl phthalate
chlordane
4-chloroaniline
1-chloronaphthalene
2-chloronaphthalene
4-chloro-3-methylphenol
chrysene
4,4'-DDD
4,4'-DDE
4,4'-DDT
dibenz(a,j)acridine
dibenz(a,h)anthracene
di-n-butyl phthalate
1,3-dichlorobenzene
1,4-dichlorobenzene
1,2-dichlorobenzene
3,3'-dichlorobenzidine
dichloroethyl ether
2,4-dichlorophenol
2,6-dichlorophenol
dieldrin
diethyl phthalate
p-dimethylaminoazobenzene
7,12-dimethylbenz(a)anthracene
alpha-, alpha-dimethylphenethylamine
2,4-dimethylphenol
dimethyl phthalate
4,6-dinitro-2-methylphenol
2,4-dinitrophenol
2,4-dinitrotoluene
2,6-dinitrotoluene
diphenylamine
1,2-diphenylhydrazine
di-n-octyl phthalate
endosulfan I
endosulfan II
endosulfan sulfate
endrin
endrin aldehyde
endrin ketone
ethyl methanesulfonate
fluoranthene
fluorene
heptachlor
heptachlor epoxide
hexachlorobenzene
hexachlorobutadiene
hexachlorocyclopentadiene
hexachloroethane
indeno(1,2,3-cd)pyrene
isophorone
methoxychlor
3-methylcholanthrene
methyl methanesulfonate
2-methylnaphthalene
2-methylphenol
4-methylphenol
naphthalene
1-naphthylamine
2-naphthylamine
3-nitroaniline
4-nitroaniline
nitrobenzene
2-nitrophenol
4-nitrophenol
N-nitrosodi-n-butylamine
N-nitrosodimethylamine
N-nitrosodiphenylamine
N-nitrosodiproylamine
N-nitrosopiperidine
pentachlorobenzene
pentachloronitrobenzene
pentachlorophenol
phenacetin
phenanthrene

phenol
2-picoline
pronamide
pyrene
1,2,4,5-tetrachlorobenzene
2,3,4,6-tetrachlorophenol
1,2,4-trichlorobenzene
2,4,5-trichlorophenol
2,4,6-trichlorophenol
toxaphene

5.0 APPLICABLE CLEAN AIR ACT LIST COMPOUNDS

acetamide
acetophenone (8270 target)
2-acetylaminofluorene
acrylamide
acrylic acid
4-aminobiphenyl (8270 target)
aniline (8270 target)
o-anisidine
benzidine (8270 target)
benzotrichloride
benzyl chloride
biphenyl
bis(2-ethylhexyl) phthalate (8270 target)
bromoform
catechol
chlordane (8270 target)
chloroacetic acid
2-chloroacetophenone
chlorobenzene
chlorobenzilate
o-cresol (8270 target)
m-cresol
p-cresol (8270 target)
cresylic acid (mixture of cresols)
cumene
2,4-D
DDE (8270 target)
1,2-dibromo-3-chloropropane
dibutyl phthalate (8270 target)
1,4-dichlorobenzene (8270 target)
3,3'-dichlorobenzidine (8270 target)
1,3-dichloropropene
dichlorvos
diethanolamine
N,N-diethylaniline
diethyl sulfate
3,3'-dimethoxybenzidine
dimethylaminoazobenzene (8270 target)
3,3'-dimethylbenzidine
dimethyl formamide
dimethyl phthalate (8270 target)
dimethyl sulfate
4,6-dinitro-o-cresol (8270 target)
2,4-dinitrophenol (8270 target)
2,4-dinitrotoluene (8270 target)
1,4-dioxane
1,2-diphenylhydrazine (8270 target)
epichlorohydrin
ethylbenzene
ethylene dibromide
ethylene glycol
glycol ethers
heptachlor (8270 target)
hexachlorobenzene (8270 target)
hexachlorobutadiene (8270 target)
hexachlorocyclopentadiene (8270 target)
hexachloroethane (8270 target)
hexamethylene-1,6-diisocyanate
hydroquinone
isophorone (8270 target)
lindane (8270 target)
maleic anhydride
methoxychlor (8270 target)
4,4'-methylene bis(2-chloroaniline)
methylene diphenyl diisocyanate
4,4'-methylenedianiline
naphthalene (8270 target)
nitrobenzene (8270 target)
4-nitrobiphenyl
4-nitrophenol (8270 target)

N-nitroso-N-methylurea
N-nitrosodimethylamine (8270 target)
N-nitrosomorpholine
parathion
pentachloronitrobenzene (8270 target)
pentachlorophenol (8270 target)
phenol (8270 target)
p-phenylenediamine
phthalic anhydride
polychlorinated biphenyls
polycyclic organic matter
1,3-propane sultone
beta-propiolactone
quinoline
quinone
styrene
styrene oxide
1,1,2,2-tetrachloroethane
tetrachloroethylene
toluene
2,4-toluenediamine
2,4-toluene diisocyanate
o-toluidine
toxaphene (8270 target)
1,2,4-trichlorobenzene (8270 target)
1,1,2-trichloroethane
2,4,5-trichlorophenol (8270 target)
2,4,6-trichlorophenol (8270 target)
trifluralin
xylenes: o-xylene, m-xylene, p-xylene

EPA METHOD 8280 - PCDDs & PCDFs By HRGC/MS Using Selected Ion Monitoring

The Analysis of PCDDs and PCDFs (Polychlorinated Dibenzo-p-Dioxins and Polychlorinated Dibenzofurans)

REFERENCE:

Test Methods for Evaluating Solid Waste, Third Edition. Report No. SW-846. U.S. Environmental Protection Agency, Office of Solid Waste and Emergency Response. Washington, DC: 1986.

1.0 SCOPE AND APPLICATION

EPA Method 8280 is used to determinate tetra-, penta-, hexa-, hepta-, and octachlorinated dibenzo-p-dioxins (PCDDs) and dibenzofurans (PCDFs) in chemical wastes, including still bottoms, fuel oils, sludges, fly ash, reactor residues, soil, and water.

2.0 SUMMARY OF METHOD

This procedure uses a matrix-specific extraction, analyte-specific cleanup, and high resolution capillary column gas chromatography/low resolution mass spectrometry techniques. An analytical system with a temperature-programmable gas chromatograph and all required accessories including syringes, analytical columns, and gases is needed. Fused silica capillary columns are required, and one of the following is recommended: (1) 50 m CP-Sil-88; (2) DB-5, 30 m x 0.25 mm ID, 0.25 μm film thickness; or (3) 30 m SP- 2250. However, any capillary column that provides separation of 2,3,7,8- tetrachlorodibenzo-p-dioxin (TCDD) from all other TCDD's with higher isomers may be used.

Two types of calibration procedures are required: An initial calibration is required before any samples are analyzed, and routine calibration conducted intermittently throughout the sample analyses. The routine calibration consists of analyzing the column performance check solution and a mid-range concentration calibration solution.

Samples are analyzed with Selected Ion Monitoring (SIM) mass spectrometry (MS) techniques. Strict identification criteria for PCDDs and PCDFs are listed in the method. The concentration of individual isomers is determined using relative response factors that were calibrated using data obtained from the analysis of multi-level calibration standards.

Because of the extreme toxicity of many of the compounds listed as analytes, the analyst must take the necessary precautions to prevent exposure to materials known or believed to contain PCDDs or PCDFs. Laboratory personnel must ensure that safe handling procedures are employed.

3.0 INTERFERENCES

The sensitivity of this method depends on the level of interferents within a given matrix. Interferents coextracted from the sample will vary considerably from source to source, depending upon the industrial process being sampled. PCDDs and PCDFs are often associated with other interfering chlorinated compounds such as PCBs and polychlorinated diphenyl ethers, which may be found at higher concentrations than that of the analytes of interest.

4.0 METHOD TARGET COMPOUNDS

tetrachlorodibenzo-p-dioxin
tetrachlorodibenzofuran
pentachlorodibenzo-p-dioxin
pentachlorodibenzofuran
hexachlorodibenzofuran
hexachlorodibenzo-p-dioxin
heptachlorodibenzo-p-dioxin
heptachlorodibenzofuran
octachlorodibenzofuran
octachlorodibenzo-p-dioxin

5.0 APPLICABLE CLEAN AIR ACT LIST COMPOUNDS

dibenzofurans
2,3,7,8-tetrachlorodibenzo-p-dioxin.

EPA METHOD 8290 - PCDDs & PCDFs By HRGC/HRMS With Labeled Standards

DRAFT METHOD

Analytical Procedures and Quality Assurance for Multimedia Analysis of Polychlorinated Dibenzo-p-dioxins and Polychlorinated Dibenzofurans by High Resolution Gas Chromatography/High Resolution Mass Spectrometry

REFERENCE:

Test Methods for Evaluating Solid Waste, Third Edition. Report No. SW-846. U.S. Environmental Protection Agency, Office of Solid Waste and Emergency Response. Washington, DC: 1986.

1.0 SCOPE AND APPLICATION

EPA Draft Method 8290 provides for the detection and quantitative measurement of 2,3,7,8-tetrachlorodibenzo-p-dioxin (2,3,7,8-TCDD), polychlorinated dibenzo-p-dioxins (PCDDs; tetra- through octachlorinated homologs), and polychlorinated dibenzofurans (PCDFs; tetra-through octachlorinated homologs) in a variety of environmental matrices and at part-per-trillion (ppt) concentrations. This method has been applied to samples from the following matrices: soil, sediment, fly ash, water, sludges, fuel oil, still bottoms, fish tissue, paper pulp, human adipose tissue, and gaseous and particulate emissions.

Because of the extreme toxicity of many of the compounds listed as analytes, the analyst must take the necessary precautions to prevent exposure to materials known or believed to contain PCDDs or PCDFs. Laboratory personnel must ensure that safe handling procedures are employed.

2.0 SUMMARY OF METHOD

This procedure uses matrix-specific extraction, analyte-specific clean-up, and high resolution capillary gas chromatography/high resolution mass spectrometry (HRGC/HRMS) techniques. A specified amount of matrix (or the entire sorbent sampling trap plus filter from a sampling train for gaseous emissions) is spiked with a solution containing specified amounts of the nine isotopically-labeled ($^{13}C_{12}$) PCDDs/PCDFs. The sample is then extracted according to a matrix-specific extraction procedure. The extracts are submitted to an acid/base washing treatment and dried. Following a solvent exchange, the residue is purified by column chromatography on neutral alumina and carbon on Celite 545 ®. Preparation of the final extract for HRGC/HRMS analysis is accomplished by adding a solution containing the isotopically-labeled recovery standards to the concentrated carbon column eluate. The recovery standard is used to determine the percent recoveries of isotopically-labeled standards spiked at the beginning of the extraction process. An aliquot of the concentrated extract is injected into a HRGC/HRMS system capable of performing selected ion monitoring at resolving powers of at least 10,000 (10 percent valley definition).

Compound identification for the compounds for which an isotopically-labeled standard is used is based upon elution at the exact retention time established by analysis of standards and simultaneous detection of the two most abundant ions in the molecular ion region. Compounds for which no isotopically-labeled standard is available are identified by their relative retention times, which must fall within the established retention windows, and the simultaneous detection of the two most abundant ions in the molecular ion region. The retention windows are established by analysis of a GC Performance Evaluation solution. Identification is confirmed by comparing the ratio of the integrated ion abundance of the molecular ion species to the theoretical abundance ratio.

Quantification of the individual congeners, total PCDDs and PCDFs, is based upon a multipoint (seven points) calibration curve for each homolog, with each calibration solution being analyzed once.

3.0 INTERFERENCES

Solvents, reagents, glassware, and other sample processing hardware may yield discrete artifacts or elevated baselines that may cause misinterpretation of the chromatographic data. All of these materials must be demonstrated to be free from interferents under the conditions of the laboratory analysis by analyzing laboratory method blanks. Analysts should avoid using polyvinyl chloride gloves. High-purity reagents and solvents help minimize interference problems. Purifying solvents by distillation in all-glass systems may be necessary.

The sensitivity of the method is dependent upon the level of interferences within a given matrix. Interferences co-extracted from the sample matrix will vary considerably from matrix to matrix. PCDDs and PCDFs are often associated with other interfering chlorinated substances such as polychlorinated biphenyls (PCBs), polychlorinated diphenyl ethers (PCDPEs), polychlorinated naphthalenes, and polychlorinated xanthenes which may be found at concentrations several orders of magnitude higher than the analytes of interest. Retention times of target analytes must be verified using reference standards, and these values must correspond to the established retention time windows. While certain purification techniques are included as part of this method, unique samples may require other additional purification steps to achieve lower detection limits.

4.0 METHOD TARGET COMPOUNDS

polychlorinated dibenzo-p-dioxins
polychlorinated dibenzofurans
2,3,7,8-tetrachlorodibenzo-p-dioxin

5.0 APPLICABLE CAA AMENDMENTS POLLUTANTS

dibenzofurans
2,3,7,8-tetrachlorodibenzo-p-dioxin

EPA METHOD 8310 - Polynuclear Aromatic Hydrocarbons By HPLC/UV Or Fluorescence

Polynuclear Aromatic Hydrocarbons

REFERENCE:

Test Methods for Evaluating Solid Waste, Third Edition. Report No. SW-846. U.S. Environmental Protection Agency, Office of Solid Waste and Emergency Response. Washington, DC: 1986.

1.0 SCOPE AND APPLICATION

EPA Method 8310 is used to determine the concentration of certain polynuclear aromatic hydrocarbons (PAH) in ground water and wastes at parts-per-billion levels. By extension, the methodology should be applicable to material extracted from a solid sorbent module of a sampling train from EPA Method 0010, which is used to sample gaseous emissions from a stationary source.

Extension of the methodology to PAH containing functional groups should be possible, depending upon the ability to adjust analytical conditions and the availability of standards for the compounds of interest.

Use of Method 8310 presupposes a high expectation of finding the specific compounds of interest. To screen samples for any or all of the method target compounds (see listed below), independent protocols for the verification of identity must be developed. One method that can be used to certify identity is gas chromatography/mass spectrometry (GC/MS).

Method detection limits are compound-dependent, ranging from 0.4 μg/L for indeno(1,2,3-cd)pyrene in ground water to 230,000 μg/L for acenaphthylene in non-water miscible waste. Detection limits for PAH in gaseous emissions have not been determined directly. This methodology has not been directly and specifically applied to the determination of polycyclic organic matter other than the PAH specifically listed in the methodology. A quantitative analysis of other PAH and functionalized PAH will require adjustment of analytical conditions and the use of appropriate standards. An additional method such as GC/MS, if applicable, may be required to identify additional compounds.

If coelution of compounds is encountered in samples taken from gaseous emissions of stationary sources, Method 8310 may not be applicable unless analytical conditions can be adjusted to achieve chromatographic resolution.

2.0 SUMMARY OF METHOD

Prior to using Method 8310, appropriate sample extraction methods must be used. A 5- to 25 μL aliquot of extract is injected into a high performance liquid chromatograph (HPLC), and compounds in the effluent are detected by ultraviolet (uv) and fluorescence detectors. If interferences prevent proper detection of the analytes of interest, the method may also be performed on extracts that have undergone purification using silica gel column cleanup (EPA Method 3630).

3.0 INTERFERENCES

The sensitivity of the method usually depends on the level of interferences rather than instrumental limitations. The limits of detection above for the liquid chromatographic approach represent sensitivities that can be achieved in the absence of interferences. When interferences are present, the level of sensitivity will be lower, if analysis is possible at all.

Solvents, reagents, glassware, and other sample processing hardware may yield discrete artifacts and/or elevated baselines, causing misinterpretation of the chromatograms. All of these materials must be demonstrated to be free from interferences under the conditions of the analysis by analyzing method blanks. Specific selection of reagents and purification of solvents by distillation in all-glass systems may be required.

Interferences coextracted from the samples will vary considerably from source to source. Although a general cleanup technique is provided as part of Method 8310, individual samples may require additional cleanup approaches to achieve the desired sensitivity.

The chromatographic conditions described in Method 8310 allow for a unique resolution of the specific PAH compounds covered by this method. Other PAH compounds, in addition to matrix artifacts, may interfere.

4.0 METHOD TARGET COMPOUNDS

acenaphthene
acenaphthylene
anthracene
benzo(a)anthracene
benzo(a)pyrene
benzo(b)fluoranthene
benzo(ghi)perylene
benzo(k)fluoranthene
chrysene
dibenzo(a,h)anthracene
fluoranthene
fluorene
indeno(1,2,3-cd)pyrene
naphthalene
phenanthrene
pyrene

5.0 APPLICABLE CLEAN AIR ACT LIST COMPOUNDS

naphthalene*
4-aminobiphenyl
benzidine
biphenyl
chlorobenzilate
3,3'-4 dichlorobenzidine
3,3'dimethoxybenzidine
dimethylaminoazobenzene
3,3'-dimethylbenzidine
4,4'-methylene bis(chloroaniline)
methylene diphenyl isocyanate
4,4'-methylenedianiline
4-nitrobiphenyl
p-phenylenediamine
2,4-toluenediamine
2,4-toluene diisocyanate
o-toluidine
polycyclic organic matter

* **Method 8310 is directly applicable only to this compound.**

EPA METHOD 8315 - Formaldehyde, Aldehydes & Ketones By HPLC Of DNPHs

DRAFT METHOD Determination of Formaldehyde by DNPH Derivatization, Solid Sorbent Extraction, and HPLC Detection

REFERENCE:

Test Methods for Evaluating Solid Waste, Third Edition. Report No. SW-846. U.S. Environmental Protection Agency, Office of Solid Waste and Emergency Response. Washington, DC: 1986.

1.0 SCOPE AND APPLICATION

EPA Draft Method 8315 uses high performance liquid chromatography (HPLC) to determinate formaldehyde in liquid environmental matrices and leachates of solid samples applicable to the determination of formaldehyde and acetaldehyde.

Extension of the methodology to HPLC determination of formaldehyde and acetaldehyde in gaseous emission samples is feasible, and the methodology can also be applied to other aldehydes and ketones. When this method is used to analyze unfamiliar sample matrices, compound identification should be supported by at least one additional qualitative technique such as gas chromatography/mass spectrometry.

Actual detection limits are compound- and matrix-dependent. However, for a list of aldehydes and ketones tested, detection limits were approximately 2 ppbv when reagent capacity (for sampling by EPA Draft Method 0011) was 60-100 ppm.

2.0 SUMMARY OF METHOD

In Draft Method 0011, the 2,4-dinitrophenylhydrazine (DNPH) derivative of aldehydes in the emission stream is formed during sampling, since the emissions are bubbled through impingers containing an aqueous acidic solution of DNPH. This solution is returned to the laboratory and extracted with methylene chloride. The methylene chloride extract is concentrated to less than 10 mL using the Kuderna-Danish procedure. Liquid chromatographic conditions described in Draft Method 8315 which permit the separation and measurement of formaldehyde (and other aldehydes and ketones) in the extract by absorbance detection at 360 nm.

3.0 INTERFERENCES

Analysis for formaldehyde is especially complicated by its ubiquitous occurrence in the environment. The volatile aldehydes, such as formaldehyde and acetaldehyde, may be contaminants in volatile organic solvents. Since formaldehyde is widely used in building insulation, great care is required to determine whether the laboratory atmosphere is contaminated with formaldehyde. Blanks and controls that are treated under the same laboratory conditions as samples are of crucial importance in assessing background levels of aldehydes. Solvent blanks for each lot of solvents used in sample preparation are important. Glassware must not be rinsed with acetone in the cleaning process.

Matrix interferences will result from contaminants that are coextracted from the sample, and will vary from source to source. Since the analytical methodology is HPLC, quantitative analysis of compounds of interest depends on the absence of coeluting interferences.

4.0 METHOD TARGET COMPOUNDS

acetaldehyde*
acetone/propionaldehyde
acrolein
benzaldehyde
butyraldehye
dimethylbenzaldehyde
formaldehyde*
hexaldehyde
isovalderaldehyde
methyl ethyl ketone
o-, m-, and p-tolualdehyde
valeraldehyde

* The target compounds actually cited in Draft Method 8315 are formaldehyde and acetaldehyde. The other listed compounds have been determined by DNPH derivatization followed by HPLC analysis.

5.0 APPLICABLE CLEAN AIR ACT LIST COMPOUNDS

acetophenone
acrolein
2-chloroacetophenone
formaldehyde
isophorone
methyl ethyl ketone
methyl isobutyl ketone
propionaldehyde
quinone

EPA METHOD 8318 - N-Methyl Carbamates By HPLC/Fluorescence Of Derivatives

DRAFT METHOD N-Methyl Carbamates by HPLC (High Performance Liquid Chromatography)

REFERENCE:

Test Methods for Evaluating Solid Waste, Third Edition. Report No. SW-846. U. S. Environmental Protection Agency, Office of Solid Waste and Emergency Response. Washington, DC: 1986.

1.0 SCOPE AND APPLICATION

EPA Draft Method 8318 is used to determine the concentration of N-methylcarbamates in soil, water, and waste matrices. Extension of the methodology is required to allow preparation and analysis of samples from stationary sources.

2.0 SUMMARY OF METHOD

N-Methylcarbamates are extracted from aqueous samples with methylene chloride, and from soils, oily solid waste, and oils with acetonitrile. The extract solvent is exchanged to methanol/ethylene glycol, and then the extract is cleaned up on a C-18 cartridge, filtered, and eluted on a C-18 analytical column. After separation, the target analytes are hydrolyzed and derivatized post-column, then quantified fluorimetrically.

Due to the specific nature of this analysis, confirmation by a secondary method is not essential. However, fluorescence due to post-column derivatization may be confirmed by substituting the NaOH and o-phthalaldehyde solutions with deionized water and reanalyzing the sample. If fluorescence is detected, then a positive interference is present and care should be taken in the interpretation of the results.

The sensitivity of the method usually depends upon the level of interferences present, rather than on the instrumental conditions. Waste samples with a high level of extractable fluorescent compounds are expected to yield significantly higher detection limits.

3.0 INTERFERENCES

Fluorescent compounds, primarily alkyl amines and compounds that yield primary alkyl amines on base hydrolysis, are potential sources of interferences. Coeluting compounds that are fluorescence quenchers may result in negative interferences. Impurities in solvents and reagents are additional sources of interferences. Before processing any samples, the analyst must demonstrate daily, through the analysis of solvent blanks, that the entire analytical system is interference-free.

4.0 METHOD TARGET COMPOUNDS

aldicarb (Temik)
aldicarb sulfone
carbaryl (Sevin)
carbofuran (Furadan)
dioxacarb
3-hydroxycarbofuran
methiocarb (Mesurol)
methomyl (Lannate)
promecarb
propoxur (Baygon).

5.0 APPLICABLE CAA AMENDMENTS POLLUTANTS

carbaryl
propoxur
ethyl carbamate

EPA METHOD 9057 - HCl & Chlorine Analysis By Ion Chromatography

Analysis of Samples from HCl/Cl_2 Emission Sampling

REFERENCE:

Test Methods for Evaluating Solid Waste, Third Edition. Report No. SW-846. U. S. Environmental Protection Agency, Office of Solid Waste and Emergency Response. Washington, DC: 1986.

1.0 SCOPE AND APPLICATION

EPA Method 9057 is used to determine hydrogen chloride (HCl) and chlorine (Cl_2) in stack gas emission samples collected from hazardous waste and municipal waste incinerators using the midget impinger HCl/Cl_2 sampling train (Method 0051) or the isokinetic HCl/Cl_2 sampling train (Method 0050).

The lower detection limit is 0.1 æg of chloride ions per mL of sample solution. Samples with concentrations that exceed the linear range of the analytical instrumentation may be diluted.

Method 9057 is recommended for use only by analysts experienced in the use of ion chromatography and in the interpretation of ion chromatograms.

2.0 SUMMARY OF METHOD

The stoichiometry of HCl and Cl_2 collection in the sampling train (Methods 0050 and 0051) is as follows: in the acidified water absorbing solution, the HCl gas is solubilized and forms chloride ions. The Cl_2 gas present in the emissions has a very low solubility in acidified water and passes through to the alkaline absorbing solution where it undergoes hydrolysis to form a proton (H+), chloride ion, and hypochlorous acid. Non-suppressed or suppressed ion chromatography (IC) is used for analysis of the chloride ion.

3.0 INTERFERENCES

Volatile materials that produce chloride ions upon dissolution during sampling are obvious interferences in the measurement of HCl. One likely interferent is Cl_2, which disproportionates to HCl and hypochlorous acid (HClO) upon dissolution in water. Cl_2 exhibits a low solubility in water, however, and the use of acidic rather than neutral or basic solutions for collection of HCl greatly reduced the dissolution of any Cl_2 present in the emissions.

Reducing agents such as SO2 may cause a positive bias in the chlorine measurement by the following reaction:

$$HClO + HSO_3^- = H_2SO_4 + Cl^-$$

4.0 METHOD TARGET COMPOUNDS

chlorine
hydrogen chloride

5.0 APPLICABLE CAA AMENDMENTS POLLUTANTS

chlorine
hydrogen chloride

EPA METHOD TO-1 - VOCs In Ambient Air Using Tenax ® & GC/MS

Determination of Volatile Organic Compounds in Ambient Air Using TENAX ® Adsorption and Gas Chromatography/Mass Spectrometry (GC/MS)

REFERENCE:

Compendium of Methods for the Determination of Toxic Organic Compounds in Ambient Air. U.S. Environmental Protection Agency. EPA-600/4-89-017 (Supplements: 600/4-87-006, 600/4-87-013).

1.0 SCOPE AND APPLICATION

Method TO-1 is used to collect and determine volatile, non-polar organics (aromatic hydrocarbons, chlorinated hydrocarbons) that can be captured on Tenax ® and determined by thermal desorption techniques. The compounds to be determined by this technique have boiling points in the range of 80-200°C.

2.0 SUMMARY OF METHOD

Air is drawn through a cartridge containing 1-2 g of Tenax ®. The cartridge is analyzed in the laboratory and purged with an inert gas into first a gas chromatograph (GC) followed by a mass spectrometer (MS). Only capillary GC techniques should be used. The GC temperature is increased through a temperature program and the compounds are eluted from the column on the basis of boiling points. The MS identifies and quantifies the compounds by mass fragmentation patterns. Compound identification is normally accomplished using a library search routine on the basis of GC retention time and mass spectral characteristics.

3.0 INTERFERENCES

The most common interferences are structural isomers.

4.0 METHOD TARGET COMPOUNDS

aromatic hydrocarbons
chlorinated hydrocarbons

5.0 APPLICABLE CAA AMENDMENTS POLLUTANTS

acrylonitrile
allyl chloride
benzene
bis(chloromethyl) ether
bromoform
1,3-butadiene
carbon disulfide
carbon tetrachloride
carbonyl sulfide
chlorobenzene
chloroform
chloromethyl methyl ether
chloroprene
cumene
1,4-dichlorobenzene
dichloroethyl ether
1,3-dichloropropene
1,1-dimethylhydrazine
1,4-dioxane
1,2-epoxybutane
ethyl acrylate
ethylbenzene
ethyl chloride
ethylene dibromide
ethylene dichloride
ethylene imine
ethylidene dichloride
hexachlorobutadiene
hexachloroethane
hexane
methyl bromide
methyl chloride
methyl chloroform
methyl ethyl ketone
methyl hydrazine
methylene chloride
methyl iodide
methyl isobutyl ketone
methyl isocyanate
methyl methacrylate
methyl tert-butyl ether
nitrobenzene
2-nitropropane
N-nitrosodimethylamine
propylene dichloride
propylene oxide
1,2-propyleneimine
styrene
1,1,2,2-tetrachloroethane
tetrachloroethylene
toluene
1,2,4-trichlorobenzene
trichloroethylene
2,2,4-trimethylpentane
vinyl acetate
vinyl bromide
vinyl chloride
vinylidene chloride
m, p, o-xylenes

EPA METHOD TO-2 - VOCs In Ambient Air By Carbon Molecular Sieve & GC/MS

Determination of Volatile Organic Compounds in Ambient Air by Carbon Molecular Sieve Adsorption and Gas Chromatography/Mass Spectrometry (GC/MS)

REFERENCE:

Compendium of Methods for the Determination of Toxic Organic Compounds in Ambient Air. U.S. Environmental Protection Agency. EPA-600/4-89-017 (Supplements: 600/4-87-006, 600/4-87-013).

1.0 SCOPE AND APPLICATION

Method TO-2 is used to collect and determine highly volatile, non-polar organics (vinyl chloride, vinylidene chloride, benzene, toluene) that can be captured on carbon molecular sieve and determined by thermal desorption techniques. The compounds to be determined by this technique have boiling points in the range of -15-120øC.

2.0 SUMMARY OF METHOD

Air is drawn through a cartridge containing 0.4 g of a carbon molecular sieve (CMS) adsorbent. The cartridge is analyzed in the laboratory by flushing with dry air to remove adsorbed moisture and purging the sample with helium while heating the cartridge to 350-400°C. The desorbed organics are collected in a cryogenic trap and flash evaporated into first a gas chromatograph (GC) followed by a mass spectrometer (MS). Only capillary GC techniques should be used. The GC temperature is increased through a temperature program and the compounds are eluted from the column on the basis of boiling points. The MS identifies and quantifies the compounds by mass fragmentation patterns. Compound identification is normally accomplished using a library search routine on the basis of GC retention time and mass spectral characteristics.

3.0 INTERFERENCES

The most common interferences are structural isomers.

4.0 METHOD TARGET COMPOUNDS

benzene
toluene
vinyl chloride Z vinylidene chloride

5.0 APPLICABLE CAA AMENDMENTS POLLUTANTS

allyl chloride
benzene
carbon tetrachloride
chloroform
ethyl chloride
ethylene dichloride
methyl bromide
methyl chloride
methyl chloroform
methylene chloride
toluene
trichloroethylene
vinyl bromide
vinyl chloride
vinylidene chloride

EPA METHOD TO-4 - Organochlorine Pesticides & PCBs In Ambient Air By GC/ECD

Determination of Organochlorine Pesticides and Polychlorinated Biphenyls (PCBs) in Ambient Air

REFERENCE:

Compendium of Methods for the Determination of Toxic Organic Compounds in Ambient Air. U.S. Environmental Protection Agency. EPA-600/4-89-017 (Supplements: 600/4-87-006, 600/4-87-013).

1.0 SCOPE AND APPLICATION

Method TO-4 is used to analyze a variety of organochlorine pesticides and polychlorinated biphenyls (PCBs) in ambient air.

2.0 SUMMARY OF METHOD

A modified high volume sampler consisting of a glass fiber filter with a polyurethane foam (PUF) backup absorbent cartridge is used to sample air at a rate of 200-280 L/min. PCBs and pesticides are recovered by Soxhlet extraction with 5 percent ether in hexane. The extracts are reduced in volume with a Kuderna-Danish concentration technique and subject to column chromatography cleanup. The extracts are analyzed using gas chromatography with electron capture detection.

3.0 INTERFERENCES

Extraneous organic compounds may interfere. It may be difficult to identify an individual pesticide or PCBs in a multiple component mixture.

4.0 METHOD TARGET COMPOUNDS

organochlorine pesticides
polychlorinated biphenyls

5.0 APPLICABLE CAA AMENDMENTS POLLUTANTS

DDE
heptachlor
lindane
methoxychlor
parathion
toxaphene

EPA METHOD TO-5 - Aldehydes & Ketones In Ambient Air By HPLC/UV Of DNPHs

Determination of Aldehydes and Ketones in Ambient Air Using High Performance Liquid Chromatography (HPLC)

REFERENCE:

Compendium of Methods for the Determination of Toxic Organic Compounds in Ambient Air. U.S. Environmental Protection Agency. EPA-600/4-89-017 (Supplements: 600/4-87-006, 600/4-87-013).

1.0 SCOPE AND APPLICATION

Method TO-5 is used to analyze individual aldehydes and ketones in ambient air.

2.0 SUMMARY OF METHOD

Air is drawn through a midget impinger containing dinitrophenylhydrazine (DNPH) reagent and isooctane where the target compounds are derivatized. The organic fraction is evaporated to dryness and dissolved in methanol. The derivatives are determined using reverse phase high performance liquid chromatography HPLC with an ultra-violet detector.

3.0 INTERFERENCES

Isomeric aldehydes or ketones may be unresolved by the HPLC system.

4.0 METHOD TARGET COMPOUNDS

aldehydes
ketones

5.0 APPLICABLE CAA AMENDMENTS POLLUTANTS

acetaldehyde
acetophenone
acrolein
formaldehyde
methyl ethyl ketone
methyl isobutyl ketone
propionaldehyde
quinone
trifluralin

EPA METHOD TO-6 - Phosgene In Ambient Air By HPLC/UV

Determination of Phosgene in Ambient Air Using High Performance Liquid Chromatography (HPLC)

REFERENCE:

Compendium of Methods for the Determination of Toxic Organic Compounds in Ambient Air. U.S. Environmental Protection Agency. EPA-600/4-89-017 (Supplements: 600/4-87-006, 600/4-87-013).

1.0 SCOPE AND APPLICATION

Method TO-6 is used to determine phosgene in ambient air at the 0.1 ppbv level.

2.0 SUMMARY OF METHOD

Air is drawn through a midget impinger containing an aniline/toluene mixture. The solution is heated to dryness and dissolved in acetonitrile. The sample is analyzed by reverse phase high performance liquid chromatography (HPLC) with an ultraviolet detector.

3.0 INTERFERENCES

Chloroformates and acidic materials may interfere.

4.0 METHOD TARGET COMPOUNDS

phosgene

5.0 APPLICABLE CAA AMENDMENTS POLLUTANTS

phosgene

EPA METHOD TO-7 - N-Nitrosodimethylamine In Ambient Air By GC/MS

Determination of N-Nitrosodimethylamine in Ambient Air Using Gas Chromatography (GC)

REFERENCE:

Compendium of Methods for the Determination of Toxic Organic Compounds in Ambient Air. U.S. Environmental Protection Agency. EPA-600/4-89-017 (Supplements: 600/4-87-006, 600/4-87-013).

1.0 SCOPE AND APPLICATION

Method TO-7 is used to determine N-nitrosodimethylamine in ambient air.

2.0 SUMMARY OF METHOD

Air is drawn through a Thermosorb/N adsorbent cartridge at a rate of 2 L/min. The cartridges are pre-eluted with dichloromethane to remove interferences. The sample is eluted with acetone and injected into a gas chromatograph (GC) that is followed by a detection system such as mass spectrometer (MS). A Carbowax 20M capillary column should be used in the GC. The GC temperature is increased through a temperature program and the compounds are eluted from the column on the basis of boiling points. The MS identifies and quantifies the compounds by mass fragmentation patterns. Compound identification is normally accomplished using a library search routine on the basis of GC retention time and mass spectral characteristics.

3.0 INTERFERENCES

Compounds with similar GC retention times and similar detectable MS ions may interfere.

4.0 METHOD TARGET COMPOUNDS

N-nitrosodimethylamine

5.0 APPLICABLE CAA AMENDMENTS POLLUTANTS

N-nitrosodiethylamine
N-nitrosomorpholine

EPA METHOD TO-8 - Phenols & Cresols In Ambient Air By HPLC

Determination of Phenol and Methyl Phenols (Cresols) in Ambient Air Using High Performance Liquid Chromatography (HPLC)

REFERENCE:

Compendium of Methods for the Determination of Toxic Organic Compounds in Ambient Air. U.S. Environmental Protection Agency. EPA-600/4-89-017 (Supplements: 600/4-87-006, 600/4-87-013).

1.0 SCOPE AND APPLICATION

Method TO-8 is used to determine cresols and phenols in ambient air at the 1-5 ppbv level.

2.0 SUMMARY OF METHOD

Air is drawn through two midget impingers, each containing sodium hydroxide. The solution is adjusted to pH 4 in the laboratory after sampling, and analyzed by reverse-phase high performance liquid chromatography (HPLC) with ultraviolet, electrochemical, or fluorescent detection.

3.0 INTERFERENCES

Compounds having the same HPLC retention times will interfere with this method. The phenolic compounds of interest may be oxidized during sampling.

4.0 METHOD TARGET COMPOUNDS

cresols
phenols

5.0 APPLICABLE CAA AMENDMENTS POLLUTANTS

cresylic acid
o-cresol
m-cresol
p-cresol

EPA METHOD TO-9 - PCDDs In Ambient Air By HRGC/HRMS

Determination of Polychlorinated Di-Benzo-p-Dioxins (PCDDS) in Ambient Air Using High Resolution Gas Chromatography/ High-Resolution Mass Spectrometry (HRGC/HRMS)

REFERENCE:

Compendium of Methods for the Determination of Toxic Organic Compounds in Ambient Air. U.S. Environmental Protection Agency. EPA-600/4-89-017 (Supplements: 600/4-87-006, 600/4-87-013).

1.0 SCOPE AND APPLICATION

Method TO-9 is used to determine p-dioxins in ambient air, especially 1,2,3,4- and 2,3,7,8-tetrachlorodibenzo-p-dioxins (TCDDs), polychlorinated dibenzo-p-dioxins (HxCDDs) and octachlorodibenzo-p-dioxin (OCDD). With careful attention to reagent purity, the method can detect PCDDs at levels below 15 pg/m^3.

2.0 SUMMARY OF METHOD

Air is drawn through a glass fiber filter with a polyurethane foam (PUF) back-up absorbent cartridge. Silica gel can be used in place of PUF to give lower detection limits, but silica gel does not always give consistent sample recoveries and will require extensive clean-up. The filters and PUF adsorbent cartridge are extracted together with benzene, diluted with hexane and cleaned up using column chromatography. The sample is injected into a high resolution gas chromatograph (HRGC) in line with a high-resolution mass spectrometer (HRMS). In the HRGC, temperature is increased through a temperature program and the sample compounds are eluted from the column on the basis of boiling points. The HRMS identifies and quantifies the compounds by mass fragmentation patterns. Compound identification is normally accomplished using a library search routine on the basis of HRGC retention time and mass spectral characteristics.

3.0 INTERFERENCES

Polychlorindated biphenyls (PCBs), methoxybiphenyls, chlorinated hydroxydiphenylethers and naphthalenes, DDE, DDT, and other compounds with similar retention times and mass fractions are analytical interferences. Inaccurate measurements can occur if PCDDs are retained or absorbed onto particulate matter, the filter, or PUF cartridge, or are chemically changed during sampling and storage in ways not traceable with isotopically labelled spikes.

4.0 METHOD TARGET COMPOUNDS

polychlorinated dibenzo-p-dioxins
polychlorinated dibenzofurans

5.0 APPLICABLE CAA AMENDMENTS POLLUTANTS

dibenzofurans
2,3,7,8-tetrachlorodibenzo-p-dioxin

EPA METHOD TO-10 - Organochlorine Pesticides In Ambient Air By PUF & GC/ECD

Determination of Organochlorine Pesticides in Ambient Air Using Low Volume Polyurethane Foam (PUF) Sampling with Gas Chromatography/Electron Capture Detector (GC/ECD)

REFERENCE:

Compendium of Methods for the Determination of Toxic Organic Compounds in Ambient Air. U.S. Environmental Protection Agency. EPA-600/4-89-017 (Supplements: 600/4-87-006, 600/4-87-013).

1.0 SCOPE AND APPLICATION

Method TO-10 is used to determine organochlorine pesticides in ambient air. This method is applicable to multi-component mixtures containing the compounds of interest in concentrations of 0.01 to 50 $\mu g/m^3$. The sampling procedure is also applicable to other pesticides that may be determined by gas chromatography coupled with a nitrogen-phosphorus detector, flame photometric detector, Hall electrolytic conductivity detector, or a mass spectrometer (MS). For some organopesticides, high performance liquid chromatography coupled with ultraviolet or electrochemical detection may be preferred.

2.0 SUMMARY OF METHOD

A low volume sampler is used to collect source vapors onto a sorbent cartridge containing polyurethane foam. Pesticides are extracted from the cartridge with 5 percent diethyl ether in hexane and recovered by Soxhlet extraction with 5 percent ether in hexane. The extracts are reduced in volume with a Kuderna-Danish concentration technique and subjected to an alumina chromatographic cleanup procedure. The sample extracts are analyzed by gas chromatography coupled with an electron capture detector (ECD).

3.0 INTERFERENCES

ECD and other detectors (except the MS) will be subject to responses from a variety of compounds other than the compounds of interest. Polychlorinated biphenyls in particular may interfere. Certain organochlorine pesticides (e.g., chlordane) are complex mixtures of individual compounds that can make accurate quantification of a particular compound in the mixture difficult.

4.0 METHOD TARGET COMPOUNDS

organochlorine pesticides

5.0 APPLICABLE CAA AMENDMENTS POLLUTANTS

captan
2,4-D salts and esters
DDE
heptachlor
lindane
methoxychlor
parathion
toxaphene
trifluralin

EPA METHOD TO-11 - Formaldehyde In Ambient Air By HPLC/UV Of Its DNPH

Determination of Formaldehyde in Ambient Air Using Adsorbent Cartridge Followed by High Performance Liquid Chromatography (HPLC)

REFERENCE:

Compendium of Methods for the Determination of Toxic Organic Compounds in Ambient Air. U.S. Environmental Protection Agency. EPA-600/4-89-017 (Supplements: 600/4-87-006, 600/4-87-013).

1.0 SCOPE AND APPLICATION

Method TO-11 is used to determine formaldehyde in ambient air. Other aldehydes and ketones can be detected with a modification of the basic procedure.

2.0 SUMMARY OF METHOD

Air is drawn through a midget impinger sampling train (without impinger) containing a silica gel cartridge coated with acidified dinitrophenylhydrazine (DNPH). The cartridge is eluted with acetonitrile in the laboratory to form a formaldehyde-DPNH derivative. The concentration of formaldehyde is determined with isocratic reverse phase high performance liquid chromatography (HPLC) with ultraviolet absorption detection.

3.0 INTERFERENCES

Isomeric aldehydes and ketones, and other compounds with the same HPLC retention times as formaldehyde may interfere with this method.

4.0 METHOD TARGET COMPOUNDS

formaldehyde

5.0 APPLICABLE CAA AMENDMENTS POLLUTANTS

acetaldehyde
acetophenone
acrolein
2-chloroacetophenone
formaldehyde
isophorone
methyl ethyl ketone
methyl isobutyl ketone
phosgene
propionaldehyde
quinone

EPA METHOD TO-13 - PAHs In Ambient Air By GC/MS Or GC/FID Or HPLC

Determination of Polynuclear Aromatic Hydrocarbons (PAHs) in Ambient Air Using High Volume Sampling with Gas Chromatography/Mass Spectrometry (GC/MS) and High Resolution Liquid Chromatographic (HRLC)

REFERENCE:

Compendium of Methods for the Determination of Toxic Organic Compounds in Ambient Air. U.S. Environmental Protection Agency. EPA-600/4-89-017 (Supplements: 600/4-87-006, 600/4-87-013).

1.0 SCOPE AND APPLICATION

Method TO-13 is used to determine benzo(a)pyrene and other polynuclear aromatic hydrocarbons (PAHs) in ambient air. Nitro-PAHs are not included with this method.

2.0 SUMMARY OF METHOD

Air is drawn through a filter and adsorbent cartridge containing XAD-2 ® or polyurethane foam. The filters and adsorbents are extracted by Soxhlet apparatus. The extract is reduced in volume with a Kuderna-Danish concentration technique and subjected to cleanup with silica gel column chromatography. The sample is further concentrated with a Kuderna-Danish evaporator and analyzed by either gas chromatography equipped with flame ionization or a mass spectrometer, or high performance liquid chromatography.

3.0 INTERFERENCES

Method interferences may be caused by contaminants in solvents, reagents, glassware, and sampling hardware. Matrix interferences may be caused by contaminants that are co-extracted with the sample. Heat, ozone, nitrogen dioxide, and ultraviolet light may cause sample degradation.

4.0 METHOD TARGET COMPOUNDS

polynuclear aromatic hydrocarbons

5.0 APPLICABLE CAA AMENDMENTS POLLUTANTS

acetamide
acetophenone
2-acetylaminofluorene
acrylamide
acrylic acid
4-aminobiphenyl
aniline
o-anisidine
benzotrichloride
benzyl chloride
biphenyl
bis(2-ethylhexyl) phthalate
caprolactam
carbaryl
catechol
chloramben
chlordane
chloroacetic acid
2-chloroacetophenone
chlorobenziliate
o,m,p-cresol, cresylic acid
cumene
1,2-dibromo-3-chloropropane
dibutyl phthalate
1,4-dichlorobenzene
3,3'-dichlorobenzidine
dichloroethyl ether
dichlorvos
diethanolamine
N,N-diethylaniline
diethyl sulfate
3,3'-dimethoxybenzidine
dimethylaminoazobenzene
3,3'-dimethylbenzidine
dimethyl carbamoyl chloride
dimethyl formamide
dimethyl phthalate
dimethyl sulfate
4,6-dinitro-o-cresol and salts
2,4-dinitrophenol
2,4-dinitrotoluene
1,2-diphenylhydrazine
epichlorohydrin
ethyl carbamate
ethylene dibromide
ethylene glycol
ethylene thiourea
glycol ethers
hexachlorobenzene
hexachlorobutadiene
hexachlorocyclopentadiene
hexachloroethane
hexamethylene-1,6-diisocyanate
hexamethylphosphoramide
hydroquinone
maleic anhydride
methoxychlor
methylene diphenyl isocyanate
4,4'-methylenedianiline
4,4'-methylene bis(2-chloroaniline)
naphthalene
4-nitrobiphenyl
4-nitrophenol
2-nitropropane
N-nitroso-N-methylurea
N-nitrosodimethylamine
N-nitrosomorpholine
pentachloronitrobenzene
pentachlorophenol
phenol
p-phenylenediamine
phthalic anhydride
polychlorinated biphenyls
1,3-propane sultone
beta-propiolactone
propoxur
quinoline
styrene oxide
2,4-toluenediamine
o-toludine
1,2,4-trichlorobenzene
2,4,5-trichlorophenol
2,4,6-trichlorophenol
triethylamine
o-, m-, p-xylenes

EPA METHOD TO-14 - VOCs In Ambient Air By Canister Sampling & HRGC

Determination of Volatile Organic Compounds (VOCs) in Ambient Air Using SUMMA ® Polished Canister Sampling and Gas Chromatography (GC)

REFERENCE:

Compendium of Methods for the Determination of Toxic Organic Compounds in Ambient Air. U.S. Environmental Protection Agency. EPA-600/4-89-017 (Supplements: 600/4-87-006, 600/4-87-013).

1.0 SCOPE AND APPLICATION

Method TO-14 is used to determine semi-volatile and volatile organic compounds in ambient air. The sample canisters can be placed above or below atmospheric pressure. Pressurized samples can be detected at the ppbv level.

2.0 SUMMARY OF METHOD

Air is drawn through a sampling train into a pre-evacuated sample SUMMA ® canister. The canister is attached to the analytical system. Water vapor is reduced in the gas stream by a Nafion dryer and VOCs are concentrated by collection into a cryogenically-cooled trap. The cryogen is removed and the temperature of the sample raised to volatilize the sample into a high resolution gas chromatogrph (HRGC). The GC temperature is increased through a temperature program and the compounds are eluted from the column on the basis of boiling points into a detector. The choice of detector depends on the specificity and sensitivity required by the analysis. Non-specific detectors include nitrogen-phosphorus detectors, flame ionization detectors, electron capture detectors, and photoionization detectors. Specific detectors include a mass spectrometer (MS) operating in the selected ion mode or the SCAN mode, or an ion trap detector. Identification errors can be reduced by employing simultaneous detection by different detectors.

3.0 INTERFERENCES

Interferences can occur because of moisture contamination in the dryer. Polar organic compounds may be lost with moisture removed in the GC/MS dryer.

4.0 METHOD TARGET COMPOUNDS

semi-volatile organic compounds
volatile organic compounds
Freon ®

5.0 APPLICABLE CAA AMENDMENTS POLLUTANTS

acetonitrile
acrylonitrile
allyl chloride
benzene
bis(choromethyl) ether
bromoform
1,3-butadiene
carbon disulfide
carbon tetrachloride
carbonyl sulfide
chlorobenzene
chloroform
chloromethyl methyl ether
chloroprene
cumene
1,4-dichlorobenzene
dichloroethyl ether
1,3-dichloropropene
1,1-dimethylhydrazine
1,4-dioxane
1,2-epoxybutane
ethyl acrylate
ethylbenzene
ethyl chloride
ethylene dibromide
ethylene dichloride
ethylene imine
ethylene oxide
ethylidene dichloride
hexachlorobutadiene
hexachloroethane
hexane
methanol
methyl bromide
methyl chloride
methyl chloroform
methyl ethyl ketone
methyl hydrazine
methylene chloride
methyl iodide
methyl isobutyl ketone
methyl isocyanate
methyl methacrylate
methyl tert-butyl ether
nitrobenzene
2-nitropropane
propylene dichloride
propylene oxide
1,2-propyleneimine
styrene
1,1,2,2-tetrachloroethane
tetrachloroethylene
toluene
trichloroethylene
2,2,4-trimethylpentane
vinyl acetate
vinyl bromide
vinyl chloride
vinylidene chloride
o-, m-, p-xylenes

California CARB METHOD 427 - Isokinetic Sampling & Analysis Of Asbestos

Determination of Asbestos Emissions from Stationary Sources

REFERENCE:

Stationary Source Test Methods, Volume III: Methods for Determining Emissions of Toxic Air Contaminants from Stationary Sources. State of California Air Resources Board, Monitoring and Laboratory Division. Sacramento, CA: 1989.

1.0 SCOPE AND APPLICATION

CARB Method 427 is used to determine asbestos emissions from stationary sources. The method describes a stack sampling method, but an alternate non-stack method (NIOSH Method 7400) can be used. The method of analysis described in NIOSH 7400 (light microscopy), and permitted under Method 427, does not distinguish asbestos from other mineral fibers.

2.0 SUMMARY OF METHOD

The method consists of a stack sampling train designed to isokinetically collect a particulate sample from a known sample of gas, corrected to a dry basis. A probe is used to withdraw the sample isokinetically from the duct. A tared filter in filter holder is placed behind the probe within the stack to collect the bulk of the particulate sample. A series of impingers or a condenser following the filter assembly outside the stack is used to both measure the moisture present in the gas and protect the dry gas meter that is located at the end of the train.

The sample is composed of the particulate-laden filter and the dried washings of the probe, nozzle, and front half of the filter holder. Transmission electron microscopy (TEM) can be used to analyze the collected sample. TEM analysis is used to classify the fibers as chrysotile, amphibole, or non-asbestos and to describe the aggregation of asbestos into single fibers, bundles, or mats.

3.0 INTERFERENCES

Phase contrast light microscopy will not differentiate between asbestos and other mineral fibers.

4.0 METHOD TARGET COMPOUNDS

asbestos

5.0 APPLICABLE CAA AMENDMENTS POLLUTANTS

asbestos
mineral fibers

California CARB Method 429 - PAHs From Stationary Sources By HRGC/MS Or HRMS

Determination of Polycyclic Aromatic Hydrocarbon (PAH) Emissions from Stationary Sources

REFERENCE:

Stationary Source Test Methods, Volume III: Methods for Determining Emissions of Toxic Air Contaminants from Stationary Sources. State of California Air Resources Board, Monitoring and Laboratory Division. Sacramento, CA: 1989.

1.0 SCOPE AND APPLICABILITY

CARB Method 429 is used to determine polycyclic aromatic hydrocarbon (PAH) emissions from stationary sources. The sensitivity that can ultimately be achieved for a given sample will depend upon the types and concentrations of other chemical compounds in the sample, as well as the original sample size and instrument sensitivity.

The limitations on extension of the methodology to compounds other than the method target compounds are the availability of standards and the ability to perform chromatographic separations on vaporized compounds.

2.0 SUMMARY OF METHOD

Particulate and gaseous-phase PAH are extracted isokinetically from the stack and collected on a filter, on XAD-2 ® resin, in the impingers, or in upstream sampling train components. Only the total amounts of each PAH in stack emissions can be determined with this method. It has not been demonstrated that the partitioning in the different parts of the sampling train is representative of the partitioning in the stack gas sample for particulate and gaseous PAH.

The analytical method entails the addition of internal standards to all samples in known quantities, matrix-specific extraction of the sample with appropriate organic solvents, preliminary fractionation and cleanup of extracts (if necessary), and analysis of the processed extract for PAH using high-resolution capillary column gas chromatography coupled with either low resolution or high resolution mass spectrometry.

3.0 INTERFERENCES

Modified Method 5 sampling train can cause artifactual formation and PAH transformation. The fact that PAH can degrade or transform on sample filters is well documented. When trapped on filters, certain reactive PAH such as benzo[a]pyrene, benzo[a]anthracene, and fluoranthene can readily react with stack gases. Low levels of nitric acid and higher levels of nitrogen oxides, ozone, and sulfur oxides have been known to react with these PAH.

PAH degradation may be of even greater concern when they are trapped in the impingers. When stack gases such as sulfur oxides and nitrogen oxides come in contact with the impinger water, they are converted into sulfuric acid and nitric acid, respectively. There is evidence that under such conditions certain PAH will be degraded. It is recommended that the levels in the impingers be used as a qualitative tool to determine if breakthrough has occurred in the resin.

In order to assess the effects of ozone, sulfur oxides, and nitrogen oxides, the tester should monitor concurrently for these gases during PAH sampling.

4.0 METHOD TARGET COMPOUNDS

naphthalene
acenaphthylene
acenaphthene
fluorene
phenanthrene
anthracene
fluoranthene
pyrene
benz[a]anthracene
chrysene
benzo[b]fluoranthene
benzo[k]fluoranthene
benzo[a]pyrene
benzo[ghi]perylene
dibenz[a,h]anthracene
indeno[1.2.3-cd]pyrene

5.0 APPLICABLE CLEAN AIR ACT LIST COMPOUNDS

polycyclic organic matter

California CARB METHOD 431 - Ethylene Oxide From Stationary Sources By GC/FID

Determination of Ethylene Oxide Emissions from Stationary Sources

REFERENCE:

Stationary Source Test Methods, Volume III: Methods for Determining Emissions of Toxic Air Contaminants from Stationary Sources. State of California Air Resources Board, Monitoring and Laboratory Division. Sacramento, CA: 1989.

1.0 SCOPE AND APPLICATION

CARB Method 431 is used to determine ethylene oxide emissions from sterilization chambers in pounds per sterlization cycle. The method can be modified to determine emissions of ethylene oxide from other stationary sources.

2.0 SUMMARY OF METHOD

A sterilization chamber is operated empty of items to be sterilized, minimizing interferences and simulating worst-case vented emission conditions. Volumetric flow of vented gas is monitored and vented gas is analyzed repeatedly by gas chromatography during chamber purging. Total emissions of ethylene oxide for the sterilization cycle are calculated from curves of flow and concentration over time.

Ethylene oxide is used in sterilizers at lethal concentrations. Contact with vented gas can cause skin burns. Inhalation can cause injury or death. Caution should be observed to avoid contact with or inhalation of vented gas.

The lower limit of sensitivity will vary according to the gas chromatography equipment and span gases used. With appropriate span gases the method is expected to achieve useful accuracy over the expected range of emissions from either controlled or uncontrolled sterilizers.

3.0 INTERFERENCES

Ethylene oxide is frequently used in sterilizers in a mixture with dichlorodifluoromethane (Freon 12). If the gas chromatographic conditions are not well selected, the ethylene oxide peak may be overwhelmed by the tail of the Freon 12 peak when testing emissions at low concentrations. This matrix problem is effectively eliminated by selecting a gas chromatographic column where ethylene oxide elutes before Freon 12.®

4.0 METHOD TARGET COMPOUNDS

ethylene oxide

5.0 APPLICABLE CAA AMENDMENTS POLLUTANTS

ethylene oxide

NIOSH METHOD 2515 - Diazomethane By GC/FID

Diazomethane

REFERENCE:

NIOSH Manual of Analytical Methods, Part 1: NIOSH Monitoring Methods, Volume 1. U.S. Department of Health, Education, and Welfare, National Institute for Occupational Safety and Health. Cincinnati, Ohio: Revised 1985.

1.0 SCOPE AND APPLICATION

NIOSH Method 2515 is used to determine 0.1 to 0.6 ppm (0.2 to 1 mg/m^3) diazomethane for a 10 L air sample.

2.0 SUMMARY OF METHOD

An air sample is obtained using a solid sorbent tube (octanoic acid-coated XAD-2 ® resin, 100 mg/50 mg) attached to a personal sampling pump with flexible tubing. The sample is collected at a known flow rate of 0.2 ñ 0.03 L/min for a total sample size of 6 to 30 L.

Analysis is by gas chromatography (GC) with flame ionization detector. A stainless steel column, 3 m x 3 mm OD with 5% SP-1000 on 100/120 mesh Chromosorb WHP preceded by a 15 cm x 3 mm OD stainless steel precolumn 80/100 mesh Gas Chrom Q is recommended.

Methyl octanoate (analyte) is desorbed in carbon disulfide with an internal standard (tridecane) for daily calibration over a range of 1 to 32 æg methyl octanoate per sample. The GC is set according to recommendations and conditions given in this method for optimum and the peak resolution samples are injected manually using the solvent flush technique or with an autosampler.

The concentration of diazomethane per sample is calculated by converting mg methyl octanoate to mg diazomethane by multiplying the corrected mg per sample by the molecular weight ratio, 41.04/158.24, and calculating the apparent concentration of diazomethane in the air volume sampled.

3.0 INTERFERENCES

The collection efficiency and reaction of diazomethane with the octanoic acid-coated resin may be strongly dependent on sample flow rates; therefore, all samples must be collected at a flow rate of 0.2 L/min only.

4.0 METHOD TARGET COMPOUNDS

diazomethane

5.0 APPLICABLE CAA AMENDMENTS POLLUTANTS

diazomethane

NIOSH METHOD 7400 - Fibers By Light Microscopy

Fibers

REFERENCE:

NIOSH Manual of Analytical Methods, Part 1: NIOSH Monitoring Methods, Volume 1. U.S. Department of Health, Education, and Welfare, National Institute for Occupational Safety and Health. Cincinnati, Ohio: Revised 1985.

1.0 SCOPE AND APPLICATION

NIOSH Method 7400 gives an index of airborne fibers but will not differentiate asbestos from other mineral fibers. Fibers less than 0.25 æm will not be detected by this method.

2.0 SUMMARY OF METHOD

A sample is withdrawn through a 0.8 to 1.2 æm ester membrane filter in a 25 mm cassette filter holder at 0.5 to 16 L/min. The sample is prepared according to the acetone/triacetin "hot block" method onto a phase-shift test slide. The fibers are counted manually using a light microscope and a Walton-Beckett graticule.

3.0 INTERFERENCES

Any airborne fibers may interfere as all particles meeting the counting criteria will be counted. Chain-like fibers may appear fibrous. High levels of non-fibrous dust particles may obscure fibers in the field of view and increase the detection limit.

4.0 METHOD TARGET COMPOUNDS

asbestos
various mineral fibers

5.0 APPLICABLE CAA AMENDMENTS POLLUTANTS

asbestos
fine mineral fibers

NIOSH METHODS 9010 AND 9012 - Colorimetric Total And Amenable Cyanide Analysis

Total and Amenable Cyanide Analysis

REFERENCE:

NIOSH Manual of Analytical Methods, Part 1: NIOSH Monitoring Methods, Volume 1. U.S. Department of Health, Education, and Welfare, National Institute for Occupational Safety and Health. Cincinnati, Ohio: Revised 1985.

1.0 SCOPE AND APPLICATION

NIOSH Methods 9010 and 9012 are used to analyze an aqueous sample for total inorganic cyanide. The methods detect simple soluble salts or complex radicals, total cyanide, and cyanide amenable to chlorination.

2.0 SUMMARY OF METHOD

The sample is refluxed with strong acid and distilled into an absorber/scrubber that contains sodium hydroxide solution. Cyanide is released as hydrogen cyanide. The cyanide ion is converted to cyanochloride by reaction with chloramin-T at a pH less than 8. Color is formed by addition of pyridine-barbituric acid reagent. The concentration of cyanide ion is determined colorimetrically (UV) either manually (Method 9010) or with an automated system (Method 9012) by comparison to known standards.

3.0 INTERFERENCES

Sulfides, nitrates, or nitrites adversely affect the colorimetric procedure but can be eliminated as interferences by pretreating the sample.

4.0 METHOD TARGET COMPOUNDS

inorganic cyanide

5.0 APPLICABLE CAA AMENDMENTS POLLUTANTS

cyanide compounds

OSHA METHOD ID-101 - Chlorine By Ion Specific Electrode Analysis

Chlorine in Workplace Atmospheres

REFERENCE:

OSHA Method ID-101

1.0 SCOPE AND APPLICATION

The method describes the collection and analysis of airborne chlorine for industries where chlorine is used as a bleaching agent or for chlorination of various organic compounds.

2.0 SUMMARY OF METHOD

An air sample is drawn into a solution of 0.1 percent sulfamic acid. An aliquot of the sample is reacted with acidic potassium iodide. Chlorine oxidizes the potassium iodide to iodine which is measured by an ion specific electrode. The detection limit is 0.4 mg/m^3 for a 15 liter air volume.

3.0 INTERFERENCES

Strong oxidizing agents including iodate, bromine, cupric ion and manganese dioxide can interfere with the analysis. High silver and mercuric concentrations (greater than 15 ppm) also can cause interference.

4.0 METHOD TARGET COMPOUNDS

Chlorine.

5.0 APPLICABLE CAA AMENDMENTS POLLUTANTS

Chlorine.

EPA METHOD 6 - Sulphur Dioxide From Stationary Sources By Titration

Determination of Sulphur Dioxide Emissions from Stationary Sources

REFERENCE:

U.S. Environmental Protection Agency. Code of Federal Regulations. Title 40, Part 60, Appendix A. Washington, D.C. Office of the Federal Register, July 1, 1987.

1.0 SCOPE AND APPLICATION

EPA Method 6 is used to determine sulphur dioxide (SO_2) emissions from stationary sources.

2.0 SUMMARY OF METHOD

A heated probe is used to collect a sample from the source. The probe is equipped with a filter (either in or out of stack) to remove particulates and sulfuric acid mist (including sulphur trioxide). The sample is collected in impingers filled with isopropanol and hydrogen peroxide. The concentration of SO_2 is determined by titration of the sample with barium perchlorate to a thorin endpoint.

The impinger solution can be modified to allow sampling/analysis of other compounds.

3.0 INTERFERENCES

Free ammonia, water soluble cations, and fluorides may interfere.

4.0 METHOD TARGET COMPOUNDS

sulphur dioxide

5.0 APPLICABLE CAA AMENDMENTS POLLUTANTS

ammonia
cyanide

EPA METHODS 7C AND 7D - NOx By Colorimetric Or Ion Chromatographic Methods

Determination of Nitrogen Oxide Emissions from Stationary Sources-Alkaline-Permanganate Methods

REFERENCE:

U.S. Environmental Protection Agency. Code of Federal Regulations. Title 40, Part 60, Appendix A. Washington, D.C. Office of the Federal Register, July 1, 1987.

1.0 SCOPE AND APPLICATION

EPA Methods 7C and 7D are used to determine nitrogen oxide (NOx) emissions from fossil-fuel fired steam generators, electric utility plants, nitric acid plants, among other sources.

2.0 SUMMARY OF METHOD

An integrated sample is collected in a heated probe packed with glass wool for particulate collection. The sample is passed through a series of impingers containing alkaline potassium permanganate solution. NOx is oxidized to nitrate ions. The nitrate is reduced to nitrite with cadmium and the nitrite is analyzed colorimetrically (Method 7C) or is analyzed as nitrate by ion chromatography (Method 7D) against known standards. The detection limits are 7 ppm NOx when sampling at 500 cc/min for 1 hour.

3.0 INTERFERENCES

Sulphur dioxide and ammonia may interfere.

4.0 METHOD TARGET COMPOUNDS

nitrogen dioxide
nitrogen oxide

5.0 APPLICABLE CAA AMENDMENTS POLLUTANTS

phosphine

EPA METHOD 12 - Lead By Atomic Absorption

Determination of Inorganic Lead Emissions from Stationary Sources

REFERENCE:

U.S. Environmental Protection Agency. Code of Federal Regulations. Title 40, Part 60, Appendix A. Washington, D.C. Office of the Federal Register, July 1, 1987.

1.0 SCOPE AND APPLICATION

This method applies to the determination of inorganic lead emissions from specified stationary sources only.

2.0 SUMMARY OF METHOD

Particulate and gaseous lead emissions are withdrawn isokinetically from the source and collected on a filter and in dilute nitric acid. The collected samples are digested in acid solution and analyzed by atomic absorption (AA) spectrometry using an air acetylene flame.

3.0 INTERFERENCES

Sample matrix effects may interfere with the analysis for lead by flame atomic absorption. If matrix interference is suspected, the analyst may confirm the presence of these matrix effects and frequently eliminate the interference by using the Method of Standard Additions.

High concentrations of copper may interfere with the analysis for lead at 217.0 nm. This interference can be avoided by analyzing the samples at 283.3 nm.

4.0 METHOD TARGET COMPOUNDS

lead compounds

5.0 APPLICABLE CLEAN AIR ACT LIST COMPOUNDS

lead compounds

EPA METHOD 13A - Spectrophotometric Determination Of Total Fluoride

Determination of Total Fluoride Emissions from Stationary Sources by the SPADNS Zirconium Lake Method

REFERENCE:

Standards of Performance for New Stationary Sources. Compilation. U. S. Environmental Protection Agency, EPA-340/1-77-015.

1.0 SCOPE AND APPLICATION

EPA Method 13A is used to determine fluoride emissions from stationary sources. It does not measure fluorocarbons.

2.0 SUMMARY OF METHOD

Gaseous and particulate fluoride are withdrawn isokinetically from the source through a heated probe. A filter (with optional heating) is placed either before the first impinger or between the third and fourth impinger. The sample is collected in impingers containing deionized water and analyzed spectrophotometrically after distillation and addition of SPADNS reagent [4,5- dihydroxy-3-(p-sulfophenylazo)-2,7-naphthalene-disulfonic acid trisodium salt]. The range of this method is 0 to 1.4 µg F /mL. The sensitivity has not been determined.

3.0 INTERFERENCES

Large quantities of chloride will interfere with the analysis. Grease on sample-exposed surfaces may cause low results because of adsorption of fluoride.

4.0 METHOD TARGET COMPOUNDS

fluoride

5.0 APPLICABLE CAA AMENDMENTS POLLUTANTS

hydrogen fluoride

EPA METHOD 13B - Specific Ion Electrode Determination Of Total Fluoride

Determination of Total Fluoride Emissions from Stationary Sources by Specific Ion Electrode Method

REFERENCE:

Standards of Performance for New Stationary Sources. Compilation. U. S. Environmental Protection Agency, EPA-340/1-77-015.

1.0 SCOPE AND APPLICATION

EPA Method 13B is used to determine fluoride emissions from stationary sources. It does not measure fluorocarbons.

2.0 SUMMARY OF METHOD

Gaseous and particulate fluoride are withdrawn isokinetically from the source through a probe heated filter and collected in impingers containing deionized water. The sample is analyzed after distillation by a specific ion electrode that is calibrated with known standards.

3.0 INTERFERENCES

Grease on sample-exposed surfaces may cause low results because of adsorption of fluoride.

4.0 METHOD TARGET COMPOUNDS

fluoride

5.0 APPLICABLE CAA AMENDMENTS POLLUTANTS

hydrogen fluoride

EPA METHOD 14 - Fluoride By Spectrophotometry Or Specific Ion Electrode

Determination of Fluoride from Roof Monitors

REFERENCE:

Standards of Performance for New Stationary Sources. Compilation. U. S. Environmental Protection Agency, EPA-340/1-77-015.

1.0 SCOPE AND APPLICATION

EPA Method 14 is used to determine fluoride emissions from potroom roof monitors for primary aluminum plants when specified by New Source Performance Standards. It does not measure fluorocarbons.

2.0 SUMMARY OF METHOD

The sample is drawn into a manifold that is connected to a duct. A sample is withdrawn from the duct isokinetically and analyzed, either spectrophotometrically or by use of a specific ion electrode.

3.0 INTERFERENCES

Large quantities of chloride will interfere with the analysis. Grease on sample-exposed surfaces may cause low results because of adsorption of fluoride.

4.0 METHOD TARGET COMPOUNDS

fluoride

5.0 APPLICABLE CAA AMENDMENTS POLLUTANTS

hydrogen fluoride

EPA METHOD 15 - Sulphur Compounds By GC/FPD

Determination of Hydrogen Sulfide, Carbonyl Sulfide, and Carbon Disulfide from Stationary Sources

REFERENCE:

Standards of Performance for New Stationary Sources. Compilation. U. S. Environmental Protection Agency, EPA-340/1-77-015.

1.0 SCOPE AND APPLICATION

EPA Method 15 is used to determine hydrogen sulfide, carbonyl sulfide, and carbon disulfide from tail gas control units of sulfur recovery plants. Any method that uses the principle of gas chromatography separation with flame photometric detection can be substituted, providing that sample-line loss and the calibration precision are met.

2.0 SUMMARY OF METHOD

Gas is drawn through a heated sample probe followed by a particulate filter (Teflon ®) outside the stack. A sulfur dioxide (SO_2) scrubber made up of impingers containing citrate buffer removes SO_2 from the sample. The sample then is diluted with clean dry air (9:1) and fed into a gas chromatograph (GC) equipped with a flame photometric detector ((FPD). In the GC, temperature is increased through a temperature program and the compounds of interest are eluted from the column on the basis of boiling point. Concentrations of the sulfur compounds of interest are determined by calibration of the GC/FPD against known standards. The minimum detectable quantity depends on the sample size and would be about 0.5 ppm for a 1 mL sample.

3.0 INTERFERENCES

Moisture, carbon monoxide, carbon dioxide, SO_2, elemental sulfur, and alkali mist are possible interferences.

4.0 METHOD TARGET COMPOUNDS

carbon disulfide
carbonyl sulfide
hydrogen sulfide

5.0 APPLICABLE CAA AMENDMENTS POLLUTANTS

carbon disulfide
carbonyl sulfide

EPA METHOD 18 - Gaseous Organics By GC

Measurement of Gaseous Organic Compound Emissions by Gas Chromatography

REFERENCE:

40 CFR Part 60, Appendix A.

1.0 SCOPE AND APPLICATION

EPA Method 18 is used to analyze approximately 90 percent of the total gaseous organics emitted from an industrial source.

Method 18 does not include techniques to identify and measure trace amounts of organic compounds such as those found in building air and fugitive emission sources. This method will not determine compounds that: (1) are polymeric (high molecular weight); (2) can polymerize before analysis; or (3) have very low vapor pressures at stack or instrument conditions.

2.0 SUMMARY OF METHOD

A presurvey must be performed on each source to be tested. The purpose of the presurvey is to obtain all information necessary to design the emission test. The most important presurvey data are the average stack temperature and temperature range, approximate particulate concentration, static pressure, water vapor content, and identity and expected concentration of each organic compound to be analyzed. Some of this information can be obtained from literature surveys, direct knowledge, or plant personnel. However, presurvey samples of the gas shall be obtained for analysis to confirm the identity and approximate concentrations of the specific compounds prior to the final testing. The presurvey samples shall be used to develop and confirm the best sampling and analysis scheme.

The major components of a gas mixture are separated with a gas chromatograph (GC) and measured with a suitable detector. The retention time of each separated component is compared with the retention time of a known compound under identical conditions. Therefore, the analyst confirms the identity and approximate concentration of the organic emission components beforehand. With this information, the analyst then prepares or purchases commercially available standard mixtures to calibrate the GC under conditions identical to those of the samples. The analyst also determines the need for sample dilution to avoid detector saturation, gas stream filtration to eliminate particulate matter, and prevention of moisture condensation. The range of this method is from about 1 part per million (ppm) to the upper limit governed by GC detector saturation or column overloading. The upper limit can be extended by diluting the stack gases with an inert gas or by using smaller gas sampling loops.

The sensitivity limit for a compound is defined as the minimum detectable concentration of that compound, or the concentration that produces a signal-to-noise ratio of three to one. The minimum detectable concentration is determined during the presurvey calibration for each compound. Gas chromatographic techniques typically provide a precision of 5 to 10 percent relative standard deviation (RSD), but an experienced GC operator with a reliable instrument can readily achieve a 5 percent RSD. For this method, the following combined GC/operator values are required:

1. Precision -- duplicate analyses are within 5 percent of their mean value.

2. Accuracy -- analysis results of prepared audit samples are within 10 percent of preparation values.

3.0 INTERFERENCES

Chromatographic resolution interferences that may occur can be eliminated by appropriate GC column and detector choice or by shifting the retention times through changes in the column flow rate and the use of temperature programming. If chromatographic resolution cannot be achieved by application of these techniques, quantitative results cannot be obtained from the application of this method.

The analytical system is demonstrated to be essentially free from contaminants by periodically analyzing blanks that consist of hydrocarbon-free air or nitrogen. Sample cross-contamination that occurs when high-level and low-level samples or standards are analyzed alternately is best dealt with by thorough purging of the GC sample loop between samples.

To ensure consistent detector response, calibration gases are contained in dry air. To eliminate errors in concentration calculations due to the volume of water vapor in the samples, moisture concentrations are determined for each sample, and a correction factor is applied to any sample with greater than 2 percent water vapor.

4.0 METHOD TARGET COMPOUNDS

gaseous organics

5.0 APPLICABLE CLEAN AIR ACT LIST COMPOUNDS

acetaldehyde
allyl chloride
1,3-butadiene
carbon disulfide
carbonyl sulfide
chloroprene
ethyl chloride
ethylene imine
ethylene oxide
formaldehyde
hydrazine
methanol
methyl bromide
methyl chloride
methyl iodide
methyl isocyanate
methylene chloride
phosgene
propylene oxide
1,2-propylene imine
vinyl bromide
vinyl chloride

EPA METHOD 23 - Polychlorinated Dioxins & Dibenzofurans By HRGC/HRMS

Determination of Polychlorinated Dibenzo-p-dioxins (PCDDs) and Polychlorinated Dibenzofurans (PCDFs) from Stationary Sources

REFERENCE:

Federal Register, February 13, 1991 (56 FR 5758). To be included in 40 CFR Part 60, Appendix A.

1.0 SCOPE AND APPLICATION

EPA Method 23 is used to determine polychlorinated dibenzo-p-dioxins (PCDDs) and polychlorinated dibenzofurans (PCDFs) from stationary sources.

2.0 SUMMARY OF METHOD

A sample is withdrawn isokinetically from the stack through a probe, a filter, and a trap packed with a solid adsorbent. The PCDDs and the PCDFs are collected in the probe, on the filter, and on the solid absorbent.

The sampling train used in this method is identical to that described in EPA Method 5 with the exceptions and modifications noted within this method.

The sample is analyzed using a gas chromatograph coupled to a mass spectrometer (GC/MS). A 1 to 5 æL aliquot of the sample extract is injected into the GC and measured with the MS. The total PCDDs and PCDFs are the sum of the individual isomers. Strict identification criteria for PCDDs and PCDFs are listed in the method. Fused silica capillary columns are required, and one of the following is recommended:

- DB-5, 0.25 æm film thickness; or
- SP-2331 column to measure the 2,3,7,8-tetrachlorodibenzofuran isomer.

Two types of calibration procedures are required: an initial calibration is required before any samples are analyzed and intermittently calibrations are performed throughout sample analyses. The routine calibration consists of analyzing the column performance check solution and a concentration calibration solution.

The peak areas for the two ions monitored for each analyte are summed to yield the total response for each analyte. Each internal standard is used to quantify the PCDDs or PCDFs in its homologous series.

[**EDITOR'S NOTE:** *This is a very complex method requiring experienced analysts and a high resolution mass spectrometer (HRMS) in addition to high resolution capillary column GC. The sample fractions that are obtained from the sampling train must be treated separately before they are combined into one sample for further treatment. Sample fractions typically include (1) nozzle probe and front half of filter holder rinse, (2) particulates from the filter and cyclone, (3) glassware and backhalf filter rinse, and (4) XAD-2 ® resin extract. Furthermore, injection of unpurified sample extracts into the GC/MS can ruin the expensive capillary GC column as well as deteriorate peak resolution and result in more complex chromatograms with extraneous compounds that coelute with the chlorinated dioxins and furans of interest. There are several clean-up procedures commonly in use and they are usually time-consuming and expensive and require great care and experience to be performed* *satisfactorily. And finally, there are strict quality assurance procedures that must satisfactorily be performed. These include the addition of stable isotope standards at various intervals in the protocol, GC relative retention time matches within 3 seconds of an authentic standard, ion mass ratios must be within plus or minus 15% of the theoritical value for positive identifications, the detection limit for 2,3,7,8-TCDD/TCDF must be 0.2 ng or lower, and various recovery efficiencies for the TCDD/TCDF homologs are specified.]*

3.0 INTERFERENCES

PCDDs and PCDFs are often associated with other interfering chlorinated compounds such as polychlorinated biphenyls (PCBs) and polychlorinated diphenyl ethers, which may be found at higher concentrations than those of the analytes of interest.

4.0 METHOD TARGET COMPOUNDS

2,3,7,8-tetrachlorodibenzo-p-dioxin
2,3,7,8-tetrachlorodibenzofuran
1,2,3,7,8-pentachlorodibenzo-p-dioxin
1,2,3,7,8-pentachlorodibenzofuran
2,3,4,7,8-pentachlorodibenzofuran
2,3,4,7,8-hexachlorodibenzo-p-dioxin
1,2,3,6,7,8-hexachlorodibenzo-p-dioxin
1,2,3,7,8,9-hexachlorodibenzo-p-dioxin
2,3,4,6,7,8-hexachlorodibenzo-p-dioxin
1,2,3,4,6,7,8-heptachlorodibenzo-p-dioxin
1,2,3,4,7,8-hexachlorodibenzofuran
1,2,3,6,7,8-hexachlorodibenzofuran
1,2,3,7,8,9-hexachlorodibenzofuran
1,2,3,4,6,7,8-hepatachlorodibenzofuran
1,2,3,4,7,8,9-hepatachlorodibenzofuran
octachlorodibenzo-p-dioxin
octachlorodibenzofuran

5.0 APPLICABLE CLEAN AIR ACT CHEMICALS

dibenzofurans
2,3,7,8-tetrachlorodibenzo-p-dioxin

EPA METHOD 101 - Mercury & Mercury Compounds By Atomic Absorption

Determination of Particulate and Gaseous Mercury Emissions from Chlor-Alkali Plants - Air Streams

REFERENCE:

U.S. Environmental Protection Agency. Code of Federal Regulations. Title 40, Part 60, Appendix B. Washington, D.C. Office of the Federal Register, July 1, 1987.

1.0 SCOPE AND APPLICATION

This method applies to the determination of particulate and gaseous mercury emissions from chlor-alkali plants and other sources (as specified in the regulations), where the carrier gas in the duct or stack is principally air.

2.0 SUMMARY OF METHOD

Particulate and gaseous mercury emissions are withdrawn isokinetically from the source and collected in acidic iodine monochloride solution. The mercury collected (in the mercuric form) is reduced to elemental mercury, which is then aerated from the solution into an optical cell and measured by atomic absorption (AA) spectrophotometry.

3.0 INTERFERENCES

Sampling SO_2 reduces iodine monochloride and causes premature depletion of the iodine monochloride.

Iodine monochloride concentrations greater than 10-4 molar inhibit the reduction of the HgII in the aeration cell. Condensation of water vapor on the optical cell windows causes a positive interference.

4.0 METHOD TARGET COMPOUNDS

mercury compounds

5.0 APPLICABLE CAA AMENDMENTS POLLUTANTS

mercury compounds

EPA METHOD 101A - Mercury & Mercury Compounds By Atomic Absorption

Determination of Particulate and Gaseous Mercury Emissions from Sewage Sludge Incinerators

REFERENCE:

U.S. Environmental Protection Agency. Code of Federal Regulations. Title 40 CFR, Part 61, Appendix B. Washington, D.C. Office of the Federal Register, July 1, 1987.

1.0 SCOPE AND APPLICATION

EPA Method 101A applies to the determination of particulate and gaseous mercury emissions from sewage sludge incinerators and other sources as specified in the regulations.

2.0 SUMMARY OF METHOD

This method is similar to EPA Method 101, except acidic potassium permanganate solution is used instead of acidic iodine monochloride for collection.

Particulate and gaseous mercury emissions are withdrawn isokinetically from the source and collected in acidic potassium permanganate solution. The mercury collected (in the mercuric form) is reduced to elemental mercury, which is then aerated from the solution into an optical cell and measured by atomic absorption (AA) spectrophotometry.

3.0 INTERFERENCES

Excessive oxidizable organic matter in the stack gas prematurely depletes the potassium permanganate solution and prevents further collection of mercury.

Condensation of water vapor on the optical cell windows causes a positive interference.

4.0 METHOD TARGET COMPOUNDS

mercury compounds

5.0 APPLICABLE CAA AMENDMENTS POLLUTANTS

mercury compounds

EPA METHOD 103 - Isokinetic Beryllium Sampling For Screening Methods

Beryllium Screening Method

REFERENCE:

U.S. Environmental Protection Agency. Code of Federal Regulations. Title 40 CFR, Part 61, Appendix B. Washington, D.C. Office of the Federal Register, July 1, 1987.

1.0 SCOPE AND APPLICABILITY

EPA Method 103 details guidelines and requirements for methods acceptable for use in determining beryllium emissions in ducts or stacks at stationary sources.

2.0 SUMMARY OF METHOD

Beryllium emissions are isokinetically sampled from three points in a duct or stack. The collected sample is analyzed for beryllium using an appropriate analytical technique.

3.0 INTERFERENCES

Not specified in Method.

4.0 METHOD TARGET COMPOUNDS

beryllium compounds

5.0 APPLICABLE CAA AMENDMENTS POLLUTANTS

beryllium compounds

EPA METHOD 104 - Beryllium & Beryllium Compounds By Atomic Absorption

Reference Method for Determination of Beryllium Emissions from Stationary Sources

REFERENCE:

U.S. Environmental Protection Agency. Code of Federal Regulations. Title 40 CFR, Part 61, Appendix B. Washington, D.C. Office of the Federal Register, July 1, 1987.

1.0 SCOPE AND APPLICATION

EPA Method 104 is applicable to the determination of beryllium emissions in ducts or stacks at stationary sources. Unless otherwise specified, this method is not intended to apply to gas streams other than those emitted directly to the atmosphere without further processing.

2.0 INTERFERENCES

Beryllium emissions are isokinetically sampled from the source, and the collected sample is digested and analyzed by atomic absorption (AA) spectrophotometry.

3.0 INTERFERENCES

Not specified in method.

4.0 METHOD TARGET COMPOUNDS

beryllium compounds

5.0 APPLICABLE CAA AMENDMENTS POLLUTANTS

beryllium compounds

EPA METHOD 106 - Vinyl Chloride By GC/FID

Determination of Vinyl Chloride from Stationary Sources

REFERENCE:

U.S. Environmental Protection Agency. Code of Federal Regulations. Title 40 CFR, Part 61, Appendix B. Washington, D.C. Office of the Federal Register, July 1, 1987.

1.0 SCOPE AND APPLICATION

EPA Method 106 is applicable to the measurement of vinyl chloride in stack gases from ethylene dichloride, vinyl chloride, and polyvinyl chloride manufacturing processes. The method does not measure vinyl chloride contained in particulate matter.

2.0 SUMMARY OF METHOD

An integrated bag sample of stack gas containing vinyl chloride is subjected to GC analysis using a flame ionization detector.

3.0 INTERFERENCES

The chromatographic columns and the corresponding operating parameters described in Method 106 normally provide an adequate resolution of vinyl chloride. However, resolution interferences may be encountered on some sources.

4.0 METHOD TARGET COMPOUNDS

vinyl chloride

5.0 APPLICABLE CAA AMENDMENTS POLLUTANTS

vinyl chloride

EPA METHOD 114 - Radionuclides

Test Methods for Monitoring Radionuclide Emissions from Stationary Sources

REFERENCE:

Standards of Performance for New Stationary Sources. Compilation. U. S. Environmental Protection Agency, EPA-340/1-77-015.

1.0 SCOPE AND APPLICATION

EPA Method 114 contains guidance on continuous stack sampling for radionuclides and radiochemical methods used to analyze the sample. Radionuclides differ in the chemical and physical forms, half-lives, and type of radiation emitted. The appropriate type of sample extraction, collection, and analysis for an individual radionuclide depends on many interrelated factors, including the mixture of other radionuclides present. Therefore the procedure for radionuclide sampling and analysis described in Method 114 is actually a series of methods based on principles of measurement that provides the user with flexibility to choose the combination of sampling and analysis schemes most applicable to the effluent stream measured.

The analysis methods described in Method 114 are for commonly found radionuclides that have the greatest potential for public harm. The analyses are grouped according to the type of radiation emitted: alpha, beta, or gamma.

2.0 SUMMARY OF METHOD

2.1 Particulate Sampling

Continuous sampling can be performed according to EPA Method 5 or SW-846 Method 0020 to isokinetically sample radionuclides present as particulate matter. A probe is equipped with a filter that has a high efficiency for submicrometer particles. Additional guidance can be found in the American National Standards Institutes' "Guide to Sampling Airborne Radioactive Materials at Nuclear Facilities" (1969).

2.2 Gaseous Sampling

Radionuclides of hydrogen, oxygen, carbon, nitrogen, the noble gases, and in some cases iodine, will be in the gaseous form. Radionuclides of these elements will require direct flow into a counter or suitable bubblers to collect the radionuclides. The following are suggested media for gaseous sample collection:

Radionuclide	Medium
tritium	silica gel, molecular sieves, ethylene glycol, water
iodine	charcoal, metal zeolites, caustic solution
argon, krypton, xenon	charcoal or metal zeolites
oxygen, carbon, nitrogen, radon	caustic solution

2.3 Analysis

2.3.1 Alpha Analysis

The element of interest is separated from the sample matrix by precipitation, ion exchange, or solvent extraction; or by carriers using electrodeposition or coprecipitation. The alpha energy is measured by alpha spectrometry or an alpha counter. Particulate samples can be analyzed from the surface of the filter directly by alpha spectrometry or an alpha counter.

2.3.1.1 Uranium Analysis. Uranium is dissolved and extracted into hexane. The extract can be analyzed colorimetrically by adding dibenzoylmethane, or fluorometrically after fusion with sodium fluoride-lithium fluoride.

2.3.1.2 Radon-222. Radon in the gaseous form can be detected continuously using a calibrated scintillation cell.

2.3.2.Beta Analysis

Gaseous samples can be continuously analyzed by flow through a ionization chamber or another beta detector such as a Geiger-M ller tube. In non-gaseous sample mixtures, the element of interest is separated from other radionuclides by precipitation, distillation, ion exchange, solvent extraction, or carriers. The extracted radionuclide is analyzed with a scintillation cell or a beta counter. Particulate samples can be analyzed directly from the filter with a beta counter when it is known that the sample contains only the radionuclide of interest.

2.3.3.Gamma Analysis

High resolution gamma spectroscopy can be used to directly measure gaseous, liquid, or particulate gamma-emitting radionuclides. Prior chemical separation is not usually necessary. For simple mixtures, low resolution gamma spectroscopy can be used. Single channel gamma spectrometry can be used when only one gamma-emitting radionuclide is present.

3.0 INTERFERENCES

Some analysis methods state that only one radionuclide can be present without interference in the analysis. Sufficient separation must occur to produce a pure sample from combined extracts and with a correction for chemical yield.

4.0 METHOD TARGET COMPOUNDS

carbon*	oxygen*
hydrogen*	polonium-210
iodine*	radon-222
nitrogen*	tritium*
noble gases (including argon)	uranium

* Usually gaseous

NOTE: Method target compounds are not limited to those listed above.

5.0 APPLICABLE CAA AMENDMENTS POLLUTANTS

radionuclides

EPA METHODS 515/615 - Chlorinated Herbicides By GC/ECD

Determination of Chlorinated Herbicides in Drinking Water (Method 515)
Determination of Chlorinated Herbicides in Industrial and Municipal Wastewater (Method 615)

REFERENCES:

Method 515 - Supplement to "Method for the Determination of Organic Compounds in Finished Drinking Water and Raw Source Water." EPA EMSL. September 1986.

Method 615 - Pressley, Thomas A., and Longbottom, James E., EPA EMSL. "The Determination of Chlorinated Herbicides in Industrial and Municipal Wastewater." January 1982.

1.0 SCOPE AND APPLICATION

EPA Methods 515 and 615 are used to determine certain chlorinated acid herbicides in drinking water (Method 515) and wastewater (Method 615). Chlorinated herbicides in air will be distributed between particulate and gas phase, depending upon temperature.

2.0 SUMMARY OF METHOD

Methods 515 and 615 need to be modified to allow extraction of solid sorbent used for air sampling. The original methods require acidifying approximately 1 L of sample. The acid herbicides and their esters and salts are then extracted with ethyl ether using a separatory funnel. The esters are hydrolyzed and converted to acid salts with potassium hydroxide solution. The aqueous phase containing the acid salts is then solvent-washed to remove extraneous organic material. After acidification, the acids are extracted into organic phase and the sample volume reduced to 5 mL in methyl t-butyl ether (MTBE) with a K-D concentrator. The acids are converted to their methyl ester using diazomethane as the derivatizing agent. Excess reagent is removed and the esters are determined by electron capture gas chromatography.

3.0 INTERFERENCES

Interferences may be caused by contaminants in solvents, reagents, and glassware. Therefore, care must be taken during sample preparation to ensure minimal interferences.

The acid forms of the herbicides are strong organic acids that react readily with alkaline substances and can be lost during analysis. Glassware and glass wool must be acid-rinsed, and the sodium sulfate must be acidified prior to use to avoid this possibility.

Organic acids and phenols, especially chlorinated compounds, cause the most direct interference with the analysis. Alkaline hydrolysis and subsequent extraction of the basic solution remove many chlorinated hydrocarbons and phthalate esters that might otherwise interfere with the electron capture analysis.

4.0 METHOD TARGET COMPOUNDS

2,4-D
2,4-DB
dalapon
dicamba
dichlorprop*
dinoseb
MCPA*
MCPP*
pentachlorophenol (PCP)
picloram
2,4,5-T
2,4,5-TP (silvex)

* Method 615 only

5.0 APPLICABLE CAA AMENDMENTS POLLUTANTS

2,4-D esters and salts
pentachlorophenol

EPA METHOD 531 - N-Methyl Carbamoyloximes & Carbamates By HPLC/Fluorescence

Measurement of N-Methyl Carbamoyloximes and N-Methyl Carbamates in Drinking Water by Direct Aqueous Injection HPLC with Post-Column Derivatization

REFERENCE:

U.S. Environmental Protection Agency, EPA/600/485/054

1.0 SCOPE AND APPLICATION

EPA Method 531 is used to identify and measure N-methylcarbamoyloximes and N-methyl carbamates in finished drinking water, raw source water, or drinking water at any treatment stage.

2.0 SUMMARY OF METHOD

Air samples are collected using EPA sampling Method 0010. The solid sorbents used in Method 0010 are extracted with methylene chloride. Extensive modification of chromatographic conditions of Method 531 will be required to perform analysis of air samples.

In Method 531, a high performance liquid chromatographic (HPLC) system capable of injecting 200 to 400 æL aliquots and performing binary linear gradients at a constant flow rate is used for sample analysis. The recommended column is a 10 cm long x 8 mm ID radially compressed HPLC column packed with 10 æm æ-Bondapak C18 or equivalent. Use of a guard column is also recommended. A post-column reactor capable of mixing reagents into the mobile phase is needed. The fluorescence detector should be capable of excitation at 230 mm and detecting emission energies greater than 419 nm. Fluorometers should have dispersive optics for excitation and be able to utilize either filters or dispersive optics at the emission detector.

The water sample is filtered and a 400-æL aliquot is injected into a reverse phase HPLC column. Separation of the analytes is achieved using gradient elution chromatography. After elution from the HPLC column, the analytes are hydrolyzed with 0.05N sodium hydroxide at 95°C. The methyl amine formed during hydrolysis is reacted with o-phthalaldehydes to form a highly fluorescent derivative which is detected using a fluorescence detector.

The analytes are identified by comparing the retention times of the unknowns to the retention times of standards, and the concentration of individual compounds in the sample is determined.

3.0 INTERFERENCES

Any matrix interferences that will interfere with the chromatography are interferences for the method. Matrix interference that will produce saturation of the chromatographic system will interfere with the analysis of target compounds even if the fluorescence detector will resolve the compounds.

4.0 METHOD TARGET COMPOUNDS

aldicarb
aldicarb sulfone
aldicarb sulfoxide
carbaryl
carbofuran
3-hydroxycarbofuran
methomyl
oxamyl

5.0 APPLICABLE CAA AMENDMENTS POLLUTANTS

carbaryl
dimethyl carbamoyl chloride

EPA METHOD 632 - Carbamate & Urea Pesticides By HPLC/UV

Determination of Carbofuran, Fluometuron, Methomyl, and Oxamyl in Wastewater

REFERENCE:

U. S. Environmental Protection Agency, EMSL, Physical and Chemical Methods Branch. November, 1985.

1.0 SCOPE AND APPLICATION

EPA Method 632 is a high performance liquid chromatographic (HPLC) method used to determine certain carbamate and urea pesticides in industrial and municipal wastewater discharges.

2.0 SUMMARY OF METHOD

Air samples may be obtained using EPA Method 0010, and the solid sorbent extracted with methylene chloride.

When Method 632 is applied to wastewater, a measured volume of sample (about 1 L) is solvent-extracted with methylene chloride using a separatory funnel. The methylene chloride extract is dried and concentrated to a volume of 10 mL or less. HPLC conditions are described that permit the separation and measurement of the compounds in the extract.

The HPLC analytical system should include high pressure syringes or sample injection loop, analytical columns, ultraviolet (UV) detector and strip chart recorder. A guard column is recommended for all applications. The recommended analytical column is a 30 cm long x 4 mm ID stainless steel packed with æ-Bondapak C18 (10 æm) or equivalent with Whatman Co. PELL ODS (30 - 38 æm) guard column, 7 cm long x 4 mm ID. The UV detector should be capable of monitoring at 254 nm and 280 nm.

The HPLC system may be calibrated using either the external or internal standard technique. The standards and extracts must be in the solvent (acetonitrile or methanol) compatible with the mobile phase.

The sample extract is injected (around 10 æL), the resulting peak size in area or peak height units is recorded, and the concentration of individual compounds in the sample is determined.

3.0 INTERFERENCES

Matrix interferences may be caused by contaminants that are coextracted from the sample. The extent of matrix interferences will vary considerably from source to source, depending upon the nature and diversity of the industrial complex or municipality sampled.

4.0 METHOD TARGET COMPOUNDS

aminocarb
barban
carbaryl
carbofuran
chlorpropham
diuron
fenuron
fenuron-TCA
fluometuron
linuron
methiocarb
monuron
monuron-TCA
neburon
oxamyl
propham
propoxur
siduron
swep
methomyl
mexacarbate

5.0 APPLICABLE CAA AMENDMENTS POLLUTANTS

caprolactam
ethylene thiourea
ethyl carbamate
propoxur

EPA METHOD 680 - Pesticides & PCBs By HRGC/MS

Determination of Pesticides and PCBs in Water and Soil/Sediment by Gas Chromatography/Mass Spectrometry

REFERENCE:

U. S. Environmental Protection Agency, EMSL, Physical and Chemical Methods Branch. November, 1985.

1.0 SCOPE AND APPLICATION

EPA Method 680 provides procedures for mass spectrometric determination of polychlorinated biphenyls (PCBs) and pesticides in water, soil, or sediment. These compounds in air will be mostly associated with particulate, although some can exist in the gas phase if the source temperature is high enough and the compound is sufficiently stable. A modification of the method in order to analyze air samples is possible.

2.0 SUMMARY OF METHOD

Sample preparation consists of placing a 1-L water sample in a separatory funnel, extracting with methylene chloride, followed by hexane exchange. Method 680 must be modified, however, to allow for extraction of solid sorbent used for air sampling. Sample extract components are then separated with capillary column gas chromatography (GC) and identified and measured with low resolution electron ionization mass spectrometry. Two surrogate compounds and two internal standards are added to each sample. Because of the multi-compound characteristics of PCBs, they are identified and measured as isomer groups. A concentration is measured for each PCB isomer group, and total PCB concentration in each sample extract is obtained by summing isomer group concentrations.

3.0 INTERFERENCES

Interferences may be caused by contaminants in solvents, reagents, and glassware. Therefore, care must be taken in sample preparation to assure minimal interferences.

With both pesticides and PCBs, interferences can be caused by the presence of much greater quantities of other sample components that overload the capillary column. Therefore, additional sample cleanup procedures may be necessary to eliminate these interferences. Capillary column GC retention times and the compound-specific characteristics of mass spectra eliminate many of the interferences that formerly were of concern with pesticide/PCB determinations with electron capture detection. The approach and identification criteria used in this method eliminate interference by most chlorinated compounds other than other PCBs. With the isomer groups approach, coeluting PCBs that contain the same number of chlorines are identified and measured together. Therefore, coeluting PCBs are a problem only if they contain a different number of chlorine atoms.

4.0 METHOD TARGET COMPOUNDS

aldrin
alpha-BHC
beta-BHC
chlordane
delta-BHC
4,4'-DDD
4,4'-DDE
4,4'-DDT
decachlorobiphenyl
dichlorobiphenyls
dieldrin
endosulfan I
endosulfan II
endosulfan sulfate
endrin
endrin aldehyde
heptachlor
heptachlor epoxide
heptachlorobiphenyls
hexachlorobiphenyls
lindane
methocychlor
monochlorobiphenyls
nonachlorobiphenyls
octachlorobiphenyls
pentachlorobiphenyls
tetrachlorobiphenyls
trichlorobiphenyls

5.0 APPLICABLE CAA AMENDMENTS POLLUTANTS

chlordane
4,4'-DDE
polychlorinated biphenyls
heptachlor
lindane
methoxychlor
toxaphene

Chapter 3

Chemical, Physical, Hazardous and Toxicological Properties of EPA's Air Toxics

Air Toxics Chemical Names

The substance names and Chemical Abstract Service (CAS) Registry Numbers of EPA's Air Toxics are listed below. There are problems with some of these names and these are listed at the bottom of the list.

CAS Number	Air Toxic Substance Name
75-07-0	Acetaldehyde
60-35-5	Acetamide
75-05-8	Acetonitrile
98-86-2	Acetophenone
53-96-3	2-Acetylaminofluorene
107-02-8	Acrolein
79-06-1	Acrylamide
79-10-7	Acrylic acid
107-13-1	Acrylonitrile
107-05-1	Allyl chloride
92-67-1	4-Aminobiphenyl
62-53-3	Aniline
90-04-0	o-Anisidine
Various	Antimony compounds
Various	Arsenic compounds
1332-21-4	Asbestos
71-43-2	Benzene
92-87-5	Benzidine
98-07-7	Benzotrichloride
100-44-7	Benzyl chloride
Various	Beryllium compounds
92-52-4	Biphenyl
75-25-2	Bromoform
106-99-0	1,3-Butadiene

CAS Number	Air Toxic Substance Name
Various	Cadmium compounds
156-62-7	Calcium cyanamide
105-60-2	Caprolactam
133-06-2	Captan
63-25-2	Carbaryl
75-15-0	Carbon disulfide
56-23-5	Carbon tetrachloride
463-58-1	Carbonyl sulfide
120-80-9	Catechol
133-90-4	Chloramben
57-74-9	Chlordane
7782-50-5	Chlorine
79-11-8	Chloroacetic acid
532-27-4	2-Chloroacetophenone
108-90-7	Chlorobenzene
510-15-6	Chlorobenzilate
67-66-3	Chloroform
542-88-1	bis(Chloromethyl) ether
107-30-2	Chloromethyl methyl ether
126-99-8	Chloroprene
Various	Chromium compounds
Various	Cobalt compounds
8007-45-2	Coke oven emissions
95-48-7	o-Cresol
108-39-4	m-Cresol
106-44-5	p-Cresol
1319-77-3	Cresols/Cresylic acid (isomers & mixture)
98-82-8	Cumene
Various	Cyanide compounds
94-75-7	2,4-D, salts and esters
72-55-9	DDE
334-88-3	Diazomethane
132-64-9	Dibenzofuran
96-12-8	1,2-Dibromo-3-chloropropane
84-74-2	Dibutyl phthalate
106-46-7	1,4-Dichlorobenzene
91-94-1	3,3'-Dichlorobenzidine
111-44-4	Dichloroethyl ether
542-75-6	1,3-Dichloropropene

CAS Number	Air Toxic Substance Name
62-73-7	Dichlorvos
111-42-2	Diethanolamine
91-66-7	N,N-Diethylaniline
64-67-5	Diethyl sulfate
119-90-4	3,3'-Dimethoxybenzidine
60-11-7	Dimethylaminoazobenzene
121-69-7	N,N-Dimethylaniline
119-93-7	3,3'-Dimethylbenzidine
79-44-7	Dimethylcarbamoyl chloride
68-12-2	N,N-Dimethylformamide
57-14-7	1,1-Dimethylhydrazine
131-11-3	Dimethyl phthalate
77-78-1	Dimethyl sulfate
534-52-1	4,6-Dinitro-o-cresol, and salts
51-28-5	2,4-Dinitrophenol
121-14-2	2,4-Dinitrotoluene
123-91-1	1,4-Dioxane
122-66-7	1,2-Diphenylhydrazine
106-89-8	Epichlorohydrin
106-88-7	1,2-Epoxybutane
140-88-5	Ethyl acrylate
100-41-4	Ethylbenzene
51-79-6	Ethyl carbamate
75-00-3	Ethyl chloride
106-93-4	Ethylene dibromide
107-06-2	Ethylene dichloride
107-21-1	Ethylene glycol
151-56-4	Ethyleneimine
75-21-8	Ethylene oxide
96-45-7	Ethylene thiourea
117-81-7	bis(2-Ethylhexyl) phthalate
75-34-3	Ethylidene dichloride
50-00-0	Formaldehyde
Various	Glycol ethers
76-44-8	Heptachlor
118-74-1	Hexachlorobenzene
87-68-3	Hexachlorobutadiene
77-47-4	Hexachlorocyclopentadiene
67-72-1	Hexachloroethane
822-06-0	Hexamethylene-1,6-diisocyanate
680-31-9	Hexamethylphosphoramide

CAS Number	Air Toxic Substance Name
110-54-3	Hexane
302-01-2	Hydrazine
7647-01-0	Hydrochloric acid
7664-39-3	Hydrogen fluoride
123-31-9	Hydroquinone
78-59-1	Isophorone
Various	Lead Compounds
58-89-9	Lindane (all isomers)
108-31-6	Maleic anhydride
Various	Manganese compounds
Various	Mercury compounds
67-56-1	Methanol
72-43-5	Methoxychlor
74-83-9	Methyl bromide
1634-04-4	Methyl tert-butyl ether
74-87-3	Methyl chloride
71-55-6	Methyl chloroform
78-93-3	Methyl ethyl ketone
60-34-4	Methylhydrazine
74-88-4	Methyl iodide
108-10-1	Methyl isobutyl ketone
624-83-9	Methyl isocyanate
80-62-6	Methyl methacrylate
101-14-4	4,4'-Methylenebis(2-chloroaniline)
75-09-2	Methylene chloride
101-68-8	Methylenediphenyl diisocyanate
101-77-9	4,4'-Methylenedianiline
None	Mineral fibers (fine)
91-20-3	Naphthalene
Various	Nickel compounds
98-95-3	Nitrobenzene
92-93-3	4-Nitrobiphenyl
100-02-7	4-Nitrophenol
79-46-9	2-Nitropropane
684-93-5	N-Nitroso-N-methylurea
62-75-9	N-Nitrosodimethylamine
59-89-2	N-Nitrosomorpholine
56-38-2	Parathion
82-68-8	Pentachloronitrobenzene
87-86-5	Pentachlorophenol

CAS Number	Air Toxic Substance Name
108-95-2	Phenol
106-50-3	p-Phenylenediamine
75-44-5	Phosgene
7803-51-2	Phosphine
7723-14-0	Phosphorus
85-44-9	Phthalic anhydride
1336-36-3	Polychlorinated biphenyls
Various	Polycyclic organic matter
1120-71-4	1,3-Propane sultone
57-57-8	beta-Propiolactone
123-38-6	Propionaldehyde
114-26-1	Propoxur
78-87-5	Propylene dichloride
75-56-9	Propylene oxide
75-55-8	1,2-Propylenimine
91-22-5	Quinoline
106-51-4	Quinone
Various	Radionuclides (including Radon)
Various	Selenium compounds
100-42-5	Styrene
96-09-3	Styrene oxide
1746-01-6	2,3,7,8-Tetrachlorodibenzo-p-dioxin
79-34-5	1,1,2,2-Tetrachloroethane
127-18-4	Tetrachloroethylene
7550-45-0	Titanium tetrachloride
108-88-3	Toluene
95-80-7	2,4-Toluenediamine
584-84-9	2,4-Toluene diisocyanate
95-53-4	o-Toluidine
8001-35-2	Toxaphene
120-82-1	1,2,4-Trichlorobenzene
79-00-5	1,1,2-Trichloroethane
79-01-6	Trichloroethylene
95-95-4	2,4,5-Trichlorophenol
88-06-2	2,4,6-Trichlorophenol
121-44-8	Triethylamine
1582-09-8	Trifluralin
540-84-1	2,2,4-Trimethylpentane
108-05-4	Vinyl acetate
593-60-2	Vinyl bromide
75-01-4	Vinyl chloride
75-35-4	Vinylidene chloride

CAS Number	Air Toxic Substance Name
1330-20-7	Xylenes (isomers and mixture)
95-47-6	o-Xylene
108-38-3	m-Xylene
106-42-3	p-Xylene

Note: CAS Numbers published in the Federal Register lacked the dashes that area part of CAS Numbers. We have added them. The first dash is always to theleft of the last digit and the second dash is always to the left of the third digit from the end of the number set. We also corrected the nomenclature of several compounds:

* the "(p)" was removed from the end of 1,4-Dichlorobenzene;

* the prime was added to 3,3'-Dichlorobenzidine;

* the prime was added to 3,3'-Dimethoxybenzidine;

* the "N,N-" was added to N,N-Dimethylformamide and the extra space before "formamide" was removed;

* the letter "s" was removed from Dibenzofuran, making it singular;

* the space before "carbamoyl" was removed from Dimethylcarbamoyl chloride

* the space before "imine" was removed from Ethyleneimine

* the space before "hydrazine" was removed from Methylhydrazine

* the space before "diphenyl" was removed from Methylenediphenyl diisocyanate

* the space before "diamine" was removed from 2,4-Toluenediamine

* the prime was added to 4,4'-Methylenebis(2-chloroaniline) and the extra space was removed; and

* the letter "m" was added to N,N-Dimethylaniline

The last correction has serious consequences because N,N-Dimethylaniline (CAS # 121-69-7) is a different compound from N,N-Diethylaniline (CAS # 91-66-7). The Clean Air Act Amendment, as published in the Federal Register lists the above compound as "N,N- Diethylaniline (N,N-Dimethylaniline)" with a CAS # of 121-69-7. Clearly the letter "m" was omitted as a typo thus changing the compound name. Because of the confusion caused by this error we have included chemical and physical properties for both of these compounds, thus increasing the number of records in this database from 189 to 190.

In addition, the CAS # for DDE as published in the Clean Air Act Amendment is wrong (it was published as CAS # 3547-04-4). The correct number, as listed in this Appendix and in the database, is CAS # 72-55-9. The EPA is in the process of making this and other nomenclature corrections. The above table lists the CAS # for DDE as it was published in the Clean Air Act Amendment and it is therefore incorrect in that volume only.

Representative Analytes

As seen in the list of Air Toxics Chemical Names above, not all of the substances listed in the 1990 Clean Air Act are individual analytes. It is unfortunate that the persons responsible for selecting these compounds had so little knowledge of environmental chemical analysis. Even a novice chemist knows that in order to be able to analyze for specific environmental chemicals there must first exist an environmental reference material. Thus, where substances were chosen that have many potential individual compounds, representative analytes are needed. The table below lists the compounds and elements that we chose as representative analytes when EPA's Air Toxics substances were nonspecific.

REPRESENTATIVE COMPOUNDS AND ELEMENTS

Analyte Group	Representative Analyte
Antimony compounds	Antimony oxide
	Antimony potassium tartrate
Arsenic compounds (inorganic, including arsine)	Arsenic
	Arsenic chloride
	Arsenic trioxide
	Arsine
	Sodium arsenite
Beryllium compounds	Beryllium
	Beryllium sulfate tetrahydrate
Cadmium compounds	Cadmium
	Cadmium chloride
	Cadmium oxide

Chromium compounds	Chromium
	t-Butyl chromate
	Calcium chromate, anhydrous
	Calcium chromate dihydrate
	Chromium carbonyl
	Chromium trioxide
	bis(Cyclopentadienyl) chromium
	Potassium dichromate
Cobalt compounds	Cobalt
	Cobaltocene
	Cobalt sulfate heptahydrate
Coke oven emissions	Ammonia
	Hydrogen sulfide
	2-Picoline
	3-Picoline
	4-Picoline
	Pyridine
Cyanide compounds	Calcium cyanide
	Copper (I) cyanide
	Hydrogen cyanide
	Nickel cyanide
	Potassium cyanide
	Silver cyanide
	Silver potassium cyanide
	Sodium cyanide
	Zinc cyanide
2,4-D, salts and esters	2,4-Dichlorophenoxyacetic acid, (2,4-D)
	2,4-D, n-Butyl ester
Glycol ethers	Ethylene glycol diethyl ether
	Ethylene glycol dimethyl ether
	Ethylene glycol monobutyl ether
	Ethylene glycol monophenyl ether
Lead compounds	Lead
	Lead acetate
	Lead dioxide
	Lead subacetate
Manganese compounds	Manganese
	Manganese (II) sulfate monohydrate
Mercury compounds	Mercury
	Mercuric chloride
	Mercury((O-carboxyphenyl)thio)ethyl, sodium salt
	Methyl mercury
	Methyl mercury (II) chloride
	Methyl mercury (II) hydroxide

Nickel compounds	Nickel
	Nickel cyanide
	Nickelocene
	Nickel oxide
	Nickel subsulfide
	Nickel sulfate
	Nickel sulfate hexahydrate
Polychlorintated biphenyls	Aroclor 1221
	Aroclor 1232
	Aroclor 1242
	Aroclor 1248
	Aroclor 1254
	Aroclor 1260
	Aroclor 1262
	Aroclor 1268
Polycyclic organic matter	2-Acetylaminofluorene*
	4-Aminobiphenyl*
	Anthracene
	Benzidine*
	Benzo(a)pyrene
	Chrysene
	3,3'-Dichlorobenzidine*
	4-Dimethylaminoazobenzene*
	4-Nitrobiphenyl*
	Phenanthrene
	Pyrene
Radionuclides (including Radon)	Radium
	Radon
Selenium compounds	Selenium
	Selenious acid
	Selenium disulfide
	Selenium sulfide

* These representative compounds of polycyclic organic matter are listed separately as Air Toxics in the 1990 Clean Air Amendment. Data for these compounds occurs under each compound's respective name in alphabetical order in this book.

Fields of Information

The information in this section describes the data fields with each individual air toxic compound or element.

Synonyms

Most compounds are known by more than one chemical name. Selected common synonyms are listed in this section for each air toxic material.

Chemical Formula

Empirical formulas show numbers of each type of atom present in the molecules.

Molecular Weight

The molecular weights given are formula weights, or molar masses in daltons, usually to the nearest one hundredth of an atomic mass unit (amu) and are based on multiples of the average atomic mass units listed below.

Elements, Symbols, and Masses Used

Element	Symbol	Atomic Mass
Antimony	Sb	121.75
Arsenic	As	74.922
Beryllium	Be	9.012
Bromine	Br	79.909
Cadmium	Cd	112.411
Calcium	Ca	40.08
Carbon	C	12.011
Chlorine	Cl	35.453
Chromium	Cr	51.996
Cobalt	Co	58.933
Fluorine	F	18.998
Hydrogen	H	1.008
Iodine	I	126.904
Lead	Pb	207.19
Manganese	Mn	54.938
Mercury	Hg	200.59
Nickel	Ni	58.71
Nitrogen	N	14.007
Oxygen	O	15.999
Phosphorous	P	30.974
Radon	Ra	226.0254
Selenium	Se	78.96
Sulfur	S	32.064
Titanium	Ti	47.90
Uranium	U	238.03

Wiswesser Line Notation

The Wiswesser Line Notation (WLN) uses numbers to express lengths of alkyl chains and the sizes of rings. These symbols then are cited in connected order from one end of the molecule to the other. The symbols used in WLNs are the 26 capital letters, the ten numerals, the blank space and four punctuation marks (&, -, /, and *). These are all available on existing computers so no special equipment is needed to computerize the data. The utility of WLNs is in computerized sorting, searching, list printing, structure-activity relationships, and data retrieval.

The source used for checking all WLNs and for generating those not available was The Wiswesser Line-Formula Chemical Notation, Third Edition, by Elbert G. Smith and Peter A. Baker, (1976), Chemical Information Management, Inc., Cherry Hill, N.J., U.S.A.

Physical Description

The physical description of the materials listed was drawn from the literature. When this was not available physical descriptions were provided from observations of the chemicals in Radian Corporation's repository of these compounds.

Specific Gravity

The specific gravity of a chemical is a ratio of the mass of the chemical, solid or liquid, to the mass of an equal volume of water at a specified temperature. The usual convention is to use the mass of water at 4øC, which is the temperature at which it is at its greatest density. The first temperature listed is that at which the mass of the chemical was measured; the second temperature listed is that at which the mass of the water was measured and it is separated from the first temperature by a slash.

Density

Density is the mass of a substance per unit volume. It is the mass in grams per milliliter (cubic centimeter). Density measurements were drawn from the available literature sources.

Melting Point

Melting points are reported as a range in degrees Centigrade (unless otherwise noted). Where two or more melting points are recorded the highest, or the one considered to be most probable, is listed in this field. Any other melting points are listed in the field named Other Physical Data.

Boiling Point

Boiling points are drawn from available literature sources and are reported in degrees Centigrade (unless otherwise noted). If no pressure notations are included with the boiling point it is assumed that the boiling point recorded was at standard pressure (760 mm Hg). If the boiling point was obtained at a reduced pressure it is so noted directly after the temperature. When two or more boiling points at a given pressure are noted in the literature the highest temperature or the one considered most probable is listed in this field. The other boiling point is then listed in the field named Other Physical Data. Likewise, if data at more than one pressure is available it will also be listed in the latter field along with the pressure at which it was obtained.

Solubility

Common solvents used in toxicity testing include, water, DMSO (dimethyl sulfoxide), 95% ethanol, acetone, methanol, and toluene. When available, solubility data is presented for each of these solvents over a specific weight per volume range using units of milligrams per liter. To be considered soluble at the referenced range all of the subject material must have dissolved in the given amount of solvent. All solubility measurements in this data field were conducted by Radian Corporation (unless referenced otherwise) and the temperature at which the measurement was made is included.

Other Solvents

Solubility data in this field is drawn from literature sources. While it sometimes significantly increases the solubility information from additional, and sometimes more diverse solvents, it also has the disadvantage of being less precise than the experimentally determined values in the above field. Literature information involving chemical solubilities is often poorly documented, highly variable, and subjective. In addition, inconsistent terms such as "insoluble," "slightly soluble," "poorly soluble," etc., are typically used. These subjective descriptions vary from source to source. Nevertheless, they are usually the best that are available and are reported as found in the literature in this data field.

Other Physical Data

A large variety of miscellaneous chemical and physical information is included in this section. Data involving refractive index, boiling points under reduced pressure, conflicting information on any of the above physical or chemical properties, multiple crystalline or other physical states, vapor pressure, vapor density, odor, pK, optical rotation, etc., are all listed in this data field as available.

Hazardous Air Pollutant Weighting Factors

The hazardous information characteristics for each Hazardous Air Pollutant (HAP) in this chapter includes EPA's proposed toxicity weighting factor, shown as "HAP Weighting Factor." Use of the proposed weighting factors in demonstrating compliance with early reductions requirements is explained in the following subsection.

For each Hazardous Air Pollutant (HAP) in this publication, the hazardous properties information includes EPA's proposed toxicity weighting factor, shown as "HAP WEIGHTING FACTOR." Use of the proposed weighting factors in demonstrating compliance with early reductions requirements is explained in the following discussion.

The Clean Air Act Amendments of 1990 establish a two-phased strategy to reduce and control emissions of Hazardous Air Pollutants (HAP's) from major sources (i.e., those which emit 10 tons per year of a single HAP or 25 tons per year for any combination of HAP's). Phase I requires the application of Maximum Achievable Control Technology (MACT) to specific source categories. Early reductions of 90% or more (95% for particulates) of HAP emissions are encouraged through the incentive of extension of compliance dates for meeting MACT standards. Proposed regulations in 56 FR 27338-27374 and EPA's "Enabling Document for Regulations Governing Compliance Extensions for Early Reductions of Hazardous Air Pollutants" of July 1991 set forth proposed guidance for early reduction demonstrations. In the proposal, actual total HAP emissions must be demonstrated to have been reduced by 90% (95% for particulates), and actual total HAP emissions adjusted for high risk pollutants must also be demonstrated to have been reduced by 90% (95% for particulates). The proposed regulations provide toxicity weighting factors for thirty-five high risk pollutants as follows:

Substances Listed as Noncarcinogens	Weighting Factor
2,4-Toluene diisocyanate	10
Acrolein	10
Acrylic acid	10
Chloroprene	10
Dibenzofuran	10
Mercury compounds	10
Methyl isocyanate	10
Methylene diphenyl diisocyanate (MDI)	10
Phosgene	10

Substances Listed as Carcinogens	Weighting Factor
2,3,7,8-Tetrachlorodibenzo-p-dioxin	100,000
Benzidene	1,000
bis(Chloromethyl) ether	1,000
Asbestos	100
Chromium compounds	100
Hydrazine	100
Arsenic compounds (inorganic including arsine)	100
Chloromethyl methyl ether	10
Cadmium compounds	10
Heptachlor	10
Beryllium compounds	10
Acrylamide	10
Coke oven emissions	10
Hexachlorobenzene	10
Chlordane	10
Dichloroethyl ether (bis(2-Chloroethyl ether)	10
1,3-Butadiene	10
Benzotrichloride	10
Ethylene dibromide (Dibromoethane)	10
Ethylene oxide	10
Vinyl chloride	10
Acrylonitrile	10
1,1,2,2-Tetrachloroethane	10
Vinylidene chloride (1,1-Dichloroethylene)	10
Benzene	10
1,2-Propylenimine (2-Methyl aziridine)	10

* HAP not on the high risk list have a weighting factor of 1.

To calculate emissions adjusted for toxicity weighting factors, the base year emission for each HAP is multiplied by its weighting factor and compared to the post-control emission also multiplied by the weighting factor. As a simple example, if the source emits benzene and toluene:

	Base Year Emissions	Post-control Emissions
Benzene	20 tons/year	3 tons/year
Toluene	200 tons/year	18 tons/year
Total	220 tons/year	21 tons/year

For the first demonstration, the total HAP emissions have been reduced by (220 - 21) divided by 220 x 100% = 90.5%. Therefore, the required 90% reduction in total HAP's has been achieved.

For the second demonstration using the proposed weighting factors of 10 for benzene and 1 for toluene, the adjusted emissions are:

	Base Year Emissions	Post-control Emissions
Benzene	20 tons/year x 10	3 tons/year x 10
Toluene	200 tons/year x 1	18 tons/year x 1
Total	400 tons/year adjusted	48 tons/year adjusted

The percent reduction using adjusted values is calculated as (400 - 48) divided by 400 x 100% = 88%, and therefore the early reduction criteria have not been met. [713]

Volatility

Information on volatility is provided in two forms whenever possible; vapor pressure and vapor density. Data on vapor pressure is usually provided in units of millimeters (mm) of mercury at a given temperature. The higher the vapor pressure the more volatile, and potentially dangerous, a chemical is. The vapor density is expressed in units relative to air. Thus a chemical with a vapor density of 2.0 is twice as dense as air and will tend to "flow" or "pool" around ground level and low places whereas chemicals with a vapor density less than 1.0 will tend to rise and more easily become dispersed in air.

Fire Hazard

This data field describes whether a compound is known or believed to be flammable (according to NFPA definitions) and thus to present a potential serious fire hazard or just to be combustible and therefore present a less potential fire hazard. Where flammability data is listed by class identifiers such as IA, IB, IC, II, etc., OSHA criteria (29 CFR 1910.106) have been used. Flash points and autoignition temperatures are provided when they could be found. In some cases flash points of liquids were measured by chemists at Radian Corporation. In addition, when available information permitted, recommended fire extinguishing methods are provided.

Fire Hazard Codes

There are several systems in use for describing the fire hazards associated with a chemical. Following is a summary of the most commonly used systems.

National Fire Protection Association codes (NFPA) represent hazards under fire conditions. They range from four to zero with four being the most flammable and zero being for materials which will not burn even at 1500 F for five minutes.

OSHA (29 CFR 1910.106)

Class IA flammable liquid	Flash point below 73 F and BP below 100 F
Class IB flammable liquid	Flash point below 73 F and BP at or above 100 F
Class IC flammable liquid	Flash point at or above 73 F and below 100 F
Class II combustible liquid	Flash point at or above 100 F and below 140 F
Class IIIA combustible liquid	Flash point at or above 140 F and below 200 F
Class IIIB combustible liquid	Flash point at or above 200 F

U.S. DOT (49 CFR 100-199)

U.S. Department of Transportation (U.S. DOT) codes are:

Extremely flammable	Flash point below 70 F
Flammable	Flash point 70-100 F
Combustible	Flash point between 100-200 F

LEL and UEL

These are fields for Lower Explosive Limits and Upper Explosive Limits (also known as lower flammability limits and upper flammability limits -- LFL and UFL) respectively. These numbers represent the lowest and highest percentages of these compounds in air that, in the presence of a spark or flame, can cause an explosive combustion.

Reactivity

This data field contains information on incompatibility of chemicals with water, air, or other chemicals.

Stability

This data field provides information on known or believed stability of chemicals. If protection from light, air, or heat is known or believed to be required or prudent for long term storage, these recommendations are provided where available. Stability in one or more solutions is often provided. In some cases stability measurements were made by chemists at Radian Corporation. In other cases an evaluation was made from data in the literature and/or from a review of the hazardous properties data. In the latter case, the chemical stability information is qualified by the words "...should be stable...".

Uses and Comments

When background information concerning the commercial uses of chemicals is available it appears in the Uses data field. This is sometimes supplemented by other information of a widely varying nature in the Comments data field.

Acute/Chronic Hazards

Toxicology is the study of poisons and the effects of their actions on the body. It includes their detection and treatment of the conditions produced. Since all chemicals can be toxic, it is necessary to define the conditions of exposure, the amount (concentration), time duration and route(s) of exposure. Environmental factors and the physiological status of the subject may also affect the seriousness of the hazard. Hence, the toxic effects of a chemical depend on the degree of exposure.

In addition, many toxic agents are hazardous only to specific cells or species. Therefore, knowledge may be limited to extrapolation or interpolation to other conditions or species and may be based on speculation and subject to errors based on the assumptions made during the assessment process.

Information provided under Acute/Chronic hazards is designed to alert users to the type of hazards that specific chemicals cause. Acute hazards have a rapid onset of symptoms of exposure and a short duration. Examples may include irritation of exposed tissues (a reversible reaction that may vary from mild spots that itch to more severe rashes and welts), corrosion (irreversible destruction) of tissues, lachrymation (tearing of the eyes) onset of coma, or death. Other acute hazards may result from heating chemicals to the point of decomposition (as, for example, a fire). Hazards from thermally decomposing chemicals may vary from acrid, noxious fumes to toxic gases such as phosgene or carbon monoxide.

Chronic hazards may include sensitization, carcinogenicity (causes cancer) or reproductive and developmental effects (including teratogenicity), among other hazards.

Symptoms

Symptoms of exposure are tremendously variable and range from irritation of the eyes, nose, and other exposed tissues to coma and, ultimately, death. Because of the thousands of different combinations and medically related grouping of symptoms, this publication is especially useful for searching for compounds that have various symptoms or combinations of symptoms.

Information given under Symptoms is designed to provide the greatest variety of observed effects that may possibly result from exposure to each chemical. This means, of course, that not all symptoms of exposure may be observed with anyone who is exposed to one or more of these Air Toxics. Actual symptoms of exposure will depend on the conditions of exposure, the concentrations of the chemicals, the duration of the exposure and the health and general well being of the person who is exposed.

The information in this section is still very useful for many purposes. In the electronic version of this Compendium these symptoms can be searched for individually or in combination with each other to help correlate potential relationships between symptoms of exposure and various types of substances that comprise EPA's Clean Air Act Air Toxics.

References for Chapter 3

The reference numbers from [001] through [700] were incorporated from the National Toxicology Program' Chemical Database (Lewis Publishers, 1992). There were gaps in the reference number sequence. This is intentional so that when new editions of the same reference were procured they could be given a new number that was sequential to earlier editions. In other cases spaces were left for potential future references of a like nature. Reference numbers higher than 700 have come from other sources and are in numeric sequence.

Note that the references below apply only to the data on the Air Toxics in this chapter. Other references for Chapters 1 and 2 are listed at the end of Chapter 2.

[001] Chemical Abstracts Service. STN International: CA and REGISTRY On-line. Columbus, OH.

[005] Christensen, H.E. and T.T. Luginbyhl, Eds. Registry of Toxic Effects of Chemical Substances. National Institute for Occupational Safety and Health. Rockville, MD. 1975.

[007] Fairchild, E.J., R.J. Lewis, Sr. and R.L. Tatken, Eds. Registry of Toxic Effects of Chemical Substances. DHEW (NIOSH) Publication No. 78-104-A. National Institute for Occupational Safety and Health. Cincinnati, OH. 1977.

[010] Lewis, R.J., Sr. and R.L. Tatken, Eds. Registry of Toxic Effects of Chemical Substances. DHEW (NIOSH) Publication No. 79-100. National Institute for Occupational Safety and Health. Cincinnati, OH. 1979.

[012] Lewis, R.J., Sr. and R.L. Tatken, Eds. Registry of Toxic Effects of Chemical Substances. DHEW (NIOSH) Publication No. 81-116. National Institute for Occupational Safety and Health. Cincinnati, OH. 1980.

[015] Lewis, R.J., Sr. and R.L. Tatken, Eds. Registry of Toxic Effects of Chemical Substances. On-line Ed. National Institute for Occupational Safety and Health. Cincinnati, OH, April 1991.

[016] Weast, R.C., D.R. Lide, M.J. Astle, and W.H. Beyer, Eds. CRC Handbook of Chemistry and Physics. 70th Ed. CRC Press, Inc. Boca Raton, FL. 1989.

[017] Weast, R.C., M.J. Astle, and W.H. Beyer, Eds. CRC Handbook of Chemistry and Physics. 67th Ed. CRC Press, Inc. Boca Raton, FL. 1986.

[018] Weast, R.C., M.J. Astle, and W.H. Beyer, Eds. CRC Handbook of Chemistry and Physics. 65th Ed. CRC Press, Inc. Boca Raton, FL. 1984.

[019] Weast, R.C. and M.J. Astle, Eds. CRC Handbook of Chemistry and Physics. 63rd Ed. CRC Press, Inc. Boca Raton, FL. 1982.

[020] Weast, R.C. and M.J. Astle, Eds. CRC Handbook of Chemistry and Physics. 60th Ed. CRC Press, Inc. Boca Raton, FL. 1979.

[021] Weast, R.C. and M.J. Astle. Eds. CRC Handbook of Chemistry and Physics. 57th Ed. CRC Press, Inc. Boca Raton, FL. 1977.

[022] Weast, R.C. and M.J. Astle, Eds. CRC Handbook of Chemistry and Physics. 56th Ed. CRC Press, Inc. Boca Raton, FL. 1976.

[025] Buckingham, J., Ed. Dictionary of Organic Compounds. 5th Ed. Chapman and Hall. New York. 1982.

[027] Edmundson, R.S. Ed. Dictionary of Organophosphorus Compounds. Chapman and Hall. New York. 1988.

[028] Buckingham, J., Ed. Dictionary of Organic Compounds. 5th Ed. Chapman and Hall. New York. 1988. Supplement 6.

[029] Buckingham, J., Ed. Dictionary of Organic Compounds. 5th Ed. and Supplements. Chapman and Hall. New York. 1988.

[030] Windholz, M., Ed. The Merck Index. 9th Ed. Merck and Co. Rahway, NJ. 1976.

[031] Windholz, M., Ed. The Merck Index. 10th Ed. Merck and Co. Rahway, NJ. 1983.

[032] Merck & Company. MRCK: The Merck Index On-line. BRS Information Technologies. Latham, New York 1991.

[033] Budavari, Susan, Ed. The Merck Index. 11th Ed. Merck and Co., Inc. Rahway, NJ. 1989.

[035] Bretherick, L., Ed. Hazards in the Chemical Laboratory. 3rd Ed. The Royal Society of Chemistry. London. 1981.

[036] Bretherick, L., Ed. Hazards in the Chemical Laboratory. 4th Ed. The Royal Society of Chemistry. London. 1986.

[038] Stull, D.R. Vapor pressure of pure substances: Organic Compounds. Industrial and Engineering Chem. 39(4):517-550. 1947.

[039] Boublik, T., V. Fried and E. Hala. The Vapor Pressures of Pure Substances. New York, Elsevier Scientific Pub. Co., 1973.

[040] Sax, N.I. Dangerous Properties of Industrial Materials. 4th Ed. Van Nostrand Reinhold. New York. 1975.

[041] Sax, N.I. Dangerous Properties of Industrial Materials. 5th Ed. Van Nostrand Reinhold. New York. 1979.

[042] Sax, N.I. Dangerous Properties of Industrial Materials. 6th Ed. Van Nostrand Reinhold. New York. 1984.

[043] Sax, N.I. and Richard J. Lewis, Sr. Dangerous Properties of Industrial Materials. 7th Ed. Van Nostrand Reinhold. New York. 1989.

[045] Sax, N.I., Ed. Industrial Pollution. Van Nostrand Reinhold. New York. 1974.

[046] Sax, N.I., Ed. Cancer Causing Chemicals. Van Nostrand Reinhold. New York. 1981.

[047] Weast, R.C. and M.J. Astle, Eds. CRC Handbook of Data on Organic Compounds. CRC Press, Inc. Boca Raton, FL. 1985.

[050] International Technical Information Institute. Toxic and Hazardous Industrial Chemicals Safety Manual for Handling and Disposal with Toxicity and Hazard Data. International Technical Information Institute. 1978.

[051] Sax, N. Irving, Ed. Dangerous Properties of Industrial Materials Report. Bimonthly Updates. Van Nostrand Reinhold Company, Inc. New York.

[052] Midwest Research Institute. NTP Analytical Chemistry Reports. Kansas City, MO.

[053] Arthur D. Little, Inc. NTP Health and Safety Packages. Arthur D. Little, Inc. Cambridge, MA.

[054] Tracor Jitco. NTP Safety and Toxicity Packages. Tracor Jitco. Bethesda, MD.

[055] Verschueren, K. Handbook of Environmental Data on Organic Chemicals. 2nd Ed. Van Nostrand Reinhold. New York. 1983.

[058] Information Handling Services. Material Safety Data Sheets Service. Microfiche Ed. Bimonthly Updates.

[060] Hawley, G.G., Ed. The Condensed Chemical Dictionary. 9th Ed. Van Nostrand Reinhold. New York. 1977.

[061] Hawley, G.G., Ed. The Condensed Chemical Dictionary. 10th Ed. Van Nostrand Reinhold. New York. 1981.

[062] Sax, N.I. and R.J. Lewis Sr., Eds. Hawley's Condensed Chemical Dictionary. 11th Ed. Van Nostrand Reinhold. New York. 1987.

[065] Bretherick, L. Handbook of Reactive Chemical Hazards. 2nd Ed. Butterworths. London. 1984.

[066] Bretherick, L. Handbook of Reactive Chemical Hazards. 3rd Ed. Butterworths. London. 1985.

[070] Proctor, N.H. and J.P. Hughes. Chemical Hazards of the Workplace. J.B. Lippincott. Philadelphia. 1978.

[071] Sax, N. Irving, Ed. Hazardous Chemicals Information Annual, No. 1. Van Nostrand Reinhold Information Services. New York. 1986.

[072] Sax, N. Irving, Ed. Hazardous Chemicals Information Annual, No. 2. Van Nostrand Reinhold Information Services. New York. 1987.

[075] Nielson, J.M., Ed. Material Safety Data Sheets. Technology Marketing Operation, General Electric Co. Schenectady, NY. 1980.

[076] Nielson, J.M., Ed. Material Safety Data Sheets. Volume II: Tradename Materials. Technology Marketing Operation, General Electric Co. Schenectady, NY. 1980.

[080] U.S. Environmental Protection Agency, Office of Toxic Substances. Toxic Substances Control Act Chemical Substances Inventory, Initial Inventory. 6 Vols. U.S. Environmental Protection Agency. Washington, D.C. 1979.

[081] U.S. Environmental Protection Agency, Office of Toxic Substances. Toxic Substances Control Act Chemical Substances Inventory, Cumulative Supplement II to the Initial Inventory. U.S. Environmental Protection Agency. Washington, DC. 1982.

[082] U.S. Environmental Protection Agency, Office of Toxic Substances. Toxic Substances Control Act Chemical Substance Inventory: 1985 Edition. 5 Vols. U.S. Environmental Protection Agency. Washington, D.C. January 1986.

[084] Commission of the European Communities. Constructing EINECS: Basic Documents. Official Journal of the European Communities. Office for Official Publications of the European Communities. Luxembourg. 1981.

[085] Commission of the European Communities. Constructing EINECS: Basic Documents, Compendium of Known Substances. Vols. 1-3. Office for Official Publications of the European Communities. Luxembourg. 1981.

[086] Commission of the European Communities. Constructing EINECS: Basic Documents, European Core Inventory. Vols. 1-4. Office for Official Publications of the European Communities. Luxembourg. 1981.

[090] Steere, N.V., Ed. Handbook of Laboratory Safety. 2nd Ed. CRC Press, Inc. Cleveland, OH. 1971.

[092] Wilhoit, Randolph C. and Bruno J. Zwolinski. Handbook of Vapor Pressures and Heats of Vaporization of Hydrocarbons and Related Compounds. Evans Press. Fort Worth, TX. 1971.

[095] Gardner, W., E.I. Cooke and R.W.I. Cooke. Handbook of Chemical Synonyms and Trade Names. 8th Ed. CRC Press. Boca Raton, FL. 1978.

[099] Grant, W. Morton, M.D. Toxicology of the Eye. 3rd Ed. Charles C. Thomas, Publisher. Springfield, IL. 1986.

[100] Grasselli, J.G., Ed. CRC Atlas of Spectral Data and Physical Constants for Organic Compounds. CRC Press, Inc. Cleveland,,OH. 1973.

[102] U.S. Department of Health and Human Services and U.S. Department of Labor. NIOSH/OSHA Occupational Health Guidelines for Chemical Hazards. 3 Vols. DHHS (NIOSH) Publication No. 81-123. January, 1981.

[105] The Society of Dyers and Colourists. Colour Index. Vols. 1-8. The Society of Dyers and Colourists. Yorkshire, England. American Association of Textile Chemists and Colorists. Research Triangle Park, NC. 1971-1987.

[107] Occupational Health Services, Inc. Hazardline. Occupational Health Services, Inc. New York.

[110] Oak Ridge National Laboratory. Environmental Mutagen Information Center (EMIC), Bibliographic Data Base. Oak Ridge National Laboratory. Oak Ridge, TN.

[120] Oak Ridge National Laboratory. Environmental Teratogen Information Center (ETIC), Bibliographic Data Base. Oak Ridge National Laboratory. Oak Ridge, TN.

[125] Shepard, T.H. Catalog of Teratogenic Agents. 3rd Ed. The Johns Hopkins University Press. Baltimore. 1980.

[130] Strauss, H.J. Handbook for Chemical Technicians. McGraw-Hill. New York. 1979.

[135] Braker, W. and A.L. Mossman. Matheson Gas Data Book. Matheson Gas Products. Secaucus, NJ. 1980.

[137] Braker, W. and A.L. Mossman The Matheson Unabridged Gas Data Book. Matheson Gas Products. Secaucus, NJ. 1974.

[140] Manufacturing Chemists Association. Guide for Safety in the Chemical Laboratory. 2nd Ed. Van Nostrand Reinhold. New York. 1972.

[141] Veterinary Medicine Publishing Co. Veterinary Pharmaceuticals and Biologicals. 6th Ed. Veterinary Medicine Publishing Co. Lenexa, KS. 1988.

[142] U.S. Department of the Interior, Bureau of Mines. Dust Explosibility of Chemicals, Drugs, Dyes and Pesticides. National Technical Information Service, U.S. Department of Commerce. Springfield, VA. May 1968.

[143] U.S. Department of the Interior, Bureau of Mines. Explosibility of Agricultural Dusts. National Technical Information Service, U.S. Department of Commerce. Springfield, VA. 1961.

[144] U.S. Department of the Interior, Bureau of Mines. Explosibility of Dusts used in the Plastics Industry. National Technical Information Service, U.S. Department of Commerce. Springfield, VA. 1962.

[145] International Working Group on the Toxicology of Rubber Additives. Rubber Chemicals Safety Data and Handling Precautions. International Working Group on the Toxicology of Rubber Additives. Belgium. 1984.

[149] Wagner, Sheldon L. Clinical Toxicology of Agricultural Chemicals. Noyes Data Corporation. Park Ridge, NJ. 1983.

[150] Gosselin, R.E., H.C. Hodge, R.P. Smith and M.N. Gleason. Clinical Toxicology of Commercial Products. 4th Ed. Williams and Wilkins, Co. Baltimore. 1976.

[151] Gosselin, R.E., H.C. Hodge, and R.P. Smith. Clinical Toxicology of Commercial Products. 5th Ed. Williams and Wilkins, Co. Baltimore. 1984.

[155] Thomas, C.L., Ed. Taber's Cyclopedic Medical Dictionary. 14th Ed. F.A. Davis Co. Philadelphia. 1981.

[157] Huff, B.B., Ed. Physicians' Desk Reference. 44th Ed. Medical Economics Co. Oradell, NJ. 1990.

[158] Huff, B.B., Ed. Physicians' Desk Reference. 43rd Ed. Medical Economics Co. Oradell, NJ. 1989.

[159] Huff, B.B., Ed. Physicians' Desk Reference. 41st Ed. Medical Economics Co. Oradell, NJ. 1987.

[160] Huff, B.B., Ed. Physicians' Desk Reference. 33rd Ed. Medical Economics Co. Oradell, NJ. 1979.

[161] Huff, B.B., Ed. Physicians' Desk Reference. 36th Ed. Medical Economics Co. Oradell, NJ. 1982.

[162] Huff, B.B., Ed. Physicians' Desk Reference. 38th Ed. Medical Economics Co. Oradell, NJ. 1984.

[163] Thompson, J.F., Ed. Analytical Reference Standards and Supplemental Data for Pesticides and Other Organic Compounds. EPA Publ. No. 600/9-76-012. U.S. Environmental Protection Agency, Office of Research and Development, Health Effects Research Laboratory. Research Triangle Park, NC. 1976.

[164] Huff, B.B., Ed. Physicians' Desk Reference for Nonprescription Drugs. 11th Ed. Medical Economics Co. Oradell, NJ. 1990.

[165] Wiswesser, W.J., Ed. Pesticide Index. Entomological Society of America. College Park, MD. 1976.

[168] Hartley, Douglas B.Sc., Ph.D., M.I.Inf.Sc. and Hamish Kidd B.Sc., Eds. The Agrochemicals Handbook. The Royal Society of Chemistry. Nottingham, England. 1983.

[169] Hartley, Douglas B.Sc., Ph.D., M.I.Inf.Sc. and Hamish Kidd B.Sc., Eds. The Agrochemicals Handbook. 2nd Ed. The Royal Society of Chemistry. Nottingham, England. 1987.

[170] Worthing, C.R., Ed. The Pesticide Manual, A World Compendium. 6th Ed. British Crop Protection Council. London, England. 1979.

[171] Worthing, C.R., Ed. The Pesticide Manual, A World Compendium. 7th Ed. British Crop Protection Council. London, England. 1983.

[172] Worthing, C.R., Ed. The Pesticide Manual, A World Compendium. 8th Ed. British Crop Protection Council. London, England. 1987.

[173] Hayes, W.J., Jr. Pesticides Studied in Man. Williams and Wilkins. Baltimore. 1982.

[175] Council for Agricultural Science and Technology. The Phenoxy Herbicides. 2nd Ed. CAST Report No. 77. Council for Agricultural Science and Technology. Iowa State University, Ames, Iowa. 1978.

[176] Advisory Committee on Pesticides. Further Review of the Safety for Use in the U.K. of the Herbicide 2,4,5-T. Advisory Committee on Pesticides. 1980.

[180] Ross, S.S., Ed. Toxic Substances Sourcebook. Series 1. Environmental Information Center, Inc., Toxic Substances Reference Department. New York. 1978.

[185] Esposito, M.P., T.O. Tiernan and F.E. Dryden. Dioxins. EPA Report No. 600/2-80-197. Industrial Environmental Research Laboratory. Environmental Protection Agency. Cincinnati, Ohio. 1980.

[186] Sittig, Marshall, Ed. Pesticide Manufacturing and Toxic Materials Control Encyclopedia. Noyes Data Corporation. Park Ridge, NJ. 1980.

[190] Packer, K., Ed. Nanogen Index, A Dictionary of Pesticides and Chemical Pollutants. Nanogens International. Freedom, CA. 1975 (Updated 1979).

[195] Estrin, F.E., P.A. Crosley and C.R. Haynes, Eds. CFTA Cosmetic Ingredient Dictionary. 3rd Ed and Supplement. The Cosmetic, Toiletry and Fragrance Assn. Inc. Washington. 1982.

[200] International Air Transport Association. Restricted Articles Regulations. 23nd Ed. International Air Transport Assn. Geneva. 1980.

[201] International Air Transport Association. Dangerous Goods Regulations and Supplement. 28th Ed. International Air Transport Association. Montreal, Quebec. 1987.

[205] Dean, John A., Ed. Lange's Handbook of Chemistry. 13th Ed. McGraw-Hill Book Company. New York. 1985.

[210] U.S. Coast Guard, Department of Transportation. Chemical Data Guide for Bulk Shipment by Water. 5th Ed. U.S. Coast Guard Publication No. CG-388. U.S. Coast Guard. Washington, DC. 1976.

[215] Rom, William N., Ed. Environmental and Occupational Medicine. Little, Brown and Company. Boston. 1983.

[220] Airline Tariff Publishing Co. Official Air Transport Restricted Articles Tariff No. 6-D Governing the Transportation of Restricted Articles by Air. Airline Tariff Publishing Co. Washington, DC. 19 April 1980 Revision.

[225] Armour, Margaret-Ann, Lois M. Browne and Gordon L. Weir. Hazardous Chemicals Information and Disposal Guide. University of Alberta. Alberta, Canada. 1984.

[230] Cross, R.J. and Mingos, D.M.P., Eds. Organometallic Compounds of Nickel, Palladium, Platinum, Copper, Silver and Gold. Chapman and Hall. London. 1985.

[231] Harrison, P.G., Ed. Organometallic Compounds of Germanium, Tin and Lead. Chapman and Hall. London. 1985.

[232] Knox, G.R., Ed. Organometallic Compounds of Iron. Chapman and Hall. London. 1985.

[233] Wardell, J.L., Ed. Organometallic Compounds of Zinc, Cadmium and Mercury. Chapman and Hall. London. 1985.

[234] White, C., Ed. Organometallic Compounds of Cobalt, Rhodium and Iridium. Chapman and Hall. London. 1985.

[240] Grayson, Martin, Ed. Kirk-Othmer Encyclopedia of Chemical Technology. Volumes 1-24 and Supplement. 3rd Ed. John Wiley & Sons. New York. 1978-1984.

[245] Kidd, Hamish and Douglas Hartley, Eds. UK Pesticides for Farmers and Growers. The Royal Society of Chemistry. Nottingham, England. 1987.

[250] Office of the Federal Register, National Archives and Records Service, General Services Administration. Code of Federal Regulations, Title 49, Transportation, Parts 100 to 199 (Revised as of October 1, 1979). Government Printing Office. Washington, DC. 1979.

[251] Office of the Federal Register, National Archives and Records Service, General Services Administration. Code of Federal Regulations, Title 49, Transportation, Parts 100 to 199 (Revised as of October 1, 1982). Government Printing Office. Washington, DC. 1982.

[252] Office of the Federal Register, National Archives and Records Service, General Services Administration. Code of Federal Regulations, Title 49, Transportation, Parts 100 to 199 (Revised as of October 1, 1983). Government Printing Office. Washington, DC. 1983.

[253] Office of the Federal Register, National Archives and Records Service, General Services Administration. Code of Federal Regulations, Title 49, Transportation, Parts 100 to 199 (Revised as of October 1, 1986). Government Printing Office. Washington, DC. 1986.

[255] Office of the Federal Register, National Archives and Records Service, General Services Administration. Code of Federal Regulations, Title 40, Protection of Environment, Parts 100 to 399 (Revised as of July 1, 1980). Government Printing Office. Washington, DC. 1979.

[260] Slein, M.W. and E.B. Sansone. Degradation of Chemical Carcinogens. Van Nostrand Reinhold. New York. 1980.

[269] Lenga, Robert E. The Sigma-Aldrich Library of Chemical Safety Data. Edition 1. Sigma-Aldrich Corporation. Milwaukee, WI. 1985.

[270] Aldrich Chemical Company. Aldrich Catalog/Handbook of Fine Chemicals. Aldrich Chemical Co., Inc. Milwaukee, WI. 1978.

[271] Aldrich Chemical Company. Aldrich Catalog/Handbook of Fine Chemicals. Aldrich Chemical Co., Inc. Milwaukee, WI. 1980.

[272] Aldrich Chemical Company. Aldrich Catalog/Handbook of Fine Chemicals. Aldrich Chemical Co., Inc. Milwaukee, WI. 1982.

[273] Aldrich Chemical Company. Aldrich Catalog/Handbook of Fine Chemicals. Aldrich Chemical Co., Inc. Milwaukee, WI. 1984.

[274] Aldrich Chemical Company. Aldrich Catalog/Handbook of Fine Chemicals. Aldrich Chemical Co., Inc. Milwaukee, WI. 1986.

[275] Aldrich Chemical Company. Aldrich Catalog/Handbook of Fine Chemicals. Aldrich Chemical Co., Inc. Milwaukee, WI. 1988.

[276] Aldrich Chemical Company. Aldrich Catalog/Handbook of Fine Chemicals. Aldrich Chemical Co., Inc. Milwaukee, WI. Unspecified edition.

[280] Aldrich Chemical Company. The Library of Rare Chemicals. 4th Ed. Aldrich Chemical Co., Inc. Milwaukee, WI. 1978.

[290] Deichmann, W.B. and H.W. Gerarde. Toxicology of Drugs and Chemicals. Academic Press. New York. 1969.

[295] Reynolds, James E.F., Ed. Martindale The Extra Pharmacopoeia. 28th Ed. The Pharmaceutical Press. London. 1982.

[300] Dreisbach, R.H. Handbook of Poisoning: Prevention, Diagnosis and Treatment. 10th Ed. Lange Medical Publications. Los Altos, CA. 1980.

[301] Dreisbach, R.H. Handbook of Poisoning: Prevention, Diagnosis and Treatment. 11th Ed. Lange Medical Publications. Los Altos, CA. 1983.

[305] Block, J.B. The Signs and Symptoms of Chemical Exposure. Charles H. Thomas. Springfield, IL. 1980.

[310] Haque, R., Ed. Dynamics, Exposure and Hazard Assessment of Toxic Chemicals, Ann Arbor Science Publishers. Ann Arbor, MI. 1980.

[315] Florey, Klaus Ed. Analytical Profiles of Drug Substances. Academic Press, Inc. Orlando, FL. 1973-1987.

[320] Occupational Safety and Health Administration. Tentative OSHA Listing of Confirmed and Suspected Carcinogens by Category. Occupational Safety and Health Administration. Washington, DC. 1979.

[321] Soderman, Jean V. Ed. CRC Handbook of Identified Carcinogens and Noncarcinogens: Carcinogenicity-Mutagenicity Database. Vols. 1-2. CRC Press, Inc. Boca Raton, FL. 1982.

[325] Office of the Federal Register National Archives and Records Administration. Code of Federal Regulations, Title 29, Labor, Parts 1900 to 1910. U.S. Government Printing Office. Washington. 1986.

[326] Office of the Federal Register National Archives and Records Administration. Code of Federal Regulations, Title 29, Labor, Parts 1900 to 1910. U.S. Government Printing Office. Washington. 1987.

[327] Office of the Federal Register National Archives and Records Administration. Code of Federal Regulations, Title 29, Labor, Parts 1900 to 1910. U.S. Government Printing Office. Washington. 1988.

[329] Office of the Federal Register National Archives and Records Administration. Code of Federal Regulations, Title 29, Labor, Part 1900.1200: Hazard Communication. U.S. Government Printing Office. Washington. 1986.

[330] Rappoport, Z.V.I., Ed. CRC Handbook of Organic Compound Identification. 3rd Ed. CRC Press, Inc. Cleveland, OH. 1975.

[335] United States Environmental Protection Agency. EPA Toxicology Handbook. Government Institutes, Inc. Rockville, Maryland. September 1986.

[340] Sittig, M. Hazardous and Toxic Effects of Industrial Chemicals. Noyes Data Corporation. Park Ridge, NJ. 1979.

[345] Sittig, M. Handbook of Toxic and Hazardous Chemicals. Noyes Publications. Park Ridge, NJ. 1981.

[346] Sittig, M. Handbook of Toxic and Hazardous Chemicals and Carcinogens. 2nd Ed. Noyes, Publications. Park Ridge, NJ. 1985.

[350] Sittig, M., Ed. Priority Toxic Pollutants, Health Impacts and Allowable Limits. Noyes Data Corporation. Park Ridge, NJ. 1980.

[355] Weiss, G., Ed. Hazardous Chemicals Data Book. Noyes Data Corporation. Park Ridge, NJ. 1980.

[360] Sunshine, I., Ed. CRC Handbook Series in Analytical Toxicology, Section A: General Data, Vol. 1. CRC Press, Inc. Boca Raton, FL. 1969.

[365] Connors, Kenneth A., Gordon L. Amidon and Valentino J. Stella. Chemical Stability of Pharmaceuticals: A Handbook for Pharmacists. 2nd Ed. John Wiley & Sons. New York. 1986.

[370] U.S. Coast Guard, Department of Transportation. Chemical Hazards Response Information System (CHRIS), A Condensed Guide to Chemical Hazards. U.S. Coast Guard Publication No. CG-446-1. U.S. Coast Guard. Washington, DC. 1974.

[371] U.S. Coast Guard, Department of Transportation. CHRIS Hazardous Chemical Data. U.S. Coast Guard. Washington, D.C. 1985.

[375] Barton, A.F.M. Solubility Data Series, Volume 15: Alcohols with Water. Pergamon Press. New York. 1984.

[380] U.S. Coast Guard, Department of Transportation. Chemical Hazards Response Information System (CHRIS), Hazardous Chemical Data. U.S. Coast Guard Publication No. CG-446-2. U.S. Coast Guard. Washington, DC. 1978.

[385] Williams, L.R., E. Calliga and R. Thomas. Hazardous Materials Spill Monitoring: Safety Handbook and Chemical Hazard Guide, Part A. U.S. EPA Report No. EPA-600/4-79-008a. U.S. Environmental Protection Agency. Washington, DC. 1979.

[386] Williams, L.R., E. Calliga and R. Thomas. Hazardous Materials Spill Monitoring: Safety Handbook and Chemical Hazard Guide, Part B - Chemical Data. U.S. EPA Report No. EPA-600/4-79-008b. U.S. Environmental Protection Agency. Washington, DC. 1979.

[390] Bureau of Explosives, Association of American Railroads. Hazardous Materials Regulations of the Department of Transportation by Air, Rail, Highway, Water and Military Explosives by Water, Including Specifications for Shipping Containers. Bureau of Explosives Tariff No. BOE-6000-A. Bureau of Explosives. Washington, DC. 1981.

[393] Student, P.J. Emergency Handling of Hazardous Materials in Surface Transportation. Bureau of Explosives. Washington, D.C. 1981.

[395] International Agency for Research on Cancer, World Health Organization. IARC Monographs on the Evaluation of Carcinogenic Risk of Chemicals to Man. International Agency for Research on Cancer. Geneva, Switzerland.

[400] U.S. Department of Transportation, Materials Transportation Bureau, Office of Hazardous Materials Operations. An Index to the Hazardous Materials Regulations, Title 49, Code of Federal Regulations, Parts 100-199 (January 3, 1977 Revision). U.S. Department of Transportation. Washington, DC. 1977.

[401] Nutt, A. R. Toxic Hazards of Rubber Chemicals. Elsevier Applied Science Publishers. New York. 1984.

[405] Goodman, L.S. and A. Gilman. The Pharmacological Basis of Therapeutics. 5th Ed. Macmillan Publishing Co. New York. 1975.

[406] Goodman, L.S., A. Gilman, F. Murad and T.W. Rall, Eds. The Pharmacological Basis of Therapeutics. 7th Ed. Macmillan Publishing Co. New York. 1985.

[410] American Conference of Governmental Industrial Hygenists. Threshold Limit Values for Chemical Substances and Physical Agents in the Work Environment with Intended Changes for 1982. American Conference of Governmental Industrial Hygenists. Cincinnati, OH. 1982.

[411] American Conference of Governmental Industrial Hygenists. Threshold Limit Values for Chemical Substances and Physical Agents in the Work Environment with Intended Changes for 1984-85. American Conference of Governmental Industrial Hygenists. Cincinnati, OH. 1984.

[412] American Conference of Governmental Industrial Hygienists. Threshold Limit Values and Biological Exposure Indices for 1985-1986. American Conference of Governmental Industrial Hygienists. Cincinnati, OH. 1985.

[413] American Conference of Governmental Industrial Hygienists. Threshold Limit Values and Biological Exposure Indices for 1986-1987. American Conference of Governmental Industrial Hygienists. Cincinnati, OH. 1986.

[414] American Conference of Governmental Industrial Hygienists. Threshold Limit Values and Biological Exposure Indices for 1987-1988. American Conference of Governmental Industrial Hygienists. Cincinnati, OH. 1987.

[415] American Conference of Governmental Industrial Hygienists. Threshold Limit Values and Biological Exposure Indices for 1988-1989. American Conference of Governmental Industrial Hygienists. Cincinnati, OH. 1988.

[416] American Conference of Governmental Industrial Hygienists. Threshold Limit Values and Biological Exposure Indices for 1990-1991. American Conference of Governmental Industrial Hygienists. Cincinnati, OH. 1990.

[420] American Conference of Governmental Industrial Hygenists. Documentation of the Threshold Limit Values. 4th ed. American Conference of Governmental Industrial Hygenists. Cincinnati, OH. 1980.

[421] American Conference of Governmental Industrial Hygienists. Documentation of the Threshold Limit Values. 5th Ed. American Conference of Governmental Industrial Hygienists. Cincinnati, OH. 1986.

[425] Schwope, A.D., P.P. Costas, J.O. Jackson and D.J. Weitzman. Guidelines for the Selection of Chemical Protective Clothing. Vol. I: Field Guide; Vol. II: Technical and Reference Manual. American Conference of Governmental Industrial Hygenists, Inc. Cincinnati, OH. 1983.

[426] Schwope, A.D., P.P. Costas, J.O. Jackson and D.J. Weitzman. Guidelines for the Selection of Chemical Protective Clothing. Vol. I: Field Guide; Vol. II: Technical and Reference Manual. American Conference of Governmental Industrial Hygenists, Inc. Cincinnati, OH. 1985.

[430] Clayton, G.D. and F.E. Clayton, Eds. Patty's Industrial Hygienc and Toxicology. Vol. 2. Third Rcvised Edition. John Wiley and Sons. New York. 1981.

[435] Lyman, Warren J., Ph.D., William F. Reehl and David H. Rosenblatt, Ph.D. Eds. Handbook of Chemical Property Estimation Methods. McGraw-Hill Book Company. New York. 1982.

[440] National Fire Protection Association. Flash Point Index of Tradc Name Liquids. 9th Ed. National Firc Protection Association. Boston, MA. 1978.

[445] Baker, Charles J. The Firefighter's Handbook of Hazardous Materials. 4th Ed. Maltese Enterprises, Inc. Indianapolis, Indiana. 1984.

[450] National Fire Protection Association. Fire Protection Guide on Hazardous Chemicals. 7th Ed. National Fire Protection Association. Boston. 1978.

[451] National Fire Protection Association. Fire Protection Guide on Hazardous Materials. 9th Ed. National Fire Protection Association. Quincy, MA. 1986.

[455] The Pharmaceutical Society of Great Britain. The Pharmaceutical Codex. 11th Edition. The Pharmaceutical Press. London. 1979.

[460] National Fire Protection Association. Hazardous Chemicals Data. National Fire Protection Association. Boston. 1975.

[465] National Toxicology Program. Chemical Registry Handbook, Parts I and II. National Toxicology Program. Research Triangle Park, NC. 1981.

[470] Meyer, E. Chemistry of Hazardous Materials. Prentiss-Hall. Englewood Cliffs, NJ. 1977.

[475] United States Environmental Protection Agency. Good Laboratory Practice Compliance Inspection Manual. Government Institutes, Inc. August 1985.

[480] Fletcher, J.H., O.C. Dermer and R.B. Fox, Eds. Nomenclature of Organic Compounds, Principles and Practice. American Chemical Society. Washington, DC. 1974.

[485] Rigaudy, J. and S.P. Klesney. Nomenclature of Organic Chemistry, Sections A, B, C, D, E, F and H. Pergamon Press. New York. 1979.

[490] Meidl, J.H. Explosive and Toxic Hazardous Materials. Glencoe Publishing Co. Encino, CA. 1970.

[495] Williams, Phillip L. and James L. Burson Eds. Industrial Toxicology. Van Nostrand Reinhold Company. New York. 1985.

[500] Mackison, F.W., R.S. Stricoff and L.J. Partridge, Eds. Occupational Health Guidelines for Chemical Hazards. DHHS (NIOSH) Publication No. 81-123. National Institute for Occupational Safety and Health. Cincinnati, OH. 1981.

[501] Mackison, F.W., R.S. Stricoff and L.J. Partridge, Eds. NIOSH/OSHA Pocket Guide to Chemical Hazards. DHEW (NIOSH) Publication No. 78-210. National Institute for Occupational Safety and Health. Cincinnati, OH. 1978.

[505] American Lung Association of Western New York. Chemical Emergency Action Manual. 2nd Ed. C.V. Mosby Co. St. Louis, MO. 1983.

[510] Alliance of American Insurers. Handbook of Organic Industrial Solvents. 5th Ed. Alliance of American Insurers. Chicago. 1980.

[515] Committee on Specifications and Criteria for Biochemical Compounds Division of Chemistry and Chemical Technology National Research Council. Specifications and Criteria for Biochemical Compounds. 3rd Ed. National Academy of Sciences. Washington, D.C. 1984.

[520] Lyman, W.J., W.F. Reehl and D.H. Rosenblatt. Handbook of Chemical Property Estimation Methods, Environmental Behavior of Organic Compounds. McGraw-Hill. New York. 1982.

[525] Hartwell, J.L. Survey of Compounds Which Have Been Tested for Carcinogenic Activity. 2nd Ed. Public Health Service Publication No. 149. National Cancer Institute, National Institutes of Heath. Bethesda, MD. 1963.

[527] Shubik, P. and J.L. Hartwell. Survey of Compounds Which Have Been Tested for Carcinogenic Activity. Supplement 2. Public Health Service Publication No. 149. National Cancer Institute, National Institutes of Heath. Bethesda, MD. 1969.

[528] National Cancer Institute, National Institutes of Health, Public Health Service, U.S. Department of Health, Education and Welfare. Survey of Compounds Which Have Been Tested for Carcinogenic Activity. 1961-1967 Vol. Sections I and II. DHEW (NIH) Publication No. 73-35. Public Health Service Publication No. 149. National Cancer Institute. Bethesda, MD. 1973.

[529] National Cancer Institute, National Institutes of Health, Public Health Service, U.S. Department of Health, Education and Welfare. Survey of Compounds Which Have Been Tested for Carcinogenic Activity. 1968-1969 Vol. DHEW (NIH) Publication No. 72-35, Public Health Service Publication No. 149. National Cancer Institute. Bethesda, MD. 1972.

[530] National Cancer Institute, National Institutes of Health, Public Health Service, U.S. Department of Health, Education and Welfare. Survey of Compounds Which Have Been Tested for Carcinogenic Activity. 1970-1971 Vol. DHEW (NIH) Publication No. 73-453, Public Health Service Publication No. 149. National Cancer Institute. Bethesda, MD. 1973.

[540] National Cancer Institute, National Institutes of Health, Public Health Service, U.S. Department of Health, Education and Welfare. Survey of Compounds Which Have Been Tested for Carcinogenic Activity. 1978 Vol. DHEW (NIH) Publication No. 80-453 (Formerly Public Health Service Publication No. 149). National Cancer Institute. Bethesda, MD. 1973.

[545] Office of the Federal Register National Archives and Records Administration. Federal Register, Dept. of Labor, Part III. U.S. Government Printing Office. Washington. January 19, 1989.

[550] U.S. Environmental Protection Agency, Office of Pesticides and Toxic Substances. TSCA Chemical Assessment Series, Chemical Hazard Information Profiles (CHIPs), August 1976 - August 1978. U.S. EPA Publication No. EPA-560/11-80-011. U.S. Environmental Protection Agency. Washington, DC. 1980.

[560] U.S. Environmental Protection Agency, Office of Pesticides and Toxic Substances. TSCA Chemical Assessment Series, Chemical Screening: Initial Evaluations of Substantial Risk Notices, Section 8(e), January 1, 1977 - June 30, 1979. U.S. EPA Publication No. EPA-560/11-80-008. U.S. Environmental Protection Agency. Washington, DC. 1980.

[565] Chemical and Pharmaceutical Press. MSDS Reference for Crop Protection Chemicals. Chemical and Pharmaceutical Press. New York, NY. 1989.

[566] Chemical and Pharmaceutical Press. MSDS Reference for Crop Protection Chemicals. Updates. Chemical and Pharmaceutical Press. New York, NY. 1989.

[570] Cone, M.V., M.F. Baldauf, F.M. Martin and J.T. Ensminger, Eds Chemicals Identified in Human Biological Media, A Data Base, First Annual Report, Vol. I, Parts 1 and 2, Records 1 - 1580. Interagency Collaborative Group on Environmental Carcinogenesis, National Cancer Institute, National Institutes of Health. Bethesda, MD. 1980.

[571] Cone, M.V., M.F. Baldauf, F.M. Martin and J.T. Ensminger, Eds. Chemicals Identified in Human Biological Media, A Data Base, Second Annual Report, Vol. II, Parts 1 and 2, Records 1581-3500. Interagency Collaborative Group on Environmental Carcinogenesis, National Cancer Institute, National Institutes of Health. Bethesda, MD. 1981.

[575] Consolidated Midland Corporation. Technical Bulletin no. 1002, Phorbol, TPA and Derivatives. Brewster, New York, 1982.

[580] Frick, G.W., Ed. Environmental Glossary, 2nd Ed. Government Institutes, Inc. Rockville, MD. 1982.

[582] Arbuckle, J.G., G.W. Frick, R.M. Hall Jr., M.L. Miller, T.F.P. Sullivan and T.A. Vanderver, Jr. Environmental Law Handbook. 7th Ed. Government Institutes, Inc. Rockville, MD. 1980.

[584] Government Institutes, Inc. Environmental Statutes. 1983 Ed. Government Institutes, Inc. Rockville, MD. 1983.

[590] Stever, Donald W. Law of Chemical Regulation and Hazardous Waste. Clark Boardman Company, Ltd. New York. 1986.

[600] Hazards Research Corporation. Vapor Pressure Determinations. HRC Report. Hazards Research Corporation. Rockaway, NJ.

[601] Safety Consulting Engineers Inc. Vapor Pressure Test Results. Safety Consulting Engineers Inc. Rosemont, IL.

[610] Clansky, Kenneth B., Ed. Suspect Chemicals Sourcebook: A Guide to Industrial Chemicals Covered Under Major Federal Regulatory and Advisory Programs. Roytech Publications, Inc. Burlingame, CA. 1990. Section 3.

[620] United States National Toxicology Program. Chemical Status Report. NTP Chemtrack System. Research Triangle Park, NC. January 4, 1991.

[650] 3M Company. 1989 Respirator Selection Guide. 3M Occupational Health and Environmental Safety Division. St. Paul, MN. 1989.

[651] 3M Company. 1990 Respirator Selection Guide. 3M Occupational Health and Environmental Safety Division. St. Paul, MN. 1990.

[700] Experimentally determined by Radian Corporation, Austin, Texas.

[701] EPA's IRIS Chemical Information Database. Adapted for publication by Lawrence H. Keith; Lewis Publishers, Inc., 1992.

[702] Screening Methods for the Development of Air Toxics Emission Factors, EPA-450/4- 91-021, September 1991.

[703] Dangerous Properties of Industrial Materials, Seventh Edition, N. Irving Sax and Richard J. Lewis, Sr., Van Nostrand Reinhold, New York, 1989.

[704] NIOSH Pocket Guide to Chemical Hazards, U.S. Department of Health and Human Services, Public Health Service, Centers for Disease Control, National Institute for Occupational Safety and Health, DHHS (NIOSH) Publication No. 90-117, June 1990.

[705] EPA's Pesticide Fact Sheet Database, Mary M.Walker and Lawrence H. Keith, Editors, Lewis Publishers, Inc., 1992.

[706] Handbook of Chemistry and Physics, Seventy-second Edition, David R. Lide, Editor in Chief, CRC Press, 1992.

[707] "Asbestos: Scientific Developments and Implications for Public Policy," B.T. Mossman, J. Bignon, M. Corn, A. Seaton, J.B.L. Gee, SCIENCE, Vol. 247, pp 294 - 301, 19 January 1990.

[708] Clean Air Act Amendments, Public Law #101-549, Section 301, November 15, 1990.

[709] Pesticide Handbook (Entoma), Twenty-ninth Edition, Robert L. Caswell, Kathleen J. DeBold, Lorraine S. Gilbert, Editors, Entomological Society of America, 1981.

[710] Bretherick, L. Handbook of Reactive Chemical Hazards. 3rd Ed. Butterworths. London. 1985.

[711] Kant, James A., Ed. Riegel's Handbook of Industrial Chemistry. 8th Ed. Van Nostrand Reinhold. New York. 1983.

[712] Gross, Paul and Daniel Brown. Toxic and Biomedical Effects of Fibers. Noyes Publications. Parkridge, NJ. 1984.

[713] 56 Federal Register 27338-27356. June 13, 1991.

[714] National Research Council, Committee on Hazardous Substances in the Laboratory, Assembly of Mathematical and Physical Sciences. Prudent Practices for Handling Hazardous Chemicals in Laboratories. National Academy Press. Washington, D.C. 1981.

[715] Budavari, Susan, Ed. The Merck Index: An Encyclopedia of Chemicals, Drugs, and Biologicals. 11th Ed. Merck & Co., Inc. 1989.

[716] McClanahan, James L. and Mary M. Walker. Guidebook on Polychlorinated Biphenyls. March, 1987.

[717] Exner, Jurgen H., Ed. Solving Hazardous Waste Problems. Learning from Dioxins. American Chemical Society. Washington, D.C. 1987.

[718] 40 CFR Part 761.

[719] Mobay Corporation, Agricultural Chemicals Division. Material Safety Data Sheet, Baygon 1.5. February 1, 1991.

[720] National Fire Protection Association. Fire Protection Guide on Hazardous Materials. Ninth Ed. National Fire Protection Association. Quincy, MA. 1986. □

[721] Radian Corporation information for the National Toxicology Program.

[722] Radian Corporation information for the National Toxicology Program.

[723] Radian Corporation information for the National Toxicology Program.

Acetaldehyde

CAS NUMBER: 75-07-0

CHEMICAL FORMULA: C2H4O

SYNONYMS: Acetic aldehyde Ethanal Ethyl aldehyde

MOLECULAR WEIGHT: 44.06

WLN: VH1

PHYSICAL DESCRIPTION: Colorless, fuming liquid

SPECIFIC GRAVITY: 0.778 @ 20/4 C [025]

DENSITY: Not available

MELTING POINT: -123.5 C [043,058,062]

BOILING POINT: 21 C [269,275,451]

SOLUBILITY: Water: 0.1-1.0 mg/mL @ 19 C [700] DMSO: >=100 mg/mL @ 18 C [700] 95% Ethanol: >=100 mg/mL @ 18 C [700] Acetone: >=100 mg/mL @ 18 C [700] Toluene: Miscible [062]

OTHER SOLVENTS: Gasoline: Miscible [062] Solvent naptha: Miscible [062] Xylene: Miscible [062] Turpentine: Miscible [062] Ether: Miscible [043,062] Benzene: Miscible [062] Most organic solvents: Miscible [025] Alcohol: Miscible [043]

OTHER PHYSICAL DATA: Pungent, fruity odor [058,062] Odor threshold: 0.21 ppm [371] Refractive index: 1.3316 @ 20 C [062,269,275] Specific gravity: 0.7827 @ 20/20 C [043] 0.7834 @ 18/4 C [047] Specific gravity: 0.788 @ 16/4 C [031] 0.8053 @ 0/4 C [205] Vapor pressure: 740 mm Hg @ 20 C [058,062] 200 mm Hg @ 10 C [058]

HAP WEIGHTING FACTOR: 1 [713]

VOLATILITY:

Vapor pressure: 400 mm Hg @ 4.9 C: 760 mm Hg @ 20.2 C [038]

Vapor density: 1.52 [043,058]

FIRE HAZARD: Acetaldehyde has a flash point of -40 C (-40 F) [058,062,269,275] and it is flammable. Fires involving this material may be controlled with a dry chemical, carbon dioxide or Halon extinguisher.

The autoignition temperature for acetaldehyde is 175 C (347 F) [043,058,451].

LEL: 4% [043,058,062,451] UEL: 57% [043,058,062]

REACTIVITY: Acetaldehyde can react vigorously with acid anhydrides, alcohols, ketones, phenols, ammonia, hydrogen cyanide, hydrogen sulfide, halogens and amines [043,058,269,451]. It can also react vigorously with phosphorous, isocyanates, strong alkalis and strong acids [043,269]. It is incompatible with oxidizing and reducing agents [269]. Polymerization may occur with acetic acid. Autoignition of vapor on corroded metals may also occur on contact with acetaldehyde [036]. Exothermic polymerization can occur with trace metals [036,043]. Acetaldehyde reacts with oxygen [043,058,066]. It also reacts with nitric acid, peroxides, caustic soda and soda ash [058]. It is reported that reactions with cobalt chloride, mercury (II) chlorate or mercury (II) perchlorate form sensitive and explosive products [043].

STABILITY: Acetaldehyde is dangerous when exposed to heat or flame [043,058]. It is sensitive to air and may undergo autopolymerization [269]. It is sensitive to moisture [058]. Upon prolonged storage, it may form unstable peroxides [269]. Solutions of acetaldehyde in water, DMSO, 95% ethanol or acetone should be stable for 24 hours under normal laboratory conditions [700].

USES: Acetaldehyde is used in the manufacture of pentaerythritol, peracetic acid, pyridines, paraldehyde, acetic acid, acetic anhydride, 2-ethylhexanol, aldol, chloral, 1,3-butylene glycol, trimethylolpropane, butanol, perfumes, aniline dyes, plastics and synthetic rubber. It is used in silvering mirrors and in hardening gelatin fibers. It is also used as a chemical intermediate and synthetic flavoring substance and adjuvant.

COMMENTS: None

NIOSH Registry Number: AB1925000

ACUTE/CHRONIC HAZARDS: Acetaldehyde may be an irritant of the skin, eyes, mucous membranes and respiratory tract [269,451]. It may also be an irritant to the throat [058]. It may be narcotic [031,062]. When heated to decomposition it emits acrid smoke and toxic fumes of carbon monoxide and carbon dioxide [043,058,269].

SYMPTOMS: Symptoms of exposure may include nausea, vomiting, headache, dermatitis and pulmonary edema. These effects may be delayed. It may also cause skin, eye, mucous membrane and upper respiratory irritation [269]. It may have a general narcotic action and large doses may cause death by respiratory paralysis [031]. It may also cause drowsiness, delirium, hallucinations and loss of intelligence [036]. Exposure may also cause slow mental response, severe damage to the mouth, throat and stomach; accumulation of fluid in the lungs, chronic respiratory disease, kidney and liver damage, throat irritation, dizziness, reddening and swelling of the skin and sensitization [058]. It may cause photophobia [099]. It may cause unconsciousness and liquid splashed in the eyes may cause a burning sensation, lacrimation and blurred vision [102]. It may also cause transient conjunctivitis [395].

Acetamide

CAS NUMBER: 60-35-5

SYNONYMS: Acetic acid amide Ethanamide Methanecarboxamide Acetimidic acid

CHEMICAL FORMULA: C2H5NO

MOLECULAR WEIGHT: 59.07

WLN: ZV1

PHYSICAL DESCRIPTION: Colorless crystals

SPECIFIC GRAVITY: 1.1590 @ 20/4 C

DENSITY: 1.16 g/mL

MELTING POINT: 82.3 C

BOILING POINT: 221.1 C @ 760 mm Hg

SOLUBILITY: Water: Very soluble (>=100 mg/mL @ 22 C) [700] DMSO: Very soluble (>=100 mg/mL @ 22 C) [700] 95% Ethanol: Very soluble (>=100 mg/mL @ 22 C) [700] Acetone: Very soluble (>=100 mg/mL @ 22 C) [700]

OTHER SOLVENTS: Chloroform: Soluble Toluene: Soluble hot Glycerol: Soluble Pyridine: Soluble (1 g/6 mL) Ether: Insoluble Benzene: Slightly soluble, soluble hot

OTHER PHYSICAL DATA: Deliquescent Mousy odor Triboluminescent

HAP WEIGHTING FACTOR: 1 [713]

VOLATILITY:

Vapor pressure: 1 mm Hg @ 65 C

FIRE HAZARD: The flash point of acetamide was not found but it is probably combustible. Fires involving this material may be controlled with a dry chemical, carbon dioxide or Halon extinguisher.

LEL: Not available UEL: Not available

REACTIVITY: Not available

STABILITY: Acetamide is stable under normal laboratory conditions.

USES: Acetamide is used in organic synthesis as a reactant, solvent and peroxide stabilizer and used in lacquers, explosives, soldering flux. It is also used as a hygroscopic agent, wetting agent, penetrating agent and in pharmaceuticals.

COMMENTS: None

NIOSH Registry Number: AB4025000

ACUTE/CHRONIC HAZARDS: Acetamide may cause skin and eye irritation and corneal damage.

SYMPTOMS: Exposure may cause irritation to the eyes, skin and mucous membranes.

Acetonitrile

CAS NUMBER: 75-05-8

SYNONYMS: Cyanomethane Ethanenitrile Ethyl nitrile Methanecarbonitrile Methyl cyanide

CHEMICAL FORMULA: C2H3N

MOLECULAR WEIGHT: 41.05

WLN: NC1

PHYSICAL DESCRIPTION: Clear colorless liquid

SPECIFIC GRAVITY: 0.7868 @ 20/20 C [042,051,053]

DENSITY: 0.7857 g/mL @ 20 C [017,047,205]

MELTING POINT: -45 C [031,042,051,421]

BOILING POINT: 81.6 C [017,031,058,371]

SOLUBILITY: Water: >=100 mg/mL @ 22.5 C [700] DMSO: >=100 mg/mL @ 22.5 C [700] 95% Ethanol: >=100 mg/mL @ 22.5 C [700] Acetone: >=100 mg/mL @ 22.5 C [700] Methanol: Miscible [031]

OTHER SOLVENTS: Petroleum ether: Immiscible [025] Ethylene chloride: Miscible [031,053] Methyl acetate: Miscible [031,053] Ethyl acetate: Miscible [031,053,295] Acetamide solutions: Miscible [031,053] Chloroform: Miscible [031,053,205,295] Carbon tetrachloride: Miscible [031,053,295] Many unsaturated hydrocarbons: Miscible [031,053] Many saturated hydrocarbons: Immiscible [031] Ether: Miscible [031,042,047,205] Benzene: Soluble [017,047,053] Most organic solvents: Miscible [025]

OTHER PHYSICAL DATA: Specific gravity: 0.7797 @ 23/22 C [052] 0.78745 @ 15/4 C [031,053] Specific gravity: 0.7138 @ 30/4 C [031,053] 0.783 @ 20/4 C [025] Refractive index: 1.3440 @ 20 C [269,275] 1.34604 @ 15 C [031] Ethereal odor [031,051,071,102] Odor threshold: 40 ppm [071,102,421] Olfactory fatigue: 2-3 hours [102] Dielectric constant: 38.8 @ 20 C [031] Evaporation rate (butyl acetate=1): 5.79 [102] Burns with a luminous flame [031] Surface tension: 29.04 dynes/cm @ 20 C [031] Lambda max: 400-250 nm, 235 nm, 220 nm, 210 nm, 193 nm (A=0.01, 0.02, 0.10, 0.30, 1.0) [274] Critical temperature: 274.7 C [371] Critical pressure: 47.7 atmospheres [371] Latent heat of vaporization: 174 cal/g [371] Heat of combustion: -7420 cal/g [371] Vapor pressure: 100 mm Hg @ 27 C [038,051,053,071] Pungent odor of vinegar [451]

HAP WEIGHTING FACTOR: 10 [713]

VOLATILITY:

Vapor pressure: 73 mm Hg @ 20 C [058,102,107,421]

Vapor density: 1.42 [051,053,055,071]

FIRE HAZARD: Acetonitrile has a flash point of 6 C (42 F) [036,066,269,371] and it is flammable. Fires involving this material can be controlled with a dry chemical, carbon dioxide or Halon extinguisher. A water spray may also be used [269].

The autoignition temperature is 524 C (975 F) [036,042,051,371].

LEL: 4.4% [042,051,102,371] UEL: 16% [036,066,107,371]

REACTIVITY: Acetonitrile is incompatible with sulfuric acid, oleum, chlorosulfonic acid, perchlorates, nitrating agents, indium, dinitrogen tetraoxide, water, steam and acids [042,053]. It is also incompatible with bases, reducing agents and alkali metals [269]. It will react with strong oxidizers [042, 058,102,269]. Violent or explosive reactions may occur with N-fluoro compounds, nitric acid and anhydrous $Fe(ClO4)3$ [036]. It forms a constant boiling mixture with water containing 84% of this chemical (with a boiling point of 76 C) [025,031]. It dissolves in some organic salts (e.g. silver nitrate, lithium nitrate and magnesium bromide) [031]. It will attack some forms of plastics, rubber and coatings [102]. A dangerous reaction may occur with chromic acid or sodium peroxide [451].

STABILITY: Acetonitrile may be sensitive to light [052,053]. It is sensitive to heat [102]. It is also sensitive to moisture, and is easily oxidized and unstable [042]. It is stable when stored for 2 weeks, protected from light, at temperatures up to 60 C [052]. Solutions of it in water, DMSO, 95% ethanol or acetone should be stable for 24 hours under normal lab conditions [700].

USES: Acetonitrile is used as a chemical intermediate in the synthesis of acetophenone, 1-naphthaleneacetic acid, thiamine and acetamidine, in pesticide manufacture, as an extractant for animal and vegetable oils, as a pharmaceutical solvent, as a solvent for inorganic salts and in organic synthesis, as a polymer solvent and in acrylic fibers. It is also used for separation of butadiene by extractive distillation, in perfumes, in nitrile rubber, in ABS resins, as a solvent in hydrocarbon extraction processes, as a specialty solvent, as a catalyst, to remove tars, phenols and coloring matter from petroleum hydrocarbons which are not soluble in acetonitrile, to recrystallize steroids, as an indifferent medium in physicochemical investigations, as a medium for promoting reactions involving ionizations and as a solvent in non-aqueous titrations.

COMMENTS: None

NIOSH Registry Number: AL7700000

ACUTE/CHRONIC HAZARDS: Acetonitrile may be toxic by ingestion, inhalation and skin absorption [036,062,269]. It may be an irritant of the skin, eyes, nose and throat [053, 371]. It may be a lachrymator [053,269,275]. When heated to decomposition it emits highly toxic fumes of carbon monoxide, nitrogen oxides and hydrogen cyanides [053,058,102,269]. It may also release carbon dioxide [053,102,269].

SYMPTOMS: Symptoms of exposure may include irritation of the eyes, nose and throat; nausea, vomiting, abdominal pain, weakness, flushing of the face, a feeling of chest tightness, respiratory depression, chest pain, hematemesis, shock, convulsions, unconsciousness and death [102,346]. Other symptoms may include dizziness, headache, drowsiness, a drop in blood pressure and rapid pulse [301]. It may cause fatigue and diarrhea [036]. It may also cause cough, bile stained emesis, dyspnea and tachypnea [151]. Exposure may also lead to asphyxia, lassitude and stupor [102]. A "cooling sensation" in the lungs and a slight feeling of bronchial tightness have been reported [421]. Other symptoms may include irritation of the skin, difficult breathing, irritability, skin eruptions, confusion, central nervous system depression, slight smarting of the eyes or respiratory system, and smarting and reddening of the skin [371]. It may cause irritation of the mucous membranes and respiratory tract, cyanosis and lacrimation [269]. Exposure may also cause rapid respiration (which becomes slow and gasping), and central nervous system damage [051]. Other symptoms may include reduced pulse rate, pale or ashen gray skin, subnormal temperature, collapse, abnormal liver and kidney function, incoordination, abnormal blood forming system function with anemia, abnormal blood clotting system function with easy bruising or bleeding and gastrointestinal bleeding [058]. Contact with the liquid has been reported to cause slight reversible eye injury [053]. It may also cause eye burns [071]. In severe exposures, it may cause delirium, paralysis and coma [036].

Acetophenone

CAS NUMBER: 98-86-2

SYNONYMS: Acetylbenzene Methyl phenyl ketone 1-Phenyl ethanone

CHEMICAL FORMULA: C_8H_8O

MOLECULAR WEIGHT: 120.16 [703]

PHYSICAL DESCRIPTION: Colorless liquid or plates [703]

DENSITY: 1.026 @ 20/4C [703]

MELTING POINT: 19.7 C [703]

BOILING POINT: 202.3 C [703]

SOLUBILITY: Water: Insoluble [703] Alcohol: Soluble [703] Ether: Soluble [703]

OTHER PHYSICAL DATA: Heat of Combustion: 991.60 kcal/mole [702] Vapor Pressure: 1 mm @ 15 C [703] Vapor Density: 4.14 [703]

HAP WEIGHTING FACTOR: 1 [713]

VOLATILITY:

Vapor Pressure: 1 mm @ 15 C [703]

Vapor Density: 4.14 [703]

FIRE HAZARD: Flash Point: 180 F (open cup) [703]

Flammable when exposed to heat, flame, or oxidizers. To fight fire use foam, CO2, dry chemical. [703] Combustion ranking: 31 [702].

Autoignition temperature: 1060 F [703]

LEL: Not available UEL: Not available

REACTIVITY: Not available

STABILITY: Emits acrid smoke and fumes on heating to decomposition [703].

UDRI Thermal Stability Class: 3 [702] UDRI Thermal Stability Ranking: 85 [702]

USES: In perfumery (orange-blossom-like odor); catalyst for the polymerization of olefins; in organic syntheses, especially as photo-sensitizer [715].

COMMENTS: None

NIOSH Registry Number: AM5250000

ACUTE/CHRONIC HAZARDS: Acetophenone is a poison by intraperitoneal and subcutaneous routes; moderately toxic by ingestion. It is a skin and severe eye irritant. Mutagenic data exists. Acetophenone is highly hypnotic and is narcotic in high concentration. [703]

SYMPTOMS: Not available

2-Acetylaminofluorene

CAS NUMBER: 53-96-3

SYNONYMS: 2-Acetaminofluorene N-Acetyl-2-aminofluorene N-fluoren-2-yl acetamide

CHEMICAL FORMULA: C15H13NO

MOLECULAR WEIGHT: 223.28

WLN: L B656 HHJ EMV1

PHYSICAL DESCRIPTION: White crystalline solid

SPECIFIC GRAVITY: Not available

DENSITY: Not available

MELTING POINT: 194 C [017,031,071,205]

BOILING POINT: Not available

SOLUBILITY: Water: <1 mg/mL @ 20.5 C [700] DMSO: >=100 mg/mL @ 20.5 C [700] 95% Ethanol: 10-50 mg/mL @ 20.5 C [700] Acetone: 33-50 mg/mL @ 20.5 C [700] Methanol: 10-50 mg/mL @ 19.0 C [700]

OTHER SOLVENTS: Alcohols: Soluble [031,205] Glycols: Soluble [031,205] Fat solvents: Soluble [031] Acetic acid: Soluble [017] Ether: Soluble [017]

OTHER PHYSICAL DATA: UV max: 285 nm [031] Forms needles when recrystallized from 50% alcohol or 50% acetic acid [017]

HAP WEIGHTING FACTOR: 1 [713]

VOLATILITY:

Vapor pressure: 0.00023 mm Hg @ 100 C [051,071]

FIRE HAZARD: Flash point data for 2-acetylaminofluorene are not available. It is probably combustible. Fires involving this material may be controlled with a dry chemical, carbon dioxide or Halon extinguisher.

LEL: Not available UEL: Not available

REACTIVITY: 2-Acetylaminofluorene is incompatible with acids, bases and oxidizing agents [269]. Ozone and chlorinating agents oxidize this type of compound [051,071].

STABILITY: 2-Acetylaminofluorene is stable under normal laboratory conditions. Solutions of it in water, DMSO, 95% ethanol or acetone should be stable for 24 hours under normal lab conditions [700].

USES: 2-Acetylaminofluorene is used in carcinogenesis studies, particularly for tumor induction as a positive control, and for the mechanism of hydroxylation by the liver. It is also used in the study of mutagenicity of aromatic amines as a positive control.

COMMENTS: 2-Acetylaminofluorene was intended to be used as a pesticide, but was never marketed because it was found to be a carcinogen [346].

NIOSH Registry Number: AB9450000

ACUTE/CHRONIC HAZARDS: 2-Acetylaminofluorene may be fatal if inhaled, ingested or absorbed through the skin [269]. It is irritating to the eyes and skin. It may be absorbed through the trachea or gastrointestinal tract [051,071]. It may also be absorbed through the skin [051,071,346]. When heated to decomposition it emits toxic fumes of carbon monoxide, carbon dioxide and nitrogen oxides [043,269].

SYMPTOMS: Symptoms of exposure may include skin and eye irritation. Symptoms of exposure in laboratory animals include central nervous system depression and orange-red urine [051,071]

Acrolein

CAS NUMBER: 107-02-8

SYNONYMS: Acraldehyde Acrylaldehyde Allyl aldehyde Propenal Propylene aldehyde

CHEMICAL FORMULA: C3H4O

MOLECULAR WEIGHT: 56.06

WLN: VH1U1

PHYSICAL DESCRIPTION: Clear, colorless or yellowish liquid

SPECIFIC GRAVITY: 0.8410 @ 20/4 C [017,047,051,395]

DENSITY: 0.8389 g/mL @ 20 C [031,205]

MELTING POINT: -87 C [025,062,275,395]

BOILING POINT: 53 C [036,058,275,395]

SOLUBILITY: Water: >=100 mg/mL @ 21 C [700] DMSO: >=100 mg/mL @ 21 C [700] 95% Ethanol: >=100 mg/mL @ 21 C [700] Acetone: >=100 mg/mL @ 21 C [700]

OTHER SOLVENTS: Alcohol: Soluble [017,031,043,205] Ether: Soluble [017,031,043,205] Ethanol: >10% [047]

OTHER PHYSICAL DATA: This compound is normally inhibited Specific gravity: 0.8427 @ 20/20 C [055,062,395] Density: 0.8621 g/mL @ 0 C; 0.8075 g/mL @ 50 C [031] Boiling point: 17.5 C @ 200 mm Hg; 2.5 C @ 100 mm Hg; -7.5 C @ 60 mm Hg [031] Boiling point: -64.5 C @ 1.0 mm Hg [031] Vapor pressure: 330 mm Hg @ 30 C [055] 100 mm Hg @ 2.5 C [038] Vapor pressure: 400 mm Hg @ 34.5 C; 760 mm Hg @ 52.5 C [038] Vapor pressure: 40 mm Hg @ -15 C; 60 mm Hg @ -7.5 C [038] Disagreeable, choking odor [043,051,062,421] Odor threshold: 2.0 ppm [107] Refractive index: 1.4017 @ 20 C [017,047,205,395] 1.4022 @ 19 C [031] Polymerizes to a white translucent solid [025,031] Burning rate: 3.8 mm/min [371] Critical temperature (estimated): 254 C [371] Critical pressure (estimated): 50.0 atmospheres [371] Liquid surface tension: 24 dynes/cm @ 20 C [371] Latent heat of vaporization: 120 cal/g [371] Heat of combustion: -6950 cal/g [371] Heat of polymerization (estimated): -28 cal/g [371] Spectroscopy data: lambda max: 207 nm (E = 2000) [395] Viscosity: 0.35 centipoise @ 20 C [395]

HAP WEIGHTING FACTOR: 10 [713]

VOLATILITY:

Vapor pressure: 200 mm Hg @ 17.5 C [038,051,395] 214 mm Hg @ 20 C [102,107]

Vapor density: 1.94 [043,055,102,371]

FIRE HAZARD: Acrolein has a flash point of -26 C (-15 F) [036,107,395,451] and it is flammable. Fires involving this material may be controlled with a dry chemical, carbon dioxide or Halon extinguisher.

The autoignition temperature of acrolein is 278 C (532 F) [036,051,062,451].

LEL: 2.8% [058,066,371,451] UEL: 31% [036,058,371,451]

REACTIVITY: Acrolein is reactive [051,062,066,151]. It is liable to polymerize violently, especially on contact with strong acids or bases [031,066,395,451]. It is incompatible with amines, sulfur dioxide, metal salts and oxidants (air) [043,051,451]. It is also incompatible with oxidizers [043,102,269,346]. It can react with ammonia [051,102,346,451]. Violent polymerization occurs with dimethylaniline [043,066]. Acrolein is incompatible with reducing agents [269]. Extremely violent polymerization can occur on contact with caustics, sodium hydroxide and potassium hydroxide. It reacts with 2-aminoethanol, 28% ammonium hydroxide, chlorosulfonic acid, ethylenediamine, ethyleneimine, 70% nitric acid, oleum and 96% sulfuric acid, causing an increase in temperature and pressure [451]. It polymerizes with release of heat on contact with thiourea [043,051,066,451]. Exposure to weakly acidic conditions (nitrous fumes or carbon dioxide) will cause exothermic and violent polymerization [043,066]. Exposure to some hydrolyzable salts will also cause polymerization [066].

STABILITY: Acrolein polymerizes readily unless an inhibitor is added [062,102, 451]. It is sensitive to heat and light [043,051,451]. It is also sensitive to air [043,051,395,451]. Uncatalyzed polymerization sets in at 200 C [066, 102,371]. The stabilizing effects of the added inhibitor may cease after a comparatively short storage time (such unstabilized material could polymerize explosively) [066]. UV spectrophotometric stability screening indicates that solutions of it in DMSO are stable for at least 24 hours [700].

USES: Acrolein is used as a lacrimogenic warning agent in methyl chloride refrigerant, as a component of military poison gases, as a synthetic reagent in the manufacture of methionine, glycerol and glutaraldehyde; as an aquatic herbicide and as an algicide for water treatment. It is also used as an intermediate for polyurethane and polyester resins, in pharmaceuticals, as an herbicide, as a biocide, in the manufacture of colloidal forms of metals, in making plastics and perfumes, to modify food starch, in the manufacture of 1,3,6-hexanetriol, as a fungicide and bactericide, as a liquid fuel, as an antimicrobial agent and as a slimicide in paper manufacture. It is an intermediate for acrylic acid and its esters and is used in the manufacture of 2-hydroxyadipaldehyde, quinoline, pentaerythritol, cycloaliphatic epoxy resins, oil-well additives and water-treatment formulae.

COMMENTS: Acrolein is often inhibited with 3% water and 200 ppm hydroquinone [275]. It was used in World War I as a tear gas under the name "Papite" [099,395].

It is produced in large amounts by overheated cooking oils and animal fats [151]. It is also a major contributor to the irritative quality of cigarette smoke and photochemical smog [406]. Worker exposure has been estimated at 7550 per year by NIOSH [346].

NIOSH Registry Number: AS1050000

ACUTE/CHRONIC HAZARDS: Acrolein may be a lachrymator [031,099,295,395]. It may be toxic by ingestion, inhalation and skin absorption [036,058,151,269]. It may be an irritant of the skin, eyes and respiratory tract [036]. When heated to decomposition it emits toxic fumes of carbon monoxide, carbon dioxide and peroxides [043,058,102,269].

SYMPTOMS: Symptoms of exposure may include irritation of the skin, eyes and respiratory tract [036,102,371,451]. Irritation of the nose and mucous membranes may also occur [051,395]. It may cause pain to the nose and unconsciousness [036]. Exposure causes lacrimation [031,058,151,295]. High concentrations may cause death [051,055,102]. Delayed hypersensitivity with multiple organ involvement, and respiratory system damage have occurred [043]. Asthmatic reactions and pulmonary edema have been reported [031]. Bronchial inflammation and bronchitis have also been reported [346]. Central nervous system effects may occur [301]. Exposure may cause a burning sensation, coughing, wheezing, laryngitis, shortness of breath, headache and nausea. Inhalation may be fatal as a result of spasm, inflammation and edema of the larynx and bronchi, and chemical pneumonitis [269]. Exposure may also cause a feeling of pressure on the chest, dizziness and decreased pulmonary function [102]. This compound may cause vomiting and lung injury [371]. It may also cause chest pain and difficult breathing [058]. Respiratory paralysis may also occur [051].

Ingestion of acrolein may cause severe gastrointestinal distress with pulmonary congestion [151]. It may also cause severe irritation of the mouth and stomach [371]. Eye contact may cause a burning sensation, blepharo conjunctivitis, lid edema, fibrinous or purulent discharge and corneal injury [099]. Eye contact may also cause redness, weeping and burns [371]. Skin contact will cause erythema and edema [099]. Sensitization may occur [031, 051,102,346]. Vesiculation may also occur [151,295,371]. Burns of the skin may occur on prolonged exposure [058,151,346,451]. Dermatitis has been reported [346]. Allergic skin reactions have also been reported [102,269]. Redness may also develop from skin contact [371].

Acrylamide

CAS NUMBER: 79-06-1

SYNONYMS: Acrylic amide Ethylenecarboxamide 2-Propenamide Propenoic acid, amide Propenamide

CHEMICAL FORMULA: C3H5NO

MOLECULAR WEIGHT: 71.08

WLN: ZV1U1

PHYSICAL DESCRIPTION: Colorless to white crystals

SPECIFIC GRAVITY: 1.122 @ 30/4 C [031,043,205]

DENSITY: 1.122 g/cm3 @ 30 C [900]

MELTING POINT: 84.5 C [031,051,205,421]

BOILING POINT: Decomposes >175 C [058]

SOLUBILITY: Water: >=100 mg/mL @ 22 C [700] DMSO: >=100 mg/mL @ 22 C [700] 95% Ethanol: 5-10 mg/mL @ 22 C [700] Acetone: 5-10 mg/mL @ 22 C [700] Methanol: 155 g/100 mL @ 30 C [031,900]

OTHER SOLVENTS: Chloroform: 2.66 g/100 mL @ 30 C [031,205,900] Ethyl acetate: 12.6 g/100 mL @ 30 C [031,900] n-Heptane: 0.0068 g/100 mL @ 30 C [031,900] Benzene: 0.346 g/100 mL @ 30 C [031,900] Alcohol: 86 g/100 mL @ 30 C [205] Ether: Very soluble [016,029,043,395] Methylene chloride: Soluble [029]

OTHER PHYSICAL DATA: Boiling point: 125 C @ 25 mm Hg [031,062,275,395] Boiling point: 103 C @ 5 mm Hg; 87 C @ 2 mm Hg [031,900] Vapor pressure: 1.6 mm Hg @ 84.5 C [043,058,107] 0.033 mm Hg @ 40 C [900] Vapor pressure: 2 mm Hg @ 87 C; 10 mm Hg @ 117 C [055] Vapor pressure: 0.07 mm Hg @ 50 C [051,900] 25 mm Hg @ 125 C [051] Heat of polymerization: 19.8 kcal/mole [900] Odorless [058,062,395,900]

HAP WEIGHTING FACTOR: 10 [713]

VOLATILITY:

Vapor pressure: 0.007 mm Hg @ 25 C [051,058,395,900]

Vapor density: 2.45 [043,058,395]

FIRE HAZARD: Acrylamide has a flash point of 84.4 C (184 F) [107]. It is combustible. Fires involving this material may be controlled with a dry chemical, carbon dioxide or Halon extinguisher. A water spray may also be used [051,058].

LEL: Decomposes [107] UEL: Decomposes [107]

REACTIVITY: Acrylamide is incompatible with acids, bases and oxidizing agents [058,269]. It is also incompatible with reducing agents, vinyl polymerization initiators, iron, copper, aluminum, brass and bronze [058].

STABILITY: Acrylamide is sensitive to air, moisture and light [269]. It polymerizes on melting or on exposure to ultraviolet light [033,058,346, 395]. Solutions of acrylamide in water, DMSO, 95% ethanol or acetone should be stable for 24 hours when protected from heat and light [700].

USES: Acrylamide is used as a polymer or copolymer in adhesives, sizing agents, soil conditioning agents, plastics, flocculants, sewage and waste treatment, chemical grouting, production of n-methylalacrylamide, permanent press fabrics, crude oil production processes and processing of paper and pulp, ore and concrete. It is also used in the synthesis of dyes.

COMMENTS: None

NIOSH Registry Number: AS3325000

ACUTE/CHRONIC HAZARDS: Acrylamide may be toxic by ingestion and inhalation [058]. It may be an irritant [033,036,043,151]. It is a potent neurotoxin [099,346]. This chemical may be absorbed through unbroken skin [033,043,099]. It may also be absorbed through the mucous membranes, lungs and gastrointestinal tract [099]. When heated to decomposition it emits acrid fumes and toxic fumes of carbon monoxide, carbon dioxide, nitrogen oxides and ammonia [043,058,269].

SYMPTOMS: Symptoms of exposure may include numbness, tingling sensation, touch tenderness, peeling of the palms, coldness of extremities, excessive sweating and bluish red skin [043,151]. Irritation of the skin and eyes may occur [043,058]. Peripheral neuropathy may also occur [099]. Other symptoms may include erythema, central peripheral distal axonopathies, skin changes, mental confusion and other psychological changes, skin ulcerations, dermatitis, gastrointestinal problems, weight loss, inability to stand, collapse, difficulty in swallowing, body tremors and hallucinations [900]. It may cause fatigue, central nervous system damage, drowsiness and slurred speech [058,900]. It may also cause ataxia, weak or absent reflexes and positive Romberg's sign [346]. Central nervous system paralysis has been reported [033]. Limb weakness has also been reported [043]. Exposure may cause peripheral nervous system damage, weakness, stumbling, shaking and redness of the skin [058]. It may also cause loss of vibration and position senses [058,346]. Midbrain disturbances have occurred [099]. Prolonged or repeated exposure may cause muscular weakness, excessive sweating of the hands and feet, cold hands, peeling of the skin and numbness [058,900]. It may also cause incoordination, skin rashes, abnormal skin or muscle sensations and central nervous system depression [058]. Persons with pre-existing skin disorders, eye problems or central or peripheral nervous system conditions may be more susceptible to the effects of this substance [058].

Acrylic acid

CAS NUMBER: 79-10-7

SYNONYMS: Acroleic acid Acrylic acid, glacial Ethylenecarboxylic acid Glacial acrylic acid Propene acid Propenoic acid 2-Propenoic acid Vinylformic acid

CHEMICAL FORMULA: C3H4O2

MOLECULAR WEIGHT: 72.06

WLN: QV1U1

PHYSICAL DESCRIPTION: Clear colorless liquid

SPECIFIC GRAVITY: 1.0511 @ 20/4 C [017,047,395]

DENSITY: 1.0511 g/mL @ 20 C [205]

MELTING POINT: 13 C [017,043,275,395]

BOILING POINT: 141 C [031,036,043,058]

SOLUBILITY: Water: >=100 mg/mL @ 17 C [700] DMSO: >=100 mg/mL @ 17 C [700] 95% Ethanol: >=100 mg/mL @ 17 C [700] Acetone: >=100 mg/mL @ 17 C [700]

OTHER SOLVENTS: Ethanol: Miscible [395,421] Alcohol: Miscible [031,043,062,205] Benzene: Miscible [043,205] Chloroform: Miscible [043,205] Ether: Miscible [031,043,205,395] Several ethers: Miscible [421]

OTHER PHYSICAL DATA: This compound is normally inhibited Specific gravity: 1.0621 @ 16/4 C [031] 1.052 @ 20/20 C [062,421] Boiling point: 122 C @ 400 mm Hg; 103.3 C @ 200 mm Hg [031] Boiling point: 86.1 C @ 100 mm Hg; 66.2 C @ 40 mm Hg; 39 C @ 10 mm Hg [031] Boiling point: 48.5 C @ 15 mm Hg [017,047] 27.3 C @ 5 mm Hg [031] Vapor pressure: 1 mm Hg @ 3.5 C; 20 mm Hg @ 52 C; 40 mm Hg @ 66.2 C [038] Vapor pressure: 60 mm Hg @ 75 C; 100 mm Hg @ 86.1 C [038] Acrid odor [031,036,395,451] Refractive index: 1.4224 @ 20 C [017,031,205,395] pKa: 4.25 @ 25 C [025] Odor threshold: 0.094 ppm [055] Spectroscopy data: Lambda max (in methanol): 252 nm (E = 13) [395] Burning rate: 1.6 mm/min [371] Critical temperature: 342 C [371] Critical pressure: 57 atm [371] Latent heat of vaporization: 151.5 cal/g [371] Heat of combustion: -4500 cal/g [371] Heat of polymerization: -257 cal/g [371] Heat of fusion: 30.03 cal/g [371] Evaporation rate (butyl acetate = 1): 1 [058]

HAP WEIGHTING FACTOR: 10 [713]

VOLATILITY:

Vapor pressure: 3.2 mm Hg @ 20 C [055] 5 mm Hg @ 27.3 C [038]

Vapor density: 2.50 [055,058,395,451]

FIRE HAZARD: Acrylic acid has a flash point of 54 C (130 F) [043,058,275,451]. It is combustible. Fires involving this material may be controlled with a dry chemical, carbon dioxide or Halon extinguisher. A water spray may also be used [036,051,371,451].

The autoignition temperature of acrylic acid is 438 C (820 F) [451].

LEL: 2.4% [371,451] UEL: 8.0% [058,421,451]

REACTIVITY: Acrylic acid is incompatible with strong oxidizers and strong bases [058,269]. It is also incompatible with strong alkalis and pure nitrogen [421]. It may polymerize (sometimes explosively) on contact with amines, ammonia, oleum and chlorosulfonic acid [051,421]. It may also polymerize with peroxides [058,421]. Mixing with 2-aminoethanol, 28% ammonium hydroxide, ethylenediamine or ethyleneimine in a closed container has caused the temperature and pressure to increase [451]. It may corrode iron and steel [395]. Polymerization may occur on contact with acids and iron salts [371]. Acrylic acid is incompatible with polymerization initiators [058].

STABILITY: Acrylic acid polymerizes readily (may be explosive) [062,269,451]. It is sensitive to heat and sunlight [058,269] and it polymerizes readily in the presence of oxygen [031,051,371,395]. Exothermic polymerization at room temperature may become explosive if confined [043]. It is normally supplied as the inhibited monomer, but because of its relatively high freezing point it often partly solidifies and the solid phase (and the vapor) will then be free of the inhibitor which remains in the liquid phase [066]. Even the uninhibited form may be stored safely below the melting point, but such material will polymerize exothermically at room temperature and may accelerate to a violent or explosive state if confined [066,451]. NMR stability screening indicates that solutions of it in DMSO are stable for at least 24 hours [700].

USES: Acrylic acid is used widely for polymerization, including production of polyacrylates. It is a monomer for polyacrylic and polymethacrylic acids and other acrylic polymers. It is used in the manufacture of plastics, as a tackifier, as a flocculant, in the production of water-soluble resins and salts, as a comonomer in acrylic emulsion and solution polymers and in molding powder for signs, construction units, decorative emblems and insignias. It is used in polymer solutions for coatings applications, in paint formulations, in leather finishings, in paper coatings, in polishes and adhesives and in general finishes and binders.

COMMENTS: Acrylic acid is often inhibited with about 200 ppm hydroquinone monomethyl ether [275].

NIOSH Registry Number: AS4375000

ACUTE/CHRONIC HAZARDS: Acrylic acid may be corrosive [031,043,275,395]. It may be an irritant of the skin, eyes and respiratory tract [031,043,269,451]. It may be extremely destructive to tissue of the mucous membranes and upper respiratory tract, eyes and skin [058,269]. It may be toxic by ingestion, inhalation and skin absorption [051,058,062,269]. When heated to decomposition it emits toxic fumes of carbon monoxide and carbon dioxide [058,269].

SYMPTOMS: Symptoms of exposure may include irritation of the skin, eyes and respiratory tract [036,051,269,451]. Corrosion and burns may occur [058]. Eye damage may also occur [058,451]. Pulmonary effects have been reported [301,346]. Exposure may cause coughing, wheezing, laryngitis, shortness of breath, headache, nausea and vomiting. Inhalation may be fatal as a result of spasm, inflammation and edema of the larynx and bronchi, chemical pneumonitis and pulmonary edema [058,269]. A burning sensation may also occur [269]. It may cause irritation of the nose and throat, and gastrointestinal tract damage [371]. Exposure may also cause irritation of the mouth, stomach and lungs, sensitization (allergic reaction) and lung damage. Ingestion may lead to pain and burning in the mouth, pharynx and stomach, diarrhea, fall in blood pressure, asphyxia due to edema of the glottis, and destruction of mucous membranes of the gastrointestinal tract. Skin contact may lead to inflammation, skin rashes and systemic poisoning (if absorbed through the skin) [058].

Acrylonitrile

CAS NUMBER: 107-13-1

SYNONYMS: Carbacryl Cyanoethylene Propenenitrile 2-Propenenitrile Propenonitrile Vinyl cyanide

CHEMICAL FORMULA: C3H3N

MOLECULAR WEIGHT: 53.06

WLN: NC1U1

PHYSICAL DESCRIPTION: Clear, colorless to pale yellow, volatile liquid

SPECIFIC GRAVITY: 0.8060 @ 20/4 C [017,031,042,205]

DENSITY: 0.8075 g/mL @ 20 C [371]

MELTING POINT: -83 C [055,269,274,326]

BOILING POINT: 77 C [036,052,269,275]

SOLUBILITY: Water: 10-50 mg/mL @ 21.6 C [700] DMSO: >=100 mg/mL @ 21.6 C [700] 95% Ethanol: >=100 mg/mL @ 21.6 C [700] Acetone: >=100 mg/mL @ 21.6 C [700]

OTHER SOLVENTS: Benzene: Soluble [017,047,395] Ether: Soluble [017,047,395] Most organic solvents: Miscible [031,173,205,421]

OTHER PHYSICAL DATA: Specific gravity: 0.8004 @ 25/4 C [031] Density: 0.8004 g/mL @ 25 C [055,062] Boiling point: 64.7 C @ 500 mm Hg; 45.5 C @ 250 mm Hg [031] Boiling point: 23.6 C @ 100 mm Hg; 8.7 C @ 50 mm Hg [031] Refractive index: 1.3888 @ 25 C [031] 1.3911 @ 20 C [017,047,205,395] Mild, pungent odor (resembling that of peach seed kernels) [371] Odor threshold: 21.4 ppm [371] Evaporation rate (butyl acetate=1): 4.54 [326] Critical temperature: 263 C [371] Critical pressure: 45 atmospheres [371] Latent heat of vaporization: 147 cal/g [371] Heat of combustion: -7930 cal/g [371] Lambda max: 203 nm (E=1163) [395] pH (5% aqueous solution): 5.5-7.5 [058]

HAP WEIGHTING FACTOR: 1 [713]

VOLATILITY:

Vapor pressure: 100 mm Hg @ 22.8 C [038,042] 200 mm Hg @ 38.7 C [038]

Vapor density: 1.83 [042,055,168,326]

FIRE HAZARD: Acrylonitrile has a flash point of 0 C (32 F) [031,036,205,395] and it is flammable. Fires involving it can be controlled with a dry chemical, carbon dioxide or Halon extinguisher.

The autoignition temperature is 481 C (898 F) [036,051,058,371].

LEL: 3.0% [036,062,066,326] UEL: 17% [036,066,326,371]

REACTIVITY: Acrylonitrile polymerizes violently in the presence of strong bases, strong acids, bromine and silver nitrate [036,042,066]. It is incompatible with strong oxidizers, copper, copper alloys, ammonia and amines [326,346]. It is also incompatible with "Redox" catalysts, peroxides and azo compounds [058]. It reacts with 1,2,3,4-tetrahydrocarbazole and with benzyltrimethylammonium hydroxide [042]. It also reacts with potassium hydroxide, sodium hydroxide and sulfuric acid [051]. It can penetrate leather and high concentrations will attack aluminum [371]. It is corrosive to metals [168]. It will also attack some forms of plastics, rubber and coatings [326]. Polymerization in a sealed tube in an oil bath @ 110 C led to a violent explosion. It has exploded with (benzyltrimethylammonium hydroxide + pyrrole) and (tetrahydrocarbazole + benzyl trimethylammonium hydroxide). At pressures above 6000 bar, free radical polymerization sometimes occurs [066].

STABILITY: Acrylonitrile may polymerize spontaneously, especially in the absence of oxygen or on exposure to visible light [031,371]. On standing it may develop a yellow color particularly after excessive exposure to light [031]. It is heat sensitive [058,269,326,371]. Even when inhibited, it will polymerize exothermically above 200 C [066]. It is easily oxidized [042]. The pure chemical may self-polymerize [326]. Solutions of it in water, DMSO, 95% ethanol or acetone should be stable for 24 hours when protected from light [700].

USES: Acrylonitrile is used in the manufacture of acrylic and modacrylic fibers, high strength whiskers, acrylostyrene plastics, acrylonitrile-butadiene-styrene plastics, synthetic rubber, nitrile rubber, chemicals, adhesives and surface coatings. It is also a chemical intermediate in the synthesis of antioxidants, pharmaceuticals, dyes, textile fibers, surface active agents and adiponitrile, etc. In organic synthesis, it is used to introduce a cyanoethyl group;

as a modifier for natural polymers and in the cyanoethylation of cotton. In synthetic soil blocks (acrylonitrile polymerized in wood pulp), it is used as a grain fumigant and a pesticide. It is a monomer for a semiconductive polymer that can be used like inorganic oxide catalysts in dehydrogenation of tert-butanol to isobutylene and water.

COMMENTS: Acrylonitrile is often inhibited with 35-45 ppm hydroquinone monomethyl ether [275]. Its usage in agriculture is now limited [168].

NIOSH Registry Number: AT5250000

ACUTE/CHRONIC HAZARDS: Acrylonitrile may be highly toxic by ingestion, inhalation and skin absorption [031,036,269,275]. It may be readily absorbed through intact skin [421]. It is a lachrymator [269]. It is an irritant of the skin, eyes and nose [051,058,099, 371]. When heated to decomposition it emits highly toxic fumes of hydrogen cyanide gas, nitrogen oxides and carbon monoxide [051, 269, 371,395].

SYMPTOMS: Symptoms of exposure may include irritation of the eyes, nausea, vomiting, headache, sneezing and weakness [326,346,371]. Other symptoms may include irritation of the skin and respiratory tract; severe lung irritation, dizziness, confusion, incoordination, nonspecific discomfort, difficult breathing, shortness of breath, reduced appetite, abdominal pain, cyanosis, convulsions, unconsciousness, abnormal liver and kidney function, reduced urine volume and edema, temporary nervous system depression with anesthetic effects, abnormal blood forming system with anemia, reduced white blood cell production, bone marrow effects and effects on the kidneys, liver, central nervous system and respiratory system [058]. It may cause lightheadedness, irritability, tiredness, mild jaundice and alcohol intolerance [151]. It may also cause diarrhea, increased salivation, flushing of the face, deepened or shallow respiration, oppressive feeling in the chest, photophobia and leucocytosis [042]. Symptoms of exposure may also include rapid respiration, drowsiness, drop in blood pressure and rapid pulse [301]. lacrimation has been reported [269]. Other symptoms may include insomnia, fatigue and irritation of the mucous membranes [051]. Exposure may lead to neurasthenic syndrome, asthenovegetative syndrome and gastritis [395]. Sore throats may also occur [173]. Skin contact may lead to discomfort, rash and temporary burning and numbness [058]. It may also lead to erythema and epidermal necrolysis or necrosis [173]. Prolonged contact with the skin may result in blisters [036,051,151,326]. It may also result in smarting and reddening of the skin [371]. Repeated skin exposure may produce scaling dermatitis [151,326]. Eye contact may lead to irritation with discomfort, tearing or blurred vision [058]. Exposure to high concentrations may result in profound weakness, asphyxia and death [346]. Chronic exposure may cause giddiness [036]. A single drop to rabbit corneas has produced transient disturbance without corneal opacification [099].

Allyl chloride

CAS NUMBER: 107-05-1

SYNONYMS: 3-Chloropropene Chlorallylene 3-Chloroprene

CHEMICAL FORMULA: C3H5Cl

MOLECULAR WEIGHT: 76.53

WLN: G2U1

PHYSICAL DESCRIPTION: Clear colorless liquid

SPECIFIC GRAVITY: 0.938 @ 20/4 C [031,043,051,205]

DENSITY: 0.9392 g/mL @ 20 C [395]

MELTING POINT: -134.5 C [025,031,102,205]

BOILING POINT: 45 C [017,036,371,451]

SOLUBILITY: Water: 1-10 mg/mL @ 19 C [700] DMSO: >=100 mg/mL @ 19 C [700] 95% Ethanol: 1-10 mg/mL @ 19 C [700] Acetone: >=100 mg/mL @ 19 C [700]

OTHER SOLVENTS: Ligroin: Soluble [017] Chloroform: Miscible [031,205,395,421] Petroleum ether: Miscible [031,205,395,421] Alcohol: Miscible [031,062,205,421] Ether: Miscible [031,205,395,421] Benzene: Soluble [017] Most organic solvents: Miscible [025]

OTHER PHYSICAL DATA: Vapor pressure: 10 mm Hg @ -42.9 C; 20 mm Hg @ -32.8 C; 400 mm Hg @ 27.5 C [038] Refractive index: 1.4157 @ 20 C [017,047] Unpleasant pungent odor [031,036,102,421] Evaporation rate (butyl acetate = 1): ~7 [102] Viscosity: 0.336 centipoise @ 20 C [395] Odor threshold: 25 ppm [430] Critical temperature: 241 C [371] Critical pressure: 47 atmospheres [371] Liquid surface tension: 28.9 dynes/cm [371] Ratio of specific heats of vapor (gas): 1.124 [371] Heat of combustion: -5416 cal/g [371]

HAP WEIGHTING FACTOR: 1 [713]

VOLATILITY:

Vapor pressure: 340 mm Hg @ 20 C; 440 mm Hg @ 30 C [055]

Vapor density: 2.64 [043,051,055]

FIRE HAZARD: Allyl chloride has a flash point of -31 C (-25 F) [031,043,058,062] and it is flammable. Fires involving this material can be controlled with a dry chemical, carbon dioxide or Halon extinguisher. A water spray may also be used [036,051,058,371].

The autoignition temperature of allyl chloride is 391 C (737 F) [102,371,430].

LEL: 3.3% [102,371,421,430] UEL: 11.2% [043,051,058,421]

REACTIVITY: Allyl chloride is incompatible with strong oxidizers, acids, amines, peroxides, chlorides of iron and aluminum [102,346]. On contact with $AlCl_3$ or BF_3, it may cause a violent exothermic polymerization [036]. A vigorous or explosive reaction occurs above -70 C with alkyl aluminum chlorides (trichlorotriethyl dialuminum, ethyl aluminum dichloride or diethyl aluminum chloride) and aromatic hydrocarbons (benzene or toluene). Violently exothermic polymerization reactions occur with Lewis acids (aluminum chloride, boron trifluoride or sulfuric acid) and metals (magnesium, zinc or galvanized metals) [043]. It undergoes hydrolysis in water to form corrosive compounds. It will attack some forms of plastics, rubber and coatings [102]. It is incompatible with caustics, ammonia and ferric chloride [058]. It is also incompatible with nitric acid, ethylene imine, ethylenediamine, chlorosulfonic acid, oleum and NaOH [043].

STABILITY: Exposure to heat and sunlight can lead to degradation of allyl chloride [051]. NMR stability screening indicates that solutions of allyl chloride in acetone are stable for at least 24 hours [700].

USES: Allyl chloride is used in the manufacture of glycerol, epichlorohydrin, allylamines, polymers, allyl alcohol, allyl silanes, allyl ethers of starch, 1,2,3-trichloropropane, sodium allylsulphonate, quaternary ammonium salts, allyl ethers of a variety of alcohols, phenols, polyols and mono-, di- and tri-allylamine. It is also used in the manufacture of allyl isothiocyanate, eugenol, 1,2-dibromo-3-chloropropane, glycerol chlorohydrins, cyclopropane and amines containing other alkyl groups (e.g., diallylmethylamine). Other uses include thermosetting resins for varnishes, plastics, adhesives, synthesis of pharmaceuticals and insecticides. It is used in a number of barbiturate hypnotic agents such as aprobarbital, butalbital, methohexital sodium, secobarbital, talbital and thiamylal sodium.

COMMENTS: None

NIOSH Registry Number: UC7350000

ACUTE/CHRONIC HAZARDS: Allyl chloride may be very toxic by inhalation, ingestion or absorption through the skin [269]. It may be readily absorbed through the skin [031, 051,102,269]. It may cause eye irritation [031,055,062,102]. It may also cause irritation of the skin, mucous membranes, lungs, nose and throat [031, 051,055,301]. It is a lachrymator [269,275]. When heated to decomposition it may emit toxic fumes of hydrogen chloride, phosgene and carbon monoxide [102].

SYMPTOMS: Symptoms of exposure may include unconsciousness, liver and kidney damage, eye, skin, respiratory tract, nose and throat irritation, local vasoconstriction, numbness and aches and pains in the bones [043,051]. It may cause mucous membrane and lung irritation [301]. It may also cause orbital pain and deep seated aches after eye or skin contact [346]. Other symptoms may include inflammation and edema of the larynx and bronchi, chemical pneumonitis, pulmonary edema, burning sensation, wheezing, laryngitis, shortness of breath and vomiting [269]. It may cause motor and sensory neurotoxic damage at the distal parts of the extremities [395]. It may also cause eye and skin burns [371]. Exposure to this compound may cause skin sensitization, lung and peripheral nerve damage, central nervous system depression evidenced by giddiness, headache, dizziness and nausea, death, changes in urine output, urine appearance or edema, chronic cough, muscular weakness and loss of sensation in the arms and legs [058]. It may cause conjunctivitis, reddening of the eyelids, corneal burn, odor of garlic on the breath and body [430]. It may also cause eye pain with photophobia [099,102].

4-Aminobiphenyl

CAS NUMBER: 92-67-1

SYNONYMS: Anilinobenzene p-Aminobiphenyl p-Aminodiphenyl 4-Aminodiphenyl (1,1'-Biphenyl)-4-amine 4-Biphenylamine p-Biphenylamine p-Diphenylamine p-Phenylaniline p-Xenylamine

CHEMICAL FORMULA: C12H11N

MOLECULAR WEIGHT: 169.23

WLN: ZR DR

PHYSICAL DESCRIPTION: Colorless to yellowish-brown crystals

SPECIFIC GRAVITY: 1.160 @ 20/20 C [043]

DENSITY: Not available

MELTING POINT: 53 C [031,043,062,395]

BOILING POINT: 302 C [016,025,043,062]

SOLUBILITY: Water: <0.1 mg/mL @ 19 C [700] DMSO: 10-50 mg/mL @ 22 C [700] 95% Ethanol: 1-5 mg/mL @ 22 C [700] Acetone: 50-100 mg/mL @ 22 C [700]

OTHER SOLVENTS: Alcohol: Soluble [016,031,205] Ether: Soluble [016,205] Chloroform: Soluble [016,031] Non-polar solvents: Soluble [395] Lipids: Soluble [395]

OTHER PHYSICAL DATA: Boiling point: 191 C @ 15 mm Hg [016,205,275,395] 166 C @ 5 mm Hg [025] pKa: 4.24 (@ 25 C, in water) [025] Volatile with steam [025,031]

HAP WEIGHTING FACTOR: 1 [713]

VOLATILITY: Not available

FIRE HAZARD: 4-Aminobiphenyl has a flash point of >112 C (>235 F) [269]. It is probably combustible. Fires involving this material may be controlled with a dry chemical, carbon dioxide or Halon extinguisher. A water spray may also be used.

The autoignition temperature of 4-aminobiphenyl is 450 C (842 F) [043].

LEL: Not available UEL: Not available

REACTIVITY: 4-Aminobiphenyl is incompatible with strong oxidizers [043,269]. It is also incompatible with acids and acid anhydrides [269]. It is a weak base that forms salts with compounds such as hydrochloric acid and sulfuric acid. It can be diazotized, acetylated and alkylated [395].

STABILITY: 4-Aminobiphenyl is oxidized by air (darkens on oxidation) [395]. It is sensitive to heat [043]. Solutions of it in water, DMSO, 95% ethanol or acetone should be stable for 24 hours under normal laboratory conditions [700].

USES: 4-Aminobiphenyl is used in organic research and in the detection of sulfates. It was formerly used as a dyestuffs intermediate and as a raw material in the manufacture of rubber antioxidants.

COMMENTS: The use of 4-aminobiphenyl is prohibited in the United Kingdom under the Carcinogenic Substances Regulations 1967 [025,036].

NIOSH Registry Number: DU8925000

ACUTE/CHRONIC HAZARDS: 4-Aminobiphenyl may be toxic by ingestion, inhalation and skin absorption [062,269]. It may cause irritation [301]. When heated to decomposition it emits toxic fumes of carbon monoxide, carbon dioxide and nitrogen oxides [043,269].

SYMPTOMS: Symptoms of exposure may include irritation, corneal damage, bladder irritation, kidney and liver damage, central nervous system effects and debility [301]. Other symptoms of exposure may include headache, lethargy, cyanosis, urinary burning and hematuria. Cystoscopy reveals diffuse hyperemia, edema and frank slough [346]. It may also cause nausea and vomiting [031].

Aniline

CAS NUMBER: 62-53-3

SYNONYMS: Aminobenzene Aminophen Benzenamine

CHEMICAL FORMULA: C6H7N

MOLECULAR WEIGHT: 93.13

WLN: ZR

PHYSICAL DESCRIPTION: Colorless to brown oily liquid

SPECIFIC GRAVITY: 1.022 @ 20/20 C [033,269,371,395]

DENSITY: 1.0217 g/mL @ 20 C [205]

MELTING POINT: -6.2 C [038,102,395,430]

BOILING POINT: 184 C [016,055,102,395]

SOLUBILITY: Water: 10-50 mg/mL @ 23 C [700] DMSO: >=100 mg/mL @ 23 C [700] 95% Ethanol: >=100 mg/mL @ 23 C [700] Acetone: >=100 mg/mL @ 23 C [700]

OTHER SOLVENTS: Chloroform: Miscible [033,295] Ligroin: Soluble [016,047] Most organic solvents: Miscible [033,395] Carbon tetrachloride: Soluble [205] Ether: Miscible [295,395] Benzene: Miscible [033,295,395] Alcohol: Miscible [033,295] Acids: Soluble [205] Most lipids: Miscible [395] Oils: Miscible [295]

OTHER PHYSICAL DATA: Refractive Index: 1.5863 @ 20 C [016,033,395,430] pKa: 9.3 @ 25 C [029] pKb: 9.30 [033,395] Boiling point: 68.3 C @ 10 mm Hg [016] 71 C @ 9 mm Hg [029] pH of 0.2 molar aqueous solution: 8.1 [033] Sweet, amine-like odor [058] Burning taste [033,295,395,430] Evaporation rate (butyl acetate = 1): <1 [058,102] Saturated concentration: 1.5 g/m3 @ 20 C, 3.4 g/m3 @ 30 C [055] Odor thresholds: lower 2 ppm, medium 70.1 ppm, upper 128 ppm [051] Critical temperature: 425.6 C [371] Critical pressure: 52.4 atm [371] Liquid surface tension: 45.5 dynes/cm [371] Liquid water interfacial tension: 5.8 dynes/cm [371] Latent heat of vaporization: 110 cal/g [371] Heat of combustion: -8320 cal/g [371]

HAP WEIGHTING FACTOR: 1 [713]

VOLATILITY:

Vapor pressure: 0.3 mm Hg @ 20 C [055] 1 mm Hg @ 34.8 C [038,043,058]

Vapor density: 3.22 [055,058,102,430]

FIRE HAZARD: The flash point for aniline is 70 C (158 F) [058,102,371,421]. It is combustible. Fires involving this material may be controlled with a dry chemical, carbon dioxide or Halon extinguisher. A water spray may also be used [269,371,451].

The autoignition temperature of aniline is 615 C (1139 F) [043,451].

LEL: 1.3% [051,058,102,451] UEL: 11.0% [058,451]

REACTIVITY: Aniline can react vigorously with some oxidizing materials (including perchloric acid, fuming nitric acid, sodium peroxide and ozone) [036,058]. It reacts violently with BCl3 [036,043,066]. It combines with acids to form salts and dissolves alkalis or alkaline earth metals with evolution of hydrogen [033]. It is incompatible with albumin, solutions of iron, zinc and aluminum, acids and alkalis [033,051] but couples readily with phenols and aromatic amines. It is easily acylated and alkylated [395]. It is corrosive to copper and copper alloys [051]. Spontaneously explosive reactions occur with benzenediazonium-2-carboxylate, dibenzoyl peroxide, fluorine nitrate, nitrosyl perchlorate, peroxodisulfuric acid and tetranitromethane. Violent reactions also may occur with peroxyformic acid, diisopropyl peroxydicarbonate, fluorine, trichloronitromethane (145 C), acetic anhydride, chlorosulfonic acid, hexachloromelamine, (HNO3 + N2O4 + H2SO4), (nitrobenzene + glycerin), oleum, (HCHO + HClO4), perchromates, K2O2, beta-propiolactone, AgClO4, Na2O2, H2SO4, trichloromelamine, acids, peroxydisulfuric acid, FO3Cl, diisopropyl peroxy-dicarbonate, n-haloimides and trichloronitromethane. It also ignites on contact with sodium peroxide + water and forms heat or shock sensitive explosive mixtures with anilinium chloride (detonates at 240 C/7.6 bar), nitromethane, hydrogen peroxide, 1-chloro-2,3-epoxypropane and peroxomonosulfuric acid. Reactions with perchloryl fluoride also form explosive products [043,066].

STABILITY: Aniline is heat sensitive [102]. It darkens on exposure to air and light [029,033,058,395]. It also polymerizes to a resinous mass on exposure to air and light [395]. It also is steam volatile [029,033,395].

USES: Aniline is used in the manufacture of rubber chemicals, agriculture chemicals and dyestuffs and in the production of MDI group isocyanates used in polyurethane. It is the parent substance for many dyes and drugs and is also used in rubber accelerators, antioxidants, photographic chemicals, explosives, petroleum refining, diphenylamine, phenolics, herbicides and fungicides, marking inks, tetryl, optical whitening agents, resins, varnishes, perfumes, shoe polishes and many organic chemicals.

COMMENTS: Aniline is readily biodegradable [051].

NIOSH Registry Number: BW6650000

ACUTE/CHRONIC HAZARDS: Aniline may be toxic by inhalation, ingestion and skin absorption [036, 058,102,346]. There is danger of cumulative effects [036], it may be highly toxic [033,269], and may be an allergin [062]. This compound may also may be readily absorbed through the skin [346]. When heated to decomposition it emits toxic fumes of carbon monoxide, carbon dioxide and oxides of nitrogen [043,269].

SYMPTOMS: Symptoms of exposure may include headache, drowsiness, cyanosis, mental confusion, convulsions, nervous system effects, blood effects, fatigue, loss of appetite and dizziness [036]. Other symptoms may include irritation of the skin, severe irritation of the eyes, methemoglobin formation, prolonged anoxemia, central nervous system depression, hemolysis of red blood cells followed by stimulation of bone marrow, liver effects and jaundice [043]. Nausea, weakness, anemia, incoordination and loss of consciousness may also occur [058]. Symptoms of exposure to this compound may also include photophobia, impairment of vision, brownish discoloration of the conjunctiva and cornea, discoloration of the vessels of the retina, disturbances of speech, tinnitus, faintness, paresthesias, muscle pain and sluggish pupillary action [099]. Other symptoms may include irritability, dyspnea and death from asphyxiation [346]. Vomiting, delirium, coma, shock and skin lesions may also occur [371]. Exposure to this compound may lead to respiratory paralysis [451]. Other symptoms may include a burning sensation, coughing, wheezing and laryngitis [269]. Insomnia and paleness may also occur [102]. Cardiac arrhythmias may result from exposure to this compound [295]. Lethargy and a drop in blood pressure may also occur [301].

---

o-Anisidine

CAS NUMBER: 90-04-0

SYNONYMS: o-Aminoanisole 2-Aminoanisole 1-Amino-2-methoxybenzene

CHEMICAL FORMULA: C7H9NO

MOLECULAR WEIGHT: 123.15

WLN: ZR BO1

PHYSICAL DESCRIPTION: Yellowish to reddish liquid

SPECIFIC GRAVITY: 1.0923 @ 20/4 C [017,047,055]

DENSITY: 1.097 g/mL @ 20 C [062,421]

MELTING POINT: 5 C [025,031,062,205]

BOILING POINT: 225 C [031,058,205,346]

SOLUBILITY: Water: <0.1 mg/mL @ 19 C [700] DMSO: >=100 mg/mL @ 23 C [700] 95% Ethanol: >=100 mg/mL @ 23 C [700] Acetone: >=100 mg/mL @ 23 C [700]

OTHER SOLVENTS: Benzene: Miscible [395] Ether: Miscible [031,205,395,421] Dilute mineral acid: Soluble [062,421]

OTHER PHYSICAL DATA: Specific gravity: 1.098 @ 15/15 C [031,205] 1.0923 @ 20/20 C [395] Density: 1.108 g/mL @ 26 C [051] Boiling point: 90 C @ 4 mm Hg [017,047] Vapor pressure: 5 mm Hg @ 88 C; 40 mm Hg @ 101-107 C [058] Volatile with steam [031,051,062,395] Refractive index: 1.5730 @ 20 C [205,269,275] pKa @ 15 C: 9.72 [025] Amine (fishy) odor [102,346]

HAP WEIGHTING FACTOR: 1 [713]

VOLATILITY:

Vapor pressure: <0.1 mm Hg @ 20 C [102,421]   1 mm Hg @ 61.0 C [058]

Vapor density: 4.25 [055,102]

FIRE HAZARD: o-Anisidine has a flash point of 98 C (210 F) [058,205,269,275]. It is combustible. Fires involving this material may be controlled with a dry chemical, carbon dioxide or Halon extinguisher. A water spray may also be used [058,269,451].

LEL: Not available   UEL: Not available

REACTIVITY: o-Anisidine is incompatible with strong oxidizers [058,102,269,346]. It is also incompatible with acids, acid chlorides, acid anhydrides and chloroformates [269]. It will attack some forms of plastics, rubber and coatings [102].

STABILITY: o-Anisidine darkens on exposure to air [031,062,395,421]. It is sensitive to heat [102]. It is also sensitive to exposure to light [031, 051]. NMR stability screening indicates that solutions of it in DMSO are stable for at least 24 hours [700].

USES: o-Anisidine is used in the manufacture of azo or triphenylmethane dyes and intermediates, in the preparation of organic compounds, in the synthesis of guaicol, in the synthesis of hair dyes, as a corrosion inhibitor for steel storage and as an antioxidant for some polymercaptan resins. In western Europe, it is used in the production of pharmaceuticals and textile processing chemicals.

COMMENTS: None

NIOSH Registry Number: BZ5410000

ACUTE/CHRONIC HAZARDS: o-Anisidine may be highly toxic by ingestion, inhalation or skin contact [036,058,269]. It may be readily absorbed through the skin [031, 051, 058,269]. It may be an irritant of the skin [031,036]. When heated to decomposition it emits toxic fumes of carbon monoxide, carbon dioxide and nitrogen oxides [269].

SYMPTOMS: Symptoms of exposure may include skin sensitization and irritation [031,301]. Other symptoms may include methemoglobinemia, corneal damage, and kidney and liver damage [301]. It may cause headache, drowsiness, blue-gray coloring of the skin and lips (cyanosis), red blood cell Heinz bodies, vertigo, nausea, vomiting, unconsciousness and death [102]. Dizziness has been reported [346].

Exposure may cause an increase in sulfhemoglobin [421]. Skin contact may result in contact dermatitis [042,051].

Antimony compounds

NOTE: For all listings which contain the word "compounds," the following applies: Unless otherwise specified, these listings are defined as including any unique chemical substance that contains the named chemical as part of that chemical's infrastructure. [708] Information for the metallic form of antimony is given below. Other compounds described are intended as selected examples of antimony compounds and do not comprise a comprehensive list.

Antimony

CAS NUMBER: 7440-36-0

CHEMICAL FORMULA: Sb

MOLECULAR WEIGHT: 121.75

WLN: SB

PHYSICAL DESCRIPTION: Silver white metal, hexagonal

SPECIFIC GRAVITY: 6.684 @ 25 C

DENSITY: Not available

MELTING POINT: 630.74 C

BOILING POINT: 1750 C

SOLUBILITY: Water: Insoluble DMSO: Insoluble 95% Ethanol: Insoluble Acetone: Insoluble

OTHER SOLVENTS: Sulfuric acid: Soluble hot

OTHER PHYSICAL DATA: Mohs hardness: 3-3.5

VOLATILITY:

Vapor pressure: 1 mm Hg @ 886 C

FIRE HAZARD: Moderate in the forms of dust and vapor, when exposed to heat or flame.

LEL: Not available UEL: Not available

REACTIVITY: Antimony can react violently with halogens, ammonia nitrate, nitric acid, potassium nitrate, potassium permanganate, sodium nitrate, BrF_3, BrN_3, $HClO_3$, ClO, ClF_3 and K_2O_2.

STABILITY: Not available

USES: Antimony is used in the manufacture of alloys, such as Britannia or Babbitt metal, hard lead, white metal, type, bullets and bearing metal; in fireworks; for thermelectric piles, blackening iron, coating metals, etc.

COMMENTS: It is element #51 on the Periodic Table. Explosion hazard: Moderate in the form of dust when exposed to heat or flame. Disaster hazard: Moderate to dangerous; when heated or in contact with acid, it emits toxic fumes.

NIOSH Registry Number: CC4025000

ACUTE/CHRONIC HAZARDS: Antimony may be highly toxic and an irritant. Its fumes are also highly toxic when inhaled.

SYMPTOMS: Symptoms of exposure may include irritation and eczematous eruption of the skin, inflammation of the mucus membrane of the nose and throat, metallic taste and stomatitis, gastrointestinal upset, with vomiting and diarrhea, and various nervous complaints, such as irritability, sleeplessness, fatigue, dizziness and muscular and neuralgic pain.

Antimony oxide

CAS NUMBER: 1309-64-4

CHEMICAL FORMULA: Sb_2O_3

MOLECULAR WEIGHT: 291.52

WLN: Not available

PHYSICAL DESCRIPTION: White crystalline powder

SPECIFIC GRAVITY: 5.67 @ 25/4 C

DENSITY: 5.2 g/mL @ 25 C

MELTING POINT: 655 C

BOILING POINT: 1550 C (sublimes); 870 C @ 210 mm Hg

SOLUBILITY: Water: <1 mg/mL @ 20 C [700] DMSO: <1 mg/mL @ 20 C [700] 95% Ethanol: <1 mg/mL @ 20 C [700] Acetone: <1 mg/mL @ 20 C [700]

OTHER SOLVENTS: Potassium hydroxide: Soluble Tartaric acid: Soluble Hydrochloric acid: Soluble Acetic acid: Soluble Dilute HNO_3: Slightly soluble Dilute H_2SO_4: Slightly soluble

OTHER PHYSICAL DATA: Odorless Tasteless Sublimes in high vacuum @ 400 C Exists in vapor phase as Sb_4O_6 Heat of vaporization: 17.82 kcal/mol Heat capacity: 24.11 cal/g/atom/C @ 21 C Specific gravity: 5.67

VOLATILITY:

Vapor pressure: 1 mm Hg @ 574 C

FIRE HAZARD: Flash point data for antimony oxide is not available. It is probably combustible. Fires involving it may be controlled using a CO_2, foam, and/or Halon extinguishers.

LEL: Not available UEL: Not available

REACTIVITY: Not available

STABILITY: Antimony oxide is stable under normal laboratory conditions.

USES: It is used in the manufacture of tartar emetic; as paint pigment; in enamels and glasses; as mordant; in flame-proofing canvas; catalyst; intermediate; staining iron and copper; phosphors.

COMMENTS: Laboratory preparation is from $SbCl_2$ and water. It is obtained from antimony ore by a volatilization (roasting) process.

NIOSH Registry Number: CC5650000

ACUTE/CHRONIC HAZARDS: Antimony oxide may be toxic. There is evidence that it may be a suspected carcinogen. It may also be a skin and eye irritant.

SYMPTOMS: Symptoms of exposure may include skin irritation and eczema; mucous membranes inflammation; sleeplessness; fatigue; dizziness; irritability; metallic taste; stomatitis; vomiting; diarrhea; muscular and neuralgic pains.

Antimony potassium tartrate

CAS NUMBER: 28300-74-5

SYNONYMS: Antimonate(2)-, bis(mu-tartrato(4-))di-, dipotassium, trihydrate

CHEMICAL FORMULA: C4H4KO7Sb

MOLECULAR WEIGHT: 324.93

WLN: OVYQYQVO & .SB-O..KA.QH

PHYSICAL DESCRIPTION: Colorless crystals or white granular powder

SPECIFIC GRAVITY: 2.607 [042,274]

DENSITY: Not available

MELTING POINT: 332-335 C [900]

BOILING POINT: Not available

SOLUBILITY: Water: 10-50 mg/mL @ 21 C [700] DMSO: <1 mg/mL @ 21 C [700] 95% Ethanol: <1 mg/mL @ 21 C [700] Acetone: <1 mg/mL @ 21 C [700]

OTHER SOLVENTS: Glycerol: 1 g/15 mL [031] Boiling water: 1 g/3 mL [031] Chloroform: Insoluble [053] Ether: Insoluble [053]

OTHER PHYSICAL DATA: Sweetish metallic taste Optical rotation: +140.69 degrees @ 20 C (c=2 in H2O); +139.25 degrees (c=2 in glycerol) [031] Aqueous solution is slightly acid Loses 0.5 mol H2O @ 100 C [017,042,053,062] Loses an additional H20 @ 210 C [031]

VOLATILITY: Not available

FIRE HAZARD: The flash point for this chemical is not available. It is probably combustible. Fires involving this material may be controlled with a dry chemical, carbon dioxide or a Halon extinguisher. A water spray may also be used [058].

LEL: Not available UEL: Not available

REACTIVITY: Antimony potassium tartrate reacts with tannic acid, alkalis and their carbonates, lead salts, astringent infusions (cinchona, rhubarb, etc.), acacia, antipyrene and mercury bichloride [031]. It also reacts with acids, salts of heavy metals, albumins, soap and tannins [053]. It can react with strong oxidizing agents [269].

STABILITY: Antimony potassium tartrate effloresces on exposure to air [031,062,173]. Solutions of it in water, DMSO, 95% ethanol or acetone should be stable for 24 hours under normal laboratory conditions [700].

USES: Antimony potassium tartrate is a mordant in the textile and leather industry and is also used as an antischistosomal, parasiticide, expectorant, cuminatoric and insecticide.

COMMENTS: None

NIOSH Registry Number: CC6825000

ACUTE/CHRONIC HAZARDS: When heated to decomposition antimony potassium tartrate emits toxic fumes of carbon monoxide, carbon dioxide, antimony and antimony oxides [269]. It may be an irritant and may be toxic via ingestion [058].

SYMPTOMS: Symptoms of exposure may include cough, metallic taste, salivation, nausea and diarrhea [042,058]. Skin rash may also occur. Large doses may cause severe damage to the liver [042]. It may be a strong irritant and emetic [151]. Other symptoms may include eye and respiratory tract irritation, headache, dizziness, dyspnea, anaphytoxis, hypotension and weakness [053].

Arsenic compounds (inorganic including Arsine)

NOTE: For all listings which contain the word "compounds," the following applies: Unless otherwise specified, these listings are defined as including any unique chemical substance that contains the named chemical as part of that chemical's infrastructure. [708] Information for the metalloid form of arsenic is given below. Other compounds described are intended as selected examples of arsenic compounds and do not comprise a comprehensive list.

Arsenic

CAS NUMBER: 7440-38-2

ATOMIC FORMULA: As

ATOMIC WEIGHT: 74.92

PHYSICAL DESCRIPTION: Silvery to black, brittle, crystalline and amorphous metalloid [703]

SPECIFIC GRAVITY: 5.73 @ 20C/4C [704]

MELTING POINT: gray arsenic, 816C @ 28 atm [706]

BOILING POINT: gray arsenic, sublimes @ 613 C [706]

SOLUBILITY: Water: Insoluble [703] HNO3: Soluble [703]

OTHER PHYSICAL DATA: Vapor Pressure: 1 mm @ 372 C (sublimes) [703] Density: black crystals 5.724 @ 14 C [703]

VOLATILITY:

Vapor Pressure: 1 mm @ 372 C (sublimes) [703]

FIRE HAZARD: Flash Point: Not available

Flammable in the form of dust when exposed to heat or flame or by chemical reaction with powerful oxidizers (see REACTIVITY). Slightly explosive in the form of dust when exposed to flame. [703]

Autoignition temperature: Not available

LEL: Not available UEL: Not available

REACTIVITY: Flammable by chemical reaction with powerful oxidizers. Emits highly toxic fumes on contact with acid or acid fumes; can react vigorously on contact with oxidizing materials. [703] Incompatible with bromine azide; rubidium acetylide; halogens; palladium; zinc; platinum; NCl3; AgNO3; CrO3; Na2O2 [710]; hexafluoroisopropylideneamino lithium. [703] Arsenical materials are incompatible with any reducing agent [706].

STABILITY: When heated, emits highly toxic fumes [703].

USES: Arsenic is used in bronzing, pyrotechny, and for hardening and improving the sphericity of shot. Arsenic is finding increasing uses as a doping agent in solid-state devices such as transistors. Some arsenic compounds such as calcium arsenate and lead arsenate have been used as agricultural insecticides and poisons. Gallium arsenide is used as a laser material to convert electricity directly into coherent light.

COMMENTS: Arsenic and its compounds are poisonous. Exposure to arsenic and its compounds should not exceed 0.2 mg/m3 as elemental As during an 8-hour work day. These values, however, are being studied, and may be lowered. [706]

NIOSH Registry Number: CG0525000

ACUTE/CHRONIC HAZARDS: Arsenic is a human carcinogen. It is a poison by subcutaneous, intramuscular, and intraperitoneal routes. Human systemic skin and gastrointestinal effects occur by ingestion. It is an experimental teratogen and tumorigen. Mutagenic data exists. When heated or on contact with acid or acid fumes, it emits highly toxic fumes. [703] Target organs include the liver, kidneys, skin, lungs and lymphatic system. [704]

SYMPTOMS: Symptoms of exposure to arsenic may include ulceration of the nasal septum, dermatitis, gastrointestinal disturbances, peripheral neuropathy, respiratory irritation, and hyperpigmentation of the skin; a carcinogen. [704]

Arsenic chloride

CAS NUMBER: 7784-34-1

SYNONYMS: Arsenic(III) chloride Arsenic chloride, liquid (DOT) Arsenic trichloride, liquid (DOT) Arsenious chloride Arsenous chloride Arsenous trichloride

CHEMICAL FORMULA: $AsCl_3$

MOLECULAR WEIGHT: 181.28

WLN: Not available

PHYSICAL DESCRIPTION: Colorless oily liquid

SPECIFIC GRAVITY: 2.163 @ 25 C

DENSITY: 2.15 g/mL @ 25 C

MELTING POINT: -16 C

BOILING POINT: 130 C

SOLUBILITY: Water: Decomposes DMSO: >=100 mg/mL @ 19 C [700] 95% Ethanol: >=100 mg/mL @ 19 C [700] Acetone: >=100 mg/mL @ 19 C (Radian)

OTHER SOLVENTS: Chloroform: Miscible Carbon tetrachloride: Miscible Iodine: Miscible Oils and fats: Miscible Hydrochloric acid: Soluble HBr: Soluble Ether: Miscible

OTHER PHYSICAL DATA: Unpleasant odor

VOLATILITY:

Vapor pressure: 10 mm Hg @ 23.5 C

Vapor density: 6.25

FIRE HAZARD: Arsenic chloride is nonflammable.

LEL: Not available UEL: Not available

REACTIVITY: Arsenic chloride reacts with water to generate hydrogen chloride and will corrode metal and it reacts violently with sodium, potassium and aluminum on impact.

STABILITY: Arsenic chloride will decompose in water and on exposure to ultraviolet light. Arsenic chloride may be stable in anhydrous acetone or anhydrous alcohol, but not in 95% alcohol [700].

USES: It is used as an intermediate for organic arsenicals (pharmaceuticals, insecticides) and in the ceramic industry.

COMMENTS: None

NIOSH Registry Number: CG1750000

ACUTE/CHRONIC HAZARDS: When heated to decomposition arsenic chloride emits highly toxic fumes; it may be also a severe irritant.

SYMPTOMS: Symptoms of exposure may include irritation of eyes, skin, stomach, nose, and throat. It may also cause irritation of the heart, liver and kidneys; nervousness, thirst, vomiting, diarrhea, cyanosis and collapse.

Arsenic trioxide

CAS NUMBER: 1327-53-3

SYNONYMS: Arsenic oxide Arsenic (III) oxide Arsenic sesquioxide Arsenious acid Arsenious oxide Arsenious trioxide Arsenous acid Arsenous oxide

CHEMICAL FORMULA: As2O3

MOLECULAR WEIGHT: 197.84

WLN: .AS2. O3

PHYSICAL DESCRIPTION: White or transparent, glassy, amorphous lumps or crystal

SPECIFIC GRAVITY: 3.738

DENSITY: Not available

MELTING POINT: 312.3 C

BOILING POINT: 465 C

SOLUBILITY: Water: Not available DMSO: Not available 95% Ethanol: Not available Acetone: Not available

OTHER SOLVENTS: Hydrochloric acid: Soluble Alkali: Soluble Alkali carbon: Soluble

OTHER PHYSICAL DATA: Not available

VOLATILITY: Not volatile

FIRE HAZARD: Not available (not expected to be flammable)

LEL: Not available UEL: Not available

REACTIVITY: Can react with oxidizing materials.

STABILITY: Arsenic trioxide is stable under normal laboratory storage conditions.

USES: Arsenic trioxide is primarily used in the manufacturing of glass, paris green, enamels, weed killers, for preserving hides, killing rodents and insects. It is also used in sheep dips and textile mordants.

COMMENTS: None

NIOSH Registry Number: CG3325000

ACUTE/CHRONIC HAZARDS: Arsenic trioxide may be toxic; emits highly toxic fumes when heated or in contact with acids or acid fumes.

SYMPTOMS: Symptoms of exposure may include marked irritation of the stomach and intestines with nausea, vomiting and diarrhea. In severe cases, the vomitus and stools may be bloody and may be accompanied by collapse and shock, with loss of appetite, cramps, constipation, liver damage, jaundice, disturbances of the blood, kidneys and nervous system, skin abnormalities such as itching, pigmentation and eruptions, constriction in the throat, difficulty in swallowing, dehydration with intense thirst, hematuria, albuminuria, glycosuria, vertigo, stupor, delirium, syncope, convulsions, ventricular fibrillation and peripheral neuritis.

Arsine

CAS NUMBER: 7784-42-1

SYNONYMS: Arsenic hydrid Arsenic hydride Arsenic trihydride Arseniuretted hydrogen Arsenous hydride Hydrogen arsenide

CHEMICAL FORMULA: AsH3

MOLECULAR WEIGHT: 77.95

WLN: Not available

PHYSICAL DESCRIPTION: Colorless gas

SPECIFIC GRAVITY: Not available

DENSITY: 2.695 g/L

MELTING POINT: -116 C

BOILING POINT: -62.5 C

SOLUBILITY: Water: 20 cc/100 cc DMSO: Not available 95% Ethanol: Slightly soluble Acetone: Not available

OTHER SOLVENTS: Chloroform: Soluble Alkalies: Slightly soluble Benzene: Soluble

OTHER PHYSICAL DATA: Mild garlic odor Dissociation pressure @ 0 C=0.806 atm Decomposes @ 300 C, depositing arsenic which volatilizes @ 400 C Aqueous solutions are neutral Transition temperature: -167.61 C Critical Temperature: 99.90 C Latent Heat of Fusion: -116.93 C Latent Heat of Vapor: -62.48 C

VOLATILITY:

Vapor pressure: 76.544 cm Hg @ -62.31 C; 14.95 cm @ 21.1 C

Vapor density: 2.66

FIRE HAZARD: Arsine is flammable. Fires involving it may be controlled with a dry chemical, carbon dioxide or Halon extinguisher.

LEL: Not available UEL: Not available

REACTIVITY: Arsine is incompatible with strong oxidizers, chlorine or nitric acid. It is strongly endothermic and capable of detonation with suitably powerful initiation.

STABILITY: Arsine will decompose on exposure to heat, light or moisture.

USES: Arsine is used for organic synthesis, as a military poison gas, and as a doping agent for solid state electronic components.

COMMENTS: None

NIOSH Registry Number: CG6475000

ACUTE/CHRONIC HAZARDS: Arsine may be extremely toxic via inhalation. When heated to decomposition it emits extremely toxic fumes.

SYMPTOMS: Symptoms of exposure via inhalation may include headache, dizziness, nausea and vomiting, epigastric pain and weakness, tea colored or bloody urine, jaundice and tenderness of the liver, anemia, chills, anorexia, paresthesia, hematomesis, hemoglobinuria and anuria, burning and stinging of the chest, dysphagia, electrocardiographic abnormalities, delirium and coma. Death by inhalation may occur as a result of anoxia, pulmonary edema, cardiac failure and renal failure.

Sodium arsenite

CAS NUMBER: 7784-46-5

CHEMICAL FORMULA: AsO2.Na

MOLECULAR WEIGHT: 129.91

WLN: .Na..As-O2

PHYSICAL DESCRIPTION: White or grayish white powder

SPECIFIC GRAVITY: 1.87 @ 20 C

DENSITY: Not available

MELTING POINT: Not available

BOILING POINT: Not available

SOLUBILITY: Water: Very soluble DMSO: Not available 95% Ethanol: Slightly soluble Acetone: Not available

OTHER SOLVENTS: Not available

OTHER PHYSICAL DATA: Odorless

VOLATILITY: Not available

FIRE HAZARD: Sodium arsenite is nonflammable.

LEL: Not available UEL: Not available

REACTIVITY: Not available

STABILITY: Sodium arsenite is somewhat hygroscopic. It absorbs carbon dioxide from the air.

USES: Sodium arsenite is used in the manufacture of dyes and soap for use on skins, for treating vines against certain scale diseases, as an insecticide, (especially against termites), as an antiseptic, a topical acaricide, hide preservative and herbicide.

COMMENTS: None

NIOSH Registry Number: CG3675000

ACUTE/CHRONIC HAZARDS: Sodium arsenite may be highly toxic via oral and inhalation routes. When heated to decomposition this compound emits toxic fumes.

SYMPTOMS: Exposure to the dust from this compound can cause eye irritation. Ingestion or excessive inhalation of the dust causes irritation of the stomach and intestines with nausea, vomiting and diarrhea; bloody stools, shock, rapid pulse and coma.

Asbestos

CAS NUMBER: 1332-21-4

NOTE: "Asbestos" is a broad commercial term for a group of naturally occurring hydrated silicates that crystallize in a fibrous habit. Asbestos minerals can be subdivided into serpentine fibers (the most common of which is chrysotile) and amphibole minerals, including crocidolite, amosite, anthophyllite, tremolite, actinolite, riebeckite (nonfibrous crocidolite) and grunerite (nonfibrous amosite). [707] The chemical composition of each form is given below under Chemical Formula.

CHEMICAL FORMULA:

Chrysotile: $Mg_6Si_4O_{10}(OH)_8$ Crocidolite: $Na_2(Fe(III))_2(Fe(II))_3Si_8O_{22}(OH)_2$
Amosite: $(Fe,Mg)_7Si_8O_{22}(OH)_2$ Anthophyllite: $(Mg,Fe)_7Si_8O_{22}(OH)_2$
Tremolite: $Ca_2Mg_5Si_8O_{22}(OH)_2$ Actinolite: $Ca_2(Mg,Fe)_5Si_8O_{22}(OH)$

MOLECULAR WEIGHT: Varies [704]

PHYSICAL DESCRIPTION: White or greenish (chrysotile), blue (crocidolite), or gray-green (amosite) fibrous, odorless solids [704]

SPECIFIC GRAVITY: Not available

MELTING POINT: 1112F (Decomposes) [704]

BOILING POINT: Decomposes [704]

SOLUBILITY: Water: Insoluble [704]

OTHER PHYSICAL DATA: Vapor Pressure: 0 mm (approx) [704]

HAP WEIGHTING FACTOR: 100 [713]

VOLATILITY:

Vapor Pressure: 0 mm (approx) [704]

FIRE HAZARD: Flash Point: Not applicable

A noncombustible solid [704].

LEL: Not applicable UEL: Not applicable

REACTIVITY: None reported [704].

STABILITY: Not available

USES: Asbestos is resistant to heat and chemicals, and has high tensile strength. It has been used as fire-proofing material, insulation, as a component of cement construction materials (such as roofing, shingles, and cement pipes), friction materials (brake linings and clutch pads), jointing and gaskets, asphalt coats and sealants, and other similar products. [712] Many uses of asbestos are now banned or restricted due to possible health hazards.

COMMENTS: None

NIOSH Registry Number: CI6475000

ACUTE/CHRONIC HAZARDS: Asbestos is a human carcinogen and experimental tumorigen. Human pulmonary system effects occur by inhalation. Usually at least 4 to 7 years of exposure are required before serious lung damage (fibrosis) results. [703] Exposure to asbestos may lead to asbestosis (scarring of the lung tissue), lung cancer (squamous cell carcinoma, adenocarcinoma, large cell carcinoma, and small cell carcinoma), pleural and peritoneal mesothelioma, and stomach, esophageal, larynx and colon cancers [721].

Occupational exposure to asbestos can cause 4 types of disorders: asbestosis; lung cancer; mesotheliomas of the pleura, pericardium, and peritoneum; and benign changes in the pleura. Asbestosis caused progressive lung damage, disability and death in many workers exposed before enforcement of occupational standards. In general, lung cancers have been found in asbestos workers who are smokers and only rarely in nonsmokers, and in most cases have occurred 20 or more years after their first exposure. [707]

SYMPTOMS: Symptoms of exposure to asbestos may include dyspnea, interstitial fibrosis, restricted pulmonary function, and finger clubbing. [704]

Benzene

CAS NUMBER: 71-43-2

NOTE: This Clean Air Act category includes benzene from gasoline. [708]

SYNONYMS: Benzol Cyclohexatriene Phenyl hydride

CHEMICAL FORMULA: C6H6

MOLECULAR WEIGHT: 78.11

WLN: RH

PHYSICAL DESCRIPTION: Clear, colorless liquid

SPECIFIC GRAVITY: 0.8765 @ 20/4 C [017,047]

DENSITY: 0.905 g/mL @ 21 C [700]

MELTING POINT: 5.5 C [017,058,395,421]

BOILING POINT: 80.1 C [017,031,055,395]

SOLUBILITY: Water: 1-5 mg/mL @ 18 C [700] DMSO: >=100 mg/mL @ 22 C [700] 95% Ethanol: >=100 mg/mL @ 22 C [700] Acetone: >=100 mg/mL @ 22 C [700]

OTHER SOLVENTS: Cyclohexane: >=100 mg/mL @ 21 C [700] Glacial acetic acid: Miscible [031,295,395] Chloroform: Miscible [031,395] Carbon disulfide: Miscible [031,062,395] Carbon tetrachloride: Miscible [031,062,395] Most organic solvents: Soluble in 100 parts [205] Light petroleum: Miscible [295] Oils: Miscible [031,295,395] Ether: Miscible [031,062,295,395] Alcohol: Miscible [031,062,295,395] Acetic acid: >10% [047]

OTHER PHYSICAL DATA: Specific gravity: 0.8787 @ 15/4 C [031,395] 0.879 @ 15/15 C [058] Density: 0.87865 g/mL @ 20 C [421] Vapor pressure: 100 mm Hg @ 26.1 C [043,301,430] 166 mm Hg @ 38 C [058] Vapor pressure: 118 mm Hg @ 30 C [055] 40 mm Hg @ 7.6 C [038,051] Refractive index: 1.5011 @ 20 C [017,047,062,430] 1.4979 @ 25 C [205] Surface tension: 28.9 dynes/cm [062,371] Gasoline-like odor [371] Vapor is heavier than air [051,371] Vapor burns with a smoky flame [062,295] Burning rate: 6.0 mm/min [371] Evaporation rate (butyl acetate = 1): >1 [058] UV max: 400-350 nm, 330 nm, 300 nm, 290 nm, 280 nm (A = 0.01, 0.02, 0.06, 0.15. 1.0) [275] Odor threshold: 0.84 ppm [051] Noncorrosive [051]

HAP WEIGHTING FACTOR: 10 [713]

VOLATILITY:

Vapor pressure: 60 mm Hg @ 15 C [038,055] 76 mm Hg @ 20 C [055]

Vapor density: 2.77 [043,051,055,058]

FIRE HAZARD: Benzene has a flash point of -11 C (12 F) [031,036,058,275] and it is flammable. Fires involving this material may be controlled with a dry chemical, carbon dioxide or Halon extinguisher.

The autoignition temperature is 562 C (1044 F) [036,043,058,062].

LEL: 1.4% [036,043,058,430] UEL: 8% [036,043,058,062]

REACTIVITY: Benzene is incompatible with oxidizers [036,058,269,346]. It is also incompatible with strong acids [058]. It can react with chlorine, ozone, permanganates, sulfuric acid, peroxides, perchlorates, nitrating acid, nitric acid, chromic acid anhydride and chromium trioxide [051]. It can also react with (bromine + iron) [346]. Reactions may also occur with interhalogens and uranium hexafluoride. The complex with silver perchlorate exploded on crushing in mortar. Mixtures with 84% nitric acid are highly sensitive to detonation. A solution with rubber exploded when ozonized [036]. Mixtures with liquid oxygen are explosive [036,051,058]. It readily undergoes substitution reactions to form halogen, nitrate, sulfonate and alkyl derivatives [430].

STABILITY: Benzene is hygroscopic [269]. It is also sensitive to heat [058]. It is stable under normal lab storage conditions [700].

USES: Benzene is used as a solvent. It is also used in the manufacture of medicines, dyes, artificial leather, linoleum, oil cloth, pesticides, plastics and resins, PCB, aviation fuel, detergents, flavors and perfumes, paints and coatings, airplane dope, varnishes, lacquers, explosives and other organics. It is used in photogravure printing and as a component of high-octane gasoline. It is used to manufacture ethylbenzene, isopropyl- benzene, cyclohexane, aniline, maleic anhydride and alkylbenzenes. It is used in veterinary medicine to destroy screwworm larvae in wounds.

COMMENTS: Benzene occurs in coal tar and petroleum naphtha and is also a constituent of gasoline. It also occurs in thermal degradation gases from high-density polystyrene and in solid waste gasification products. It has also been identified in condensates of tobacco smoke. Its use is declining due to its toxicity [430] and carcinogenicity.

NIOSH Registry Number: CY1400000

ACUTE/CHRONIC HAZARDS: Benzene may be highly toxic by ingestion, inhalation and skin contact [036,165,295,395]. It may be an irritant of the skin, eyes, nose and throat [058, 346,371,430]. When heated to decomposition it emits toxic fumes of carbon monoxide and carbon dioxide [058,269]. There is evidence that it may be also a carcinogen.

SYMPTOMS: Symptoms of exposure may include staggering gait, vomiting, loss of consciousness, central nervous system depression, dizziness, headache, constriction of the chest and euphoria [051,301,406,430]. It may cause somnolence, shallow and rapid pulse, delirium, chemical pneumonia, fear of impending death, breathlessness, visual disturbances, hilarity, weariness, fatigue, vertigo, death following coma, dyspnea, inebriation with euphoria, hemorrhaging in the brain, pericardium, urinary tract, mucous membranes and skin, blood dyscrasias, hyperbilirubinemia, splenomegaly, adrenomegaly, abnormal caloric labyrinth irritability, impairment of hearing, erythema, vesiculation, chromosome aberrations and tinnitus [430]. In addition, it may cause irritation of the mucous membranes, restlessness, depression and aplasia [031]. Edema and blistering of the skin, blood changes, intestinal irritation, cataract formation and embryotoxic activity have been reported [058]. It may cause weakness, anemia, anorexia, bone marrow hyperplasia with leucocytosis, epistaxis, menstrual disturbances, mental confusion, mydriasis, pulmonary edema, purpura, rapid respiration, speech problems, skin and eye irritation, scaling, urine color change and weight loss [051]. Other symptoms may include tremors, visual blurring, sleepiness, loss of appetite, nervousness, shallow and rapid breath, aplastic anemia and nausea [301,406]. This material may cause pallor, excitation, respiratory failure, liver and kidney damage, ventricular arrhythmias and irritability [295]. It has caused bone marrow depression, reduced platelets and red and white blood cells and leukemia [058, 301,421]. Occasionally hemorrhages in the retina and in the conjunctiva have occurred and on rare instances neuroretinal edema and papilledema have accompanied retinal hemorrhages [099]. Petechiae, coma, fibrillation and necrosis or fatty degeneration of the heart, liver and adrenals are possible [301]. Also, it may cause leukopenia, thrombocytopenia, irregular heart beat and inflammation of the respiratory tract [395].

Benzidine

CAS NUMBER: 92-87-5

SYNONYMS: 4,4'-Diamino-1,1'-biphenyl p,p'-Dianiline 4,4'-Diphenylenediamine p,p'-Bianiline 4,4'-Bianiline 4,4'-Biphenyldiamine 4,4'-Biphenylenediamine p,p'-Diaminobiphenyl

CHEMICAL FORMULA: C12H12N2

MOLECULAR WEIGHT: 184.23

WLN: ZR DR DZ

PHYSICAL DESCRIPTION: Grayish-yellow, white or reddish-gray

SPECIFIC GRAVITY: 1.250 @ 20/4 C [043,051,055,395]

DENSITY: Not available

MELTING POINT: 128 C [017,025,031,205]

BOILING POINT: 400 C @ 740 mm Hg [017,025,047,205]

SOLUBILITY: Water: <1 mg/mL @ 22 C [700] DMSO: >=100 mg/mL @ 20 C [700] 95% Ethanol: 10-50 mg/mL @ 20 C [700] Acetone: >=100 mg/mL @ 20 C [700]

OTHER SOLVENTS: Ether: 1 g/50 mL [031] Absolute alcohol: 1 g/13 mL [900] Less polar solvents: Readily soluble [395]

OTHER PHYSICAL DATA: Melting point: 115-120 C (when slowly heated) [031,051] Odorless [902] Can be sublimed [395] Spectroscopy data: lambda max 287 nm (in ethanol) [395]

HAP WEIGHTING FACTOR: 1000 [713]

VOLATILITY:

Vapor density: 6.36 [055]

FIRE HAZARD: Benzidine is combustible [051,062, 901]. Fires involving this material may be controlled with a dry chemical, carbon dioxide or Halon extinguisher. A water spray may also be used [902].

LEL: Not available UEL: Not available

REACTIVITY: Benzidine is a weak base that forms insoluble salts with sulfuric acid. It can be diazotized, acetylated and alkylated [395]. It is hypergolic with red fuming nitric acid [066].

STABILITY: Benzidine darkens on exposure to air and light [031,395,421,901]. UV spectrophotometric stability screening indicates that solutions of this compound in acetone are stable for at least 24 hours under normal lab conditions [700].

USES: Benzidine is used as a precursor in the synthesis of dyes and pigments used to color textiles, rubber, plastic products, printing inks, paints, lacquers, leathers and paper products, manufacture of a wide variety of organic chemicals, rubber compounding agent, as a reagent for hydrogen peroxide (H2O2) in milk, for detection of blood stains, as a stain in microscopy, in the manufacture of plastic films, in the production of security paper and as a laboratory reagent in determining hydrogen cyanide, sulfate, nicotine and certain sugars.

COMMENTS: Its use in the United Kingdom is prohibited under The Carcinogenic Substances Regulations 1968 [025].

NIOSH Registry Number: DC9625000

ACUTE/CHRONIC HAZARDS: Benzidine may be toxic by ingestion, inhalation and skin contact [062]. There is evidence it may be a carcinogen. It may be rapidly absorbed through the skin [031,036,051,346]. It may be a severe irritant [902] and when heated to decomposition it emits toxic fumes of nitrogen oxides [043, 051].

SYMPTOMS: Symptoms of exposure may include irritation of the skin and eyes [346,902]. Other symptoms may include damage to the blood, including hemolysis and bone marrow depression [043,051]. If ingested, it may cause nausea and vomiting. This is sometimes followed by liver and kidney damage [031,043,051]. It may cause skin sensitization and contact dermatitis [051,346]. It may also cause bladder irritation [301]. Long term exposure may result in an increase in urination, hematuria and urinary tract tumors [346,902]. It may also cause pain on urination [346]. Cystitis has been reported [051].

Benzotrichloride

CAS NUMBER: 98-07-7

SYNONYMS: alpha, alpha, alpha-Trichlorotoluene 1-(Trichloromethyl)benzene Phenylchloroform

CHEMICAL FORMULA: $C_7H_5Cl_3$

MOLECULAR WEIGHT: 195.48

WLN: GXGGR

PHYSICAL DESCRIPTION: Colorless to yellowish liquid

SPECIFIC GRAVITY: 1.38 @ 15/15 C

DENSITY: 1.3723 g/mL @ 20 C

MELTING POINT: -4.8 C

BOILING POINT: 220.6 C @ 760 mm Hg

SOLUBILITY: Water: Decomposes DMSO: >=100 mg/mL @ 15 C [700] 95% Ethanol: >=100 mg/mL @ 15 C [700] Acetone: >=100 mg/mL @ 15 C [700]

OTHER SOLVENTS: Ether: Soluble Benzene: Soluble

OTHER PHYSICAL DATA: Refractive index: 1.5580 @ 20 C Pungent odor Boiling point: 129 C @ 60 mm Hg; 105 C @ 25 mm Hg; 89 C @ 10 mm Hg

HAP WEIGHTING FACTOR: 10 [713]

VOLATILITY:

Vapor pressure: 1 mm Hg @ 45.8 C; 40 mm Hg @ 119.8 C; 760 mm Hg @ 213.5 C

Vapor density: 6.77

FIRE HAZARD: Benzotrichloride has a flash point of 97 C (207 F). It is combustible. Fires involving it can be controlled with a carbon dioxide, dry chemical and Halon extinguisher.

LEL: Not available UEL: Not available

REACTIVITY: Benzotrichloride hydrolyzes in the presence of moisture. It reacts with water, lime, ammonia, strong alkalis, organic amines, chlorates and acids.

STABILITY: Benzotrichloride is sensitive to moisture. UV spectrophotometric stability screening indicates that solutions of this compound in ethanol are stable for at least 24 hours [700].

USES: Synthetic dyes, organic synthesis

COMMENTS: None

NIOSH Registry Number: XT9275000

ACUTE/CHRONIC HAZARDS: Benzotrichloride may be toxic by inhalation. It may be corrosive and an irritant. When heated to decomposition it emits toxic fumes.

SYMPTOMS: Benzotrichloride may be corrosive to the skin, eyes and mucous membranes. It has caused central nervous system depression in experimental animals.

Benzyl chloride

CAS NUMBER: 100-44-7

SYNONYMS: (Chloromethyl)benzene omega-Chlorotoluene a-Chlorotoluene alpha-Chlorotoluene Tolyl chloride Chloromethylbenzene Chlorophenylmethane

CHEMICAL FORMULA: C_7H_7Cl

MOLECULAR WEIGHT: 126.59

WLN: G1R

PHYSICAL DESCRIPTION: Colorless to brown-yellow liquid

SPECIFIC GRAVITY: 1.100 @ 20/20 C

DENSITY: Not available

MELTING POINT: -43 C

BOILING POINT: 179.3 C @ 760 mm Hg (dec)

SOLUBILITY: Water: Reaction DMSO: >=100 mg/mL @ 22 C [700] 95% Ethanol: >=100 mg/mL @ 22 C [700] Acetone: >=100 mg/mL @ 22 C [700]

OTHER SOLVENTS: Chloroform: Miscible Ether: Miscible

OTHER PHYSICAL DATA: Unpleasant, irritating odor Steam-volatile Refractive index: 1.5380 @ 20 C; 1.5415 @ 15 C Boiling point: 66 C @ 11 mm Hg; 99 C @ 62 mm Hg Viscosity: 1.3 centistokes @ 25 C Critical temperature: 411 C Odor threshold: 0.047 ppm

HAP WEIGHTING FACTOR: 1 [713]

VOLATILITY:

Vapor pressure: 1 mm Hg @ 22 C; 20 mm Hg @ 75 C; 100 mm Hg @ 114.2 C

Vapor density: 4.36

FIRE HAZARD: Benzyl chloride has a flash point of 73 C (165 F). It is combustible. Fires involving it can be controlled with a dry chemical, carbon dioxide or Halon extinguisher.

The autoignition temperature for benzyl chloride is 525 C (1085 F).

LEL: 1.1% UEL: Not available

REACTIVITY: Benzyl chloride can react vigorously with oxidizing agents, iron and iron salts. It also reacts with sodium acetate, pyridine, aluminum, brass, copper, zinc, magnesium and other active metals. Contact with water and/or steam will produce toxic fumes.

STABILITY: Gas chromatography stability screening indicates that solutions of benzyl chloride in 95% ethanol are stable for at least 24 hours [700]. The neat material is moisture-sensitive. Upon standing this compound will pick up the water vapor from the air and turn cloudy.

USES: Benzyl chloride is used in the manufacture of benzyl compounds, perfumes, pharmaceutical products, dyes, synthetic tannins, artificial resins, photography, gasoline gum inhibitors and formerly used as irritant gas in chemical warfare.

COMMENTS: None

NIOSH Registry Number: XS8925000

ACUTE/CHRONIC HAZARDS: Benzyl chloride may be corrosive and extremely irritating and may cause burns on contact with the skin, eyes and mucous membranes. It may also be a lachrymator and an experimental carcinogen. When heated to decomposition it emits toxic fumes.

SYMPTOMS: Symptoms of exposure may include severe irritation of the skin, eyes, and upper respiratory tract; coughing, wheezing, laryngitis, shortness of breath, headache, nausea and vomiting, central nervous system depression, a burning sensation and sore throat. Inhalation may be fatal as a result of spasm, inflammation and edema of the larynx and bronchi, chemical pneumonitis and pulmonary edema.

Beryllium compounds

NOTE: For all listings which contain the word "compounds," the following applies: Unless otherwise specified, these listings are defined as including any unique chemical substance that contains the named chemical as part of that chemical's infrastructure. [708] Information for the metallic form of beryllium is given below. Other compounds described are intended as selected examples of beryllium compounds and do not comprise a comprehensive list.

Beryllium

CAS NUMBER: 7440-41-7

CHEMICAL FORMULA: Be

MOLECULAR WEIGHT: 9.01 [703]

PHYSICAL DESCRIPTION: A grayish-white, hard, light metal [703]

SPECIFIC GRAVITY: 1.85 @ 20C/4C [704]

MELTING POINT: 1278 C [703]

BOILING POINT: 2970 C [703]

SOLUBILITY: Water: Insoluble [704]

OTHER PHYSICAL DATA: Not available

VOLATILITY:

Vapor Pressure: 0 mm (approx) [704]

FIRE HAZARD: Flash Point: Not applicable

A moderate fire hazard in the form of dust or powder, or when exposed to flame or by spontaneous chemical reaction.

Autoignition temperature: Not available

LEL: Not applicable UEL: Not applicable

REACTIVITY: Incompatible with halocarbons. Warm beryllium reacts incandescently with fluorine or chlorine. Mixtures of the powder with carbon tetrachloride or trichloroethylene will flash or spark on heavy impact. [703, 710] Incompatible or reacts with acids, caustics, chlorinated hydrocarbons, oxidizers, molten lithium [704].

STABILITY: Slight explosion hazard in the form of powder or dust. When heated to decomposition in air it emits very toxic fumes of BeO. [703]

USES: Beryllium has many desirable properties for industrial uses. These include light weight, high melting point, and modules of elasticity that are greater than that of steel. Beryllium also resists attack by concentrated nitric acid, it hasexcellent thermal conductivity, and it is non-magnetic. At ordinary temperatures beryllium resists oxidation in air. Beryllium is used as an alloying agent in producing beryllium copper. It slso finds application as a structural material for high-speed aircraft, missiles, spacecraft, and communication satellites. Since beryllium is relatively transparent to X-rays, ultra-thin foil made from beryllium finds industial use in applications such as X-ray lithography (which is used for reproduction of microminiature integrated circuits. Another very important use of beryllium is in nuclear reactors as a reflector or moderator. This use is because beryllium has a low thermal neutron absorption cross section. It is also used in gyroscopes, computer parts, and instruments where lightness, stiffness, and dimensional stability are required. Beryllium oxide has a very high melting point and therefore it finds use in nuclear work and ceramic applications. [706]

COMMENTS: None

NIOSH Registry Number: DS1750000

ACUTE/CHRONIC HAZARDS: Beryllium is a deadly poison by intravenous route and a suspected human carcinogen. It is an experimental carcinogen, neoplastigen and tumorigen. Human systemic effects by inhalation include lung fibrosis, dyspnea and weight loss. Human mutagenic data exists. When heated to decomposition it emits very toxic fumes of BeO. [703] Targeted organs include the lungs, skin, eyes and mucous membranes. [704]

SYMPTOMS: Symptoms of exposure may include respiratory symptoms, weakness, fatigue and weight loss. [704]

Beryllium sulfate tetrahydrate

CAS NUMBER: 7787-56-6

SYNONYMS: Sulfuric acid, beryllium salt (1:1), tetrahydrate Beryllium sulfate tetrahydrate (1:1:4)

CHEMICAL FORMULA: BeO4S.4H2O

MOLECULAR WEIGHT: 177.14

WLN: BE S-O4 &QH 4

PHYSICAL DESCRIPTION: Colorless crystals

SPECIFIC GRAVITY: 1.713 [058]

DENSITY: 1.713 g/mL @ 11 C [205]

MELTING POINT: 270 C [205]

BOILING POINT: 580 C (Decomposes) [062,205]

SOLUBILITY: Water: >=100 mg/mL @ 23 C [700] DMSO: <1 mg/mL @ 23 C [700] 95% Ethanol: <1 mg/mL @ 23 C [700] Acetone: <1 mg/mL @ 23 C [700]

OTHER SOLVENTS: Concentrated H2SO4: Slightly soluble [016,395] Alcohol: Insoluble [016,062]

OTHER PHYSICAL DATA: Partial dehydration (loses 2 waters) occurs @ 100 C and total dehydration (loses 4 waters) occurs @ 400 C [016,395] Refractive index:

1.472, 1.440 [016,205] Sinks and mixes with water [371] Heat of solution: -6 cal/g [371] Odorless [058,371]

VOLATILITY: Not available

FIRE HAZARD: Beryllium sulfate tetrahydrate is probably nonflammable [371]. Fires involving it may be controlled with a dry chemical, carbon dioxide or Halon extinguisher.

LEL: Not available UEL: Not available

REACTIVITY: Beryllium sulfate tetrahydrate is incompatible with acids [269].

STABILITY: Solutions of beryllium sulfate tetrahydrate should be stable under normal lab conditions.

USES: Beryllium sulfate tetrahydrate is used as a chemical intermediate in the processing of beryl and bertrandite ores.

COMMENTS: None

NIOSH Registry Number: DS5000000

ACUTE/CHRONIC HAZARDS: Beryllium sulfate tetrahydrate may be fatal if inhaled, swallowed or absorbed through the skin [269]. It may be an irritant and when heated to decomposition it may emit toxic fumes of sulfur oxides [043,269].

SYMPTOMS: Symptoms of exposure may include lung disease with cough, chest pain, shortness of breath, weight loss, weakness and fatigue. It may cause eye and skin irritation, rash and ulcers. Implantation may lead to wart like granulomas. It may cause irritation or burns to the mouth, esophagus and stomach [058]. It may also cause irritation to the respiratory system and cyanosis [036]. It may cause conjunctivitis, hyperemia of the conjunctiva, edema of the lids, burning sensation and photophobia [099]. Other symptoms of exposure may include difficulty breathing, anorexia, malaise, pneumonitis, nasopharyngitis, tracheobronchitis, dyspnea, conjunctival inflammation and dermatitis [371]. It may cause rhinitis, death, pulmonary fibrosis, coronary pulmonale, respiratory insufficiency, cardiac arrest, virus pneumonia and renal insufficiency [395].

Biphenyl

CAS NUMBER: 92-52-4

SYNONYMS: Diphenyl Bibenzene Phenylbenzene Xenene

CHEMICAL FORMULA: $C_{12}H_{10}$

MOLECULAR WEIGHT: 154.22

WLN: RR

PHYSICAL DESCRIPTION: White crystals

SPECIFIC GRAVITY: Not available

DENSITY: 1.041 g/mL

MELTING POINT: 69-71 C

BOILING POINT: 254-255 C

SOLUBILITY: Water: Insoluble DMSO: Not available 95% Ethanol: Not available Acetone: Not available

OTHER SOLVENTS: Ether: Soluble Benzene: Very soluble

OTHER PHYSICAL DATA: Pleasant but peculiar odor

HAP WEIGHTING FACTOR: 1 [713]

VOLATILITY: Not available

FIRE HAZARD: The flash point of biphenyl is 109 C (235 F). It is combustible. Fires involving it should be controlled using a dry chemical, carbon dioxide or Halon extinguisher.

The autoignition temperature is 540 C (1004 F).

LEL: Not available UEL: Not available

REACTIVITY: Biphenyl is incompatible with oxidizers.

STABILITY: Biphenyl is stable under normal laboratory conditions.

USES: Biphenyl is used in organic synthesis, as a heat transfer fluid, as a preservative in food, and as a fungicide.

COMMENTS: None

NIOSH Registry Number: DU8050000

ACUTE/CHRONIC HAZARDS: Not available

SYMPTOMS: Symptoms of exposure may include fatigue, headache, insomnia, sensory impairment and anorexia. Biphenyl also may irritate skin and eyes when they come into contact with it.

Bromoform

CAS NUMBER: 75-25-2

SYNONYMS: Tribromomethane Methenyl tribromide

CHEMICAL FORMULA: CHBr3

MOLECULAR WEIGHT: 252.75

WLN: EYEE

PHYSICAL DESCRIPTION: Colorless, heavy liquid

SPECIFIC GRAVITY: 2.89 @ 20/4 C [025,042,055,058]

DENSITY: 2.887 g/mL [062]

MELTING POINT: 9 C [062,421]

BOILING POINT: 149.5 C [025,042,047,346]

SOLUBILITY: Water: <0.1 mg/mL @ 22.5 C [700] DMSO: >=100 mg/mL @ 18 C [700] 95% Ethanol: >=100 mg/mL @ 18 C [700] Acetone: >=100 mg/mL @ 18 C [700]

OTHER SOLVENTS: Chloroform: Soluble [062] Ligroin: Soluble [017] Solvent naphtha: Soluble [062] Petroleum ether: Miscible [031] Ether: Soluble [062] Benzene: Soluble [062]

OTHER PHYSICAL DATA: Odor and taste similar to chloroform Specific gravity: 2.90 @ 15/15 C [025] 2.9035 @ 15/4 C [031] Boiling point: 46 C @ 15 mm Hg [047] Refractive index: 1.5976 @ 20 C; 1.6005 @ 15 C Freezes to hexagonal crystals

HAP WEIGHTING FACTOR: 1 [713]

VOLATILITY:

Vapor pressure: 5 mm Hg @ 20 C [058] 5.6 mm Hg @ 25 C [055]

Vapor density: 8.7 [055,058]

FIRE HAZARD: Bromoform is nonflammable [042,062, 205,275].

LEL: Not available UEL: Not available

REACTIVITY: Bromoform reacts with chemically active metals such as sodium, potassium, calcium, powdered aluminum, zinc and magnesium; and with strong caustics [058, 269,346]. It also reacts with lithium and NaK alloy [042]. It reacts with acetone when catalyzed by potassium hydroxide or another base [036,066].

STABILITY: Bromoform gradually decomposes, acquiring a yellow color. This decomposition is accelerated by exposure to air and light [031]. Protect from freezing [058]. Solutions of it in water, DMSO, 95% ethanol or acetone should be stable for 24 hours when protected from light and air [700].

USES: Bromoform is used as an intermediate in organic synthesis; in geological assaying; solvent for waxes, greases and oils; in medicines (as a sedative); and as an ingredient in fire resistant chemicals and gauge fluid.

COMMENTS: Bromoform is sometimes stabilized with 1% ethanol.

NIOSH Registry Number: PB5600000

ACUTE/CHRONIC HAZARDS: Bromoform may be a lachrymator, an irritant and it has narcotic effects. When heated to decomposition it emits toxic fumes of HBr [042]. It also may be absorbed through the skin [058].

SYMPTOMS: Symptoms of exposure may include irritation of the eyes, skin and respiratory tract [346]. Ingestion may cause respiratory difficulties, tremors and unconsciousness [036]. Dizziness, disorientation, slurred speech, and death may also result from ingestion [058,102]. Inhalation may be fatal as a result of spasm, inflammation and edema of the larynx and bronchi; chemical pneumonitis, and pulmonary edema. Other symptoms may include liver and kidney damage; and central nervous system depression [269]. Inhalation may cause irritation of the respiratory tract, pharynx and larynx producing lacrimation and salivation. It may also cause skin and eye irritation. Repeated or prolonged contact may cause dermatitis [058].

1,3-Butadiene

CAS NUMBER: 106-99-0

SYNONYMS: Biethylene Bivinyl Butadiene Buta-1,3-diene alpha-gamma-Butadiene Divinyl Erythrene Pyrrolylene Vinylethylene

CHEMICAL FORMULA: C4H6

MOLECULAR WEIGHT: 54.09

WLN: 1U2U1

PHYSICAL DESCRIPTION: Colorless gas

SPECIFIC GRAVITY: 0.6211 @ 20/4 C [017,047,062,395]

DENSITY: Not available

MELTING POINT: -108.9 C [017,047,062,395]

BOILING POINT: -4.4 C [017,205,395,421]

SOLUBILITY: Water: Insoluble [036,062,205,430] DMSO: Not available 95% Ethanol: Soluble [017,047,062,395] Acetone: Very soluble [395]

OTHER SOLVENTS: Most organic solvents: Soluble [031,395] Ether: Soluble [017,047,062,395] Benzene: Soluble [017,047,395]

OTHER PHYSICAL DATA: Critical temperature: 161.8 C Critical pressure: 32376 mm Hg Refractive index: 1.4292 @ -25 C Mild, aromatic odor Specific gravity: 0.650 @ -6/4 C [031,205] Boiling point: 15.3 C @ 1520 mm Hg; 47 C @ 3800 mm Hg [031] Odor threshold: 0.16 ppm (4 mg/m3) Evaporation rate: >25

HAP WEIGHTING FACTOR: 10 [713]

VOLATILITY:

Vapor pressure: 1840 mm Hg @ 21 C [042,395,430] 760 mm Hg @ -4.5 C [395]

Vapor density: 1.87 [042,102,395,430]

FIRE HAZARD: 1,3-Butadiene has a flash point of -76 C (-105 F) [042,062,371,430] and it is flammable. Fires involving it can be controlled with a dry chemical, carbon dioxide or Halon extinguisher.

The autoignition temperature of 1,3-butadiene is 450 C (842 F) [102].

LEL: 2% [036,042,062,137] UEL: 11.5% [036,042,137,430]

REACTIVITY: When not inhibited, 1,3-butadiene polymerizes and copolymerizes easily, e.g. under the influence of sodium, thereby forming synthetic rubbers [031]. It can react with oxidizing materials [042,346,395]. It can also react with aluminum tetrahydroborate, buten-3-yne, cobalt, crotonaldehyde, nitrogen oxides, oxygen and sodium nitrite. When stored under an inert atmosphere (oxygen up to 1.8%), it polymerized and then decomposed violently [066]. It may react with copper and copper alloys [102,346]. It is a high explosion hazard if mixed with phenol and ClO2 [042]. It will attack some forms of plastics, rubber and coatings [102]. It can also react with acetylides [058].

STABILITY: 1,3-Butadiene is usually inhibited and then it is stable under normal laboratory conditions [371]. When not inhibited, it may form acrolein and peroxides upon exposure to air [395]. Peroxides formed on long contact with air are explosive, and may also initiate polymerization. It may react violently when heated under pressure [036].

USES: 1,3-Butadiene is used as polymer component in the manufacture of synthetic rubber, synthetic polymeric elastomers, rocket fuels, plastics, resins, ABS resins, chemical intermediates and latex paints.

COMMENTS: It is often inhibited with p-tert-butylcatechol.

NIOSH Registry Number: EI9275000

ACUTE/CHRONIC HAZARDS: 1,3-Butadiene may be an irritant and narcotic in high concentrations [031,036]. When heated to decomposition it emits acrid fumes [042].

SYMPTOMS: Symptoms of exposure may include nausea, blurred vision, prickling and dryness of the mouth, throat and nose and decreased blood pressure and pulse rate [430]. It may also cause bone marrow damage and central nervous system effects [301]. Dermatitis and frostbite may result from exposure to the liquid and evaporating gas. In high concentrations, it may cause cough, narcotic effects, fatigue, drowsiness, headache, vertigo, loss of consciousness, respiratory paralysis and death [346]. It may also cause hematological disorders, liver enlargement, liver and bile duct diseases, kidney malfunctions, laryngitis, upper respiratory tract irritation, conjunctivitis, gastritis, various skin disorders and a variety of neuraesthenic symptoms [395].

Cadmium compounds

NOTE: For all listings which contain the word "compounds," the following applies: Unless otherwise specified, these listings are defined as including any unique chemical substance that contains the named chemical as part of that chemical's infrastructure. [708] Information for the metallic form of cadmium is given below. Other compounds described are intended as selected examples of cadmium compounds and do not comprise a comprehensive list.

Cadmium

CAS NUMBER: 7440-43-9

CHEMICAL FORMULA: Cd

MOLECULAR WEIGHT: 112.41

WLN: CD

PHYSICAL DESCRIPTION: Silver-white, blue tinged and lustrous metal

SPECIFIC GRAVITY: 8.642

DENSITY: Not available

MELTING POINT: 320.9 C

BOILING POINT: 765 C

SOLUBILITY: Water: Insoluble DMSO: Not available 95% Ethanol: Not available Acetone: Not available

OTHER SOLVENTS: Acid: Soluble NH4NO3: Soluble Sulfuric Acid (hot): Soluble

OTHER PHYSICAL DATA: Refractive index: 1.13 Specific gravity: 8.642 Hexagonal crystals, silver-white, malleable metal [703]

VOLATILITY:

Vapor pressure: 1 mm Hg @ 394 C

FIRE HAZARD: Cadmium is flammable in powder a form.

LEL: Not available UEL: Not available

REACTIVITY: Cadmium reacts readily with dilute nitric acid and reacts slowly with hot hydrochloric acid but it does not react with alkalis. Cadmium dust can react vigorously with oxidizing materials.

STABILITY: Cadmium slowly oxidizes by moist air to form cadmium oxide (CdO).

USES: Cadmium is used as a constituent of easily fusible alloys; soft solder and solder for aluminum. Electroplating is its major use. It is also used as a deoxidizer in nickel plating, in process engraving, in electrodes for cadmium vapor lamps, in photoelectric cells, with photometry of ultraviolet sun-rays, and in Ni-Cd storage batteries. The powder is also used as an amalgam (1 Cd : 4 Hg) in dentistry.

COMMENTS: None

NIOSH Registry Number: EU9800000

ACUTE/CHRONIC HAZARDS: Cadmium and its salts may be highly toxic and there is evidence that it is a suspected carcinogen.

SYMPTOMS: Ingestion of cadmium metal and solvated compounds may cause increased salivation, choking, vomiting, abdominal pain, anemia, renal dysfunction, diarrhea and tenesmus. Inhalation (dust or fumes) may cause throat dryness, cough, headache, vomiting, chest pain, extreme restlessness and irritability, pneumonitis and possibly bronchopneumonia.

Cadmium chloride

CAS NUMBER: 10108-64-2

CHEMICAL FORMULA: CdCl2

MOLECULAR WEIGHT: 183.32

WLN: CD G2

PHYSICAL DESCRIPTION: Colorless to white hexagonal crystals

SPECIFIC GRAVITY: Not available

DENSITY: 4.047 g/mL @ 25 C [017,043,053,430]

MELTING POINT: 568 C [017,033,205,275]

BOILING POINT: 960 C [017,033,043,058]

SOLUBILITY: Water: >=100 mg/mL @ 20 C [700] DMSO: 50-100 mg/mL @ 20 C [700] 95% Ethanol: <1 mg/mL @ 21.5 C [700] Acetone: <1 mg/mL @ 20 C [700] Methanol: 1.7 g/100 mL @ 15.5 C [017,053]

OTHER SOLVENTS: Ether: Insoluble [017,051,173,430] Water (hot): 150 g/100 mL @ 100 C [017] Dilute acids: Dissolved [421] Ammonia solutions: Dissolved [421] Alcohol: 1.52 g/100 mL @ 15 C [017]

OTHER PHYSICAL DATA: Odorless [058,062,173,371] Refractive index: 1.65 @ 20 C (@ 589.3 nm) [395] Heat of fusion: 28.8 cal/g [371]

VOLATILITY:

Vapor pressure: 10 mm Hg @ 656 C [043,051,053,058] 0 mm Hg @ 20 C [053,102]

FIRE HAZARD: Cadmium chloride is nonflammable [051,053, 371].

LEL: Not available UEL: Not available

REACTIVITY: Cadmium chloride reacts violently with bromine trifluoride and potassium [043,051,058,269]. It is incompatible with oxidizing materials and acids [269]. It is also incompatible with elemental sulfur, selenium and tellerium [053,102,346].

STABILITY: Cadmium chloride is hygroscopic [033,051,269]. Solutions of it in water, DMSO, 95% ethanol or acetone should be stable for 24 hours under normal lab conditions [700].

USES: Cadmium chloride is used in photography, in dyeing and calico printing, in the vacuum tube industry, in the manufacture of cadmium yellow, in galvanoplasty, in the manufacture of special mirrors, as an ice-nucleating agent, as a lubricant, in analysis of sulfides to absorb hydrogen sulfide, in testing for pyridine bases, as a fungicide, in the preparation of cadmium sulfide, in analytical chemistry, as an ingredient of electroplating baths, in additions of tinning solutions, in paints and electronic components, as a pesticide, as an insecticide, as a nematocide, as a polymerization catalyst, and in pigments, glass and glazes.

COMMENTS: None

NIOSH Registry Number: EV0175000

ACUTE/CHRONIC HAZARDS: Cadmium chloride may be highly toxic by ingestion and inhalation [036,051, 058,269]. It is an irritant and when heated to decomposition it emits toxic fumes of chlorine [043]. It may also emits oxides of the contained metal and halogen, possibly also free, or ionic halogen [058].

SYMPTOMS: Symptoms of exposure may include tightness in the chest, emesis, tubular dysfunction and necrosis of the kidneys and liver, irritation, redness and discomfort of the skin and eyes, decrease in bone density, kidney stones and vomiting [058]. In addition, it may cause nausea, diarrhea, shortness of breath, headache, dyspnea, shock, metallic taste in the mouth, cough with foamy or bloody sputum, emphysema, pulmonary edema, renal failure, gastrointestinal inflammation, inflammation of the pulmonary epithelium, muscular aches, weakness, pain in the legs, bubbling rales, diminished urine formation, fever, lung consolidation, loss of sense of smell, chest pain, weight loss, anemia, yellow stained teeth, irritability, hematuria, proteinuria, abdominal pain, lowered red and white blood cell counts, elevated erythrocyte sedimentation rate, increased lung density and bronchial pneumonia [301]. Other symptoms may include irritation and possible damage of the lungs, increased salivation, choking and stomach pains [036]. It may cause cramps, dark urine, inflammation of the mucous membranes, death, bone damage and blood changes [051]. It may cause irritation to the mucous membranes and upper respiratory tract [269]. Exposure can lead to tenesmus, thickening of the alveolar septa, osteomalacia, skeletal deformity, severe pain, mild hypochromic anemia and mild jaundice [173]. It may also lead to convulsions, gastroenteric distress, prostration, sensory disturbances, smarting of skin, first degree burns (on short exposure) and second degree burns (on long exposure) [371]. Other symptoms may include sweating, chills and cor pulmonale [346]. This compound may cause irritation to the nose and throat, and ulceration of the nose

[102]. It may cause hypertension [395]. Symptoms may also include peribronchial and perivascular fibrosis and dizziness [406].

Cadmium oxide

CAS NUMBER: 1306-19-0

SYNONYMS: Cadmium monoxide Cadmium(II) oxide

CHEMICAL FORMULA: CdO

MOLECULAR WEIGHT: 128.40

WLN: .CD..O

PHYSICAL DESCRIPTION: Red-brown crystals or colorless

SPECIFIC GRAVITY: 6.95 @ 20/4 C

DENSITY: Not available

MELTING POINT: 900 C (decomposes)

BOILING POINT: 1559 C (sublimes)

SOLUBILITY: Water: <1 mg/mL @ 20 C [700] DMSO: <1 mg/mL @ 20 C [700] 95% Ethanol: <1 mg/mL @ 20 C [700] Acetone: <1 mg/mL @ 20 C [700]

OTHER SOLVENTS: Acids: Soluble Alkalies: Insoluble Ammonium salt solutions: Soluble

OTHER PHYSICAL DATA: There are two forms of this compound. The alternate form is dark brown crystals with a density of 8.15. The melting and boiling points are the same

VOLATILITY:

Vapor pressure: 1 mm Hg @ 1000 C

FIRE HAZARD: Flash point data for cadmium oxide are not available. It is probably combustible. This material may be controlled with a dry chemical, carbon dioxide, foam or Halon extinguisher.

LEL: Not available UEL: Not available

REACTIVITY: Cadmium oxide reacts violently with magnesium.

STABILITY: Cadmium oxide is stable under normal laboratory conditions.

USES: Cadmium oxide is used for electroplating, in storage battery electrodes, as a catalyst, in semiconductors, in the manufacture of silver alloys, in ceramic glazes, as a nematocide, as an anthelmintic, in phosphors, in glass, for cadmium electroplating, and an ascaricide in swine.

COMMENTS: None

NIOSH Registry Number: EV1925000

ACUTE/CHRONIC HAZARDS: Cadmium oxide is an irritant and when heated to decomposition it emits toxic fumes.

SYMPTOMS: Symptoms of exposure may include nausea, vomiting, diarrhea, head and muscular aches, salivation, abdominal pain, metallic taste, shortness of breath, chest pain, cough with foamy or bloody sputum, and weakness. Inhalation may cause severe or fatal lung irritation, proteinuria, and elevated blood pressure. It affects the sense of smell, the kidneys, and heart rate. Ingestion causes kidney and liver injuries.

Caprolactam

CAS NUMBER: 105-60-2

SYNONYMS: Aminocaproic lactam 6-Aminocaproic acid lactam 6-Aminohexanoic acid cyclic lactam 6-Caprolactam gamma-Caprolactam

CHEMICAL FORMULA: C6H11NO

MOLECULAR WEIGHT: 113.16

WLN: T7MVTJ

PHYSICAL DESCRIPTION: White crystalline solid

SPECIFIC GRAVITY: 1.02 @ 75/4 C [031,205,421]

DENSITY: Not available

MELTING POINT: 69 C [043,346]

BOILING POINT: 266.9 C @ 760 mm Hg [395]

SOLUBILITY: Water: >=100 mg/mL @ 20.5 C [700] DMSO: >=100 mg/mL @ 20.5 C [700] 95% Ethanol: >=100 mg/mL @ 20.5 C [700] Acetone: >=100 mg/mL @ 20.5 C [700] Methanol: Freely soluble [031]

OTHER SOLVENTS: Cyclohexane: <1 mg/mL @ 21 C [700] Chloroform: Soluble [017,395] Chlorinated solvents: Soluble [031,062,395] Petroleum distillates: Soluble [031,051,062,395] Tetrahydrofurfuryl alcohol: Freely soluble [031] Dimethyl formamide: Freely soluble [031,395] Ether: Freely soluble [031] Benzene: Soluble [017,395] Cyclohexene: Soluble [031,051,062,395]

OTHER PHYSICAL DATA: Boiling point: 100 C @ 3 mm Hg [421] 139 C @ 12 mm Hg [017,029,055] 180 C @ 50 mm Hg [062,205] Refractive index: 1.4965 @ 31 C; 1.4935 @ 40 C [062] Density: 1.05 g/mL @ 25/4 C (70% aqueous soln.) [031,062]

HAP WEIGHTING FACTOR: 1 [713]

VOLATILITY:

Vapor pressure: 0.001 mm Hg @ 20 C [055,395] 3 mm Hg @ 100 C [062]

Vapor density: 3.91 [055,395]

FIRE HAZARD: The flash point for caprolactam is 125 C (257 F) [031,421]. It is combustible. Fires involving this compound may be controlled using a dry chemical, carbon dioxide or Halon extinguisher.

The autoignition temperature for this compound is 375 C (707 F) [107].

LEL: 1.4% [107] UEL: 8% [107]

REACTIVITY: Caprolactam can react with strong oxidizing agents and strong bases [269]. It can also react with chlorinated hydrocarbons and nitro compounds [107]. A potentially explosive reaction occurs with acetic acid + dinitrogen trioxide [043].

STABILITY: Caprolactam is hygroscopic [031,036,269,421]. Solutions of it in water, DMSO, 95% ethanol or acetone should be stable for 24 hours under normal lab conditions [700].

USES: Caprolactam is used in the manufacture of synthetic fibers, plastics, bristles, film, coatings, synthetic leather, plasticizers and paint vehicles. It is a cross-linking agent for polyurethanes and it is also used in the synthesis of lysine.

COMMENTS: None

NIOSH Registry Number: CM3675000

ACUTE/CHRONIC HAZARDS: Caprolactam is an irritant and when heated to decomposition it emits toxic fumes of carbon monoxide, carbon dioxide and nitrogen oxides [269].

SYMPTOMS: Symptoms of exposure may include eye irritation, visual disturbance, cough, respiratory distress, dry throat, nausea, dyspnea, eczema, erythema, dermatitis, skin sensitization, skin burns and vomiting [107]. It may also cause contact dermatitis, eczema of the hands and light sensitivity [346]. It may also have central nervous system effects [301].

Captan

CAS NUMBER: 133-06-2

SYNONYMS: N-Trichloromethylmercapto-4-cyclohexene-1,2-dicarboximide

CHEMICAL FORMULA: $C9H8Cl3NO2S$

MOLECULAR WEIGHT: 300.59

WLN: T56 BVNV GUTJ CSXGGG

PHYSICAL DESCRIPTION: White crystals

SPECIFIC GRAVITY: 1.74 [031,169,395,421]

DENSITY: 1.74 g/mL [025,047]

MELTING POINT: 178 C [031,169,172,395]

BOILING POINT: Decomposes

SOLUBILITY: Water: <1 mg/mL @ 20 C [700] DMSO: 50-100 mg/mL @ 20 C [700] 95% Ethanol: <1 mg/mL @ 20 C [700] Acetone: 1-5 mg/mL @ 20 C [700] Toluene: 0.69 g/100 mL @ 26 C [031]

OTHER SOLVENTS: Chloroform: 7.78 g/100 mL @ 26 C [031,169] Tetrachloroethane: 8.15 g/100 mL @ 26 C [031] Heptane: 0.04 g/100 mL @ 26 C [031] Ethylene dichloride: Slightly soluble [062] Ether: 0.25 g/100 mL @ 26 C [31] Benzene: 21 g/kg @ 26 C [169] Cyclohexanone: 23 g/kg @ 25 C [172,395] Dioxane: 47 g/kg @ 26 C [169] Isopropanol: 1.7 g/kg @ 26 C [169,395] Methylene chloride: Soluble [025] Petroleum oils: Insoluble [169,172,173,395] Xylene: 20 g/kg @ 25 C [169,172,395]

OTHER PHYSICAL DATA: Melting point also reported as 158-164 C [062] Odorless [031,043]

HAP WEIGHTING FACTOR: 1 [713]

VOLATILITY: Not available

FIRE HAZARD: Flash point data for this chemical are not available. It is probably combustible. Fires involving this material may be controlled with a dry chemical, carbon dioxide or Halon extinguisher.

LEL: Not available UEL: Not available

REACTIVITY: Captan is incompatible with strong alkaline and oxidizing materials, sulfur and (sulfur + moisture) [058].

STABILITY: Captan is slowly hydrolyzed in aqueous neutral and acidic media. It is rapidly hydrolyzed in alkaline media. It decomposes at or near the melting point [169]. UV spectrophotometric stability screening indicates that solutions of it in ethanol are stable for at least 24 hours [700].

USES: Captan is used as a fungicide in paints, plastics, leather and fabrics, in fruit preservation and as a bacteriostat. It is used to control diseases of many fruit, ornamentals and vegetable crops. It is also used as a spray, root dip or seed treatment to protect young plants against rot and damping off. Captan controls a wide range of fungal diseases such as pome fruit rot, shot-hole of stone fruit, peach leaf curl, brown rot of cherries, apricots, peaches, plums and citrus fruit, downy mildew and black rot of vines, early and late blights of potatoes and tomatoes, blight and leaf spot in carrots, anthracnose and downy mildew of cucurbits, leaf spot diseases in ornamentals, anthracnose and leaf spot disease of tomatoes and brown patch on turf. It is used in the topical treatment of fungal infections of the skin. It is also used in pastes for wallpaper, in paint for greenhouses, in medical facilities, in food packaging, in vinyl coated fabrics and vinyl car roofs, in lacquers, in paper, in rubber stabilizers and in polyethylene for garbage bags and pond liners.

COMMENTS: None

NIOSH Registry Number: GW5075000

ACUTE/CHRONIC HAZARDS: Captan is an irritant and when heated to decomposition it emits toxic fumes of chlorine, sulfur oxides and nitrogen oxides [043].

SYMPTOMS: Symptoms of exposure may include vomiting and diarrhea [031], and irritation of the eyes, skin, and mucous membranes.

Carbaryl

CAS NUMBER: 63-25-2

SYNONYMS: Carbamic acid, methyl-, 1-naphthyl ester Methylcarbamic acid-1-naphthyl ester 1-Naphthyl methylcarbamate Sevin

CHEMICAL FORMULA: C12H11NO2

MOLECULAR WEIGHT: 201.24 [702]

PHYSICAL DESCRIPTION: White crystalline solid, essentially odorless [705]

SPECIFIC GRAVITY: 1.23 @ 20C/4C [704]

MELTING POINT: 142 C [705]

BOILING POINT: Decomposes [702]

SOLUBILITY: Water: 40 mg/L [702] Water: 40 ppm @ 30 C [709] Soluble in most organic solvents [709]

OTHER PHYSICAL DATA: Vapor pressure: <0.005 mmHg @ 26 C [705] Vapor pressure: 0.002 mmHg @ 40 C [709] Melting Point: 145 C [702]

HAP WEIGHTING FACTOR: 1 [713]

VOLATILITY:

Vapor Pressure: 0.002 mmHg @ 40 C [709]

FIRE HAZARD: Flash Point: Not applicable

Flammability: Noncombustible solid, but may be dissolved in flammable liquids [704].

Autoignition temperature: Not available

LEL: Not applicable UEL: Not applicable

REACTIVITY: Incompatible or reacts with strong oxidizers such as chlorates, bromates, and nitrates [704].

STABILITY: When heated to decomposition it emits toxic fumes of NOx [703].

USES: Insecticide (trade name "SEVIN") [709].

COMMENTS: Noncorrosive; hydrolyzes rapidly in alkaline solutions [709].

NIOSH Registry Number: FC5950000

ACUTE/CHRONIC HAZARDS: A poison by ingestion, intravenous, intraperitoneal, and possibly other routes; human mutagenic data exists. An experimental carcinogen, teratogen, and tumorigen; produces experimental reproductive effects. Carbaryl is also a severe skin irritant. Skin absorption is slow with no accumulation in tissue. A reversible cholinesterase inhibitor. [703] Targeted organs include the respiratory system, central nervous system, cardiovascular system, and skin. [704] When heated to decomposition, emits toxic fumes of NOx. [703] Carbaryl is a primary eye (conjuntival) irritant at 24 hours but it cleared at 48 hours. [705]

SYMPTOMS: Symptoms of exposure to carbaryl (trade name "Sevin") may include miosis, blurred vision, tearing; nasal discharge, salivation; sweat; abdominal cramps, nausea, vomiting, diarrhea; tremor; cyanosis; convulsions; and skin irritation. [704] Symptoms include blurred vision, headache, stomach ache, and vomiting. Symptoms similar to but less severe than those due to parathion. [703]

Carbon disulfide

CAS NUMBER: 75-15-0

SYNONYMS: Carbon bisulfide Carbon sulfide Dithiocarbonic anhydride

CHEMICAL FORMULA: Carbon disulfide

MOLECULAR WEIGHT: 76.14

WLN: C S2

PHYSICAL DESCRIPTION: Colorless to yellow liquid

SPECIFIC GRAVITY: 1.2632 @ 20/4 C [016,031,047]

DENSITY: 1.2632 g/mL [016,031,047]

MELTING POINT: -111.5 C [016,047,102]

BOILING POINT: 46.5 C [025,031,043,051]

SOLUBILITY: Water: <1 mg/mL @ 20 C [700] DMSO: >=100 mg/mL @ 20 C [700] 95% Ethanol: >=100 mg/mL @ 20 C [700] Acetone: >=100 mg/mL @ 20 C [700]

OTHER SOLVENTS: Alcohol: Soluble [016,062,421] Benzene: Miscible [025,031] Chloroform: Miscible [031,172,295] Ether: Miscible [025,031,172,295] Carbon tetrachloride: Miscible [031]

OTHER PHYSICAL DATA: Specific gravity: 1.29272 @ 0/4 C, 1.27055 @ 15/4 C, 1.24817 @ 20/20 C [031] Specific gravity: 1.260 @ 25/25 C [062] 1.2570 @ 25/21 C [052] 1.263 @ 20/4 C [051,055] Boiling point: -73.8 C @ a mm Hg, -44.7 C @ 10 mm Hg, -5.1 C @ 100 mm IIg, 28.0 C @ 400 mm Hg, 69.1 @ 2 atm, 104.8 @ 5 atm [031] Boiling point: 20 C @ 297 mm Hg [025] Vapor pressure: 1 mm Hg @ -73.8 C, 5 mm Hg @ -54.3 C, 10 mm Hg @ -44.7 C, 20 mm Hg @ -34.3 C, 40 mm Hg @ -22.5 C, 60 mm Hg @ -15.3 C, 100 mm Hg @ -5.1 C, 200 mm Hg @ 10.4 C, 760 mm Hg @ 46.5 C [038] Forms a hemihydrate which decomposes @ -3 C [025] Critical temperature: 280.0 C [031] Critical pressure: 72.9 atm [031] Latent heat of vaporization: 85 cal/g [371] Heat of fusion: 13.80 cal/g [371] Heat capacity: 18.17 cal/mol/deg @ 24.3 C [031] Liquid surface tension: 0.032 N/m @ 20 C [371] Liquid water interfacial tension: 0.0484 N/m @ 20 C [371] Heat of combustion: -3230 cal/g [371] Nearly odorless when pure [043,051,062] Refractive index: 1.6232 @ 25 C [062] 1.6319 @ 20 C [016,047] Refractive index: 1.6295 @ 18 C [172] 1.6315 @ 15 C [025]

HAP WEIGHTING FACTOR: 1 [713]

VOLATILITY:

Vapor pressure: 300 mm Hg @ 20 C [102,421); 400 mm Hg @ 28 C [038,043,051]

Vapor density: 2.64 [043,051,055]

FIRE HAZARD: The flash point for carbon disulfide is -30 C (-22 F) [031,058,395,451] and it is flammable. Fires involving carbon disulfide may be controlled using a dry chemical, carbon dioxide or Halon extinguisher. A water spray may also be used [043,051,269].

The autoignition temperature is 90 C (194 F) [051,102,451].

LEL: 1% [036,062,173,421] UEL: 50% [031,062,421,451]

REACTIVITY: Carbon disulfide reacts violently with fluorine, azide solutions, zinc dust and liquid chlorine in the presence of iron. It also reacts violently with aluminum, potassium, sodium and their alloys [038]. It is incompatible with rust or iron, permanganic acid, nitrogen oxide, CsN3, ClO, ethylamine diamine, ethylene imine, Pb(N3)2, LiN3, (sulfuric acid + permanganates), KN3, RbN3, NaN3, phenylcopper-triphenylphosphine complexes and metals. Mixtures with dinitrogen tetraoxide are heat, spark and shock sensitive explosives [043]. This compound is also incompatible with oxidants [043,051,269].

STABILITY: Carbon disulfide may be sensitive to prolonged exposure to air and light [051]. Solutions of it in water, DMSO, 95% ethanol or acetone should be stable for 24 hours under normal lab conditions [700].

USES: Carbon disulfide is used as a disinfectant, an insecticide, in the manufacture of rayon and carbon tetrachloride, in electronic vacuum tubes and as a solvent for sulfur, phosphorus, iodine, bromine, rubber intermediates, selenium, fats and resins. It also used in the manufacture of xanthogenates, cellophane, optical glass, matches and paper. This compound is also used as a bactericide, a wood preservative, a nematocide, in the synthesis of dyes, pharmaceuticals and flotation agents, as a solvent in dry spinning PVC and oil wells, in the extraction and processing of oils, fats, resins and waxes and in veterinary medicine as a parasiticide.

COMMENTS: None

NIOSH Registry Number: FF6650000

ACUTE/CHRONIC HAZARDS: Carbon disulfide may be highly toxic via all routes of exposure [051,102,269]. It is an irritant, a nerve poison, and it can be absorbed through the skin. When heated to decomposition it emits toxic fumes of sulfur oxides and sulfur dioxides [043,051,269].

SYMPTOMS: Symptoms of exposure may include narcotic effects leading to unconsciousness, eye irritation, severe central nervous system damage, failure of vision, mental disturbances and paralysis [036]. Symptoms of acute poisoning may include euphoria, restlessness, mucous membrane irritation, nausea, vomiting, unconsciousness and terminal convulsions. Symptoms of chronic poisoning may include hallucinations, tremors, auditory disturbances, visual disturbances, weight loss and blood dyscrasias. Dermal contact may cause burning pain, erythema and exfoliation [031]. Exposure to this compound may cause death from respiratory failure [043,051,151].

Other symptoms may include peripheral sensory neuritis with paresthesias, pain (mostly in the legs), impairment of peripheral tendon reflexes, weakening of the legs, central scotoma, decreased visual activity, impaired recognition of red and green, permanent blindness, disturbed pupillary reaction to light, abnormal adaptation to darkness and retinal microaneurysms [099]. Chromosome aberrations, thyroid hypofunction, excitation of the central nervous system, depression, stupor, headache, neuritis, crawling sensation in the skin, cold sensation, irritability, Parkinsonian paralysis, insanity, insomnia, loss of memory and personality changes may also occur [051]. Reproductive disorders and kidney, liver and heart damage may occur. Depending on the intensity and duration of exposure, effects ranging from mild irritation to severe destruction of tissue can occur [269].

Additional symptoms may include a drop in blood pressure, dizziness, diarrhea, numbness, unsteady gait, difficulty swallowing, increased arteriosclerosis, stomach problems, spontaneous abortion, aspermia and menstrual irregularities [102]. Exposure to this compound may lead to narcosis, nervousness, indigestion, excessive fatigue, loss of appetite, polyneuritis, sleepiness and emotional disturbances [421]. Mental disturbances and motor nerve damage may also occur [186,401]. Other symptoms may include irritation of the skin, nose and throat, difficult breathing, skin and eye burns, garlicky breath, abdominal pain, weak pulse, palpitations, fatigue, vertigo, mania, hallucinations of sight, hearing, taste and smell, central nervous system depression, respiratory paralysis, coma and convulsions [371]. Exposure to this compound may also result in damage to the peripheral nerves and the hemopoietic system, congestion and edema of the gastrointestinal tract, changes in the brain and spinal cord, cerebrovascular changes, blurred vision, muscle spasms, psychotic behavior, reddening, cracking and peeling of the skin, cyanosis, respiratory depression, bizarre sensations in the extremities, sensory loss, muscular weakness, partial blindness, Parkinsonian tremor, vascularization of the retina, dilation of retinal arterioles, blanching of the optic disk and diminished or lost corneal and pupillary reflexes [301]. Other symptoms may include disturbance of peripheral nerve function, emotional disorders, psychosis and Parkinsonism [295]. In addition, bronchitis, emphysema, anger, suicidal tendencies, nightmares, impotency, chronic gastritis, impairment of endocrine activity and abnormal erythrocytic development with hypochromic anemia may occur.

Carbon tetrachloride

CAS NUMBER: 56-23-5

SYNONYMS: Carbon chloride Methane tetrachloride Perchloromethane Tetrachloromethane

CHEMICAL FORMULA: CCl4

MOLECULAR WEIGHT: 153.82

WLN: GXGGG

PHYSICAL DESCRIPTION: Clear, heavy, colorless liquid

SPECIFIC GRAVITY: 1.5940 @ 20/4 C [017,047]

DENSITY: 1.597 g/mL @ 20 C [043,051]

MELTING POINT: -23 C [047,055,172,275]

BOILING POINT: 76.7 C [031,055,205,395]

SOLUBILITY: Water: <1 mg/mL @ 21 C [700] DMSO: >=100 mg/mL @ 21 C [700] 95% Ethanol: >=100 mg/mL @ 21 C [700] Acetone: >=100 mg/mL @ 21 C [700]

OTHER SOLVENTS: Chloroform: Miscible [031,062,295,395] Carbon disulfide: Miscible [031,173] Petroleum ether: Miscible [031,173] Solvent naphtha: Miscible [062,395] Ether: Miscible [031,062,295,455] Benzene: Miscible [031,062,173,430] Most organic solvents: Miscible [172,421] Oils: Miscible [031,173] Dehydrated alcohol: Miscible [031,173,295,455] Light petroleum: Miscible [295] Absolute ethanol: Soluble [052] Fat solvents: Miscible [395]

OTHER PHYSICAL DATA: Ethereal odor [043,051,058,451] Refractive index: 1.4607 @ 20 C [062,172,395] 1.4631 @ 15 C [029] Specific gravity: 1.585 @ 25/4 C [062,395,430] 1.632 @ 0/4 C [029] Specific gravity: 1.589 @ 25/25 C [031] Vapor pressure: 40 mm Hg @ 4.3 C; 60 mm Hg @ 12.3 C; 200 mm Hg @ 38.3 C [038] Vapor pressure: 56 mm Hg @ 10 C; 137 mm Hg @ 30 C [055] Vapor pressure: 100

mm Hg @ 23 C [038,043,051] Critical temperature: 283 C [058] Critical pressure: 45 atmospheres [058] Evaporation rate (butyl acetate=1): 12.8 [421] Freezes to trimorphous solid having melting points of -28.6 C, -23.8 C and -21.2 C [029] Burning taste [295,455] Odor threshold (lower): 21.4 ppm [051]

HAP WEIGHTING FACTOR: 1 [713]

VOLATILITY:

Vapor pressure: 91 mm Hg @ 20 C [107,301,421] 113 mm Hg @ 25 C [055,430]

Vapor density: 5.3 [051,058,371,421]

FIRE HAZARD: Carbon tetrachloride is nonflammable [031, 107, 295, 451].

LEL: Not available UEL: Not available

REACTIVITY: Carbon tetrachloride can react violently with allyl alcohol, triethylaluminum sesquichlorides, bromine trifluoride, lithium, disilane, potassium-tert-butoxide, liquid oxygen, sodium-potassium alloy, (silver perchlorate + hydrochloric acid), tetrasilane, trisilane, uranium, zirconium and burning wax [043,051,451]. It can also react violently with aluminum trichloride, chlorine trifluoride, dimethylformamide, 1,2,3,4,5,6-hexachlorocyclohexane and dinitrogen tetraoxide [043]. It may react violently with fluorine, aluminum, barium, beryllium, potassium, sodium and zinc [036,043,066]. It is reduced by acid [395]. A cobalt/molybdenum-alumina catalyst will generate a substantial exotherm on contact with its vapor in the presence of air. It is incompatible with triethyl dialuminum trichloride, borane, carbaboranes or their derivatives, calcium hypochlorite and 1,11-diamino-3,6,9-triazaundecane [066]. Magnesium and zinc may react with a mixture of this compound and methanol [043, 066]. It is also incompatible with other chemically active metals such as magnesium [058,107,346,421]. It may react with alkali metals and oxidizing agents [058,269]. It reacts slowly with copper and lead. It reacts, sometimes explosively, with aluminum alloys. It forms telomers with ethylene and vinyl compounds under pressure in the presence of a peroxide inhibitor [395]. When mixed with ethylene and initiated by dibenzoyl peroxide, it may explode. In the presence of iron, and at temperatures well below 100 C, a potentially dangerous reaction with dimethylformamide may occur [036,066]. It may explode with calcium disilicide [036,043,066]. It may attack some forms of plastics, rubber and coatings [053,421]. It ignites in oxygen at 100 C [066]. Violent reactions may result when mixed with decaborane or diborane. A potentially explosive reaction may occur with dimethylacetamide when iron is present as a catalyst [043]. It reacts with plutonium [451].

STABILITY: Carbon tetrachloride is sensitive to light [053,269,295,455] and it is also sensitive to heat and moist air [058]. It is slowly decomposed by light and various metals if moisture is present [053,295]. It is decomposed by water at high temperatures [172,173]. Solutions of carbon tetrachloride in water, DMSO, 95% ethanol or acetone should be stable for 24 hours under normal lab conditions [700].

USES: Carbon tetrachloride has been used in fire extinguishers, in refrigerants, in mixtures with potent fumigants to reduce the fire hazard, to render benzene nonflammable and separate xylene isomers as components to reduce flammability. It is also used as a metal degreaser, as an agricultural fumigant and as a solvent for lacquers, varnishes, waxes, resins, fats, oils, rubber, organic compounds and rubber cement. It is used as a dry cleaning solvent, in cable manufacture, in the production of semiconductors, in blowing agents, in fluorocarbon propellants and chlorofluoromethanes. In veterinary medicine, it is used as an anthelmintic and to treat liver fluke infections in sheep. It is also used as an azeotropic drying agent for wet spark plugs in automobiles, to extract oil from flowers and seeds, as an extractant and intermediate in many industrial processes, in polymer technology as a reaction medium, catalyst and chain transfer agent, as a solvent for resins and in organic synthesis for chlorination of organic compounds used in soap perfumery and insecticidal industries.

COMMENTS: Carbon tetrachloride is not recommended for extinguishing fires, as phosgene is liable to be formed [036]. It has been banned for household use by the FDA [043]. Humans appear to be unusually susceptible to poisoning by this compound [151].

NIOSH Registry Number: FG4900000

ACUTE/CHRONIC HAZARDS: Carbon tetrachloride may be toxic by ingestion, inhalation or skin absorption [031,036,062,451]. It is an irritant and may cause lacrimation [043,051]. It may also be narcotic [043,051,062,102]. When heated to decomposition it emits irritating fumes and toxic fumes of chlorine, carbon monoxide, carbon dioxide, hydrogen chloride and phosgene [043,058,395,451], and it may also emit other hydrocarbon products [051].

SYMPTOMS: Symptoms of exposure may include headache, mental confusion, central nervous system depression, fatigue, anorexia, nausea, vomiting, coma, abdominal cramps, dizziness, unconsciousness, weakness, amnesia, paresthesia, tremors, jaundice, and liver and kidney damage [107, 301]. It may also cause depression, loss of appetite, bronchitis, internal irritation, stupor and damage to the heart and nervous system [036]. Skin contact may remove the natural lipid cover of the skin [031,346]. It may also lead to a dry, scaly, fissured dermatitis. Other symptoms may include gastrointestinal disturbances, abdominal pain, diarrhea, enlarged and tender liver, toxic hepatitis, diminished urinary volume, red and white blood cells in the urine and albuminuria [186,346]. It may also cause narcosis, lung damage, acute nephrosis of the kidney, polyneuritis, narrowing of visual fields and other neurological changes, cirrhosis of the liver, lacrimation, burning of the eyes, malaise, dark urine, renal casts, uremia, epigastric distress, visual disturbances (such as blind spots, spots before the eyes, visual "haze" and restriction of the visual fields) and death [043,051]. Exposure to this compound depresses and injures almost all cells of the body, including the central nervous system, liver, kidney and blood vessels. Depression of the heart muscle may result in ventricular arrhythmias. Damage to the kidneys may result in marked edema and fatty degeneration of the tubules.

Other symptoms may include slowed respiration, slowed or irregular pulse, fall of blood pressure, sudden weight gain, azotemia, anemia, blurred vision and loss of peripheral color vision [301]. It may also cause drowsiness, giddiness, oliguria, cellular necrosis of the liver, acute nephritis and aplastic anemia [295]. Eye and skin irritation, dyspnea, hematemesis, hematuria, proteinuria, weight loss, cyanosis and miosis have also been reported [107]. Ingestion of this compound with alcohol will intensify the effects of the chemical [031,051,186,295]. Other symptoms may include hepatomegaly, optic atrophy, optic neuritis and pulmonary edema [051]. It may also cause a permanent reduction in vision, deafness and retrobulbar neuritis [099]. Mucous membrane irritation and anesthesia have also been reported [172]. Hepatic nodular hyperplasia may also occur [186].

Additional symptoms may include sleepiness, increased peristalsis, erythema, gastroenteritis and death from ventricular fibrillation [173]. It may cause disorientation [058]. Alveolitis has occurred [395]. It may cause irritation of the nose and throat, a sense of fullness in the head, convulsions, hepatic steatosis, hypertension, acidosis and sudden death from depression of vital medullary centers [406]. Other symptoms may include flatulence, fatty liver, elevated SGOT and elevated serum bilirubin [421]. It may also cause incoordination, vertigo and increased nitrogen retention [102]. Other symptoms are pupillary constriction, unspecified respiratory system and gastrointestinal system effects, somnolence, severe gastrointestinal upset and liver enlargement [043].

Carbonyl sulfide

CAS NUMBER: 463-58-1

SYNONYMS: Carbon oxysulfide Oxycarbon sulfide

CHEMICAL FORMULA: COS

MOLECULAR WEIGHT: 60.07 [703]

PHYSICAL DESCRIPTION: Gas or liquid [703]

MELTING POINT: -138 C [703]

BOILING POINT: 49.9 C [703]

SOLUBILITY: Water: 1000 mL/L [702]

OTHER PHYSICAL DATA: Vapor Density: 2.1 [703] Heat of Combustion: 130.5 kcal/mole [702]

HAP WEIGHTING FACTOR: 1 [713]

VOLATILITY:

Vapor Density: 2.1 [703]

FIRE HAZARD: Flash point: Not available

A very dangerous fire hazard and moderate explosion hazard when exposed to heat or flame. [703] To fight fire, stop flow of gas or use CO2 dry chemical or water spray. [703]

Autoignition temperature: Not available

LEL: 12% [703] UEL: 28.5% [703]

REACTIVITY: Can react vigorously with oxidizing materials. [703]

STABILITY: When heated to decomposition it emits toxic fumes of CO. Moderate explosion hazard when exposed to heat or flame. [703]

USES: Not available

COMMENTS: None

NIOSH Registry Number: FG6400000

ACUTE/CHRONIC HAZARDS: Carbonyl sulfide is a poison by intraperitoneal route. It is mildly toxic by inhalation; it is an irritant. It is a narcotic in high concentration. Carbonyl sulfide may liberate highly toxic hydrogen sulfide upon decomposition. When heated to decomposition it emits toxic fumes of CO. [703]

SYMPTOMS: Not available

Catechol

CAS NUMBER: 120-80-9

SYNONYMS: Pyrocatechol o-Benzenediol 1,2-Benzenediol o-Dihydroxybenzene 1,2-Dihydroxybenzene o-Dioxybenzene o-Diphenol o-Hydroxyphenol 2-Hydroxyphenol Oxyphenic acid

CHEMICAL FORMULA: C6H6O2

MOLECULAR WEIGHT: 110.11

WLN: QR BQ

PHYSICAL DESCRIPTION: Colorless crystals

SPECIFIC GRAVITY: 1.371 @ 15/4 C [055,395]

DENSITY: 1.1493 g/mL @ 21 C [017,047]

MELTING POINT: 105 C [025,031,047,055]

BOILING POINT: 245 C (sublimes) [062]

SOLUBILITY: Water: >=100 mg/mL @ 21.5 C [700] DMSO: >=100 mg/mL @ 21.5 C [700] 95% Ethanol: >=100 mg/mL @ 21.5 C [700] Acetone: >=100 mg/mL @ 21.5 C [700]

OTHER SOLVENTS: Chloroform: Soluble [031,042,062,205] Pyridine: Soluble [062] Aqueous alkalies: Soluble [062] Ether: Soluble [017,031,047,205] Benzene: Soluble [031,042,062,205] Carbon tetrachloride: Soluble [395]

OTHER PHYSICAL DATA: Sublimes [025,031,062,395] Boiling point: 221.5 C @ 400 mm Hg; 197.7 C @ 200 mm Hg [031] Steam volatile Refractive index (20 C): 1.604 pKa1: 9.12, pKa2: 12.08 Lambda max: 214 nm (epsilon = 614.6) in cyclohexane

HAP WEIGHTING FACTOR: 1 [713]

VOLATILITY:

Vapor pressure: 5 mm Hg @ 104 C [038] 10 mm Hg @ 118.3 C [038,042]

Vapor density: 3.79

FIRE HAZARD: Catechol has a flash point of 127 C (261 F) [042,062,430]. It is combustible. Fires involving this material may be controlled with a dry chemical, carbon dioxide or Halon extinguisher.

LEL: Not available UEL: Not available

REACTIVITY: Catechol can react with acid chlorides, acid anhydrides, bases and oxidizing agents [269]. It reacts violently on contact with concentrated nitric acid [025] and it also acts as a reducing agent [395].

STABILITY: Catechol discolors to brown on exposure to air and light, especially when moist [062]. Its aqueous solutions soon turn brown [031].

USES: Catechol is used as an antiseptic, and in photography, dyestuffs, electroplating, specialty inks, antioxidants and light stabilizers, organic synthesis, cosmetics and pharmaceuticals.

COMMENTS: None

NIOSH Registry Number: UX1050000

ACUTE/CHRONIC HAZARDS: Catechol may be toxic, corrosive and a severe irritant. It causes burns and may be readily absorbed through the skin [269,036,025]. When heated to decomposition it emits toxic fumes of carbon monoxide and carbon dioxide [042,269].

SYMPTOMS: Symptoms of dermal exposure may include convulsions, injury to the blood and dermatitis [042]. Depending on the intensity and duration of exposure, effects may vary from mild irritation to severe destruction of tissue [269]. Other symptoms may include central nervous system depression and prolonged rise of blood pressure [430]. It may also cause photosensitivity and stomatitis [301]. Blood dyscrasias has also been described [151]. Inhalation of this compound may cause digestive disturbances including vomiting, difficulty in swallowing, diarrhea and loss of appetite. Ingestion may cause burning pain in the mouth and throat, abdominal pain, headache, dizziness, muscular weakness, irregular breathing, coma and possibly death. This compound may be absorbed through the skin and systemic poisoning effects may follow. Discoloration and severe burns may also occur. Eye contact may cause redness, pain and blurred vision. It may also cause severe damage and blindness [058].

Chloramben

CAS NUMBER: 133-90-4

SYNONYMS: 3-Amino-2,5-dichlorobenzoic acid Amiben

CHEMICAL FORMULA: C7H5Cl2NO2

MOLECULAR WEIGHT: 206.03

WLN: ZR BG EG CVQ

PHYSICAL DESCRIPTION: Purplish white powder

SPECIFIC GRAVITY: Not available

DENSITY: Not available

MELTING POINT: 194-197 C

BOILING POINT: Not available

SOLUBILITY: Water: <0.1 mg/mL @ 22 C [700] DMSO: >=100 mg/mL @ 21 C [700] 95% Ethanol: >=100 mg/mL @ 21 C [700] Acetone: >=100 mg/mL @ 21 C [700]

OTHER SOLVENTS: Methanol: 22.3 g/100 mL Chloroform: 0.09 g/100 mL Benzene: 0.02 g/100 mL

OTHER PHYSICAL DATA: Not available

HAP WEIGHTING FACTOR: 1 [713]

VOLATILITY:

Vapor pressure: 0.007 mm Hg @ 100 C

FIRE HAZARD: Flash point data for this chemical is not available. It is probably combustible. Fires involving this material may be controlled using a dry chemical, carbon dioxide or Halon extinguisher.

LEL: Not available UEL: Not available

REACTIVITY: Chloramben reacts with sodium hypochlorite solutions.

STABILITY: Chloramben is sensitive to prolonged exposure to air and light. Solutions should be stable for 24 hours under normal laboratory conditions [700].

USES: Chloramben is used for a weed control herbicide especially in soybean, navy bean, groundnut, maize, sweet potato, asparagus, squash and pumpkin plantings.

COMMENTS: None

NIOSH Registry Number: DG1925000

ACUTE/CHRONIC HAZARDS: Chloramben emits toxic fumes when heated to decomposition.

SYMPTOMS: Not available

Chlordane

CAS NUMBER: 57-74-9

SYNONYMS: 1,2,4,5,6,7,8,8-octachloro-2,3,3a,4,7,7a-hexahydro-4,7-methano-indene Octachlorodihydrodicyclopentadiene

CHEMICAL FORMULA: C10H6Cl8

MOLECULAR WEIGHT: 409.80

WLN: L C555 A IUTJ AG AG BG DG EG HG IG JG

PHYSICAL DESCRIPTION: Off-white powder (pure); amber solid (impure)

SPECIFIC GRAVITY: 1.57-1.63 @ 15.5/15.5 C [043]

DENSITY: 1.59-1.63 g/mL @ 25 C [033,172,395,430]

MELTING POINT: 106-107 C (cis); 104-105 C (trans) [051,172,173,395]

BOILING POINT: 175 C @ 2 mm Hg (decomposes) [102]

SOLUBILITY: Water: <1 mg/mL @ 23 C [700] DMSO: >=100 mg/mL @ 23 C [700] 95% Ethanol: 50-100 mg/mL @ 23 C [700] Acetone: >=100 mg/mL @ 23 C [700]

OTHER SOLVENTS: Aliphatic hydrocarbon solvents: Miscible [033,169,430] Aromatic hydrocarbon solvents: Miscible [033,169,430] Deodorized kerosene: Miscible [033,062,169,430] Petroleum solvents: Soluble [151] Petroleum hydrocarbons: Soluble [395] Cyclohexanone: Miscible [172] Propan-2-ol: Miscible [172] Trichloroethylene: Miscible [172] Most organic solvents: Soluble [062,173,395,421]

OTHER PHYSICAL DATA: Chlorine-like odor [102,371] Viscosity: 69 poises @ 25 C [033] Refractive index: 1.56-1.57 @ 25 C [062,172,395,430] Liquid surface tension (estimated): 0.025 N/m @ 20 C [051,371] Heat of combustion (estimated): -2200 cal/g [051,371]

HAP WEIGHTING FACTOR: 10 [713]

VOLATILITY: Not available

FIRE HAZARD: Chlordane is noncombustible [102, 421]. However, the technical grade material has a flash point of 56 C (133 F) [051]. Fires involving this material can be controlled with a dry chemical, carbon dioxide or Halon extinguisher.

The autoignition temperature of technical grade material is 210 C (410 F) [051,371].

LEL: 0.7% [051,371] UEL: 5% [051,371]

REACTIVITY: Chlordane is incompatible with strong oxidizers [051,102,346]. It decomposes in weak alkalis [033,062,421,430]. It is also decomposed by sodium in isopropyl alcohol [186]. It is corrosive to iron, zinc and various protective coatings [169] and it attacks some forms of plastics and rubber [102].

STABILITY: Chlordane is stable under normal laboratory conditions. Solutions of it in water, DMSO, 95% ethanol or acetone should be stable for 24 hours under normal lab conditions [700].

USES: Chlordane is used as a non-systemic contact and stomach insecticide with some fumigant action. It is also used as an acaricide, a pesticide and a wood preservative. Chlordane is used in termite control and as a protective treatment for underground cables.

COMMENTS: The EPA has cancelled registrations of pesticides containing Chlordane with the exception of its use through subsurface ground insertion for termite control and the dipping of roots or tops of non-food plants [033].

NIOSH Registry Number: PB9800000

ACUTE/CHRONIC HAZARDS: Chlordane may be toxic by ingestion, inhalation or skin absorption [033, 062]. It is an irritant and may be absorbed through the skin [051, 102, 346, 371]. Effects at higher dosage levels may be cumulative [421]. When heated to decomposition it emits toxic fumes of carbon monoxide, hydrogen chloride gas, chlorine and phosgene [043,051,102,371].

SYMPTOMS: Symptoms of exposure may include skin irritation, irritability, convulsions, deep depression and degenerative changes in the liver [033]. Tremors, excitement, ataxia, gastritis, central nervous system stimulation, respiratory failure, fatty degeneration and death may also occur [043]. Other symptoms may include nervousness, loss of coordination, blood dyscrasias and acute leukemia [395]. Exposure to this compound may lead to diplopia, blurring of vision, twitching of the extremities, nausea and vomiting [099]. Additional symptoms may include headache, dizziness and mild clonic jerking [406]. Symptoms may also include severe cough, dyserythropoiesis, eosinophilia, megaloblastosis, aplastic anemia, neuroblastoma, hyperexcitability of the central nervous system, hyperactive reflexes, muscle twitching, coma, anorexia, weight loss and severe gastroenteritis [151]. Exposure to this compound may also cause confusion, oliguria, proteinuria, hematuria, mild hypertension, albuminuria, paresthesia, unconsciousness, weakness, enteritis, diffuse pneumonia, lower nephron syndrome and lateral nystagmus [173]. Delirium, abdominal pain, diarrhea and anuria may also occur [102,346]. Other symptoms may include scattered petechial hemorrhages in the lungs, kidneys and brain, damage to tubular cells of the kidneys, degenerative changes in hepatic cells and renal tubules, hyperexcitability, central nervous system depression, cognitive and emotional deterioration, impairment of memory and impairment of visual motor coordination [301]. Symptoms may also include giddiness, fatigue, pulmonary edema, liver, kidney and myocardial toxicity, hypothermia, accelerated respiration followed by depressed respiration, loss of appetite, muscular weakness and apprehensive mental state [295]. Symptoms may also include light headedness, tightness of the chest, arthralgias, sore throat, eye irritation, neurological symptoms and dry and red skin [051]. Shaking, staggering and mania may also occur [102]. Symptoms may also include gastrointestinal tract irritation [371].

Chlorine

CAS NUMBER: 7782-50-5

CHEMICAL FORMULA: Cl2

MOLECULAR WEIGHT: 70.90 [703]

PHYSICAL DESCRIPTION: Greenish-yellow gas, liquid, or rhombic crystals [703] pungent, irritating odor [704]

DENSITY (LIQUID): 1.47 @ 0 C (3.65 atm) [703]

MELTING POINT: -101 C [703]

BOILING POINT: -34.5 C [703]

SOLUBILITY: Water: 14.6 g/L @ 0 C [702]

OTHER PHYSICAL DATA: Vapor Pressure: 4800 mm @ 20 C [703] Vapor Density: 2.49 [703] Freezing Point: -150 F [704] Density: 3.214 g/L [706] Specific Gravity: 1.56 (-33.6 C) [706] Refractive Index (gas): 1.000768 [706] Refractive Index (liquid): 1.367 [706]

HAP WEIGHTING FACTOR: 1 [713]

VOLATILITY:

Vapor Pressure: 4800 mm @ 20 C [703]

Vapor Density: 2.49 [703]

FIRE HAZARD: Flash Point: Not applicable

Noncombustible gas, but a strong oxidizer [704].

Autoignition temperature: Not applicable

LEL: Not applicable UEL: Not applicable

REACTIVITY: Chlorine should be kept away from easily oxidized materials. Chlorine reacts with many organic chemicals, sometimes with explosive violence. [714] Chlorine is incompatible with ammonia, acetylene, butadiene, butane, methane, propane (or other petroleum gases), hydrogen, sodium carbide, benzene, finely divided metals, turpentine. [706]

STABILITY: Not available

USES: Chlorine is widely used in making everyday products. It is used for producing safe drinking water the world over. It is also extensively used in the production of paper products, dyestuffs, textiles, petroleum products, medicines, antiseptics, insecticides, foodstuffs, solvents, paints, plastics, and many other consumer products. Most of the chlorine produced is used in the manufacture of chlorinated compounds for sanitation, pulp bleaching, disinfectants, and textile processing. Further use is in the manufacture of chlorates, chloroform, carbon tetrachloride, and in the extraction of bromine. Organic chemistry demands much from chlorine, both as an oxidizing agent and in substitution, since it often brings desired properties in an organic compound when substituted for hydrogen, as in one form of synthetic rubber. [706]

COMMENTS: None

NIOSH Registry Number: FO2100000

ACUTE/CHRONIC HAZARDS: Moderately toxic and very irritating to humans by inhalation; human respiratory system effects include changes in trachea or bronchi, emphysema, chronic pulmonary edema or congestion. Strong irritant to eyes, mucous membranes and respiratory tract; extremely irritating to mucous membranes of the eyes @ 3 ppm. Combines with moisture to liberate O2 and forms HCl which in quantity cause inflammation of the tissues contacted. Detectable odor at concentration of 3.5 ppm; 15 ppm causes immediate irritation of the throat; 50 ppm are dangerous and 1000 ppm may be fatal, even when exposure is brief. Human mutagenic data exists. [703]

SYMPTOMS: Symptoms of exposure to chlorine may include a burning of the eyes, nose, and mouth; lacrimation, rhinorrhea; cough, choking, substernal pain; nausea, vomiting; headache, dizziness; syncope; pulmonary edema; pneumonia; hypoxemia; dermatitis; eye and skin burns. [704] Because of its intensely irritating properties, severe industrial exposure is rare, as the worker attempts to leave the exposure area. On continued exposure, initial irritation of the eyes and mucous membranes is followed by coughing, a feeling of suffocation, and later, pain and a feeling of constriction in the chest. In severe exposure, pulmonary edema may follow with rales. [703]

Chloroacetic acid

CAS NUMBER: 79-11-8

SYNONYMS: Monochloroacetic acid Chloroethanoic acid

CHEMICAL FORMULA: C2H3ClO2

MOLECULAR WEIGHT: 94.50

WLN: QV1G

PHYSICAL DESCRIPTION: Colorless crystals

SPECIFIC GRAVITY: Not available

DENSITY: 1.58 g/mL @ 20 C

MELTING POINT: 61-63 C

BOILING POINT: 187-189 C

SOLUBILITY: Water: >=100 mg/mL @ 20 C [700] DMSO: >=100 mg/mL @ 20 C [700] 95% Ethanol: >=100 mg/mL @ 20 C [700] Acetone: >=100 mg/mL @ 20 C [700]

OTHER SOLVENTS: Chloroform: Soluble Carbon disulfide: Soluble Ether: Soluble Benzene: Soluble

OTHER PHYSICAL DATA: Occurs in three crystalline forms Melting point (alpha-form): 63 C Melting point (beta-form): 56.2 C Melting point (gamma-form): 52.5 C Deliquescent

HAP WEIGHTING FACTOR: 1 [713]

VOLATILITY:

Vapor pressure: 1 mm Hg @ 43 C

Vapor density: 3.26

FIRE HAZARD: Chloroacetic acid has a flash point of 126 C (259 F). It is combustible. Fires involving this material may be controlled with a dry chemical, carbon dioxide or Halon extinguisher.

LEL: Not available UEL: Not available

REACTIVITY: Chloroacetic acid will corrode common metals.

STABILITY: Chloroacetic acid is stable under normal laboratory conditions.

USES: Chloroacetic acid is used as an herbicide, preservative, bacteriostat, and intermediate in the production of carboxymethylcellulose, ethyl chloroacetate, glycine, synthetic caffeine, sarcosine, thioglycolic acid, EDTA, 2,4-D, and 2,4,5-T.

COMMENTS: None

NIOSH Registry Number: AF8575000

ACUTE/CHRONIC HAZARDS: Chloroacetic acid may be toxic and corrosive and will cause severe irritation and burns of tissue. Toxic decomposition products may evolve in a fire.

SYMPTOMS: Inhalation may cause irritation of the mucous membranes. Contact with the liquid can cause severe irritation and burns of tissue. Ingestion can cause burns of the mouth, throat and stomach.

2-Chloroacetophenone

CAS NUMBER: 532-27-4

SYNONYMS: alpha-Chloroacetophenone omega-Chloroacetophenone Phenacyl chloride Phenylchloromethylketone

CHEMICAL FORMULA: C8H7ClO

MOLECULAR WEIGHT: 154.60

WLN: GR BV1

PHYSICAL DESCRIPTION: White to light yellow powder

SPECIFIC GRAVITY: 1.32 @ 15/4 C [371,421]

DENSITY: Not available

MELTING POINT: 56.5 C [017]

BOILING POINT: 227-228 C [017]

SOLUBILITY: Water: <1 mg/mL @ 19 C [700] DMSO: <1 mg/mL @ 19 C [700] 95% Ethanol: <1 mg/mL @ 19 C [700] Acetone: <1 mg/mL @ 19 C [700] Toluene: <1 mg/mL @ 18 C [700]

OTHER SOLVENTS: Carbon disulfide: Soluble [062,421] Ether: Soluble [017] Benzene: Soluble [062,421]

OTHER PHYSICAL DATA: Refractive Index: 1.685 @ 25 C; 1.5438 @ 20 C Saturated Concentrations: 0.034 g/m3 @ 20 C; 0.11 g/m3 @ 30 C Heat of Combustion: -5,190 cal/g Sharp, floral odor

HAP WEIGHTING FACTOR: 1 [713]

VOLATILITY:

Vapor pressure: 0.0054 mm Hg @ 20 C [031]

Vapor density: 5.2 [042]

FIRE HAZARD: 2-Chloroacetophenone has a flash point of 118 C (244 F) [371,421]. It is combustible. Fires involving this material may be controlled with a dry chemical, carbon dioxide or Halon extinguisher.

LEL: Not available UEL: Not available

REACTIVITY: 2-Chloroacetophenone is incompatible with bases, amines and alcohols [269]. It is also incompatible with water and steam [369,346]. When mixed with water, HCl is produced, although this reaction is slow and generally not hazardous [371]. It reacts slowly with metals, causing mild corrosions [355].

STABILITY: 2-Chloroacetophenone is sensitive to moisture [274]. Gas chromatography stability screenings indicate that solutions of it in acetone are stable for at least 24 hours [700].

USES: 2-Chloroacetophenone is used as a chemical warfare agent with lacrimatory properties (it is referred to as CN in military circles). It is also the principal ingredient of the riot control gas MACE, also called "Chemical Mace". It is also used as an intermediate for the synthesis of pharmaceuticals and other chemicals.

COMMENTS: The name MACE is derived from Methyl chloroACEtophenone.

NIOSH Registry Number: AM6300000

ACUTE/CHRONIC HAZARDS: 2-Chloroacetophenone is a lachrymator [031,062,269,274]. It also may be a strong eye, throat and skin irritant. It may cause corneal damage [031]. High concentrations are extremely destructive to tissues [062,269] and when heated to decomposition, it emits toxic fumes [071].

SYMPTOMS: Exposure causes irritation of the skin and mucous membranes [301]. It may cause tingling in the nose, with rhinorrhea; and may result in development of a skin rash [421]. It may also cause laryngitis, headache and vomiting [269,301]. Other symptoms may include: coughing, wheezing and nausea [269]. There may also be a burning sensation and shortness of breath [269, 371]. Possible systemic manifestations may include agitation, coma, contraction of pupils and loss of reflexes [371]. More severe exposure produces pulmonary edema [371,421]. Higher concentrations may also cause pigmentation and second degree burns [301]. There may also be denaturing reactions on the eyes, such as corneal burns [099,301].

Chlorobenzene

CAS NUMBER: 108-90-7

SYNONYMS: Benzene chloride Chlorbenzol Phenyl chloride

CHEMICAL FORMULA: C_6H_5Cl

MOLECULAR WEIGHT: 112.56

WLN: GR

PHYSICAL DESCRIPTION: Clear, colorless liquid

SPECIFIC GRAVITY: 1.107 @ 20/4 C [031,269,275,421]

DENSITY: Not available

MELTING POINT: -45 C [031,043,055,269]

BOILING POINT: 132 C [017,036,047,269]

SOLUBILITY: Water: <1 mg/mL @ 20 C [700] DMSO: Reacts [043] 95% Ethanol: >=100 mg/mL @ 20 C [700] Acetone: >=100 mg/mL @ 20 C [700]

OTHER SOLVENTS: Chloroform: Freely soluble [031,205,421] Ether: Freely soluble [031,205,421] Benzene: Freely soluble [031,205,421] Alcohol: Freely soluble [031,205,421]

OTHER PHYSICAL DATA: Refractive index: 1.5241 @ 20 C [017,047,269,275] Faint, not unpleasant, almond-like odor [036] Solidifies @ -55 C [031] Odor threshold: 0.21 ppm [371] Specific gravity: 1.113 @ 15.5/15.5 C [043] 1.105 @ 25/25 C [062] Boiling point: 22 C @ 10 mm Hg [017,047] 45 C @ 30 mm Hg [025]

HAP WEIGHTING FACTOR: 1 [713]

VOLATILITY:

Vapor pressure: 8.8 mm Hg @ 20 C [055,102] 11.8 mm Hg @ 25 C [055,058]

Vapor density: 3.88 [043,055]

FIRE HAZARD: The flash point for chlorobenzene is 23 C (75 F) [205,269] and it is flammable. Fires involving chlorobenzene may be controlled with a dry chemical, carbon dioxide or Halon extinguisher. A water spray can also be used [058,371].

The autoignition temperature for chlorobenzene is 637 C (1180 F) [062,102].

LEL: 1.3% [043,058,102,451] UEL: 7.1% [043,058,102]

REACTIVITY: A potentially explosive reaction may occur with powdered sodium or phosphorus trichloride + sodium. A violent reaction may occur with dimethyl sulfoxide. It reacts vigorously with oxidizers [043]. It will attack some forms of plastic, rubber and coatings [102]. Silver perchlorate will form a solvated salt with this compound that is shock sensitive [058].

STABILITY: Chlorobenzene is stable under normal laboratory conditions. Solutions of chlorobenzene in 95% ethanol should be stable for 24 hours under normal lab conditions [700].

USES: Chlorobenzene is used in the manufacture of phenol, chloronitrobenzene, aniline and DDT. It is a solvent for paints. It is also used as a heat transfer medium, pesticide intermediate and a solvent carrier for methylene diisocyanate.

COMMENTS: None

NIOSH Registry Number: CZ0175000

ACUTE/CHRONIC HAZARDS: Chlorobenzene may be harmful if swallowed, inhaled or absorbed through skin and it is a strong narcotic [043]. The vapor or mist may also be an irritant [269]. When heated to decomposition this compound emits toxic fumes of hydrogen chloride gas, carbon monoxide and carbon dioxide [269].

SYMPTOMS: Symptoms of exposure may include nausea, dizziness, headache, liver damage and kidney damage [269]. Other symptoms may include central nervous system depression, skin irritation, defatting, dermatitis, skin burns, drowsiness, cyanosis, spastic contractions of extremities and loss of consciousness [058]. It may also cause pallor and collapse [151]. Incoordination and lung damage may also occur [102]. Exposure may cause incoherence [346]. Coughing may result [371].

Chlorobenzilate

CAS NUMBER: 510-15-6

SYNONYMS: 4,4'-Dichlorobenzilic acid ethyl ester Ethyl p,p'-Dichlorobenzilate Ethyl 4,4'-Dichlorobenzilate

CHEMICAL FORMULA: $C_{16}H_{14}Cl_2O_3$

MOLECULAR WEIGHT: 325.20

WLN: GR DXQR DG&VO2

PHYSICAL DESCRIPTION: Viscous yellow liquid or pale yellow crystals

SPECIFIC GRAVITY: Not available

DENSITY: 1.2816 g/mL @ 20 C [169]

MELTING POINT: 35-37 C [071,395]

BOILING POINT: 156-158 C @ 0.07 mm Hg [169,172,395]

SOLUBILITY: Water: <0.1 mg/mL @ 22 C [700] DMSO: >=100 mg/mL @ 23 C [700] 95% Ethanol: >=100 mg/mL @ 23 C [700] Acetone: >=100 mg/mL @ 23 C [700] Methanol: 1000 g/kg @ 20 C [169,172] Toluene: 1000 g/kg @ 20 C [169,172]

OTHER SOLVENTS: Dichloromethane: 1000 g/kg @ 20 C [169,172] Most organic solvents: Soluble [031,169,395] Benzene: Soluble [062] Absolute ethanol: Soluble

[052] Hexane: 600 g/kg @ 20 C [169,172] Octan-1-ol: 700 g/kg @ 20 C [172] Petroleum oils: Soluble [395]

OTHER PHYSICAL DATA: Refractive index: 1.5727 @ 20 C [031,169,395] Boiling point: 141-142 C @ 0.06 mm Hg [062,071,395] Specific gravity (technical grade): 1.2816 @ 20/4 C [062,071,395]

HAP WEIGHTING FACTOR: 1 [713]

VOLATILITY: Not available

FIRE HAZARD: Flash point data for chlorobenzilate are not available. It is probably combustible. Fires involving this material may be controlled with a dry chemical, carbon dioxide or Halon extinguisher.

LEL: Not available UEL: Not available

REACTIVITY: Chlorobenzilate is hydrolyzed by alkalis and strong acids [051,071,169,395]. It is incompatible with lime [031].

STABILITY: Chlorobenzilate is stable under normal laboratory conditions. Solutions of it in water, DMSO, 95% ethanol or acetone should be stable for 24 hours under normal lab conditions [700].

USES: Chlorobenzilate is used as an insecticide, miticide, nonsystemic acaricide and synergist for DDT. It is used to control many species of phytophagous mites on pome fruit, citrus fruit, vines, soybeans, cotton, teat and vegetables. It is also used to control bee mites in beehives.

COMMENTS: The use of chlorobenzilate has been restricted in the United States because it is oncogenic in rats and mice and causes adverse testicular effects in male rats after repeated exposures [151].

NIOSH Registry Number: DD2275000

ACUTE/CHRONIC HAZARDS: Chlorobenzilate may be toxic by inhalation [025]. It may be absorbed through the respiratory tract, gastrointestinal tract and skin [295]. It is an irritant and when heated to decomposition it emits toxic fumes of chlorides [043].

SYMPTOMS: Symptoms of exposure may include central nervous system stimulation, vomiting, diarrhea, paresthesia, excitement, giddiness, fatigue, tremors, convulsions, pulmonary edema, hypothermia, headache, loss of appetite, muscular weakness, apprehensive mental state, myocardial toxicity, impotence, infertility and coma [295]. It may also cause hyperexcitability, narcosis, central nervous system depression, testicular damage kidney [062] damage and liver damage [071]. Other symptoms may include muscle pains, ataxia, mild delirium and fever [395]. It may also cause skin irritation [051,071].

Chloroform

CAS NUMBER: 67-66-3

SYNONYMS: Formyl trichloride Methane trichloride Trichloromethane Methenyl trichloride Methyl trichloride Trichloroform

CHEMICAL FORMULA: CHCl3

MOLECULAR WEIGHT: 119.38

WLN: GYGG

PHYSICAL DESCRIPTION: Clear colorless liquid

SPECIFIC GRAVITY: 1.4832 @ 20/4 C [017,047]

DENSITY: 1.49845 g/mL @ 15 C [043,051,430]

MELTING POINT: -63.5 C [017,038,043,430]

BOILING POINT: 61.2 C [062,395,421,451]

SOLUBILITY: Water: <1 mg/mL @ 19 C [700] DMSO: >=100 mg/mL @ 19 C [700] 95% Ethanol: >=100 mg/mL @ 19 C [700] Acetone: >=100 mg/mL @ 19 C [700]

OTHER SOLVENTS: Carbon tetrachloride: Miscible [031,062] Carbon disulfide: Miscible [031,062] Solvent naphtha: Miscible [395] Ether: Miscible [031,295,395,455] Benzene: Miscible [031,062,395] Petroleum ether: Miscible [031] Ligroin: >10% [047] Ethanol: Miscible [395] Alcohol: Miscible [031,295,421,455] Oils: Miscible [031,395,421] Light petroleum: Miscible [295] Most organic solvents: Miscible [295,395,421,455]

OTHER PHYSICAL DATA: Sweet, burning taste [295,395,455] Heavy, ethereal odor [043,051,371] Forms a constant boiling mixture with 7% alcohol (boiling @ 59 C) [031] Refractive index: 1.4459 @ 20 C [017,047,395] 1.4422 @ 25 C [062,395] Refractive index: 1.4486 @ 15 C [205] Specific gravity: 1.485 @ 20/20 C [062,395]

1.481 @ 25/4 C [029] Density: 1.474-1.479 g/mL @ 20 C [295,455] Vapor pressure: 100 mm Hg @ 10.4 C [038,043,051] 246 mm Hg @ 30 C [395] Vapor pressure: 60 mm Hg @ 0.5 C; 400 mm Hg @ 42.7 C; 760 mm Hg @ 61.3 C [038] Odor threshold: 0.3 mg/m3 [421] Evaporation rate (butyl acetate = 1): 11.6 [102] Evaporation rate (ether = 1): 0.56 [058] UV max (in water): 400-290 nm, 270 nm, 255 nm, 245 nm (A = 0.01, 0.02, 0.15, 1.0) The addition of a small percentage of alcohol greatly retards the gradual oxidation which occurs when it is exposed to air and light [295,455] Critical temperature: 263.2 C [371] Critical pressure: 54 atmospheres [371] Liquid surface tension: 27.1 dynes/cm [371] Latent heat of vaporization: 59.3 cal/g [371] Heat of fusion: 17.62 cal/g [371] Spectroscopy data: Lambda (@ vaporization point): <200 nm [395]

HAP WEIGHTING FACTOR: 1 [713]

VOLATILITY:

Vapor pressure: 160 mm Hg @ 20 C [055,058,102] 200 mm Hg @ 25.9 C [038,395]

Vapor density: 4.12 [043,051,055]

FIRE HAZARD: Chloroform is nonflammable [031,058, 275,451].

LEL: Not available UEL: Not available

REACTIVITY: Chloroform reacts violently with acetone in the presence of KOH or Ca(OH)2. It may react explosively with fluorine, N2O4, sodium and (sodium + methanol) [036,043,066]. It may also react explosively with aluminum, lithium and sodium methoxide [036,043,051]. Explosive reactions also occur with (sodium hydroxide + methanol) [036,066,451]. It is incompatible with bases [058,269,346,395]. Gentle heating of a 1:1 solution with bis(dimethylamino)dimethyl stannane led to a mild explosion. Mixtures with nitromethane are detonable [066]. It undergoes an explosive interaction with potassium. Contact with potassium tert-butoxide has caused ignition. It may react explosively with (sodium methoxide + methanol). It reacts vigorously with triisopropylphosphine [043,066]. It reacts violently with (acetone + alkali (e.g. NaOH)), disilane, magnesium, (perchloric acid + phosphorous pentoxide) and NaK [043,051]. It also reacts violently with (methanol + alkali) [043]. It is incompatible with oxidizers [051,058,395]. It reacts readily with halogens or halogenating agents, primary amines and phenols [051,395]. It reacts violently with (KOH + methanol). It will corrode iron and certain other metals in contact with water. In the presence of strong alcohols and water, it may become violently explosive [051]. It will attack some forms of plastics, rubber and coatings [102].

STABILITY: When chloroform is exposed to air and light, gradual oxidation occurs [029,295,395,455]. Solutions of chloroform in water, DMSO, 95% ethanol or acetone should be stable for 24 hours if protected from light and air [700].

USES: Chloroform is used in the manufacture of fluorocarbon plastics, resins, refrigerants and propellants. It is used as a solvent for fats, oils, plastics, dyes, rubber, alkaloids, waxes, gutta-percha, resins and dry cleaning. It is also an extraction solvent in the manufacture of flavors and sterols. Other uses include fumigants, insecticides, manufacture of anesthetics and pharmaceuticals, primary source for chlorodifluoromethane, sweetener, manufacture of fire extinguishers, electronic circuitry manufacture and analytical chemistry. It is used as a cleansing agent, analgesic, muscle relaxant, carminative, flavoring agent, preservative, bactericide, heat transfer medium, and as a counter-irritant in liniments (has rubefacient action). It is also used in the rubber industry, in veterinary medicine as an antispasmodic, in the extraction and purification of penicillin and other pharmaceuticals, in the manufacture of artificial silk and plastics and in the sterilization of catgut.

COMMENTS: Chloroform has been prohibited by the FDA from use in drugs, cosmetics and food packaging (including cough medicines, toothpastes, etc.) [062]. It is classified as a carcinogen by EPA.

NIOSH Registry Number: FS9100000

ACUTE/CHRONIC HAZARDS: Chloroform may be toxic by ingestion or inhalation [036,062]. It may also be harmful by skin absorption [269]. It is an irritant and when heated to decomposition it emits toxic fumes of hydrogen chloride, carbon monoxide, chlorine and phosgene [043,058,269,395].

SYMPTOMS: Symptoms of exposure may include nausea, vomiting, eye and skin irritation, unconsciousness and death [043,058,102]. Other symptoms may include drowsiness, giddiness, headache, anesthesia and conjunctivitis [036]. Central nervous system depression may occur [102,151,301,406]. Respiratory failure may also occur [301]. Exposure may cause lassitude, digestive disturbances, dizziness and mental dullness [102,346]. It may also cause coma and an enlarged liver [346]. Other symptoms may include hepatotoxicity, nephrotoxicity, reduced cardiac output, cardiac arrhythmias and abdominal pain [295]. It may also cause cardiac sensitization, gastrointestinal upset, sensation of fainting, salivation, intracranial pressure and fatigue [430]. Severe anoxia has been reported [455]. Exposure may lead to chest pain, serious disorientation, hepatomegaly and, in high concentrations, blepharospasm [102]. It may also lead to ventricular tachycardia, bradycardia, necrosis and hepatomas of the liver, cardiovascular depression and ventricular fibrillation [421]. Irritation of the upper respiratory tract, nervous system disturbances and heart damage may occur [269]. Renal damage may also occur [395]. Nervous aberration and profound toxemia may follow exposure [051].

Inhalation may cause hallucinations, distorted perceptions, unspecified gastrointestinal effects, dilation of the pupils, reduced reaction to light, reduced intraocular pressure, feeling of warmth of the face and body, irritation of the mucous membranes, excitation, loss of reflexes and loss of sensation. Prolonged inhalation may cause paralysis followed by cardiac respiratory failure [043]. Other symptoms of exposure via inhalation may include irritation of the nose and throat, drunkenness and narcosis [371]. Inhalation of large doses may result in hypotension, respiratory depression and myocardial depression [031].

Inhalation or ingestion of this compound may cause cardiac irregularities, cardiac arrest, convulsions, reduced blood pressure and uncontrollable hyperthermia (rare) [301]. Ingestion may cause burning of the throat and mouth [058,102].

Skin contact may result in smarting and reddening [371]. Prolonged skin contact may result in burns [295,346,455]. Prolonged or repeated skin contact may cause dermatitis through defatting of the skin [058].

Eye contact may cause immediate burning pain, tearing, reddening of the conjunctiva and injury of the corneal epithelium [099].

Chronic exposure may result in jaundice, cirrhosis and damage of the liver and kidney damage [301]. Alcoholics seem to be affected sooner and more severely [346].

---

bis(Chloromethyl) ether

CAS NUMBER: 542-88-1

SYNONYMS: BCME Chloro(chloromethoxy)methane sym-Dichlorodimethyl ether

CHEMICAL FORMULA: C2H4Cl2O

MOLECULAR WEIGHT: 114.97 [701]

PHYSICAL DESCRIPTION: Colorless liquid (Merck, 1976) with extremely suffocating odor (Sittig,, 1981) [701]

SPECIFIC GRAVITY: 1.315 @ 20 C (Merck, 1976) [701]

MELTING POINT: -42.7 F, -41.5 C (Weast, 1979) [701]

BOILING POINT: 223 F, 106 C (Merck, 1976) [701]

SOLUBILITY: Water: Not soluble; decomposes (Weast, 1979, p. C-300) [701]

OTHER PHYSICAL DATA: Vapor Pressure: 30 mmHg @ 22 C (Callahan, 1979) [701] Vapor Density: 4.0 [703] Boiling Point: 105 C [703] Freezing Point: -43 F [704]

HAP WEIGHTING FACTOR: 1,000 [713]

VOLATILITY:

Vapor Pressure: 30 mm @ 72 F [704]

Vapor Density: 4.0 [703]

FIRE HAZARD: Flash Point: < 19 C [703]

A dangerous fire hazard [703]. Class IA Flammable Liquid [704].

Autoignition temperature: Not available

LEL: Not available UEL: Not available

REACTIVITY: Reacts with water to form hydrochloric acid and formaldehyde. Incompatible or reacts with acids. [704]

STABILITY: Unstable in moist air; decomposed by water into HCl and formaldehyde [715]. When heated to decomposition it emits very toxic fumes of Cl- [703].

USES: Not available

COMMENTS: None

NIOSH Registry Number: KN1575000

ACUTE/CHRONIC HAZARD:

A poison by inhalation, ingestion, and skin contact; a human and experimental carcinogen, neoplastigen and tumorigen. Human systemic effects by inhalation include irritation of the conjunctiva and unspecified nasal and respiratory effects. Human mutagenic data exists. When decomposed it emits very toxic fumes of Cl-. [703]

SYMPTOMS: Symptoms of exposure to bis(Chloromethyl) ether may include irritation of eyes, skin, and mucous membranes of the respiratory system; pulmonary congestion, edema; corneal damage, necrosis; reduced pulmonary function, cough, dyspnea, wheezing; blood-stained sputum; bronchial secretions. [704]

Chloromethyl methyl ether

CAS NUMBER: 107-30-2

SYNONYMS: Dimethylchloro ether Methylchloromethyl ether

CHEMICAL FORMULA: C2H5ClO

MOLECULAR WEIGHT: 80.52 [701]

PHYSICAL DESCRIPTION: Colorless liquid with an irritating odor (IARC, 1972-1985, CHRIS, 1978) [701]

SPECIFIC GRAVITY: 1.0605 at 20/4C (IARC, 1972-1985) [701]

MELTING POINT: -154.3 F, -103.5 C (Encyc. Occupat. Health and Safety, 1971) [701]

BOILING POINT: 138 F, 59 C (IARC, 1972-1985) [701]

SOLUBILITY: Water: Decomposes (Weast, 1979, p. C-300) [701]

OTHER PHYSICAL DATA: Freezing Point: -154 F [704]

HAP WEIGHTING FACTOR: 10 [713]

VOLATILITY:

Vapor Pressure: Not available

Vapor Density: Not available

FIRE HAZARD: Flash Point: < 73.4 F [703]

A very dangerous fire hazard when exposed to heat or flame [703]. Class IV flammable liquid [704].

Autoignition temperature: Not available

LEL: Not available UEL: Not available

REACTIVITY: Reaction with divalent metals forms a very reactive product [703]. Reacts with water to form hydrochloric acid and formaldehyde [704].

STABILITY: When heated to decomposition it emits toxic fumes of Cl- [703].

UDRI Thermal Stability Class: 5 [702] UDRI Thermal Stability Ranking: 218 [702]

USES: Used in synthesis of chloromethylated compounds [715].

COMMENTS: None

NIOSH Registry Number: KN6650000

ACUTE/CHRONIC HAZARDS: Chloromethyl methyl ether is a poison by inhalation and moderately toxic by ingestion. It is a suspected human carcinogen, an experimental carcinogen, tumorigen, and neoplastigen. Human mutagenic data exists. When heated to decomposition it emits toxic fumes of Cl-. [703] Targeted organs include the respiratory system, skin, eyes and mucous membranes. [704]

SYMPTOMS: Symptoms of exposure may include irritation of eyes, skin, mucous membranes; pulmonary edema, pulmonary congestion, pneumonia; burns, necrosis; coughing, wheezing, pulmonary congestion; blood-stained sputum; weight loss; bronchial secretions. [704]

Chloroprene

CAS NUMBER: 126-99-8

SYNONYMS: 2-Chloro-1,3-butadiene beta-Chloroprene

CHEMICAL FORMULA: C_4H_5Cl

MOLECULAR WEIGHT: 88.54 [703]

PHYSICAL DESCRIPTION: Colorless liquid with a pungent, ether-like odor [704]

SPECIFIC GRAVITY: 0.96 @ 20C/4C [704]

MELTING POINT: Not available

BOILING POINT: 59.4 C [703]

SOLUBILITY: Water: Slightly soluble [703] Ether: Soluble [706] Acetone: Soluble [706] Benzene: Soluble [706]

OTHER PHYSICAL DATA: Vapor density: 3.0 [703] Freezing Point: -153 F [704] Polymerizes at room temperature unless inhibited with antioxidants [704]

Polymerizes to Neoprene

HAP WEIGHTING FACTOR: 10 [713]

VOLATILITY:

Vapor Pressure: 188 mm [704]

Vapor Density: 3.0 [703]

FIRE HAZARD: Flash Point: -4 F [703]

A very dangerous fire hazard when exposed to heat or flame. Explosive in the form of vapor when exposed to heat or flame. To fight fire, use alcohol foam. [703] Class IV flammable liquid [704].

Autoignition temperature: Not available

LEL: 4.0% [703] UEL: 20% [703]

REACTIVITY: Incompatible with liquid or gaseous fluorine [703]. Incompatible or reacts with peroxides and other oxidizers (Note: polymerizes at room temperature unless inhibited with antidioxants) [704].

STABILITY: Autooxidizes in air to form an unstable peroxide which catalyzes exothermic polymerization of the monomer. When heated to decomposition it emits toxic fumes of Cl-. [703]

USES: Neoprene is an oil-resistant synthetic rubber made by the polymerization of chloroprene. Chloroprene is polymerized to neoprene, and a number of different neoprene polymers are available including the sulfur-modified polymers (GN, GNR, GRT), and unmodified types (W and WRT). Neoprene can also be vulcanized. [711]

COMMENTS: None

NIOSH Registry Number: EI9625000

ACUTE/CHRONIC HAZARDS: A poison by ingestion, intravenous, and subcutaneous routes; moderately toxic by inhalation. An experimental teratogen; produces experimental reproductive effects; human mutagenic data exists. [703]

SYMPTOMS: Symptoms of exposure to chloroprene may include irritation of the eyes and respiratory system; nervousness, irritability; dermatitis; and alopecia. [704] Human exposure also caused corneal necrosis and anemia. Exposure to the vapor can lead to asphyxia. Other effects are central nervous system depression, drop in blood pressure, and degenerative changes in vital organs. [703]

Chromium compounds

NOTE: For all listings which contain the word "compounds," the following applies: Unless otherwise specified, these listings are defined as including any unique chemical substance that contains the named chemical as part of that chemical's infrastructure. [708] Information for the metallic form of chromium is given below. Other compounds described are intended as selected examples of chromium compounds and do not comprise a comprehensive list.

Chromium

CAS NUMBER: 7440-47-3

CHEMICAL FORMULA: Cr

MOLECULAR WEIGHT: 52.00

WLN: CR

PHYSICAL DESCRIPTION: Very hard metal, cubic steel, gray crystals

SPECIFIC GRAVITY: 7.2

DENSITY: Not available

MELTING POINT: 1900 C

BOILING POINT: 2642 C

SOLUBILITY: Water: Insoluble DMSO: Insoluble 95% Ethanol: Insoluble Acetone: Insoluble

OTHER SOLVENTS: Dilute sulfric acid: Soluble Dilute hydrochloric acid: Soluble

OTHER PHYSICAL DATA: Not available

VOLATILITY:

Vapor pressure: 1 mm Hg @ 1616 C

FIRE HAZARD: Chromium metal is not flammable.

LEL: Not available UEL: Not available

REACTIVITY: Chromium reacts violently with ammonium nitrate, dinitrogen dioxide, nitrogen oxide, lithium, potassium chlorate and sulfur dioxide.

STABILITY: Chromium metal is stable under normal ambient storage conditions.

USES: It is used for chrome plating of other metals, in the manufacture of alloys, and an isotope is used for a blood disease tracer, in nuclear and high-temperature research.

COMMENTS: None

NIOSH Registry Number: GB4200000

ACUTE/CHRONIC HAZARDS: Chromium is a human poison by ingestion with gastrointestinal effects. It is an experimental tumorigen and suspected carcinogen. [703]

SYMPTOMS: Symptoms of exposure to chromium metal via inhalation may include histologic fibrosis of the lungs. [704] Symptoms of poisoning by ingestion are not available.

---

t-Butyl chromate

CAS NUMBER: 1189-85-1

SYNONYMS: bis(1,1-Dimethylethyl) ester chromic acid

CHEMICAL FORMULA: $C8H18CrO4$

MOLECULAR WEIGHT: 230.22

WLN: Not available

PHYSICAL DESCRIPTION: Liquid

SPECIFIC GRAVITY: Not available

DENSITY: Not available

MELTING POINT: -5 to 0 C

BOILING POINT: Not available

SOLUBILITY: Water: Not available DMSO: Not available 95% Ethanol: Not available Acetone: Not available

OTHER SOLVENTS: Not available

OTHER PHYSICAL DATA: Red crystals (in petroleum ether)

VOLATILITY: Not available

FIRE HAZARD: t-Butyl chromate is flammable and fires involving it may be controlled with a dry chemical, carbon dioxide or Halon extinguisher.

LEL: Not available UEL: Not available

REACTIVITY: t-Butyl chromate is hydrolyzed by water.

STABILITY: t-Butyl chromate is unstable except in solution.

USES: t-Butyl chromate is used as an oxidizing agent, as a catalyst for alkene polymerization, and as a corrosion inhibitor.

COMMENTS: None

NIOSH Registry Number: GB2900000

ACUTE/CHRONIC HAZARDS: t-Butyl chromate may be highly toxic and emit acrid smoke and irritating fumes when heated to decomposition.

SYMPTOMS: Symptoms of exposure may include skin burns and necrotic skin ulcers, drowsiness and narcosis, nausea, vomiting, diarrhea, corrosion of mucous membranes, allergic reactions, lung damage, liver damage, kidney damage, and increased bleeding tendency.

Calcium chromate, anhydrous

CAS NUMBER: 13765-19-0

SYNONYMS: Chromic acid, calcium salt (1:1) Calcium chromium oxide (CaCrO4) Calcium monochromate

CHEMICAL FORMULA: CaCrO4

MOLECULAR WEIGHT: 156.08

WLN: .CA..CR-Q2-O2

PHYSICAL DESCRIPTION: Yellow crystals

SPECIFIC GRAVITY: Not available

DENSITY: 2.89 g/mL

MELTING POINT: Not available

BOILING POINT: Not available

SOLUBILITY: Water: <0.1 mg/mL @ 22 C [700] DMSO: <1 mg/mL @ 22 C [700] 95% Ethanol: <1 mg/mL @ 22 C [700] Acetone: <1 mg/mL @ 19.5 C [700]

OTHER SOLVENTS: Dilute acids: Soluble

OTHER PHYSICAL DATA: Not available

VOLATILITY: Not available

FIRE HAZARD: Flash point data for calcium chromate are unavailable but it is probably not flammable.

LEL: Not available UEL: Not available

REACTIVITY: Calcium chromate is a strong oxidizer and reacts with acids and ethanol.

STABILITY: Calcium chromate is stable under normal laboratory conditions. Solutions of it should be stable for 24 hours under normal lab conditions [700].

USES: Calcium chromate is used as a pigment and corrosion inhibitor, in the manufacture of metallic chromium, for oxidizing reactions and for battery depolarization.

COMMENTS: None

NIOSH Registry Number: GB2750000

ACUTE/CHRONIC HAZARDS: Calcium chromate, anhydrous may be toxic, corrosive to and may be absorbed by the skin and mucous membranes. When heated to decomposition it emits toxic fumes.

SYMPTOMS: Symptoms of exposure may include respiratory irritation, cough, dyspnea, wheezing, chest pain, irritation, ulceration and perforation of the nasal septum; jaundice, kidney damage, conjunctivitis, penetrating ulcers of the skin, sensitization dermatitis, and signs of lung cancer.

Calcium chromate dihydrate

CAS NUMBER: 8012-75-7

SYNONYMS: Chromic acid, calcium salt (1:1), dihydrate

CHEMICAL FORMULA: CaCrO4*2H2O

MOLECULAR WEIGHT: 192.12

WLN: CA CR-O2-Q2 QH2

PHYSICAL DESCRIPTION: Yellow powder

SPECIFIC GRAVITY: Not available

DENSITY: Not available

MELTING POINT: 200 C (Loses 2H2O)

BOILING POINT: Not available

SOLUBILITY: Water: Soluble DMSO: Soluble 95% Ethanol: Soluble Acetone: Insoluble

OTHER SOLVENTS: Dilute acids: Soluble

OTHER PHYSICAL DATA: Not available

VOLATILITY: Not available

FIRE HAZARD: Not available

LEL: Not available UEL: Not available

REACTIVITY: Not available

STABILITY: Not available

USES: Not available

COMMENTS: None

NIOSH Registry Number: GB2800000

ACUTE/CHRONIC HAZARDS: Calcium chromate dihydrate may be toxic, an irritant and there is evidence that it is an animal positive carcinogen.

SYMPTOMS: Symptoms of exposure may include respiratory irritation; irritation of eyes, skin and mucous membranes, ulceration, and perforation of the nasal septum; jaundice; kidney damage; conjunctivitis from exposure to chromate dust; penetrating skin ulcers; sensitization dermatitis;and signs of lung cancer.

Chromium carbonyl

CAS NUMBER: 13007-92-6

SYNONYMS: Chromium hexacarbonyl Hexacarbonyl chromium

CHEMICAL FORMULA: C6CrO6

MOLECULAR WEIGHT: 220.06

WLN: Not available

PHYSICAL DESCRIPTION: White crystals

SPECIFIC GRAVITY: 1.770 @ 18/4 C

DENSITY: Not available

MELTING POINT: 150-151 C (decomposes)

BOILING POINT: 210 C (violent decomposition)

SOLUBILITY: Water: <1 mg/mL @ 20 C [700] DMSO: <1 mg/mL @ 20 C [700] 95% Ethanol: <1 mg/mL @ 20 C [700] Acetone: 1-5 mg/mL @ 20 C [700] Methanol: Insoluble

OTHER SOLVENTS: Diethyl ether: Insoluble Chloroform: Slightly soluble Nonpolar organic solvents: Slightly soluble Iodoform: Slightly soluble Carbon tetrachloride: Slightly soluble Acetic acid: Insoluble Ether: Insoluble

OTHER PHYSICAL DATA: Sublimes @ room temperature; sinters @ 90 C; decomposes @ 130 C; explodes @ 210 C Burns with a luminous flame Odorless Resists decomposition by bromine, iodine, water or cold concentrated nitric acid

VOLATILITY:

Vapor pressure: 1 mm Hg @ 36 C, 760 mm Hg @ 151 C [038]

FIRE HAZARD: Flash point data for chromium carbonyl are not available. It may be combustible. Fires involving this material may be controlled with a dry chemical, carbon dioxide or Halon extinguisher.

LEL: Not available UEL: Not available

REACTIVITY: Chromium carbonyl is decomposed by chlorine and fuming nitric acid [053,062]. It is incompatible with oxidizing agents [269].

STABILITY: UV spectrophotometric stability screening indicates that solutions of chromium carbonyl are stable for less than four hours in water but are stable for at least 24 hours in 95% ethanol [700]. Solutions of it decompose photochemically when exposed to light [033,053]. It also decomposes violently at 210 C [016,062,066,269].

USES: Chromium carbonyl is used in catalysts for olefin polymerization and isomerization, as a gasoline additive to increase octane number and in the preparation of chromous oxide (CrO). It is also used as a catalyst for hydrogenation, isomerization, water gas shift reaction and alkylation of aromatic hydrocarbons. It is used as an intermediate and in the synthesis of "sandwich" compounds from aromatic hydrocarbons.

COMMENTS: None

NIOSH Registry Number: GB5075000

ACUTE/CHRONIC HAZARDS: When heated to decomposition chromium carbonyl emits toxic fumes. This compound is an irritant and it may be fatal if inhaled, swallowed or absorbed through the skin [269].

SYMPTOMS: Symptoms of exposure to this type of compound upon ingestion may include dizziness, intense thirst, abdominal pain, vomiting and shock. Skin contact with this type of compound may result in incapacitating eczematous dermatitis with edema and ulceration. Inhalation of this type of compound may cause painless ulceration, bleeding and perforation of the nasal septum accompanied by a foul nasal discharge, conjunctivitis, lacrimation, acute hepatitis with jaundice, nausea, vomiting, loss of appetite and an enlarged, tender liver [053,301]. Ingestion of this type of compound may also lead to oliguria or anuria and death from uremia [301].

Chromium trioxide

CAS NUMBER: 1333-82-0

SYNONYMS: Chromic acid, solid Chromic acid, solution Chromic anyhdride Chromic trioxide Chromic VI acid Chromium oxide Chromium (6+) trioxide Chromium VI oxide

CHEMICAL FORMULA: CrO_3

MOLECULAR WEIGHT: 100.00

WLN: Not available

PHYSICAL DESCRIPTION: Dark, red crystals, flakes or powder

SPECIFIC GRAVITY: Not available

DENSITY: 2.70 g/mL

MELTING POINT: 196 C

BOILING POINT: 250 C (decomposes)

SOLUBILITY: Water: Very soluble DMSO: Not available 95% Ethanol: Soluble Acetone: Not available

OTHER SOLVENTS: Sulfuric acid: Soluble Nitric acid: Soluble

OTHER PHYSICAL DATA: Not available

VOLATILITY: Not available

FIRE HAZARD: Flash point data for chromium trioxide are not available. It is probably combustible. Fires involving this material may be controlled with a dry chemical, carbon dioxide or Halon extinguisher.

LEL: Not available UEL: Not available

REACTIVITY: Chromium trioxide is a powerful oxidizer and reacts with most organic substances in a violent and possibly explosive manner. Pharmaceutical incompatibilities include: alcohol, ether, glycerol, spirit nitrous ether; bromides; chlorides; iodides; hypophosphites, sulfites and sulfides. Violent or explosive reactions may occur with acetic acid, acetic anhydride, phosphorous and selenium. Incandescence may result when chromium trioxide is placed in contact with potassium, sodium, ammonia, butyric acid, hydrogen sulfide. It also may ignite acetone, methanol, ethanol, propan-2-ol, butanol, cyclohexanol, N,N-dimethylformamide, pyridine or sulfur.

STABILITY: Chromium trioxide is sensitive to moisture.

USES: Chromium trioxide is used in chromium plating: copper stripping; aluminum anodizing; corrosion inhibitor; photography; purifying oil and acetylene; hardening microscopical preparations; and also as an oxidant in organic chemistry.

COMMENTS: None

NIOSH Registry Number: GB6650000

ACUTE/CHRONIC HAZARDS: Chromium trioxide may be toxic and a severe irritant. There is evidence it may be carcinogenic.

SYMPTOMS: Dermal contact can cause primary irritation as well as ulceration and allergic eczema. Inhalation can cause nasal irritation and septal perforation. Pulmonary irritation, bronchogenic carcinoma may result from inhalation of chromate dust. Ingestion causes violent gastrointestinal irritation with vomiting and diarrhea.

---

bis(Cyclopentadienyl) chromium

CAS NUMBER: 1271-24-5

SYNONYMS: Chromocene

CHEMICAL FORMULA: C10H10Cr

MOLECULAR WEIGHT: 182.19

WLN: Not available

PHYSICAL DESCRIPTION: Red needles

SPECIFIC GRAVITY: Not available

DENSITY: Not available

MELTING POINT: 173 C

BOILING POINT: 75-90 C under vacuum (sublimes)

SOLUBILITY: Water: Decomposes DMSO: <1 mg/mL @ 22 C [700] 95% Ethanol: <1 mg/mL @ 22 C [700] Acetone: <1 mg/mL @ 22 C [700]

OTHER SOLVENTS: Carbon tetrachloride: Decomposes Carbon disulfide: Decomposes

OTHER PHYSICAL DATA: Paramagnetic

VOLATILITY: Not available

FIRE HAZARD: bis(Cyclopentadienyl)chromium is flammable. Fires involving this compound may be controlled using a dry chemical, carbon dioxide or Halon extinguisher.

LEL: Not available UEL: Not available

REACTIVITY: bis(Cyclopentadienyl)chromium reacts with water and alcohols.

STABILITY: Thin layer chromatography stability screening indicates that solutions of this compound in DMSO are stable for half an hour [700]. UV spectrophotometric stability screening indicates that solutions in water are stable for less than half an hour [700]. Fresh solutions should be prepared before each use. bis(cyclopentadienyl)chromium is sensitive to exposure to air and moisture.

USES: It is used as a catalyst, in polymer synthesis, as an ultraviolet light absorber, as a reducing agent, a free radical scavenger, and as an antiknock agent.

COMMENTS: None

NIOSH Registry Number: Not available

ACUTE/CHRONIC HAZARDS: Not available

SYMPTOMS: Not available

Potassium dichromate

CAS NUMBER: 7778-50-9

SYNONYMS: Potassium dichromate (VI)

CHEMICAL FORMULA: K2[OCrO2OCrO2O]

MOLECULAR WEIGHT: 294.20

WLN: Not available

PHYSICAL DESCRIPTION: Bright orange-red prismatic crystals

SPECIFIC GRAVITY: 2.676 @ 25/4 C

DENSITY: 2.68 g/mL

MELTING POINT: 398 C

BOILING POINT: Decomposes @ 500 C

SOLUBILITY: Water: 10-50 mg/mL @ 20 C [700] DMSO: <1 mg/mL @ 20 C [700] 95% Ethanol: <1 mg/mL @ 20 C [700] Acetone: <1 mg/mL @ 20 C [700]

OTHER SOLVENTS: Acids: Soluble

OTHER PHYSICAL DATA: Changes from triclinic to monoclinic prisms at 241.6 C 1% aqueous solution: pH=4.04 10% aqueous solution: pH=3.57 Bitter, metallic taste Odorless Heat of fusion: 29.8 cal/g Heat of solution: -62.5 cal/g Specific heat: 0.186 @ 16-98 C

VOLATILITY: Not available

FIRE HAZARD: Potassium dichromate is nonflammable.

LEL: Not available UEL: Not available

REACTIVITY: Potassium dichromate is a strong oxidizer. It is incompatible with acetone, sulfuric acid, hydrazine, hydroxylamine, slaked lime, mercury cyanides, reducing agents and many organic materials.

STABILITY: Potassium dichromate is stable under normal laboratory conditions. Solutions of it should be stable for 24 hours under normal laboratory conditions [700].

USES: Potassium dichromate is used for tanning leather, dyeing, painting, decorating porcelain, printing, photolithography, pigment-prints, staining wood, pyrotechnics, safety matches, and for blending palm oil, wax, & sponges. It is also used for waterproofing fabrics, as oxidizer in manufacture of organic compounds, in electric batteries, and as a corrosion inhibitor.

COMMENTS: Potassium dichromate usually prepared by the action of potassium chloride on sodium dichromate.

NIOSH Registry Number: HX7680000

ACUTE/CHRONIC HAZARDS: Potassium dichromate may be toxic if inhaled, swallowed or absorbed through the skin. It is corrosive and a severe irritant. When heated to decomposition it emits toxic fumes.

SYMPTOMS: Symptoms of exposure may include rash, ulcers, burning sensation, coughing, wheezing, laryngitis, shortness of breath, headache, nausea, vomiting and possible allergic reactions. Ingestion may cause violent gastroenteritis, peripheral vascular collapse, vertigo, muscle cramps, coma, and toxic nephritis with glycosuria. Inhalation may be fatal as a result of spasm, inflammation, edema or the larynx and bronchi, chemical pneumonitis and pulmonary edema.

Cobalt compounds

NOTE: For all listings which contain the word "compounds," the following applies: Unless otherwise specified, these listings are defined as including any unique chemical substance that contains the named chemical as part of that chemical's infrastructure. [708] Information for the metallic form of cobalt is given below. Other compounds described are intended as selected examples of cobalt compounds and do not comprise a comprehensive list.

Cobalt

CAS NUMBER: 7440-48-4

CHEMICAL FORMULA: Co

MOLECULAR WEIGHT: 58.93

WLN: -CO-

PHYSICAL DESCRIPTION: Silver-gray solid

SPECIFIC GRAVITY: 8.9 @ 20 C

DENSITY: 8.92 g/mL

MELTING POINT: 1495 C

BOILING POINT: 2870 C

SOLUBILITY: Water: <1 mg/mL @ 19 C [700] DMSO: <1 mg/mL @ 19 C [700] 95% Ethanol: <1 mg/mL @ 19 C [700] Acetone: <1 mg/mL @ 19 C [700] Toluene: <1 mg/mL @ 20 C [700]

OTHER SOLVENTS: Nitric acid: Readily soluble Acid: Soluble

OTHER PHYSICAL DATA: Hard, magnetic, ductile, somewhat malleable metal Exists in 2 allotropic forms, @ room temperature the hexagonal form is more stable than the cubic form

VOLATILITY:

Vapor pressure: 0.0 mm Hg @ 20 C

FIRE HAZARD: Literature sources indicate that the dust of cobalt is flammable. Fires involving this material may be controlled with a dry chemical, carbon dioxide or Halon extinguisher. Dry sand, dry dolomite or dry graphite powder may also be used.

LEL: Not available UEL: Not available

REACTIVITY: Cobalt metal readily resists oxidation but when finely divided it is pyrophoric in air. It will react with acetylene, hydrazinium nitrate, oxidizing agents and 1,3,4,7-tetramethylisoindole. Cobalt reacts with acids, but becomes passive in concentrated nitric acid.

STABILITY: Cobalt metal is stable under normal laboratory conditions.

USES: Cobalt is used in chemical manufacturing, electroplating, ceramics, lamp filaments, as a trace element in fertilizers, in glass, and as a dryer in printing inks, paints and varnishes. Its principal use is in alloys, especially, steels for permanent and soft magnets and high speed tool steels and jet engines. It is also used as a catalyst in sulfur synthesis, as a coordination and complexing agent, in the treatment of cyanide poisoning, as a pigment in enamels and glazes, and in photographic and electrical industries. Its radioactive isotope is used in medicine for diagnostic aid, biological and medical research, radiation therapy and cancer treatment.

COMMENTS: Cobalt is widely distributed in nature; its abundance in earth's crust is .001%-.002%. Cobalt appears to be essential to life, and is contained in Vitamin B12.

NIOSH Registry Number: GF8750000

ACUTE/CHRONIC HAZARDS: Cobalt may be a local irritant.

SYMPTOMS: Exposure to this element may result in wheezing, coughing, shortness of breath, and burning in the mouth, throat or chest develop.

Cobaltocene

CAS NUMBER: 1277-43-6

SYNONYMS: bis(Cyclopentadienyl) cobalt

CHEMICAL FORMULA: C10H10Co

MOLECULAR WEIGHT: 189.13

WLN: L50J 0-Co-- 0L50J

PHYSICAL DESCRIPTION: Black-purple crystals

SPECIFIC GRAVITY: Not available

DENSITY: Not available

MELTING POINT: 173-174 C [025,234]

BOILING POINT: Not available

SOLUBILITY: Water: <1 mg/mL @ 21.5 C [700] DMSO: 5-10 mg/mL @ 21.5 C [700] 95% Ethanol: <1 mg/mL @ 21.5 C [700] Acetone: 50-100 mg/mL @ 21.5 C [700]

OTHER SOLVENTS: Hydrocarbons: Soluble [062]

OTHER PHYSICAL DATA: Paramagnetic Sublimes @ 40 C and 0.1 mm Hg [025,234]

VOLATILITY: Not available

FIRE HAZARD: Cobaltocene is flammable [900]. Fires involving this material can be controlled with a dry chemical, carbon dioxide or Halon extinguisher.

LEL: Not available UEL: Not available

REACTIVITY: Cobaltocene is a highly reactive compound which is readily oxidized by water and dilute acids [062].

STABILITY: Cobaltocene is sensitive to air and is readily oxidized [025,234]. It is also sensitive to exposure to light [900]. UV spectrophotometric stability screening indicates that solutions of this compound in water are stable for less than 30 minutes and that solutions of this compound in 95% ethanol are stable for at least 2 hours [700].

USES: Cobaltocene is used to catalyze synthesis of pyridines from alkynes and nitriles; as a polymerization inhibitor of olefins up to 200 C; in Diels-Alder reactions; and as a paint dryer and oxygen stripping agent.

COMMENTS: It is a metallocene.

NIOSH Registry Number: GG0350000

ACUTE/CHRONIC HAZARDS: Cobaltocene may be toxic by ingestion [062].

SYMPTOMS: Not available

Cobalt sulfate heptahydrate

CAS NUMBER: 10026-24-1

SYNONYMS: Cobalt (II) sulfate (1:1) heptahydrate Cobalt monosulfate heptahydrate Cobalt monosulphate heptahydrate Cobaltous sulfate heptahydrate Cobaltous sulphate heptahydrate Sulfuric acid, cobalt (2+) salt (1:1), heptahydrate

CHEMICAL FORMULA: CoO4S*7H2O

MOLECULAR WEIGHT: 281.13

WLN: Not available

PHYSICAL DESCRIPTION: Pink to red monoclinic, prismatic crystals

SPECIFIC GRAVITY: 2.03 @ 25/4 C [031]

DENSITY: 1.948 [062,205] g/mL

MELTING POINT: 96.8 C [017,058,205]

BOILING POINT: 735 C (decomposes) [058]

SOLUBILITY: Water: >=100 mg/mL @ 18 C [700] DMSO: 10-50 mg/mL @ 18 C [700] 95% Ethanol: <1 mg/mL @ 18 C [700] Acetone: <1 mg/mL @ 18 C [700] Methanol: 54.5 g/mL [017]

OTHER SOLVENTS: Not available

OTHER PHYSICAL DATA: Odorless [058,371] On heating, it dehydrates @ 41.5 C and @ 71 C [031] Becomes anhydrous @ 420 C [017,058] Refractive index: 1.477, 1.483, 1.489 [205] Specific gravity: 1.948 @ 25/25 C [017]

VOLATILITY: Not available

FIRE HAZARD: Cobalt sulfate heptahydrate is nonflammable [371].

LEL: Not available UEL: Not available

REACTIVITY: Not available

STABILITY: Cobalt sulfate heptahydrate is hygroscopic [269,275]. Solutions of Cobalt sulfate heptahydrate in water, DMSO, 95% ethanol or acetone should be stable for 24 hours under normal lab conditions [700].

USES: Cobalt sulfate heptahydrate is used in storage batteries, cobalt electroplating baths, drying lithographic inks, varnishes, ceramics, enamels, and glazes to prevent discoloring. It is also used in cobalt pigments for decorating porcelain.

COMMENTS: None

NIOSH Registry Number: GG3200000

ACUTE/CHRONIC HAZARDS: Cobalt sulfate heptahydrate may be toxic and an irritant and it may also be readily absorbed through the skin [269]. When heated to decomposition it emits toxic fumes of sulfur oxides [058,269] and cobalt oxides [371].

SYMPTOMS: Symptoms of exposure via inhalation may include coughing and shortness of breath [058,371]. Inhalation may cause irritation of the respiratory tract and mucous membranes [058,269]. It may also cause permanent disability [371]. Ingestion of this compound may result in nausea, vomiting, abdominal pain, vasodilatation, mild hypotension, rash and ringing in the ears [058]. It can cause irritation, redness and pain on skin or eye contact [058]. Skin contact may also result in allergic reactions. Other symptoms may include headache and nervous system disturbances [269]. Repeated ingestion may produce a goiter and reduce thyroid activity. Prolonged or repeated skin contact may cause dermatitis. Lung disease sometimes occurs on inhalation of this type of compound [058].

Coke oven emissions

CAS NUMBER: 8007-45-2 (10/01/91) [701]

NOTE: Emissions from coking operations cover a large spectrum from smoke and particulate matter to carbon monoxide, carbon dioxide, hydrocarbons, hydrogen sulfide, and phenols. Coke plants sometimes have by-product processing systems which recover a range of organic compounds including pyridine and quinoline bases, phenols, creosote, tars containing polycyclic aromatic hydrocarbons (PAH's), and tar oils containing light aromatics.

The compounds described here are intended as selected examples of coke oven related emissions and do not comprise a comprehensive list.

Ammonia

CAS NUMBER: 7664-41-7

SYNONYMS: Ammonia anhydrous

CHEMICAL FORMULA: NH3

MOLECULAR WEIGHT: 17.04 [703]

PHYSICAL DESCRIPTION: Colorless gas, extremely pungent odor, liquefied by compression [703]

DENSITY: 0.771 g/L @ 0 C, 0.817 g/L @ -79 C [703]

MELTING POINT: -77.7 C [703]

BOILING POINT: -33.35 C [703]

SOLUBILITY: Water: Very soluble [703] Alcohol: Moderately soluble [703]

OTHER PHYSICAL DATA: Vapor Pressure: 10 atm @ 25.7 C [703] Vapor Density: 0.6 [703] Freezing Point: -108 F [704]

VOLATILITY:

Vapor Pressure: 10 atm @ 25.7 C [703]

Vapor Density: 0.6 [703]

FIRE HAZARD: Flash Point: Not applicable (gas) [704]

Difficult to ignite. Explosion hazard when exposed to flame or in a fire. NH3 + air in a fire can detonate. [703]

Autoignition temperature: 1204 F [703]

LEL: 16% [703] UEL: 25% [703]

REACTIVITY: Anhydrous ammonia is not compatible with mercury, chlorine, chlorine dioxide, calcium hypochlorite, or bromine. Anhydrous hydrofluoric acid and iodine are each incompatible with ammonia in either its aqueous or its anhydrous form. Ammonia and its aqueous solutions are potentially explosive with chlorine, bromine, or iodine. [706] Ammonia reacts violently or forms explosive products with all four halogens and some of the interhalogens. It also reacts violently or explosively with boron halides, chloroformamidnium nitrate, ethylene oxide, magnesium perchlorate, nitrogen trichloride, oxygen and platinum, strong oxidants, heavy metals and their compounds, tellurium halides and pentaborane(9), and chlorine azides. In addition, liquid ammonia reacts violently with 1,2-dichloroethane. Ammonia also reacts with a mixture of iodine and potassium (it is unclear if the reference means potassium iodide). [710] Ammonia is also reported to form sensitive explosive mixtures with air and many organic compounds such as hydrocarbons; chlorodinitrobenzenes; and stibine. [703] Ammonia is incompatible with acetaldehyde; acrolein; boron; boron triiodide; perchloric acid; chlorites; chlorosilane; ethylene dichloride and liquid ammonia mixtures; hexachloromelamine; hydrazine and alkali metals; potassium ferricyanide; potassium mercuric cyanide; tellurium hydropentachloride; trichloromelamine; tetramethylammonium amide; and thiotrihiazylchloride. [703] Ammonia used to be used as a refrigerataion gas but violent gas-air explosions were common so it is not generally used for this purpose any more. Bretherick reports that ammoniacal silver nitrate (Gomari solutions) have exploded when disturbed after they have been stored and once it was reported to explode violently when disturbed by removing a glass rod. [710]

STABILITY: Emits toxic fumes of NH3 and NOx when exposed to heat. [703]

USES: The uses for ammonia are varied. The greatest use is in the manufacture of fertilizers or as a direct application for fertilizer. [711] Ammonia is used in the manufacture of ammonium nitrate, urea, ammonium sulfate, nitric acid, ammonium phosphates, and other industrial and agricultural products.

COMMENTS: None

NIOSH Registry Number: BO0875000

ACUTE/CHRONIC HAZARDS: Ammonia is a human poison by an unspecified route and a poison experimentally by inhalation, ingestion, and possibly other routes. It is an eye, mucous membrane, and systemic irritant by inhalation. Mutagenic data exists. It emits toxic fumes of NH3 and NOx when exposed to heat. [703] Target organs include the eyes and respiratory system. [704]

SYMPTOMS: Symptoms of exposure to ammonia may include eye, nose and throat irritation; dyspnea, bronchospasm, chest pain; pulmonary edema; pink, frothy sputum; skin burns, and vesiculation. [704]

Hydrogen sulfide

CAS NUMBER: 7783-06-4

SYNONYMS: Sulfur hydride

CHEMICAL FORMULA: H2S

MOLECULAR WEIGHT: 34.08 [703]

PHYSICAL DESCRIPTION: Colorless, flammable gas; offensive odor [703]

DENSITY: 1.539 @ g/L @ O C [703]

MELTING POINT: -85.5 C [703]

BOILING POINT: -60.4 C [703]

SOLUBILITY: Water: 0.4% [704]

OTHER PHYSICAL DATA: Vapor Pressure: 20 atm @ 25.5 C [703] Vapor Density: 1.189 [703] Freezing Point: -122 F [704]

VOLATILITY:

Vapor Pressure: 20 atm 25.5 C

Vapor Density: 1.18

FIRE HAZARD: Flash Point: 260 C

Flammable. Fires involving this chemical should be extinguished with water, dry chemical, or halon extinguishers.

Autoignition temperature: 260 C (500 F)

LEL: 4.3% UEL: 46

%REACTIVITY: Hydrogen sulfide reacts violently with Na202, NI3, NF3, OF2, HNO3, PbO2, F2, Cu, CrO3, ClF3, ClO, BrF5, acetaldehyde, Na, hydrated iron oxide and absorbed Oxygen.

STABILITY: Aqueous solutions are not stable.

USES: Hydrogen sulfide is used in the manufacturing of chemicals, in metallurgy; as analytical reagent; purification of hydrochloric and sulfuric acids. It is a source of sulfur.

COMMENTS: None

NIOSH Registry Number: MX1225000

ACUTE/CHRONIC HAZARDS: Hydrogen sulfide is a gas and may be toxic if inhaled. It is an insidious poison, sense of smell may be fatigued and fail to give warning of high concentrations.

SYMPTOMS: Symptoms of exposure to high concentrations may results in collapse, coma and death from respiratory failure after one or two inspirations. Low concentrations may produce irritation of conjunctiva and mucous membranes. Headaches, dizziness, nausea, lassitude also may appear after exposure.

2-Picoline

CAS NUMBER: 109-06-8

SYNONYMS: 2-Methylpyridine

CHEMICAL FORMULA: C6H7N

MOLECULAR WEIGHT: 93.14 [703]

PHYSICAL DESCRIPTION: Colorless liquid; strong unpleasant odor [703]

DENSITY: 0.9443 @ 20C/4C [706]

MELTING POINT: -66.8 C [706]

BOILING POINT: 128.8 C [706]

SOLUBILITY: Water: Very soluble [703] Alcohol: Miscible [703] Ether: Miscible [703]

OTHER PHYSICAL DATA: Vapor Pressure: 10 mm @ 24.4 C [703] Vapor Density: 3.2 [703]

VOLATILITY:

Vapor Pressure: 10 mm @ 24.4 C [703]

Vapor Density: 3.2 [703]

FIRE HAZARD: Flash Point: 102 F (open cup) [703]

Flammable when exposed to heat and flame. To fight fire, use CO2, dry chemical. DOT Classification: Flammable or Combustible Liquid. [703]

Autoignition temperature: 1000 F [703]

LEL: Not available UEL: Not available

REACTIVITY: Mixtures with hydrogen peroxide, iron (II) sulfate, and sulfuric acid may ignite and explode. [710]

STABILITY: When heated to decomposition it emits toxic fumes of NOx. [703]

USES: Not available

COMMENTS: None

NIOSH Registry Number: TJ4900000

ACUTE/CHRONIC HAZARDS: 2-Picoline is a poison by intraperitoneal route. It is moderately toxic by ingestion and skin contact, mildly toxic by inhalation. It is a skin and severe eye irritant. [703]

SYMPTOMS: Clinical signs of intoxication caused by alkyl derivatives of pyridine include weight loss, diarrhea, weakness, ataxia, and unconsciousness. Exposures to the picolines may give rise to flushing of the face, skin rash, an increase in heart and respiration rates, headache, giddiness, nausea, and vomiting. It has been alleged that workmen exposed to picoline vapors may develop diplopia as a result of disturbance of the eye muscles. [723]

3-Picoline

CAS NUMBER: 108-99-6

SYNONYMS: 3-Methylpyridine

CHEMICAL FORMULA: C6H7N

MOLECULAR WEIGHT: 93.14

PHYSICAL DESCRIPTION: Colorless liquid; sweetish odor.

DENSITY: 0.9566 @ 20C/4C [706]

MELTING POINT: -18.3C [706]

BOILING POINT: 144.1C [706]

SOLUBILITY: Water: Miscible [703] Alcohol: Miscible [703] Ether: Miscible [703]

OTHER PHYSICAL DATA: Not available

VOLATILITY: Not available

FIRE HAZARD: Flash Point: Not available

Flammable when exposed to heat or flame. DOT Classification: Flammable or Combustible Liquid. [703]

Autoignition temperature: Not available

LEL: Not available UEL: Not available

REACTIVITY: Can react vigorously with oxidizing materials [703].

STABILITY: When heated to decomposition it emits toxic fumes of NOx and SOx [703].

USES: Not available

COMMENTS: None

NIOSH Registry Number: TJ5000000

ACUTE/CHRONIC HAZARDS: 3-Picoline is a poison by intravenous and intraperitoneal routes. It is moderately toxic by ingestion. When heated to decomposition it emits toxic fumes of NOx. [703]

SYMPTOMS: Clinical signs of intoxication caused by alkyl derivatives of pyridine include weight loss, diarrhea, weakness, ataxia, and unconsciousness. Exposures to the picolines may give rise to flushing of the face, skin rash, an increase in heart and respiration rates, headache, giddiness, nausea, and vomiting. It has been alleged that workmen exposed to picoline vapors may develop diplopia as a result of disturbance of the eye muscles. [723]

4-Picolene

CAS NUMBER: 108-89-4

SYNONYMS: 4-Methylpyridine

CHEMICAL FORMULA: C6H7N

MOLECULAR WEIGHT: 93.14

PHYSICAL DESCRIPTION: Colorless liquid; disagreeable odor

DENSITY: 0.9548 @ 20C/4C [706]

MELTING POINT: 3.6 C

BOILING POINT: 144.9 C [706]

SOLUBILITY: Water: Soluble [706] Alcohol: Soluble [706] Ether: Soluble [706] Acetone: Soluble [706]

OTHER PHYSICAL DATA: Not available

VOLATILITY

Vapor Density: 3.21 [703]

FIRE HAZARD: Flash Point: 134 F (open cup) [703]

Flammable when exposed to heat, flames, oxidizers. To fight fire, use alcohol foam. [703]

Autoignition temperature: Not available

LEL: Not available UEL: Not available

REACTIVITY: Flammable when exposed to oxidizers [703].

STABILITY: When heated to decomposition it emits toxic fumes of NOx [703].

USES: Not available

COMMENTS: None

NIOSH Registry Number: UT5425000

ACUTE/CHRONIC HAZARDS: 4-Picoline is a poison by skin contact and intraperitoneal routes. It is moderately toxic by ingestion, mildly toxic by inhalation. [703] 4-Picoline is a severe irritant to the eyes, skin and respiratory tract. It penetrates the skin rapidly and causes intense skin and eye irritation. [723]

SYMPTOMS: Clinical signs of intoxication caused by alkyl derivatives of pyridine include weight loss, diarrhea, weakness, ataxia, and unconsciousness. Exposures to the picolines may give rise to flushing of the face, skin rash, an increase in heart and respiration rates, headache, giddiness, nausea, and vomiting. It has been alleged that workmen exposed to picoline vapors may develop diplopia as a result of disturbance of the eye muscles. [723]

Pyridine

CAS NUMBER: 110-86-1

SYNONYMS: Azabenzene Azine

CHEMICAL FORMULA: C5H5N

MOLECULAR WEIGHT: 79.11 [703]

PHYSICAL DESCRIPTION: Colorless liquid; sharp, penetrating, empyreumatic odor; burning taste [703] colorless to yellow liquid with a nauseating, fish-like odor [704]

SPECIFIC GRAVITY: 0.98 @ 20C/4C [704]

MELTING POINT: -42 C [706]

BOILING POINT: 115.3 C [703]

SOLUBILITY: Water: Miscible [703] Alcohol: Miscible [703] Ether: Miscible [703]

OTHER PHYSICAL DATA: Vapor Pressure: 10 mm @ 13.2 C [703] Vapor Density: 2.73 [703] Freezing Point: -42 C [703] Volatile with steam [703]

HAP WEIGHTING FACTOR: 1 [713]

VOLATILITY:

Vapor pressure: 18 mm Hg @ 20 C [058,102,421] 20 mm Hg @ 25 C [055,430]

Vapor density: 2.72 [058,102,430]

FIRE HAZARD: Pyridine has a flash point of 20 C (68 F) [036,058,275,371] and it is flammable. Fires involving this material can be controlled with a dry chemical, carbon dioxide or Halon extinguisher.

The autoignition temperature of pyridine is 482 C (900 F) [036,058,102,371].

LEL: 1.8% [036,058,371,430] UEL: 12.4% [036,058,371,430]

REACTIVITY: Pyridine is a weak base [031,099]. It is incompatible with strong oxidizers and strong acids [058,102,269,346]. It is also incompatible with acid chlorides and chloroformates [269]. It dissolves inorganic salts [025]. It reacts violently with dinitrogen tetroxide [036,066]. It also reacts violently with chlorosulfonic acid, nitric acid, oleum, perchlorates, beta-propiolactone, silver perchlorate, sulfuric acid, formamide, sulfur trioxide and iodine [043]. Maleic anhydride decomposes exothermically in the presence of this compound. The solid obtained by reaction with bromine trifluoride ignites when dry. The complex with chromium trioxide is unstable. It is incandescent on contact with fluorine [036,066]. A highly explosive by-product is formed when it is used as an acid-acceptor in reactions involving trifluoromethyl hypofluorite [066]. It will attack some forms of plastics, rubber and coatings [102].

STABILITY: Pyridine is hygroscopic [025,269]. It slowly darkens on exposure to light [058]. It is sensitive to heat [058,102,346]. Some decomposition has been indicated when stored as the bulk chemical for 2 weeks at temperatures up to 60 C [052]. Solutions of it in water, DMSO, 95% ethanol or acetone should be stable for 24 hours under normal lab conditions [700].

USES: Pyridine is used as a solvent for anhydrous mineral salts and many organic and inorganic compounds. It is used as an organic intermediate, in analytical chemistry, as an intermediate for pesticides production, in pharmaceuticals, in waterproofing, as a solvent in Karl Fischer reactions for determination of water, as a denaturant for industrial ethanol, in the manufacture of paints, in explosives, in dyestuffs, in rubber, in vitamins, in sulfa drugs and disinfectants, as a denaturant for antifreeze mixtures, as a dyeing assistant in textiles and in fungicides.

COMMENTS: This compound is extracted in quantity from coal tar [025].

NIOSH Registry Number: UR8400000

ACUTE/CHRONIC HAZARDS: Pyridine may be toxic by ingestion, inhalation and skin contact [036, 269, 371]. High concentrations cause narcosis [058,102,346]. It can be absorbed by the skin [346] and it is an irritant [036,269,371,451]. When heated to decomposition it emits toxic fumes of carbon monoxide, carbon dioxide and nitrogen oxides [058,102,269]. It may also emit toxic fumes of cyanides and possibly ammonia [058].

SYMPTOMS: Symptoms of exposure may include irritation of the skin, eyes and respiratory tract, headache and nausea [036,058,102,346]. It can irritate the nose, throat [371] and the mucous membranes [269,346]. Other symptoms of exposure may include giddiness, vomiting, conjunctivitis, dermatitis and central nervous system effects [036]. It can cause central nervous system depression [031,043,058,151]. It can also cause dizziness, insomnia, nervousness, anorexia, diarrhea, lower back pain, urinary frequency and liver and kidney damage [102,346]. Large doses can cause gastrointestinal disturbances [031,058,102,269]. Exposure can lead to fatigue, mental depression, hyperpyrexia, delirium, hepatorenal damage and death from pulmonary edema or membranous tracheobronchitis [151]. It can also lead to central nervous system damage, abdominal pain or discomfort, weakness and temporary vertigo [102]. Other symptoms of exposure may include coughing, chest pains and difficulty breathing [269]. Skin sensitization and photosensitization have been reported [346]. Large doses act as a heart poison [036]. Damage to the corneas and heart also have occurred [301]. High concentrations may cause narcosis [058,102,346]. Eye contact may cause burns. Skin contact may cause smarting and burns [371]. Mental dullness may occur [430]. Gastrointestinal irritation and systemic effects of the liver and kidneys may also occur [058].

---

o-Cresol

CAS NUMBER: 95-48-7

SYNONYMS: o-Toluol 2-Cresol 2-Hydroxytoluene o-Cresylic acid o-Methylphenol 2-Methylphenol 1-Hydroxy-2-methylbenzene

CHEMICAL FORMULA: C7H8O

MOLECULAR WEIGHT: 108.14

WLN: QR B1

PHYSICAL DESCRIPTION: Colorless to pale brown crystals

SPECIFIC GRAVITY: 1.047 @ 20/4 C [031,043,205]

DENSITY: 1.05 g/mL @ 20 C [371]

MELTING POINT: 30.9 C [047,062,205,430]

BOILING POINT: 191 C [058,062,275,451]

SOLUBILITY: Water: <1 mg/mL @ 19 C [700] DMSO: >=100 mg/mL @ 19 C [700] 95% Ethanol: >=100 mg/mL @ 19 C [700] Acetone: >=100 mg/mL @ 19 C [700]

OTHER SOLVENTS: Chloroform: Miscible [031,205] Mineral oil: 2.0% [430] Ether: Miscible [031,205] Benzene: >10% [047] Glycol: Soluble [421] Fixed alkali hydroxides: Soluble [031] Dilute alkalis: Soluble [205,421] Alcohol: Miscible [031,205]

OTHER PHYSICAL DATA: Phenol-like odor [031,058,062,430] Boiling point: 120 C @ 76 mm Hg; 70 C @ 6 mm Hg [025] Specific gravity: 1.0397 @ 24/22 C [052] Refractive index: 1.5361 @ 20 C [047,205] Saturation concentration: 1.2 g/cu m @ 20 C; 2.8 g/3 m @ 30 C [055] Liquid surface tension: 40.3 dynes/cm [371] Liquid water interfacial tension 32.7 dynes/cm [371] Latent heat of vaporization: 99.12 cal/g [371] Heat of combustion: -7774 cal/g [371] Percent in saturated air: 0.0323 @ 25 C [430] Density of saturated air: 1.00089 @ 25 C [430]

HAP WEIGHTING FACTOR: 1 [713]

VOLATILITY:

Vapor pressure: 1 mm Hg @ 38.2 C [043,058] 5 mm Hg @ 64 C [051,055]

Vapor density: 3.72 [043,058,371,430]

FIRE HAZARD: o-Cresol has a flash point of 81 C (178 F) [043,058,275,430]. It is combustible. Fires involving o-cresol may be controlled using a dry chemical, carbon dioxide or Halon extinguisher. A water spray may also be used [451].

The autoignition temperature of o-cresol is 599 C (1110 F) [058,451].

LEL: 1.4 @ 149 C [043,451] UEL: Not available

REACTIVITY: o-Cresol is incompatible with oxidizing agents and bases [058,269]. Mixing it with chlorosulfonic acid, nitric acid and oleum in a closed container caused the temperature and pressure to increase [451].

STABILITY: o-Cresol is sensitive to light and air [031,043].

USES: o-Cresol is used as a disinfectant, solvent, resins, metal cleaner, food antioxidant, ore flotation, textile scouring agent, organic intermediate, surfactant, cresylic acid constituent, additives to lubricating oil and insecticide. It is also used in the manufacturing of perfumes, dyes, plastics, herbicides, tricresyl phosphate, salicylaldehyde and coumarin.

COMMENTS: It is a component of cresylic acid.

NIOSH Registry Number: GO6300000

ACUTE/CHRONIC HAZARDS: o-Cresol may be very toxic when ingested, inhaled or absorbed through the skin [451]. It may be readily absorbed through the skin [269] and it is corrosive to body tissues [275,346,430]. When heated to decomposition it emits toxic fumes of carbon dioxide and carbon monoxide [058].

SYMPTOMS: Symptoms of exposure may include skin eruptions, liver and kidney damage [051]. It may cause burns to the eyes, headache, dizziness, nausea, vomiting, stomach pain, exhaustion and possibly coma [036]. It may also cause painless blanching or erythema, corrosion, profuse sweating, intense thirst, diarrhea, cyanosis from methemoglobinemia, hemolysis, convulsions, pulmonary edema, death from respiratory failure, jaundice, oliguria and anuria [301]. When it comes into contact with the skin it may cause prickling, intensive burning, loss of feeling, wrinkling, white discoloration, softening and later gangrene. Other symptoms may include extensive damage to the eyes, blindness, weakness of muscles, dimness of vision, ringing of ears, rapid breathing, mental confusion and loss of consciousness [346]. It may cause skin burns, central nervous system depression and gastroenteric disturbances [371, 430]. It may also cause severe skin and eye irritation [043].

---

m-Cresol

CAS NUMBER: 108-39-4

SYNONYMS: 3-Cresol 3-Hydroxytoluene m-Cresylic acid m-Methylphenol 3-Methylphenol 1-Hydroxy-3-methylbenzene m-Toluol

CHEMICAL FORMULA: C7H8O

MOLECULAR WEIGHT: 108.14

WLN: QR C1

PHYSICAL DESCRIPTION: Colorless to yellowish liquid

SPECIFIC GRAVITY: 1.034 @ 20/4 C [031,043,051,205]

DENSITY: 1.0336 g/mL @ 20 C [371]

MELTING POINT: 11.5 C [016,047,058,371]

BOILING POINT: 203 C [062,269,275,371]

SOLUBILITY: Water: 10-50 mg/mL @ 20 C [700] DMSO: >=100 mg/mL @ 20 C [700] 95% Ethanol: >=100 mg/mL @ 20 C [700] Acetone: >=100 mg/mL @ 20 C [700]

OTHER SOLVENTS: Ether: Soluble [016,047,062] Benzene: Soluble [016,047] Most organic solvents: Soluble [295,455] Solutions of fixed alkali hydroxides: Soluble [031] Chloroform: Miscible [031,205] Dilute alkalies: Soluble [205,421] Mineral oil: 2.5% [430] Glycol: Soluble [421] Alcohol: Soluble [016,062,421]

OTHER PHYSICAL DATA: Phenolic odor [031,036,062,102] Boiling point: 86 C @ 10 mm Hg [016,047] Boiling point: 80 C @ 6 mm Hg [025] Specific gravity: 1.04 @ 22/4 C [025] Vapor pressure: 5 mm Hg @ 76.0 C; 10 mm Hg @ 87.8 C [038]

Steam-volatile [025] Refractive index: 1.5438 @ 20 C [016,047,205] Critical temperature: 432 C [371] Critical pressure: 45.0 atmospheres [371] pH: 5.5 [058] Odor threshold: 5 ppm [058] Taste threshold: 0.002 ppm [051,072] Pungent taste [295] Liquid surface tension: 41.7 dynes/cm @ 20 C [371] Liquid water interfacial tension: 31.3 dynes/cm @ 20 C [371] Ratio of specific heats of vapor (gas): 1-1.05 (estimated) [371] Latent heat of vaporization: 100.6 cal/g [371] Heat of combustion: -7798 cal/g [371] Saturation concentration: 0.24 g/m3 @ 20 C, 0.68 g/m3 @ 30 C [055] Evaporation rate (butyl acetate = 1): 0.015 [102]

HAP WEIGHTING FACTOR: 1 [713]

VOLATILITY:

Vapor pressure: 0.04 mm Hg @ 20 C [055] 1 mm Hg @ 52 C [038,043,051]

Vapor density: 3.72 [043,051,055,072]

FIRE HAZARD: m-Cresol has a flash point of 86 C (187 F) [031,062,205,275]. It is combustible. Fires involving this material may be controlled with a dry chemical, carbon dioxide or Halon extinguisher. A water spray may also be used [051,058,371,451].

The autoignition temperature of m-cresol is 558 C (1038 F) [043,062,371,451].

LEL: 1.06% [051,072,371] UEL: 1.35% [051,072,371]

REACTIVITY: m-Cresol can react vigorously with strong oxidizers and strong bases [058,102,269]. It reacts violently with nitric acid, oleum, chlorosulfonic acid, metals and strong acids. If the water content is below approximately 0.3% and the temperature above 120 C, the corrosion of aluminum and its alloys may occur violently [058]. It will attack some forms of plastics, coatings and rubber [102].

STABILITY: m-Cresol is sensitive to air, light and heat and it is also hygroscopic [269].

USES: m-Cresol is used in disinfectants, in resins, as a raw material for photographic developers, in ore flotation, in fumigation compounds, in explosives, in phenol, as an insecticide, as a wood preservative, in degreasing compounds, in paintbrush cleaners and as an additive to lubricating oils. It is also used as an intermediate in the manufacture of chemicals, dyes, plastics and antioxidants. It is used in the manufacture of antiseptics, phosphate esters, herbicides and perfumes, as a solvent, as an engine and metal cleaner and in the textile industry.

COMMENTS: m-Cresol is obtained from coal tar [031,062] and it is a component of cresylic acid.

NIOSH Registry Number: GO6125000

ACUTE/CHRONIC HAZARDS: m-Cresol may be very toxic and a severe irritant. It causes burns and may be readily absorbed through the skin [269]. It is corrosive [036, 058, 295, 346]. When heated to decomposition it emits toxic fumes of carbon monoxide and formaldehyde [102].

SYMPTOMS: Symptoms of exposure via skin contact may include prickling, intensive burning and loss of feeling. The affected skin may be wrinkled, white, discolored and softened. Later, gangrene may occur. Repeated or prolonged exposure may cause a rash [346]. Eye contact may cause burning, tearing, reddening, swelling of the eyes and surrounding tissue and, in some cases, blindness. It may cause irritation of the nose and throat, sleeplessness, damage to the lungs, liver, kidneys, blood, nervous system and respiratory system; corrosion of the skin and eyes, skin sensitization, weak pulse, tinnitus, emphysema, edema, congestion in the brain, dermatitis, disturbances of the gastrointestinal and vascular systems, difficulty in swallowing, salivation, loss of appetite, fainting, dizziness, anorexia, hypertension, slightly enlarged heart, facial muscle spasms and tremors [058]. Systemic poisoning results in weakness of the muscles, headache, dimness of vision, rapid breathing, mental confusion, unconsciousness and death [346]. It can also cause respiratory system irritation, nausea, vomiting, stomach pains, exhaustion and possibly coma [036]. It can also cause hemorrhages, destruction of cellular protoplasm of the gastroenteric tract and bronchopneumonia with petechial hemorrhages in the pleura [430]. Optic atrophy and opacification may occur [099]. It may cause a burning sensation in the mouth and esophagus, corneal damage, gastroenteric disturbances, severe depression, collapse and injury of the spleen and pancreas [371]. Allergic reactions have also been reported [051].

---

p-Cresol

CAS NUMBER: 106-44-5

SYNONYMS: 4-Cresol p-Cresylic acid 1-Hydroxytoluene 4-Hydroxytoluene 1-Methyl-4-hydroxybenzene p-Methylphenol 4-Methylphenol p-Toluol

CHEMICAL FORMULA: C7H8O

MOLECULAR WEIGHT: 108.14

WLN: QR D1

PHYSICAL DESCRIPTION: Colorless to pink crystals

SPECIFIC GRAVITY: 1.0341 @ 20/4 C [031,043,205]

DENSITY: 1.034 g/mL @ 20 C [371]

MELTING POINT: 34.8 C [016,047,055,205]

BOILING POINT: 201.8 C @ 760 mm Hg [031]

SOLUBILITY: Water: <1 mg/mL @ 21 C [700] DMSO: >=100 mg/mL @ 21 C [700] 95% Ethanol: >=100 mg/mL @ 21 C [700] Acetone: >=100 mg/mL @ 21 C [700]

OTHER SOLVENTS: Dilute alkalies: Soluble [205,421] Mineral oil: 0.7% [430] Glycol: Soluble [421] Benzene: Soluble [016] Ether: Soluble [016,062,455] Chloroform: Soluble [062] Aqueous solutions of alkali hydroxides: Soluble [031] Alcohol: Soluble [016,062,421,455]

OTHER PHYSICAL DATA: Boiling point: 179.4 C @ 200 mm Hg; 140 C @ 100 mm Hg [031] Boiling point: 117.7 C @ 40 mm Hg; 102.3 C @ 20 mm Hg [031] Boiling point: 90 C @ 11 mm Hg [025] 85.7 C @ 10 mm Hg [016,047] Vapor pressure: 1 mm Hg @ 53 C [038,043,055] 5 mm Hg @ 76.5 C [038] Vapor pressure: 10 mm Hg @ 88.6 C; 20 mm Hg @ 102.3 C [038] Refractive index: 1.5312 @ 20 C [016,047] Phenolic odor [036,043,058,295] Odor threshold: <1 ppm [102] Critical temperature: 431.4 C [371] Critical pressure: 50.8 atmospheres [371] Steam-volatile [025,031] Pungent taste [295] pH: 5.5 [058] Liquid surface tension: 41.8 dynes/cm @ 40 C [371] Liquid water interfacial tension: 31.2 dynes/cm @ 40 C [371] Latent heat of vaporization: 104.85 cal/g [371] Heat of combustion: -7786 cal/g [371] Heat of fusion: 26.28 cal/g [371] Evaporation rate (butyl acetate = 1): 0.011 [102] Saturation concentration in air: ~0.014% @ 25 C [102]

HAP WEIGHTING FACTOR: 1 [713]

VOLATILITY:

Vapor pressure: 0.04 mm Hg @ 20 C; 0.11 mm Hg @ 25 C [055]

Vapor density: 3.72 [043,055,102,371]

FIRE HAZARD: p-Cresol has a flash point of 86 C (187 F) [031,062,205,451]. It is combustible. Fires involving this material may be controlled with a dry chemical, carbon dioxide or Halon extinguisher. A water spray may also be used [058,371,451].

The autoignition temperature of this compound is 558 C (1038 F) [043, 062, 371,451].

LEL: 1.06% [371] UEL: 1.4% [371]

REACTIVITY: p-Cresol is incompatible with strong oxidizers and strong alkalis [058,102,269]. It will attack some forms of plastics, coatings and rubber [102].

STABILITY: p-Cresol is sensitive to heat [102]. It is also sensitive to light [058,455].

USES: p-Cresol is used in disinfectants, in degreasing compounds, in paintbrush cleaners, as an additive in lubricating oils, in the manufacture of antiseptics, phosphate esters, antioxidants, resins, herbicides, perfumes, explosives and photographic developers, as a solvent, as an engine and metal cleaner and in the textile industry.

COMMENTS: p-Cresol is found in a score of essential oils, including ylang-ylang and oil of jasmine [043]. It is obtained from coal tar [031] and is a component of cresylic acid.

NIOSH Registry Number: GO6475000

ACUTE/CHRONIC HAZARDS: p-Cresol may be very toxic when ingested, inhaled or absorbed through the skin. It may be readily absorbed through the skin [269], it is an irritant [036,058,371] and it is corrosive [058,275,346]. When heated to decomposition it emits toxic fumes of carbon monoxide and formaldehyde [102].

SYMPTOMS: Symptoms of exposure may include irritation of the eyes, nose and throat, severe eye damage, vomiting, sleeplessness, and damage to the lungs, liver, kidneys, blood, nervous system and respiratory system [058]. Other symptoms may include eye burns, headache, dizziness, nausea, stomach pains, exhaustion, dermatitis and coma [036]. It can cause corrosion of all tissues. Skin contact causes prickling, intense burning and loss of feeling. The skin shows wrinkling, white discoloration and softening. Gangrene may occur. Other symptoms may include blindness, skin rash, weakness of the muscles, dimness of vision, ringing of the ears, rapid breathing, mental confusion, difficulty in swallowing, salivation, diarrhea, loss of appetite, fainting, loss of consciousness and death [346]. Eye contact can result in pain, swelling of the conjunctiva and corneal damage. Exposure can cause burning sensation in the mouth and esophagus, gastroenteric disturbance, severe depression, collapse, edema of the lungs and injury of the spleen and pancreas [371]. It can also cause hemorrhages, destruction of cellular protoplasm of the gastroenteric tract, emphysema and bronchopneumonia with petechial hemorrhages in the pleura [430].

Cresols/Cresylic acid (isomers and mixture)

CAS NUMBER: 1319-77-3

NOTE: See data for the individual cresol isomers that, together, make up the mixture known as cresylic acid. The individual compounds are:

o-Cresol   m-Cresol   p-Cresol

SYNONYMS: Cresol Mixture of ortho-, meta-, or para-Cresol Mixture of 2-, 3-, or 4-Methyl phenol

---

Cumene

CAS NUMBER: 98-82-8

SYNONYMS: Isopropyl benzene 2-Phenylpropane

CHEMICAL FORMULA: C9H12

MOLECULAR WEIGHT: 120.21 [703]

PHYSICAL DESCRIPTION: Colorless liquid with a sharp, penetrating aromatic odor [704]

SPECIFIC GRAVITY: 0.86 @ 20C/4C [704]

MELTING POINT: -96 C [703]

BOILING POINT: 152 C [703]

SOLUBILITY: Water: Insoluble [704]

OTHER PHYSICAL DATA: Vapor Pressure: 10 mm @ 38.3 C [703] Vapor Density: 4.1 [703] Freezing Point: -141 F [704]

HAP WEIGHTING FACTOR: 1 [713]

VOLATILITY:

Vapor Pressure: 5 mm @ 77 F [704]

Vapor Density: 4.1 [703]

FIRE HAZARD: Flash Point: 111 F [703] 96 F [704]

Flammable when exposed to heat or flame [703]. Class IC flammable liquid [704]. To fight fire, use foam, CO_2, dry chemical [703].

Autoignition temperature: 795 F [703]

LEL: 0.9% [703] UEL: 6.5% [703]

REACTIVITY: Can react with oxidizing materials. Violent reaction with HNO_3; oleum; chlorosulfonic acid. [703]

STABILITY: Not available

USES: Cumene accounts for about 10% of the chemical usage of propylene. Almost all of the cumene is used to produce phenol and acetone by the cumene hydroperoxide process. Smaller amounts go into à-methylstyrene, acetophenone, and various solvents. [711]

COMMENTS: None

NIOSH Registry Number: GR8575000

ACUTE/CHRONIC HAZARDS: Cumene is moderately toxic by ingestion and mildly toxic by inhalation and skin contact. Other hazards include an eye and skin irritant; potential narcotic action; central nervous system depressant and antipsychotic reactions. [703]

SYMPTOMS: Symptoms of exposure to cumene may include irritation of the eyes and mucous membranes; headache; dermatitis; narcosis and coma. [704]

Cyanide compounds

NOTE: The Clean Air Act defines this group as: X'CN where X = H' or any other group where a formal dissociation may occur. For example KCN or $Ca(CN)_2$. [708] The compounds described below are intended as selected examples of cyanide compounds and do not comprise a comprehensive list.

Calcium cyanide

CAS NUMBER: 592-01-8

CHEMICAL FORMULA: C2CaN2

MOLECULAR WEIGHT: 92.19

WLN: CA CN

PHYSICAL DESCRIPTION: Rhombohedric crystals or powder

SPECIFIC GRAVITY: Not available

DENSITY: Not available

MELTING POINT: Decomposes > 350 C

BOILING POINT: Not available

SOLUBILITY: Water: Soluble (with gradual liberation of HCN) DMSO: Not available 95% Ethanol: Not available Acetone: Not available

OTHER SOLVENTS: Alcohol: Soluble

OTHER PHYSICAL DATA: Not available

VOLATILITY: Not available

FIRE HAZARD: Not available

LEL: Not available UEL: Not available

REACTIVITY: Not available

STABILITY: Decomposes in moist air liberating hydrogen cyanide [715]. When heated to decomposition it emits toxic fumes of Nox and CN-. [703]

USES: Calcium cyanide is used as a fumigant, rodenticide, in stainless-steel manufacturing, in leaching ores of precious metals, and as a stabilizer for cement [715].

COMMENTS: Very poisonous. Keep dry. [715]

NIOSH Registry Number: EW0700000

ACUTE/CHRONIC HAZARDS: A deadly poison by ingestion and probably other routes. When heated to decomposition it emits toxic fumes of NOx and CN-. [703]

SYMPTOMS: Not available.

Copper(I)cyanide

CAS NUMBER: 544-92-3

SYNONYMS: Cuprous cyanide

CHEMICAL FORMULA: CCuN

MOLECULAR WEIGHT: 89.56

WLN: CU CN

PHYSICAL DESCRIPTION: White monoclinic prisms

SPECIFIC GRAVITY: 2.92

DENSITY: Not available

MELTING POINT: 473 C in N2

BOILING POINT: Not available

SOLUBILITY: Water: Not available DMSO: Not available 95% Ethanol: Insoluble Acetone: Not available

OTHER SOLVENTS: Hydrochloric acid: Soluble Ammonium hydroxide: Soluble Ammonia: Slightly soluble

OTHER PHYSICAL DATA: Not available

VOLATILITY: Not available

FIRE HAZARD: Not available

LEL: Not available UEL: Not available

REACTIVITY: Not available

STABILITY: Not available

USES: It is used inn electroplating copper or iron, in insecticides and fungicides, as an antifouling agent in marine paints, and as a polymerization catalyst.

COMMENTS: None

NIOSH Registry Number: GL7150000

ACUTE/CHRONIC HAZARDS: Copper (I) cyanide may be toxic by ingestion.

SYMPTOMS: Not available

Hydrogen cyanide

CAS NUMBER: 74-90-8

SYNONYMS: Hydrocyanic acid; Formonitrile

CHEMICAL FORMULA: HCN

PHYSICAL DESCRIPTION: Colorless or pale blue liquid or gas (above 78 F) with a strong, irritating odor [704] odor of bitter almonds [703]

SPECIFIC GRAVITY: 0.69 @ 20C/4C [703]

MELTING POINT: -14 C [706]

BOILING POINT: 26 C [706]

SOLUBILITY: Water: Miscible [703], Soluble [706]; Alcohol: Miscible [703], Soluble [706]; Ether: Miscible [703], Soluble [706].

OTHER PHYSICAL DATA: Vapor Pressure: 400 mm @ 9.8 C [703] Vapor Density: 0.932 [703] Freezing Point: 8 F [704]

VOLATILITY:

Vapor Pressure: 400 mm @ 9.8 [703]

Vapor Density: 0.932 [703]

FIRE HAZARD: Flash Point: 0 F (Closed cup) [703]

Very dangerous fire hazard when exposed to heat, flame or oxidizers. Severe explosion hazard when exposed to heat or flame or by chemical reaction with oxidizers. Gas forms explosive mixtures with air. To fight fire, use CO2, non-alkaline dry chemical, foam. [703]

Autoignition temperature: 1000 F [703]

LEL: 5.6% [703] UEL: 40% [703]

REACTIVITY: Severe explosion hazard when exposed to heat or flame or by chemical reaction with oxidizers. Reacts violently with acetaldehyde. [703] Self reactive and dangerous with a water content of 2 - 5%. [720]

STABILITY: Can polymerize explosively at 50-60 C or in the presence of traces of alkali. The anhydrous liquid is stabilized at or below temperature by the addition of acid. The gas forms explosive mixtures with air. When heated to decomposition or in reaction with water, steam, acid or acid fumes it produces highly toxic fumes of CN-. [703]

USES: HCN is the building block for adiponitrile, methyl methacrylate, cyanuric chloride, amino carbonylic acid, sodium cyanide, methionine, and other intermediates. These intermediates serve as raw materials for the manufacture of synthetic fibers, plastics, herbicides, chelating agents, dyestuffs, pharmaceuticals, explosives, surfactants, and metal treatment agents. [711] The compressed gas is used for exterminating rodents and insects in ships and for killing insects on trees, etc. [715]

COMMENTS: None

NIOSH Registry Number: MW6825000

ACUTE/CHRONIC HAZARDS: A deadly poison by all routes. Hydrogen cyanide and other cyanides combine in the tissues with the enzymes associated with cellular oxidation and usually cause rapid death through asphyxia. When hydrogen cyanide is no longer present, normal function is restored provided death has not already occurred. Exposure to 100-200 ppm for 30-60 minutes can cause death. [703]

SYMPTOMS: Symptoms of exposure may include asphyxia and death at high levels; weakness, headache, confusion; nausea, vomiting, increased rate and depth of respiration or respiration slow and gasping. [704] Venous blood may be cherry-red. In cases of acute cyanide poisoning death is extremely rapid, although sometimes breathing may continue for a few minutes. In less acute cases, there is cyanosis, headache, dizziness, unsteadiness of gait, a feeling of suffocation, and nausea. Where the patient recovers, there is rarely any disability. [703]

Nickel cyanide

Data for this compound is listed under Nickel compounds.

Potassium cyanide

CAS NUMBER: 151-50-8

CHEMICAL FORMULA: CKN

MOLECULAR WEIGHT: 65.12

WLN: Not available

PHYSICAL DESCRIPTION: White granular powder

SPECIFIC GRAVITY: 1.52 @ 16 C

DENSITY: Not available

MELTING POINT: 634.5 C

BOILING POINT: Not available

SOLUBILITY: Water: Not available DMSO: Not available 95% Ethanol: Not available Acetone: Not available

OTHER SOLVENTS: Not available

OTHER PHYSICAL DATA: Inorganic compound with the odor of HCN

VOLATILITY: Not available

FIRE HAZARD: Not available

LEL: Not available UEL: Not available

REACTIVITY: Potassium cyanide may react violently with chlorates, nitrites, NCl_3, and $NaClO_3$.

STABILITY: On exposure to air potassium cyanide is gradually decomposed.

USES: Potassium cyanide is used for the extraction of gold and silver from ores, in analytical reagents, and as an insecticide.

COMMENTS: None

NIOSH Registry Number: TS876000

ACUTE/CHRONIC HAZARDS: Potassium cyanide may be highly toxic and it is a violent poison! Poisoning may occur by ingestion, absorption through injured skin or inhalation of hydrogen cyanide (liberated by action of carbon dioxide or other acids).

SYMPTOMS: Exposure may result in rapid death.

Silver cyanide

CAS NUMBER: 506-64-9

CHEMICAL FORMULA: AgCN

MOLECULAR WEIGHT: 133.84

WLN: AG CN

PHYSICAL DESCRIPTION: White or grayish powder

SPECIFIC GRAVITY: 3.95

DENSITY: Not available

MELTING POINT: 320 C

BOILING POINT: Not available

SOLUBILITY: Water: Not available DMSO: Not available 95% Ethanol: Not available Acetone: Not available

OTHER SOLVENTS: Nitric acid: Soluble Ammonium hydroxide: Soluble Potassium cyanide: Soluble Sodium thiosulfate: Soluble

OTHER PHYSICAL DATA: Decomposes @ 320 C degrees Odorless

VOLATILITY: Not available

FIRE HAZARD: Not available

LEL: Not available UEL: Not available

REACTIVITY: Silver cyanide can react violently with fluorine.

STABILITY: Silver cyanide is stable in dry air but it darkens on exposure to light.

USES: Silver cyanide is used for silver plating. It was formerly used for extemporaneous preparation of dilute hydrocyanic acid by treatment with hydrochloric acid.

COMMENTS: None

NIOSH Registry Number: VW3850000

ACUTE/CHRONIC HAZARDS: Silver cyanide is toxic.

SYMPTOMS: Not available

Silver potassium cyanide

CAS NUMBER: 506-61-6

CHEMICAL FORMULA: C2AgKN2

MOLECULAR WEIGHT: 199.01

WLN: AG KA CN2

PHYSICAL DESCRIPTION: White crystals

SPECIFIC GRAVITY: 2.36 @ 25 C

DENSITY: Not available

MELTING POINT: Not available

BOILING POINT: Not available

SOLUBILITY: Water: Soluble DMSO: Not available 95% Ethanol: Not available Acetone: Not available

OTHER SOLVENTS: Alcohol: Soluble Acids: Insoluble

OTHER PHYSICAL DATA: Not available

VOLATILITY: Silver potassium cyanide is not volatile.

FIRE HAZARD: Silver potassium cyanide is nonflammable.

LEL: Not available UEL: Not available

REACTIVITY: Silver potassium cyanide is sensitive to light

STABILITY: Silver potassium cyanide is stable under ambient storage conditions when protected from light.

USES: Silver potassium cyanide is used for silverplating, as a bactericide, and as an antiseptic.

COMMENTS: None

NIOSH Registry Number: TT5775000

ACUTE/CHRONIC HAZARDS: Silver potassium cyanide may be toxic by all routes of exposure but especially by inhalation or ingestion.

SYMPTOMS: Not available

Sodium cyanide

CAS NUMBER: 143-33-9

CHEMICAL FORMULA: CNNa

MOLECULAR WEIGHT: 49.01

WLN: .NA..CN

PHYSICAL DESCRIPTION: White crystalline powder

SPECIFIC GRAVITY: Not available

DENSITY: 1.6 g/mL @ 25 C [058,371]

MELTING POINT: 563.7 C [043,275]

BOILING POINT: 1496 C [043,051,058]

SOLUBILITY: Water: >=100 mg/mL @ 20 C [700] DMSO: 1-5 mg/mL @ 20 C [700] 95% Ethanol: 5-10 mg/mL @ 20 C [700] Acetone: <1 mg/mL @ 20 C [700]

OTHER SOLVENTS: Alcohol: Slightly soluble [062]

OTHER PHYSICAL DATA: Odorless when dry [031,051] Somewhat deliquescent in damp air [031] Vapor pressure: 3.34 mm Hg @ 900 C; 36 mm Hg @ 1000 C [038] Vapor pressure: 100 mm Hg @ 1214 C [038]

VOLATILITY:

Vapor pressure: ~0 mm Hg @ 20 C [102,421] 0.76 mm Hg @ 800 C [051]

FIRE HAZARD: Sodium cyanide is nonflammable [051, 058, 371].

LEL: Not available UEL: Not available

REACTIVITY: Sodium cyanide reacts with acids, releasing a highly flammable and toxic hydrogen cyanide gas [036, 043, 058, 451]. It is incompatible with strong oxidizers [051, 058, 269, 346]. This includes nitrates, chlorates and acid salts [346]. It explodes if melted with nitrite or chlorate at about 450 C [043,451]. Carbon monoxide from the air is sufficiently acidic to liberate toxic gas on contact with sodium cyanide [043]. It undergoes a mild reaction with water or steam [043, 051, 371]. Weak alkaline solutions can produce dangerous gases [058]. Sodium cyanide undergoes violent reactions with fluorine, magnesium and nitric acid [043]. A reaction with ethyl chloroacetate (upon heating) may erupt suddenly [051, 066]. Aqueous solutions readily dissolve gold and silver in the presence of air [033].

STABILITY: Sodium cyanide is stable when dry [058]. It will decompose on exposure to moisture [043, 058, 269]. It is also sensitive to heat [043]. Aqueous solutions are strongly alkaline and rapidly decompose [033,062]. Solutions of it in water (at pH greater than or equal to 7), DMSO, 95% ethanol or acetone should be stable for 24 hours under normal lab conditions [700].

USES: Sodium cyanide is used in extracting gold and silver from ores, in electroplating baths, in fumigating citrus and other fruit trees, ships, railway cars and warehouses, in the manufacture of hydrocyanic acid and many other cyanides, and in the case-hardening of steel. It is also used in insecticides, in cleaning metals, in the manufacture of dyes and pigments, in nylon intermediates, in chelating compounds, in ore flotation, in rodenticides, in metal polishes (especially silver polish), in metallurgical and photographic processes, and in fumigating rabbit burrows and rat runs.

COMMENTS: None

NIOSH Registry Number: VZ7525000

ACUTE/CHRONIC HAZARDS: Sodium cyanide may be extremely toxic by ingestion, inhalation or skin absorption [036,269,295,371]. It may be readily absorbed through the skin [036, 051,058,371]. It is an irritant and corrosive [051,151]. On contact with acids or when heated to decomposition it emits toxic fumes of carbon monoxide, carbon dioxide, nitrogen oxides and hydrogen cyanide [036,043,058,269].

SYMPTOMS: Symptoms of exposure may include hallucinations, distorted perceptions, muscle weakness and gastritis [043]. Other symptoms may include cyanosis and irritation of skin, eyes and lungs [269]. Eye contact may cause burns [058,371]. Skin contact may result in corrosion [051,151]. It may also result in deep ulcers, severe irritation, pain and second degree burns [371]. It can cause rapid death

Symptoms of exposure to this class of compounds may include weakness, headache, dizziness, nausea, vomiting, unconsciousness and death [036,058,102]. Other symptoms may include heaviness of the arms and legs, increased difficulty in breathing, pallor and cessation of breathing [036]. It can cause dilated pupils, transient blindness, central retinal edema, convulsions, hemianopia, optic neuropathies and optic neuritis [099]. It can also cause asphyxia, a rash characterized by itching and by macular, papular and vesicular eruptions, secondary infection, loss of appetite and irritation of the upper respiratory tract [043]. Exposure can lead to central nervous system stimulation followed by central nervous system depression, cardiac irregularities including bradycardia, respiratory arrest, bitter, acrid and burning taste, feeling of constriction or numbness in the throat, salivation, anxiety, confusion, vertigo, giddiness, sensation of stiffness in the lower jaw, hyperpnea, dyspnea, rapid respirations which then become slow and irregular, short inspiration with prolonged expiration, odor of bitter almonds on the breath or vomitus, rise in blood pressure, reflex slowing of the heart rate, rapid, weak and irregular pulse, palpitations, sensation of constriction in the chest, bright pink coloration of the skin, opisthotonos, trismus, involuntary micturition and defecation, paralysis, sweating, protruding eyeballs, unreactive pupils and foam-covered mouth (sometime blood-stained) indicating pulmonary edema. The skin may also be brick red [151]. Exposure may also lead to blood pressure fall, coma, flushing, drowsiness, hoarseness, conjunctivitis, weight loss and mental deterioration [301]. Other symptoms may include collapse and nervousness [173]. Inhalation may lead to nosebleed and nasal ulceration [421]. Staggering, hypotension and tachycardia may occur [295]. Other symptoms may include tearing, blurred vision and possible permanent eye damage [058].

Zinc cyanide

CAS NUMBER: 557-21-1

CHEMICAL FORMULA: C2N2Zn

MOLECULAR WEIGHT: 117.41

WLN: ZN CN2

PHYSICAL DESCRIPTION: White powder

SPECIFIC GRAVITY: 1.852

DENSITY: Not available

MELTING POINT: 800

BOILING POINT: Not available

SOLUBILITY: Water: Insoluble DMSO: Not available 95% Ethanol: Not available Acetone: Not available

OTHER SOLVENTS: Not available

OTHER PHYSICAL DATA: Not available

VOLATILITY: Not available

FIRE HAZARD: Not available

LEL: Not available UEL: Not available

REACTIVITY: Zinc cyanide can react violently with magnesium.

STABILITY: Not available

USES: Zinc cyanide is used for electroplating and for removing NH3 from producer gas.

COMMENTS: None

NIOSH Registry Number: ZH1575000

ACUTE/CHRONIC HAZARDS: Zinc cyanide may be toxic and also is an irritant.

SYMPTOMS: Symptoms of acute exposure may include headache, lack of appetite, dizziness, rapid respiration, rapid pulse, nausea and vomiting, unconsciousness, convulsion and death. Symptoms of chronic exposure may include headache, lack of appetite, weakness, inflammation of the skin with small pimples or small blistery spots.

2,4-D, salts and esters

2,4-Dichlorophenoxyacetic acid (2,4-D)

CAS NUMBER: 94-75-7

SYNONYMS: 2,4-D Dichlorophenoxyacetic acid

CHEMICAL FORMULA: C8H6Cl2O3

MOLECULAR WEIGHT: 221.04 [703]

PHYSICAL DESCRIPTION: White to yellow crystalline, odorless powder (herbicide) [704]

SPECIFIC GRAVITY: (86 F) 1.57 [704]

MELTING POINT: 141 C [703]

BOILING POINT: 160 C @ 0.4 mm [703]

SOLUBILITY: Water: 540 ppm [702]

OTHER PHYSICAL DATA: Vapor Density: 7.63 [703] Vapor Pressure: 0.4 mm Hg @ 160 C [705]

HAP WEIGHTING FACTOR: 1 [713]

VOLATILITY:

Vapor Pressure: Low [704]

Vapor Density: 7.63 [703]

FIRE HAZARD: Flash Point: 88 C [702]

Noncombustible solid, but may be dissolved in flammable liquids [704].

Autoignition temperature: Not available

LEL: Not applicable UEL: Not applicable

REACTIVITY: Incompatible or reacts with strong oxidizers [704].

STABILITY: When heated to decomposition it emits toxic fumes of Cl- [703].

UDRI Thermal Stability Class: 5 [702] UDRI Thermal Stability Ranking: 211 [702]

USES: A defoliant and herbicide [703]. Used to increase latex output of old rubber trees [715].

COMMENTS: None

NIOSH Registry Number: AG6825000

ACUTE/CHRONIC HAZARDS: Poison by ingestion, intravenous, and intraperitoneal routes. Moderately toxic by skin contact; experimental carcinogen and teratogen; suspected human carcinogen. Human systemic effects by ingestion: somnolence, convulsions, coma, nausea or vomiting. Can cause liver and kidney injury. A skin and severe eye irritant. Human mutagenic data exists; produces experimental reproductive effects. When heated to decomposition, emits toxic fumes of Cl-. [703]

SYMPTOMS: Symptoms of exposure may include weakness, stupor, hyporeflexia, muscle twitch; convulsions; dermatitis; in animals: liver and kidney damage. [704] It has also caused somnolence, convulsions, coma, and nausea or vomiting. It is a skin and severe eye irritant; it can cause liver and kidney injury. [703]

2,4-D, n-Butyl ester

CAS NUMBER: 94-80-4

SYNONYMS: 2,4-Dichlorophenoxyacetic acid, butyl ester

CHEMICAL FORMULA: C12H14Cl2O3

MOLECULAR WEIGHT: 277.16

WLN: 4OV1OR BG DG

PHYSICAL DESCRIPTION: Light brown liquid

SPECIFIC GRAVITY: 1.235-1.245 @ 20/20 C

DENSITY: Not available

MELTING POINT: Not available

BOILING POINT: 146-147 C @ 1 mm Hg

SOLUBILITY: Water: <1 mg/mL @ 21 C [700] DMSO: >=100 mg/mL @ 21 C [700] 95% Ethanol: >=100 mg/mL @ 21 C [700] Acetone: >=100 mg/mL @ 21 C [700]

OTHER SOLVENTS: Not available

OTHER PHYSICAL DATA: Not available

HAP WEIGHTING FACTOR: 1 [713]

VOLATILITY: Not available

FIRE HAZARD: Flash point data for 2,4-D, n-butyl ester are not available. It is probably combustible. Fires involving 2,4-D, n-butyl ester should be controlled using a dry chemical, carbon dioxide or Halon extinguisher.

LEL: Not available UEL: Not available

REACTIVITY: Not available

STABILITY: 2,4-D, n-butyl ester is stable under normal laboratory conditions.

USES: 2,4-D, n-butyl ester is a selective herbicide and defoliant.

COMMENTS: None

NIOSH Registry Number: AG8050000

ACUTE/CHRONIC HAZARDS: 2,4-D, n-butyl ester is toxic and an irritant. When heated to decomposition it emits toxic fumes.

SYMPTOMS: Symptoms of exposure may include local irritation of the skin, eyes and nasal passages, anorexia, diarrhea, nausea, vomiting, weakness, stupor, muscle twitching, convulsions, decrease in body temperature and coma.

DDE

CAS NUMBER: 72-55-9

SYNONYMS: 2,2-bis(p-Chlorophenyl)-1,1-dichloroethene 2,2-bis(4-Chlorophenyl)-1,1-dichloroethene 2,2-bis(p-Chlorophenyl)-1,1-dichloroethylene 2,2-bis(4-Chlorophenyl)-1,1-dichloroethylene P,P'-DDE

CHEMICAL FORMULA: $C_{14}H_8Cl_4$

MOLECULAR WEIGHT: 318.03

WLN: GYGUYR XG&R XG

PHYSICAL DESCRIPTION: White crystalline solid

SPECIFIC GRAVITY: Not available

DENSITY: Not available

MELTING POINT: 88-90 C [269,275]

BOILING POINT: 316.5 C [025]

SOLUBILITY: Water: <0.1 mg/mL @ 22 C [700] DMSO: 50-100 mg/mL @ 21 C [700] 95% Ethanol: 10-50 mg/mL @ 21 C [700] Acetone: 50-100 mg/mL @ 21 C [700]

OTHER SOLVENTS: Fats: Soluble [395] Most organic solvents: Soluble [395]

OTHER PHYSICAL DATA: Adsorption capacity: 232 mg/g

HAP WEIGHTING FACTOR: 1 [713]

VOLATILITY: Not available

FIRE HAZARD: Flash point data for DDE are not available. It is probably combustible. Fires involving DDE may be controlled with a dry chemical, carbon dioxide or Halon extinguisher.

LEL: Not available UEL: Not available

REACTIVITY: DDE is incompatible with strong oxidizing agents and strong bases [269]. Oxidation is catalyzed by UV radiation [395].

STABILITY: DDE is sensitive to exposure to light. Solutions of it in water, DMSO, 95% ethanol or acetone should be stable for 24 hours under normal lab conditions [700].

USES: DDE is used as an insecticide and military product.

COMMENTS: DDE is a metabolite of DDT and also a degradation product of DDT.

NIOSH Registry Number: KV9450000

ACUTE/CHRONIC HAZARDS: DDE is harmful if ingested, inhaled or absorbed through the skin and is an irritant. There is evidence that this compound is an animal carcinogen [015]. When heated to decomposition it emits toxic fumes of carbon monoxide and carbon dioxide [269]. It may also emit toxic fumes of hydrogen chloride gas [042,269].

SYMPTOMS: Symptoms of exposure to DDE may include liver and kidney damage [052]. Based on data for a similar compound, symptoms may also include vomiting, headache, fatigue, malaise, numbness and partial paralysis of the extremities, moderate ataxia, exaggeration of part of the reflexes, mild convulsions, loss of proprioception and vibratory sensation of the extremities, hyperactive knee-jerk reflexes, excitement, confusion and increased respiration [215]. It may also cause nausea and diarrhea [042]. Other symptoms may include tremors of the head and neck muscles, cardiac and respiratory failure and even death [031]. It may also cause paresthesias of the tongue, lips and face, irritability and dizziness [406]. It may cause tonic and clonic convulsions [031,406]. Other symptoms include apprehension and hyperesthesia of the mouth and face [215,406]. It may also cause "yellow vision" [099].

Diazomethane

CAS NUMBER: 334-88-3

SYNONYMS: Azimethylene Diazirine

CHEMICAL FORMULA: CH2N2

MOLECULAR WEIGHT: 42.05 [703]

PHYSICAL DESCRIPTION: Yellow gas at ordinary temperature [703] yellow gas with a musty odor [704]

DENSITY: 1.45 [703]

MELTING POINT: -145 C [703]

BOILING POINT: -23 C [703]

SOLUBILITY: Water: Reacts [704]

OTHER PHYSICAL DATA: Vapor Pressure: > 1 atm [704] Freezing Point: -229 F [704]

HAP WEIGHTING FACTOR: 1 [713]

VOLATILITY:

Vapor Pressure: >1 atm [704]

FIRE HAZARD: Flash Point: Not applicable (gas) [704]

Flammable gas; EXPLOSIVE [704]

Autoignition temperature: Not available

LEL: Not available UEL: Not available

REACTIVITY: Undiluted liquid or gas may explode on contact with alkali metals. On contact with acid or acid fumes it emits highly toxic fumes of NOx. Incompatible with alkali metals; calcium sulfate. [703] Incompatible or reacts with water, drying agents such as calcium arsenate [704]. Interaction with dimethylaminodimethylarsine and trimethyltin chloride to produce diazomethyldimethylarsine is accompanied by violent foaming; eye protection is essential [710].

STABILITY: Highly explosive when shocked, exposed to heat or by chemical reaction. Undiluted liquid or gas may explode on contact with rough surfaces, heat (100 C), high intensity light or shock [703]. May explode violently on exposure to sunlight or contact with rough edges such as ground glass [704]. When heated to decomposition it emits highly toxic fumes of NOx. [703] Highly endothermic; boils at -23 C; liquid or concentrated solutions may explode if impurities or solids are present, including freshly crystallized products. Explosive intermediates may also be formed during its use as a reagent, but cold dilute solutions have often been used uneventfully. Many explosions are attributed to uncontrolled or unsuitable conditions of contact between concentrated alkali and undiluted nitroso precursors. [710]

USES: Powerful methylating agent for acidic compounds such as carboxylic acids, phenols, enols [715].

COMMENTS: Caution: Explosive (use safety screen), insidious poison (a well-ventilated hood is absolutely necessary), avoid vapor. Strong irritant. Does not cause discernible reaction at time of contact, but later, even in minute amounts, produces inflammatory reaction. Hypersensitivity results which makes it impossible to work with diazomethane without attacks of asthma and associated symptoms. [715]

Shipped as a liquified compressed gas [704].

NIOSH Registry Number: PA7000000

ACUTE/CHRONIC HAZARDS: Diazomethane is an experimental tumorigen and carcinogen. It is a poison irritant by inhalation and a powerful allergen. It can cause pulmonary edema and frequently causes hypersensitivity leading to asthmatic symptoms. Mutagenic data exists. When heated to decomposition or on contact with acid or acid fumes it emits highly toxic fumes of NOx. [703]

SYMPTOMS: Symptoms of exposure may include coughing, shortness of breath; headache; flushed skin, fever; chest pain, pulmonary edema, pneumonitis; irritation of the eyes; asthma; fatigue. [704]

Dibenzofuran

NOTE: The form of this name in the Clean Air Act Amendments Section 112(b) list is Dibenzofurans. Although there are multiple isomers of chlorinated dibenzofuran and other substituted dibenzofurans there is only one dibenzofuran. The information presented here applies only to the single compound dibenzofuran.

CAS NUMBER: 132-64-9

SYNONYMS: Diphenylene oxide

CHEMICAL FORMULA: C12H8O

MOLECULAR WEIGHT: 168.19

WLN: T B656 HOJ

PHYSICAL DESCRIPTION: Colorless crystals

SPECIFIC GRAVITY: 1.0886 @ 99/4 C

DENSITY: Not available

MELTING POINT: 86-87 C

BOILING POINT: 287 C

SOLUBILITY: Water: <1 mg/mL @ 20 C [700] DMSO: >=100 mg/mL @ 20 C [700] 95% Ethanol: 10-50 mg/mL @ 20 C [700] Acetone: >=100 mg/mL @ 20 C [700]

OTHER SOLVENTS: Acetic acid: Soluble Ether: Soluble Benzene: Slightly soluble

OTHER PHYSICAL DATA: Refractive index: 1.6079 @ 99 C

HAP WEIGHTING FACTOR: 10 [713]

VOLATILITY:

Vapor density: 5.8

FIRE HAZARD: Flash point data for 1,4-dibenzofuran are not available. It is probably combustible. Fires involving this material may be controlled with a dry chemical, carbon dioxide or Halon extinguisher.

LEL: Not available UEL: Not available

REACTIVITY: Not available

STABILITY: 1,4-Dibenzofuran is sensitive to prolonged exposure to light.

USES: 1,4-Dibenzofuran is used as an insecticide and as an intermediate in the synthesis of pharmaceuticals and other chemicals.

COMMENTS: None

NIOSH Registry Number: Not available

ACUTE/CHRONIC HAZARDS: Not available

SYMPTOMS: Not available

1,2-Dibromo-3-chloropropane

CAS NUMBER: 96-12-8

SYNONYMS: 3-Chloro-1,2-dibromopropane Dibromochloropropane 1-Chloro-2,3-dibromopropanc

CHEMICAL FORMULA: C3H5Br2Cl

MOLECULAR WEIGHT: 236.35

WLN: G1YE1E

PHYSICAL DESCRIPTION: Amber to colorless liquid

SPECIFIC GRAVITY: 2.08 @ 20/20 C [055,395]

DENSITY: 2.05 g/mL @ 20 C [051,062]

MELTING POINT: 6 C [107,327]

BOILING POINT: 196 C [031,047,395,430]

SOLUBILITY: Water: <0.1 mg/mL @ 18 C [700] DMSO: >=100 mg/mL @ 20 C [700] 95% Ethanol: >=100 mg/mL @ 20 C [700] Acetone: >=100 mg/mL @ 20 C [700] Methanol: Miscible [173,395]

OTHER SOLVENTS: Dichloropropane: Miscible [031,173] Oils: Miscible [031,051,062,205] Isopropyl alcohol: Miscible [031,173,395] Liquid hydrocarbons: Miscible [395] Halogenated hydrocarbons: Miscible [395] Alcohols: Soluble [430]

OTHER PHYSICAL DATA: Pungent odor [031,173,327,395] Refractive index: 1.5518 @ 25 C [395,430] 1.553 @ 14 C [031,047,205] Boiling point: 78 C @ 16 mm Hg [031,047] 21 C @ 0.8 mm Hg [025,031] Specific gravity: 2.09 @ 20/4 C [430] Density: 2.093 g/mL @ 14 C [031,047,205] Evaporation rate (butyl acetate = 1): Very much less than 1 [327]

HAP WEIGHTING FACTOR: 1 [713]

VOLATILITY:

Vapor pressure: 0.8 mm Hg @ 21 C [031,055,173,395]

FIRE HAZARD: 1,2-Dibromo-3-chloropropane has a flash point of 76.6 C (170 F) [043,051,062,107]. It is combustible. Fires involving this material may be controlled with a dry chemical, carbon dioxide or Halon extinguisher.

LEL: Not available UEL: Not available

REACTIVITY: 1,2-Dibromo-3-chloropropane reacts with chemically active metals such as aluminum, magnesium, tin and their alloys [173,327]. It will attack some rubber materials and coatings [327,395].

STABILITY: 1,2-Dibromo-3-chloropropane is stable under normal laboratory conditions. It is stable in neutral and acidic media but it is hydrolyzed in alkali [173,395]. Solutions of it in water, DMSO, 95% ethanol or acetone should be stable for 24 hours under normal lab conditions [700].

USES: 1,2-Dibromo-3-chloropropane is used as a soil fumigant, nematocide, pesticide and intermediate in organic synthesis.

COMMENTS: None

NIOSH Registry Number: TX8750000

ACUTE/CHRONIC HAZARDS: 1,2-Dibromo-3-chloropropane is highly toxic via all routes of exposure [107]. It is a severe irritant [043,031]. It is narcotic in high concentrations [031,043]. When heated to decomposition it emits toxic fumes of hydrogen bromide, hydrogen chloride and carbon monoxide [043,326].

SYMPTOMS: Symptoms of exposure may included severe eye and skin irritation and irritation of the mucous membranes [031]. It is narcotic in high concentrations [031,043] and has been implicated in sterility in males [043,062,186,395]. It may also cause diminished renal function and degeneration and cirrhosis of the liver [186]. Other symptoms include nausea, conjunctivitis, respiratory irritation, pulmonary congestion, pulmonary edema, central nervous system depression, apathy, sluggishness and ataxia. Upon repeated exposure, erythema, inflammation and dermatitis may occur [327]. Exposure may also cause drowsiness, vomiting, liver and kidney damage, respiratory distress, testicular atrophy, spleen necrosis and sperm count depression [107].

Dibutyl phthalate

CAS NUMBER: 84-74-2

SYNONYMS: Phthalic acid, dibutyl ester o-benzenedicarboxylic acid, dibutyl ester n-butyl phthalate Di-n-butyl phthalate

CHEMICAL FORMULA: C16H22O4

MOLECULAR WEIGHT: 278.35

WLN: 4OVR BVO2

PHYSICAL DESCRIPTION: Clear, colorless, viscous liquid

SPECIFIC GRAVITY: 1.047-1.049 @ 20/20 C [043]

DENSITY: 1.05 g/mL @ 25 C [058,102]

MELTING POINT: -35 C [025,205,275,371]

BOILING POINT: 340 C [017,058,275,451]

SOLUBILITY: Water: <1 mg/mL @ 20 C [700] DMSO: >=100 mg/mL @ 20 C [700] 95% Ethanol: >=100 mg/mL @ 20 C [700] Acetone: >=100 mg/mL @ 20 C [700]

OTHER SOLVENTS: Alcohol: Soluble [017] Ether: Soluble [017,031] Benzene: Soluble [017,031] Most organic solvents: Miscible [062]

OTHER PHYSICAL DATA: Very weak, ammoniacal odor [102,346] Viscosity: 0.203 poise @ 20 C [062] Refractive index: 1.4911 @ 20 C [017,047] 1.4915 @ 25 C [062] Critical temperature: 500 C [371] Critical pressure: 17 atmospheres [371] Liquid surface tension: 34 dynes/cm [371] Liquid water interfacial tension: 27 dynes/cm [371] Heat of combustion: -7400 cal/g [371] Vapor pressure: 0.1 mm Hg @ 115 C [055] 760 mm Hg @ 340 C [038] Vapor pressure: 5 mm Hg @ 182.1 C; 10 mm Hg @ 198.2 C [038] Vapor pressure: 20 mm Hg @ 216.2 C; 40 mm Hg @ 235.8 C [038] Vapor pressure: 60 mm Hg @ 247.8 C; 100 mm Hg @ 263.7 C [038]

HAP WEIGHTING FACTOR: 1 [713]

VOLATILITY:

Vapor pressure: 1 mm Hg @ 147 C; 1.1 mm Hg @ 150 C [058]

Vapor density: 9.58 [043,071]

FIRE HAZARD: Di-n-butyl phthalate has a flash point of 157 C (315 F) [043,058,071,102]. It is combustible. Fires involving this material may be controlled with a dry chemical, carbon dioxide or Halon extinguisher.

The autoignition temperature of di-n-butyl phthalate is 403 C (757 F) [043,071,102,451].

LEL: 0.5% [058,102] UEL: Not available

REACTIVITY: Di-n-butyl phthalate can react violently with chlorine [036,043,066,451]. It is incompatible with nitrates, strong oxidizers, strong alkalis and strong acids [102].

STABILITY: Di-n-butyl phthalate is stable under normal laboratory conditions. Solutions of it in water, DMSO, 95% ethanol or acetone should be stable for 24 hours under normal lab conditions [700].

USES: Di-n-butyl phthalate is used in plasticizers, cosmetics, safety glass, insecticides, printing inks, paper coatings, adhesives, elastomers and explosives. It is also used as a solvent in polysulfide dental impression materials, solvent for perfume oils, perfume fixative, textile lubricating agent and solid rocket propellent.

Comments: None

NIOSH Registry Number: TI0875000

ACUTE/CHRONIC HAZARDS: Dibutyl phthalate is an irritant and may be toxic at sufficient concentrations. It is also a lachrymator. When heated to decomposition it emits toxic fumes of carbon dioxide and carbon monoxide [058].

SYMPTOMS: Symptoms of exposure may include irritation of the eyes, nasal passages, throat and upper respiratory tract. It may cause nausea, conjunctivitis, profuse lacrimation, vomiting, dizziness, headache, tearing of eyes and photophobia [058]. It may also cause liver and kidney damage [301].

1,4-Dichlorobenzene

NOTE: The form of this name in the Clean Air Act Amendments Section 112(b) list is 1,4-Dichlorobenzene(p). The correct notation for the compound is 1,4-Dichlorobenzene.

CAS NUMBER: 106-46-7

SYNONYMS: p-Dichlorobenzene p-Chlorophenyl chloride Paradichlorobenzene

CHEMICAL FORMULA: $C_6H_4Cl_2$

MOLECULAR WEIGHT. 147.00

WLN: GR DG

PHYSICAL DESCRIPTION: Colorless or white, volatile crystals

SPECIFIC GRAVITY: 1.2475 @ 20/4 C [017,047,395]

DENSITY: 1.2417 g/mL @ 60 C [205]

MELTING POINT: 53 C [031,036,038,205]

BOILING POINT: 174 C [017,058,395,451]

SOLUBILITY: Water: <1 mg/mL @ 23 C [700] DMSO: >=100 mg/mL @ 23 C [700] 95% Ethanol: >=100 mg/mL @ 23 C [700] Acetone: >=100 mg/mL @ 23 C [700]

OTHER SOLVENTS: Alcohol: Soluble [017,031,205,430] Ether: Soluble [017,031,205,395] Benzene: Soluble [017,031,205,395] Carbon disulfide: Soluble [031,051,205,395] Chloroform: Soluble [031,051,205,395] Most organic solvents: Soluble [173,295]

OTHER PHYSICAL DATA: This chemical readily sublimes [031,051,062,071] Specific gravity: 1.4581 @ 20.5/4 C [043,058] Density: 1.458 g/mL @ 21 C [051] Boiling point: 55 C @ 10 mm Hg [017,047] Vapor pressure: 1 mm Hg @ 25 C [173] 10 mm Hg @ 54.8 C [038,043,058] Vapor pressure: 20 mm Hg @ 69.2 C; 40 mm Hg @ 84.8 C; 60 mm Hg @ 95.2 C [038] Vapor pressure: 100 mm Hg @ 108.4 C [038] Refractive index: 1.5285 @ 60 C [017,031,205] Aromatic, mothball-like odor [051,071] Odor threshold: 15-30 ppm [051,071,102,371] Lambda max (in ethanol): 258 nm, 266 nm, 273 nm, 280 nm [395] Dielectric constant: 2.41 @ 50 C [051] Surface tension: 31.4 dynes/cm @ 60 C; 27.2 dynes/cm @ 100 C [051] Heat of fusion: 29.07 cal/g [051] Burning rate: ~1.3 mm/min [371]

HAP WEIGHTING FACTOR: 1 [713]

VOLATILITY:

Vapor pressure: 0.6 mm Hg @ 20 C [055] 1.8 mm Hg @ 30 C [055,395]

Vapor density: 5.08 [043,051,058,071]

FIRE HAZARD: 1,4-Dichlorobenzene has a flash point of 65.5 C (150 F) [033,043,062,451]. It is combustible. Fires involving this material may be controlled with a dry chemical, carbon dioxide or Halon extinguisher. A water spray may also be used [043,058,371,451].

LEL: 2.5% [051,071,102] UEL: Not available

REACTIVITY: 1,4-Dichlorobenzene is incompatible with oxidizing agents [043,051,058,269]. It is also incompatible with aluminum and its alloys [269]. It liquifies when mixed with camphor, phenol and salol [295]. It will attack some forms of plastics, rubber and coatings [102].

STABILITY: 1,4-Dichlorobenzene is stable under normal laboratory conditions. Solutions of it in water, DMSO, 95% ethanol or acetone should be stable for 24 hours under normal lab conditions [700].

USES: 1,4-Dichlorobenzene is used as a moth repellant, general insecticide, pesticide, fumigant, germicide, miticide, space odorant, air deodorant, chemical intermediate for dyes and organic chemicals, mildew control agent, disintegrating paste for molding concrete and stoneware, lubricant and disinfectant. It is also used in the manufacture of 2,5-dichloroaniline, pharmaceutical manufacture, agriculture (to fumigate soil), manufacture of polyphenylene sulfide resins (used for surface coatings and molding resins) and organic synthesis.

COMMENTS: None

NIOSH Registry Number: CZ4550000

ACUTE/CHRONIC HAZARDS: 1,4-Dichlorobenzene is toxic by ingestion [043,062], inhalation, and skin absorption [036,058,269]. It is also an irritant [031,036,058,151] and may cause severe destruction of tissues [269]. When heated to decomposition it emits toxic fumes of carbon monoxide, carbon dioxide, hydrogen chloride gas and phosgene [058,102,269,371].

SYMPTOMS: Symptoms of exposure include irritation of the skin, eyes and throat [031,051,058,151], the nose [036, 058,102,295], and irritation of the mucous membranes and lungs [301]. Other symptoms of exposure include weakness, dizziness, weight loss and liver injury [031]. Exposure may lead to drowsiness, allergic purpura, acute glomerulonephritis, dermatitis and lens opacities [295]. It may also lead to hemolytic anemia, liver necrosis, headaches, periorbital swelling and profuse rhinitis [346]. It can cause anorexia, nausea, vomiting, jaundice and cirrhosis [051,071,346]. Allergic skin reactions and damage to the liver and kidneys occur [269]. Unspecified changes may occur in the eyes, lungs, thorax and respiratory system [043]. Other symptoms of exposure are hypersensitivity, oliguria, refractory anemia, dyspnea, petechial and purpuric rash, swollen hands and feet, cataracts, coughing, hypochromic anemia, granulocytopenia, aplastic anemia, congestive heart failure, nephrosis, inflammation of the mucosa, methemoglobinemia, leukemia, myeloblastic leukemia, myeloproliferative syndrome and death [173]. Central nervous system effects have been reported [301]. Pulmonary granulocytosis has also been reported [395]. Hepatic enlargement may occur [099]. It may cause pain to the eyes and nose, swelling and inflammation of the nose [051 071]. It may also cause irritation of the upper respiratory tract, a slight burning sensation and red blotching of the skin due to allergic reactions [058]. Vertigo may result from exposure [421]. Other symptoms may include asthenia, hypogranulocytosis, hyperleukocytosis, numbness, clumsiness, ascites, elevated serum bilirubin and elevated alkaline phosphatase [430].

3,3'-Dichlorobenzidine

NOTE: The form of this name in the Clean Air Act Amendments Section 112(b) list is 3,3-Dichlorobenzidine. The correct notation for the compound is 3,3'-Dichlorobenzidine.

CAS NUMBER: 91-94-1

SYNONYMS: 4,4'-Diamino-3,3'-dichlorobiphenyl o,o'-Dichlorobenzidine 3,3'-Dichloro-4,4'-biphenyldiamine

CHEMICAL FORMULA: C12H10Cl2N2

MOLECULAR WEIGHT: 253.13

WLN: ZR BG DR DZ CG

PHYSICAL DESCRIPTION: Colorless to grayish-purple crystals

SPECIFIC GRAVITY: Not available

DENSITY: Not available

MELTING POINT: 132-133 C

BOILING POINT: Not available

SOLUBILITY: Water: Insoluble DMSO: Not available 95% Ethanol: Not available Acetone: Not available

OTHER SOLVENTS: Alcohol: Very soluble Glacial acetic acid: Very soluble Ether: Soluble Benzene: Very soluble

OTHER PHYSICAL DATA: Not available

HAP WEIGHTING FACTOR: 1 [713]

VOLATILITY: Not available

FIRE HAZARD: Flash point data for 3,3'-dichlorobenzidine are not available. It is probably combustible. Fires involving it should be extinguished with dry chemical, carbon dioxide, and/or Halon extinguishers.

LEL: Not available UEL: Not available

REACTIVITY: Not available

STABILITY: 3,3'-Dichlorobenzidine is likely to be unstable to light, heat and air.

USES: 3,3'-Dichlorobenzidine is used as an intermediate for dyes and pigments and as a curing agent for isocyanate-terminated resins for urethane plastics.

COMMENTS: None

NIOSH Registry Number: DD0525000

ACUTE/CHRONIC HAZARDS: 3,3'-Dichlorobenzidine is a known carcinogen. It is also an allergen and an irritant and can be absorbed by the skin. When heated to decomposition it emits toxic fumes.

SYMPTOMS: 3,3'-Dichlorobenzidine causes tumors in animal experiments.

Dichloroethyl ether

CAS NUMBER: 111-44-4

SYNONYMS: bis(2-Chloroethyl) ether 2,2'-Dichlorodiethyl ether 2,2'-Dichlorethyl ether

CHEMICAL FORMULA: C4H8Cl2O

MOLECULAR WEIGHT: 143.02

WLN: G2O2G

PHYSICAL DESCRIPTION: Clear, colorless liquid

SPECIFIC GRAVITY: 1.2199 @ 20/4 C

DENSITY: Not available

MELTING POINT: -24.5 C

BOILING POINT: 178 C

SOLUBILITY: Water: Reaction DMSO: >=100 mg/mL @ 21 C [700] 95% Ethanol: >=100 mg/mL @ 21 C [700] Acetone: >=100 mg/mL @ 21 C [700]

OTHER SOLVENTS: 10% Ethanol: Soluble Most organic solvents: Soluble Ether: Soluble Benzene: Soluble

OTHER PHYSICAL DATA: Boiling point: 67 C @ 15 mm Hg; 75 C @ 20 mm Hg; 82-83 C 23 mm Hg Refractive index: 1.4575 @ 20 C Specific gravity: 1.2220 @

20/20 C Weight/gallon: 10.2 pounds @ 20 C Fruity pungent odor Dissolves greases, oils, fats

HAP WEIGHTING FACTOR: 10 [713]

VOLATILITY:

Vapor pressure: 0.71 mm Hg @ 20.0 C; 1.4 mm Hg @ 25 C; 5 mm Hg @ 49.3 C

Vapor density: 4.93

FIRE HAZARD: The flash point of dichloroethyl ether is 55 C (131 F). It is combustible. Fires involving this compound may be controlled using a dry chemical, carbon dioxide or Halon extinguisher.

The autoignition temperature for dichloroethyl ether is 368 C (696 F).

LEL: Not available UEL: Not available

REACTIVITY: Dichloroethyl ether reacts vigorously with oleum, chlorosulfonic acid and oxidizing materials and it reacts with water or steam to produce toxic and corrosive fumes. It may form highly reactive peroxides on exposure to air and light.

STABILITY: Gas chromatography stability screening indicates that solutions of dichloroethyl ether in 10% ethanol in water are stable for at least 24 hours [700]. Dichloroethyl ether is hydrolyzed slowly in aqueous dimethylformamide at pH 7 (Kw 30 C <0.0005/minute). The neat material is sensitive to exposure to air and light.

USES: Dichloroethyl ether is used as a general solvent, a selective solvent for production of high-grade lubricating oils, for textile scouring and cleansing, for wetting and penetrating compounds, in organic synthesis, and as a component in paints, varnishes, lacquers, and finish removers. It is also used in spotting and dry cleaning formulations, as a soil fumigant, in medicinals and pharmaceuticals, and for the manufacture of insecticides and acaricides.

COMMENTS: None

NIOSH Registry Number: KN0875000

ACUTE/CHRONIC HAZARDS: Dichloroethyl ether is a strong irritant and a lachrymator. It may be toxic by inhalation and ingestion and also may be absorbed through the skin. When heated to decomposition it emits toxic fumes. There is evidence that this compound is an experimental carcinogen.

SYMPTOMS: Symptoms of exposure may include irritation of the eyes, skin, eyes, stomach and respiratory tract; coughing, nausea, retching, lung lesions, pulmonary edema, dermatitis, conjunctivitis, lacrimation and liver and kidney injuries.

1,3-Dichloropropene

CAS NUMBER: 542-75-6

SYNONYMS: 1,3-Dichloropropylene alpha-Chloroallyl chloride gamma-Chloroallyl chloride

CHEMICAL FORMULA: C3H4Cl2

MOLECULAR WEIGHT: 110.97

WLN: G2U1G

PHYSICAL DESCRIPTION: Clear, colorless to light straw-colored liquid

SPECIFIC GRAVITY: 1.217 @ 20/4 C [017]

DENSITY: 1.220 g/mL @ 20 C [169]

MELTING POINT: -84 C [107]

BOILING POINT: 104 C [107,430,451]

SOLUBILITY: Water: <0.1 mg/mL @ 16.5 C [700] DMSO: >=100 mg/mL @ 20 C [700] 95% Ethanol: >=100 mg/mL @ 20 C [700] Acetone: >=100 mg/mL @ 20 C [700] Methanol: Miscible [172,173] Toluene: Soluble [062,395]

OTHER SOLVENTS: Chloroform: Soluble [017,047,205,395] Octane: Soluble [062,395] Carbon tetrachloride: Miscible [172,173] Heptane: Miscible [172,173] Ether: Soluble [017,047,205,395] Benzene: Soluble [017,047,395] Hydrocarbons: Miscible [169] Halogenated solvents: Miscible [169] Ketones: Miscible [169] Esters: Miscible [169] Most organic solvents: Miscible [430]

OTHER PHYSICAL DATA: Specific gravity: 1.217 @ 25/4 C [172] 1.225 @ 10/4 C [062] Specific gravity: ~1.2 @ 20/20 C [430] Density: 1.217 g/mL @ 25 C [173] Vapors form explosive mixtures with air [451] Chloroform-like odor [031,173,421,451] Odor is also described as sharp, sweet, penetrating and irritating [055,395, 421] Odor threshold: 1-3 ppm [173,421] Refractive index: 1.4735 @ 22 C [107,169] 1.4701 @

20 C [269,275] Burning rate (estimated): 3.4 mm/min [371] Liquid surface tension: 31.2 dynes/cm @ 24 C [371] Liquid water interfacial tension: 23.8 dynes/cm @ 24 C [371] Latent heat of vaporization (estimated): 62.8 cal/g [371] Heat of combustion (estimated): 3900 cal/g [371] This compound is a mixture of isomers

HAP WEIGHTING FACTOR: 1 [713]

VOLATILITY:

Vapor pressure: 27.9 mm Hg @ 20 C [169,172] 28 mm Hg @ 25 C [107,421,430]

Vapor density: 3.83 [055,395]

FIRE HAZARD: 1,3-Dichloropropene has a flash point of 35 C (95 F) [043,107,371,451] and it is flammable. Fires involving this material can be controlled with a dry chemical, carbon dioxide or Halon extinguisher.

LEL: 5.3% [451] UEL: 14.5% [421,451]

REACTIVITY: 1,3-Dichloropropene reacts vigorously with oxidizing materials [043,269]. It reacts with aluminum, active metals and halogenated compounds [107,421]. It also reacts with acids and thiocyanates [107]. It is corrosive to magnesium, magnesium alloys and aluminum alloys [169]. It is also incompatible with some metal salts [421].

STABILITY: 1,3-Dichloropropene is sensitive to heat [043,107]. Solutions of it in water, DMSO, 95% ethanol or acetone should be stable for 24 hours under normal lab conditions [700].

USES: 1,3-Dichloropropene is used as a soil fumigant, nematocide, pesticide and chemical intermediate. It is also used in organic synthesis.

COMMENTS: Epichlorohydrin (1-2%) is often added as a stabilizer and corrosion inhibitor [055,421].

NIOSH Registry Number: UC8310000

ACUTE/CHRONIC HAZARDS: 1,3-Dichloropropene is toxic by ingestion, inhalation and skin absorption [062,151,269,451]. It is a lachrymator, a strong irritant and is extremely destructive to tissues of the mucous membranes and upper respiratory tract, eyes and skin. When heated to decomposition it emits toxic fumes of carbon monoxide, carbon dioxide and hydrogen chloride gas [043,269,451].

SYMPTOMS: Symptoms of exposure may include severe irritation of the skin, eyes and mucous membranes [031,099,151,301]. Corrosion and extreme destruction of the tissues of the mucous membranes and upper respiratory tract, eyes and skin may also occur [269]. Other symptoms may include respiratory distress, abdominal cramps, nasal irritation [107], and lacrimation [107,151,269]. Headache and nausea may occur [107, 269,395] and unconsciousness may also result [031,151,173]. Vomiting, severe lung injury, abdominal pain, muscle twitching, fatigue, irritability, chest discomfort, malaise, heart injury, kidney damage, liver damage and coma [151] have been reported as well as dizziness, dyspnea, difficulty in concentrating and decreased libido [173]. Other symptoms may include weakness and neck pain [395].

Inhalation may be fatal as a result of spasm, inflammation and edema of the larynx and bronchi, chemical pneumonitis and pulmonary edema. Symptoms of inhalation may include burning sensation, coughing, wheezing, laryngitis and shortness of breath [269]. Other symptoms of inhalation may include gasping, refusal to breathe, substernal pain and central nervous system depression [151] and lung irritation may occur [301]. Ingestion may cause acute gastrointestinal distress and pulmonary congestion [151]. Symptoms of skin contact may include a marked inflammatory response of epidermis and underlying tissues [151] and skin injuries may occur [395] that may also include vesication [165]. Prolonged skin contact may lead to severe burns [169,371,395]. Eye contact may cause conjunctivitis [395] and it may also result in eye damage [165].

Dichlorvos

CAS NUMBER: 62-73-7

SYNONYMS: Dimethyl dichlorovinyl phosphate 2,2-Dichlorovinyl dimethyl phosphoric acid ester

CHEMICAL FORMULA: $C_4H_7Cl_2O_4P$

MOLECULAR WEIGHT: 220.98 [701]

PHYSICAL DESCRIPTION: Oily colorless to amber liquid with an aromatic chemical odor (Patty, 1963; NIOSH/OSHA, 1981) [701]

SPECIFIC GRAVITY: 1.415 @ 25C/4C (Merck, 1976) [701]

MELTING POINT: NA

BOILING POINT: 284 F, 140 C at 20 mmHg (Merck, 1976) [701]

SOLUBILITY: Water: Approximately 1 g/100 mL (Merck, 1976) [701] Kerosene: Slightly soluble [705] Most organic solvents: readily soluble [705] Glycerin: Slightly soluble [703]

OTHER PHYSICAL DATA: Vapor Pressure: 0.01 mmHg at 30 C (Sunshine, 1969) [701] Boiling Point: 120 C @ 14 mm, 77 C @ 1 mm [703]

HAP WEIGHTING FACTOR: 1 [713]

VOLATILITY:

Vapor Pressure: 0.01 mm @ 86 F [704]

FIRE HAZARD: Flash Point: >175 F (open cup) [704]

Class III combustible liquid [704].

Autoignition temperature: Not available

LEL: Not available UEL: Not available

REACTIVITY: Incompatible or reacts with strong acids, strong alkalis [704].

CORROSIVITY:

Corrosive to iron and mild steel [704]. Non-corrosive to stainless-steel, aluminum, nickel, Hostelloz 13, Teflon [705].

STABILITY: Stable in the presence of hydrocarbon solvents; undergoes hydrolysis in the presence of water and is readily decomposed by strong acids and bases. [705] When heated to decomposition it emits very toxic fumes of Cl- and POx. [703]

USES: Organophosphate insecticide used against flies, mosquitos, gnats, cockroaches, and other nuisance pests. Applied in and around dwellings, on food animals and domestic animals, crops, food packaging, and food storage areas. [705]

COMMENTS: None

NIOSH Registry Number: TC0350000

ACUTE/CHRONIC HAZARDS: A poison by ingestion, inhalation, skin contact, subcutaneous, intravenous, intraperitoneal, and possibly other routes. An experimental teratogen and suspected carcinogen; produces experimental reproductive effects; human mutagenic data exists. A cholinesterase inhibitor, used in flea collars for pets. No neurotoxicity has been observed. Very rapidly metabolized and excreted. When heated to decomposition, emits very toxic fumes of Cl- and POx. [703]

SYMPTOMS: Symptoms of exposure may include miosis, aching eyes; rhinorrhea; headache; tight chest, wheezing, laryngeal spasm, salivation; cyanosis; anorexia, nausea, vomiting, diarrhea; sweat; muscle fatigue, paralysis, giddiness, ataxia; convulsions; low blood pressure, cardiovascular irregularities; irritation of the skin and eyes. [704]

Diethanolamine

CAS NUMBER: 111-42-2

SYNONYMS: 2,2'-Dihydroxydiethylamine 2,2'-Iminodiethanol bis-2-Hydroxyethylamine Diethylolamine Di(2-hydroxyethyl)amine

CHEMICAL FORMULA: C4H11NO2

MOLECULAR WEIGHT: 105.14

WLN: Q2M2Q

PHYSICAL DESCRIPTION: Viscous, oily liquid or colorless crystals

SPECIFIC GRAVITY: 1.0966 @ 20/4 C [017,047]

DENSITY: 1.097 g/mL [269,275]

MELTING POINT: 28 C [031,055,421,451]

BOILING POINT: 269.1 C (decomposes) [043,055]

SOLUBILITY: Water: >=100 mg/mL @ 14 C [700] DMSO: >=100 mg/mL @ 14 C [700] 95% Ethanol: >=100 mg/mL @ 14 C [700] Acetone: >=100 mg/mL @ 14 C [700] Methanol: Miscible [031]

OTHER SOLVENTS: Alcohol: Miscible [031,295] Carbon tetrachloride: <0.1% @ 25 C [031] n-Heptane: <0.1% @ 25 C [031] Ether: 0.8% @ 25 C [031] Benzene: 4.2% @ 25 C [031] Chloroform: Miscible [295]

OTHER PHYSICAL DATA: Boiling point: 154-155 C @ 10 mm Hg [017,025,047] Refractive index: 1.4770 @ 20 C [269,275] A 5% solution in water has a pH of 10.2 to 11.4 [295] Viscosity: 351.9 centipoise @ 30 C; 53.85 centipoise @ 60 C [031] Faint ammoniacal odor [031,058,295] Specific gravity: 1.092 @ 30/20 C [055,062] Dipole moment: 2.81 [031] Latent heat of vaporization: 148 cal/g [371] Heat of combustion: -6000 cal/g [371] Heat of solution: -7 cal/g [371] Reid vapor pressure: 0.97 psia [371]

HAP WEIGHTING FACTOR: 1 [713]

VOLATILITY:

Vapor pressure: 5 mm Hg @ 138 C [043] <0.01 mm Hg @ 20 C [055,421]

Vapor density: 3.65 [043,055,058]

FIRE HAZARD: Diethanolamine has a flash point of 137 C (279 F) [205,269]. It is combustible. Fires involving this material may be controlled with a dry chemical, carbon dioxide or Halon extinguisher. A water spray may also be used [043,058,371].

The autoignition temperature of diethanolamine is 662 C (1224 F) [043,371,451].

LEL: 1.6% (calculated) [371] UEL: 9.8% (estimated) [371]

REACTIVITY: Diethanolamine can react with oxidizing materials, acids, CO2, copper alloys, aluminum, zinc, galvanized iron and copper [269].

STABILITY: Diethanolamine is hygroscopic [025,031]. It may be sensitive to exposure to air and light [295].

USES: Diethanolamine is used as an emulsifying and dispersing agent, to solubilizing fusidic acid, for the preparation of salts of iodinated organic acids, to scrub toxic gases from smoke screen, a cation in many water soluble salts of drugs, pesticides, industrial basic solvents, as rubber chemicals, in surface active agents used in textile specialties, herbicides, petroleum demulsifiers, cosmetics, pharmaceuticals, production of lubricants for textile industry, in organic synthesis, cutting oils, shampoos, cleaners and polishers, chemical intermediate for resins, plasticizers, absorbent for acid gases, solubilizing 2,4-D and humectant.

COMMENTS: None

NIOSH Registry Number: KL2975000

ACUTE/CHRONIC HAZARDS: Diethanolamine may be corrosive [269], toxic, and may cause severe eye and mild skin irritation [043]. When heated to decomposition it may emit toxic fumes of carbon dioxide, carbon monoxide and nitrogen oxides.

SYMPTOMS: Symptoms of exposure may include sedation, ataxia, loss of righting reflex, cyanosis and death [151]. It may cause headaches, nausea, vomiting, coma, coughing, blindness, skin and eye irritation, abdominal pain and liver and kidney damage [058]. Other symptoms include wheezing, shortness of breath, laryngitis and damage to the eyes [269]. It may also cause erythema and blistering [301].

N,N-Diethylaniline

NOTE: Although N,N-diethylaniline is listed in the Clean Air Act ammendment [708], it lists N,N-dimethylaniline as a synonym and further lists the CAS number for N,N-dimethylaniline (121-69-7). This is clearly a typographical error where the letter "m" was ommitted. However, since this has not been noted, many people are still making the assumption that the compound in the Clean Air Act ammendment is the above compound when it is not. Therefore, we have included data for both this compound and N,N-dimethylaniline.

CAS NUMBER: 91-66-7

SYNONYMS: N,N-Diethylaminobenzene N,N-Diethylbenzenamine Diethylphenylamine

CHEMICAL FORMULA: C10H15N

MOLECULAR WEIGHT: 149.24

WLN: 2N2&R

PHYSICAL DESCRIPTION: Clear, light yellow liquid

SPECIFIC GRAVITY: 0.93507 @ 20/4 C

DENSITY: Not available

MELTING POINT: -38.8 C

BOILING POINT: 215.5 C; 92 C @ 10 mm Hg

SOLUBILITY: Water: <1 mg/mL @ 24.5 C [700] DMSO: >=100 mg/mL @ 24.5 C [700] 95% Ethanol: >=100 mg/mL @ 24.5 C [700] Acetone: >=100 mg/mL @ 24.5 C [700]

OTHER SOLVENTS: Chloroform: Slightly soluble Ether: Soluble

OTHER PHYSICAL DATA: Refractive Index [20 D]: 1.5409

HAP WEIGHTING FACTOR: 1 [713]

VOLATILITY:

Vapor pressure: 1 mm Hg @ 49.7 C

It is volatile with steam.

FIRE HAZARD: N,N-Diethylaniline has a flash point of 85 C (185 F). It is combustible. Fires involving this material can be controlled with a dry chemical, carbon dioxide, foam or Halon extinguisher.

LEL: Not available UEL: Not available

REACTIVITY: Not available

STABILITY: N,N-Diethylaniline is stable under normal laboratory conditions.

USES: N,N-Diethylaniline is used as a dye intermediate and also as an intermediate for the synthesis of other compounds.

COMMENTS: None

NIOSH Registry Number: BX3400000

ACUTE/CHRONIC HAZARDS: N,N-Diethylaniline is an irritant, may be absorbed through the skin, and can be toxic by ingestion. Repeated exposure may have a cumulative effect.

SYMPTOMS: Symptoms of exposure may include weakness, tremors, cyanosis, drowsiness and central nervous system depression.

Diethyl sulfate

CAS NUMBER: 64-67-5

SYNONYMS: Ethyl sulfate Sulfuric acid diethyl ester

CHEMICAL FORMULA: C4H10O4S

MOLECULAR WEIGHT: 154.20 [703]

PHYSICAL DESCRIPTION: Colorless, oily liquid; peppermint odor [703]

DENSITY: 1.172 @ 15C/4C [703]

MELTING POINT: -25 C [703]

BOILING POINT: 209.5 C (decomposes) [703]

SOLUBILITY: Water: Insoluble, decomposed by hot water [703]

OTHER PHYSICAL DATA: Vapor Pressure: 1 mm @ 47 C [703] Vapor Density: 5.31 [703] Miscible with alcohol and ether [703]

HAP WEIGHTING FACTOR: 1 [713]

VOLATILITY:

Vapor Pressure: 1 mm @ 47 C [703]

Vapor Density: 5.31 [703]

FIRE HAZARD: Flash Point: 220 F [703]

Combustible when exposed to heat or flame. To fight fire use alcohol foam, water, foam, CO2. [703]

Autoignition temperature: Not available

LEL: Not available UEL: Not available

REACTIVITY: Can react vigorously with oxidizing materials. Ignites on contact with potassium tert-butoxide. Reacts violently with 3,8-dinitro- 6-phenylphenanthridine + water. Reaction with iron + water forms the explosive hydrogen gas. [703] This compound is very reactive with chloride bases [702].

STABILITY: Combustible when exposed to heat or flame. When heated to decomposition it emits toxic fumes of SOx. [703]

USES: Used chiefly as an ethylating agent; as an accelerator in the sulfation of ethylene; in some sulfonations [715].

COMMENTS: None

NIOSH Registry Number: WS7875000

ACUTE/CHRONIC HAZARDS: Poison by subcutaneous route; moderately toxic by ingestion, skin contact and inhalation. A suspected human carcinogen; an experimental carcinogen, tumorigen and teratogen. Produces experimental reproductive effects; human mutagenic data exist. A severe skin and eye irritant. When heated to decomposition it emits toxic fumes of Sox. [703] There is sufficient evidence of carcinogenicity in animals, limited evidence of carcinogenicity in humans; the agent is probably carcinogenic to humans. [723]

SYMPTOMS: Not available.

3,3'-Dimethoxybenzidine

NOTE: The form of this name in the Clean Air Act Amendments Section 112(b) list is 3,3-Dimethoxybenzidine. The correct notation for the compound is 3,3'-Dimethoxybenzidine.

CAS NUMBER: 119-90-4

SYNONYMS: o-Dianisidine 4,4'-Diamino-3,3'-dimethoxybiphenyl

CHEMICAL FORMULA: $C_{14}H_{16}N_2O_2$

MOLECULAR WEIGHT: 244.29

WLN: ZR BO1 DR DZ CO1

PHYSICAL DESCRIPTION: Colorless crystals

SPECIFIC GRAVITY: Not available

DENSITY: Not available

MELTING POINT: 137-138 C [025,031,047,346,395]

BOILING POINT: Not available

SOLUBILITY: Water: <0.1 mg/mL @ 20 C [700] DMSO: >=100 mg/mL @ 20 C [700] 95% Ethanol: <1 mg/mL @ 20 C [700] Acetone: 5-10 mg/mL @ 20 C [700]

OTHER SOLVENTS: Ether: Soluble [017,031,047,062,395] Benzene: Soluble [017,031,047,395] Chloroform: Soluble [017,395] Most lipids: Probably soluble [395] Most organic solvents: Probably soluble [395]

OTHER PHYSICAL DATA: Turns violet on standing

HAP WEIGHTING FACTOR: 1 [713]

VOLATILITY:

Vapor density: 8.5

FIRE HAZARD: 3,3'-Dimethoxybenzidine has a flash point of 206 C (403 F) [042,062,274]. It is combustible. Fires involving this material may be controlled with a dry chemical, carbon dioxide or Halon extinguisher. A water spray may also be used [355].

LEL: Not available UEL: Not available

REACTIVITY: 3,3'-Dimethoxybenzidine is a weak base and has general characteristics of primary aromatic amines [395].

STABILITY: 3,3'-Dimethoxybenzidine is sensitive to heat, air and prolonged exposure to light.

USES: Not available.

COMMENTS: None

NIOSH Registry Number: DD0875000

ACUTE/CHRONIC HAZARDS: 3,3'-Dimethoxybenzidine is toxic by inhalation and skin contact [025,036] or by ingestion [036]. It can be absorbed through the skin [346]. It is a positive animal carcinogen [015,346,396] and an irritant of the skin and eyes [025,355]. When heated to decomposition it emits toxic fumes of nitrogen oxides [042].

SYMPTOMS: Symptoms of exposure may include irritation of the skin and eyes; headaches, drowsiness, and cyanosis [355]. Other symptoms may include severe irritation of the nose, sneezing, and cancer of the bladder [036].

Dimethylaminoazobenzene

CAS NUMBER: 60-11-7

SYNONYMS: N,N-Dimethyl-p-phenylazooaniline P-Dimethylaminoazobenzene N,N-Dimethyl-4-aminoazobenzene 4-Dimethylaminoazobenzene

CHEMICAL FORMULA: C14H15N3

MOLECULAR WEIGHT: 225.30

WLN: 1N1&R DNUNR

PHYSICAL DESCRIPTION: Yellow crystalline leaflets

SPECIFIC GRAVITY: Not available

DENSITY: Not available

MELTING POINT: 111 C (decomposes) [055,269,274]

BOILING POINT: Not available

SOLUBILITY: Water: <1 mg/mL @ 22 C [700] DMSO: 5-10 mg/mL @ 22 C [700] 95% Ethanol: <1 mg/mL @ 22 C [700] Acetone: 50-100 mg/mL @ 22 C [700] Toluene: 1.2-3 g/100 mL @ room temperature [105]

OTHER SOLVENTS: Benzene: 2.8 g/100 mL @ room temperature [105] Ether: Soluble [031,062,205,395] Mineral acids: Soluble [031,042,062,395] Oils: Soluble [031,042,062,395] Chloroform: Soluble [031,205,395] Petroleum ether: Soluble [031,205,395] Cellosolve: Soluble [105] Ethyl acetate: 2 g/100 mL @ room temperature [105] Linseed oil: 2.6-5 g/100 mL @ room temperature [105] Mineral oil: 0.1-1 g/100 mL @ room temperature [105] Oleic acid: 1.5-4.2 g/100 mL @ room temperature [105] Paraffin wax: 0.5-2 g/100 mL @ room temperature [105] Stearic acid: 1.5-10 g/100 mL @ room temperature [105] Turpentine: Soluble [105] White spirit: 0.4-1.5 g/100 mL @ room temperature [105] Xylene: 0.75-1.65 g/100 mL @ room temperature [105]

OTHER PHYSICAL DATA: Yellow solution in concentrated sulfuric acid; on dilution-red solution Red solution in hydrochloric acid, from which the hydrochloride crystallizes in purple-red, hair-like needles; the base separates as an orange precipitate on addition of sodium hydroxide

HAP WEIGHTING FACTOR: 1 [713]

VOLATILITY:

Vapor density: 7.78

FIRE HAZARD: Flash point data for 4-dimethylaminoazobenzene are not available. It is probably combustible. Fires involving this material may be controlled with a dry chemical, carbon dioxide or Halon extinguisher.

LEL: Not available UEL: Not available

REACTIVITY: 4-Dimethylaminoazobenzene is incompatible with strong oxidizing agents and strong acids [269].

STABILITY: 4-Dimethylaminoazobenzene is sensitive to heat and light [051]. Solutions of it in water, DMSO, 95% ethanol or acetone should be stable for 24 hours under normal lab conditions [700].

USES: 4-Dimethylaminoazobenzene is used as a dye (for polishes and other wax products, polystyrene, petrol and soap); indicator in volumetric analysis (yellow @ pH 4.0; red @ pH 2.9); for the determination of free hydrochloric acid in gastric juice; spot test identification of peroxidized fats; and formerly as a food dye.

COMMENTS: The FDA prohibits its use in foods or beverages.

NIOSH Registry Number: BX7350000

ACUTE/CHRONIC HAZARDS: Dimethylaminoazobenzene is harmful by ingestion, inhalation or skin absorption [269]. There is also evidence that it is carcinogenic in animals [015,269,395] and it is an OSHA carcinogen [015,051,269,325]. When heated to decomposition it emits toxic fumes of carbon monoxide, carbon dioxide and nitrogen oxides [042,269].

SYMPTOMS: Symptoms of exposure may include contact dermatitis [051, 395]. Mammals exposed to lethal doses have developed damage to the liver and kidneys. Methemoglobinemia is caused by most of the compounds in this class and this can produce asphyxia with resultant damage to cells of the central nervous system [051].

N,N-Dimethylaniline

NOTE: Although N,N-diethylaniline is listed in the Clean Air Act ammendment [708], it lists N,N-dimethylaniline as a synonym and further lists the CAS number for

N,N-dimethylaniline (121-69-7). This is clearly a typographical error where the letter "m" was ommitted. However, since this has not been noted, many people are still making the assumption that the compound in the Clean Air Act ammendment is N,N-diethylaniline when it is not. Therefore, we have included data for both N,N-diethylaniline and N,N-dimethylaniline.

CAS NUMBER: 121-69-7

SYNONYMS: Aminodimethylbenzene N,N-Dimethylbenzenamine (Dimethylamino)benzene Dimethylaniline Dimethylphenylamine N,N-Dimethylphenylamine

CHEMICAL FORMULA: C8H11N

MOLECULAR WEIGHT: 121.20

WLN: 1N1&R

PHYSICAL DESCRIPTION: Yellowish to brownish, oily liquid

SPECIFIC GRAVITY: 0.9557 @ 20/4 C

DENSITY: Not available

MELTING POINT: 2.45 C

BOILING POINT: 194 C

SOLUBILITY:

Water: <1 mg/mL @ 21 C [700] DMSO: >=100 mg/mL @ 20 C [700] 95% Ethanol: >=100 mg/mL @ 20 C [700] Acetone: >=100 mg/mL @ 20 C [700]

OTHER SOLVENTS:

Chloroform: Soluble Ether: Soluble Benzene: Soluble

OTHER PHYSICAL DATA: Boiling point: 77 C @ 13 mm Hg Refractive index: 1.5582 @ 20 C

HAP WEIGHTING FACTOR: 1 [713]

VOLATILITY:

Vapor Pressure: 1 mm Hg @ 29.5 C

Vapor Density: 4.17

FIRE HAZARD: N,N-Dimethylaniline has a flash point of 62.7 C (145 F). It is combustible. Fires involving this material can be controlled with a dry chemical, carbon dioxide or Halon extinguisher.

The autoignition temperature for this compound is 370 C (700 F).

LEL: Not available UEL: Not available

REACTIVITY: N,N-Dimethylaniline can react with oxidizing materials such as benzoyl peroxide.

STABILITY: N,N-Dimethylaniline may be sensitive to prolonged exposure to air or light.

USES: Used in the manufacture of basic dyes, vanillin and Michler's ketone. Also used as reagent for methanol, methyl furfural, H2O2, nitrate, alcohol and formaldehyde. It is also used as a stabilizer and with MBTH in a colorimetric peroxidase determination. Reagent in a sensitive procedure using p-anisidene-N,N-dimethylaniline for the catalytic determination of micro amounts of ferric and ferrous ions in as little as 10 E-7 M.

NIOSH Registry Number: BX4725000

ACUTE/CHRONIC HAZARDS: N,N-Dimethylaniline is toxic, a severe irritant, and it may be absorbed through the skin. When heated to decomposition this compound emits highly toxic fumes.

SYMPTOMS: Symptoms of exposure may include central nervous system (CNS) depression, methemoglobinemia, collapse, visual disturbances, intense abdominal pain, weakness, tremors, cyanosis, and drowsiness. Chronic exposure may cause fatigue, loss of appetite, headache and dizziness.

3,3'-Dimethylbenzidine

NOTE: The form of this name in the Clean Air Act Amendments Section 112(b) list is 3,3'-dimethyl benzidine. The correct notation for the compound is 3,3'-Dimethylbenzidine.

CAS NUMBER: 119-93-7

SYNONYMS: 4,4'-Diamino-3,3'-Dimethylbiphenyl 3,3'-Dimethyl-4,4'-biphenyldiamine 4,4'-Di-o-toluidine 3,3'-Dimethylbenzidine

CHEMICAL FORMULA: $C_{14}H_{16}N_2$

MOLECULAR WEIGHT: 212.29

WLN: ZR B1 DR DZ C1

PHYSICAL DESCRIPTION: White to reddish crystals

SPECIFIC GRAVITY: 1.0 [421]

DENSITY: Not available

MELTING POINT: 129-131 C [033,043,055,062]

BOILING POINT: 200 C [421]

SOLUBILITY: Water: <1 mg/mL @ 19 C [700] DMSO: >=100 mg/mL @ 19 C [700] 95% Ethanol: <1 mg/mL @ 19 C [700] Acetone: >=100 mg/mL @ 19 C [700]

OTHER SOLVENTS: Alcohol: Soluble [016,033,043,062] Ether: Soluble [016,033,043,062] Dilute acids: Soluble [033,421] Acetic acid: Soluble [043]

OTHER PHYSICAL DATA: Not available

HAP WEIGHTING FACTOR: 1 [713]

VOLATILITY: Not available

FIRE HAZARD: Flash point data for 3,3'-dimethylbenzidine are not available. It is probably combustible. Fires involving this material may be controlled with a dry chemical, carbon dioxide or Halon extinguisher.

LEL: Not available UEL: Not available

REACTIVITY: 3,3'-Dimethylbenzidine is a weak base that forms salts with hydrochloric acid or sulfuric acid. It can be acetylated [395]. It is incompatible with strong oxidizers [058].

STABILITY: 3,3'-Dimethylbenzidine is sensitive to exposure to light [033,062]. It may be sensitive to prolonged exposure to air. UV spectrophotometric stability screening indicates that solutions of the chemical in 95% ethanol are stable for at least 24 hours [700].

USES: 3,3'-Dimethylbenzidine is used in the manufacture of dyes, as a sensitive reagent for gold (1:10 million detectable) and for free chlorine in water, and in the production of polyurethane-based high-strength elastomers, coatings and rigid plastics.

COMMENTS: The use of 3,3'-dimethylbenzidine in the United Kingdom is controlled by the Carcinogenic Substances Regulations 1967 [029,036,401].

NIOSH Registry Number: DD1225000

ACUTE/CHRONIC HAZARDS: 3,3'-Dimethylbenzidine is harmful by ingestion, inhalation or skin absorption [036,058,395]. It is rapidly absorbed through intact skin [421] and it is an irritant [036,346]. When heated to decomposition it emits toxic fumes of nitrogen oxides [043].

SYMPTOMS: Symptoms of exposure may include irritation of the skin and eyes [036] and bladder irritation [301].

Dimethylcarbamoyl chloride

NOTE: The form of this name in the Clean Air Act Amendments Section 112(b) list is Dimethyl carbamyl chloride. The correct notation for the compound is Dimethylcarbamoyl chloride.

CAS NUMBER: 79-44-7

SYNONYMS: Chloroformic acid dimethylamide Dimethylcarbamyl chloride

CHEMICAL FORMULA: C3H6ClNO

MOLECULAR WEIGHT: 107.55

WLN: GVN1&1

PHYSICAL DESCRIPTION: Clear colorless liquid

SPECIFIC GRAVITY: 1.678 @ 20/4 C

DENSITY: Not available

MELTING POINT: -33 C

BOILING POINT: 165-167 C

SOLUBILITY: Water: Decomposes DMSO: >=100 mg/mL @ 21 C [700] 95% Ethanol: >=100 mg/mL @ 21 C [700] Acetone: >=100 mg/mL @ 21 C [700]

OTHER SOLVENTS: Not available

OTHER PHYSICAL DATA: Not available

HAP WEIGHTING FACTOR: 1 [713]

VOLATILITY:

Vapor density: 3.73

FIRE HAZARD: Dimethylcarbamoyl chloride has a flash point of 68 C (155 F). It is combustible. Fires involving this material can be controlled with a dry chemical, carbon dioxide or Halon extinguisher.

LEL: Not available UEL: Not available

REACTIVITY: Dimethylcarbamoyl chloride can react with water, steam and acids.

STABILITY: Dimethylcarbamoyl chloride is unstable in water and should be protected from moisture. NMR stability screening indicates that solutions of this compound in DMSO are stable for less than two hours but solutions in acetone are stable for at least 24 hours [700].

USES: Dimethylcarbamoyl chloride is used as a chemical intermediate in the production of drugs and pesticides.

COMMENTS: None

NIOSH Registry Number: FD4200000

ACUTE/CHRONIC HAZARDS: Dimethylcarbamoyl chloride is a severe local irritant and a lachrymator. It produces toxic and corrosive fumes on contact with water, steam, acids or acid fumes; or when heated to decomposition.

SYMPTOMS: This compound can cause severe local irritation of the skin on contact, as well as irritation of the eyes and lacrimation.

N,N-Dimethylformamide

NOTE: The form of this name in the Clean Air Act Amendments Section 112(b) list is Dimethyl formamide. The correct notation for the compound is N,N-Dimethylformamide.

CAS NUMBER: 68-12-2

SYNONYMS: N-Formyldimethylamine

CHEMICAL FORMULA: C3H7NO

MOLECULAR WEIGHT: 73.09

WLN: VHN1&1

PHYSICAL DESCRIPTION: Colorless to very slightly yellow liquid

SPECIFIC GRAVITY: 0.9487 @ 20/4 C [017,053]

DENSITY: 0.95 g/mL [055]

MELTING POINT: -61 C [043,102,269,275]

BOILING POINT: 149-156 C @ 760 mm Hg [017]

SOLUBILITY: Water: >=100 mg/mL @ 22 C [700] DMSO: >=100 mg/mL @ 22 C [700] 95% Ethanol: >=100 mg/mL @ 22 C [700] Acetone: >=100 mg/mL @ 22 C [700]

OTHER SOLVENTS: Chloroform: Soluble [017] Ether: Soluble [017] Benzene: Soluble [017] Alcohol: Soluble [017]

OTHER PHYSICAL DATA: pH (0.5 M solution in H2O): 6.7 [031] Evaporation rate (butyl acetate = 1): 0.17 [102] Faint amine odor [031,036,058,102] Refractive index: 1.4305 @ 20 C [017,047,269,275] 1.42803 @ 25 C [031] Saturation concentration: 12 g/m3 @ 20 C [055] log P oct: - 0.87/ - 0.59 [055] Odor threshold: 100 ppm [102] Burning rate: 2.2 mm/min [371]

HAP WEIGHTING FACTOR: 1 [713]

VOLATILITY:

Vapor pressure: 3.7 mm Hg @ 25 C [043,051,421] 2.7 mm Hg @ 20 C [055,102]

Vapor density: 2.51 [043,051,055]

FIRE HAZARD: The flash point for N,N-dimethylformamide is 67 C (153 F) [031,421]. It is combustible. Fires involving this material may be controlled with a dry chemical or carbon dioxide extinguisher. Do not use halogenated extinguishing media on this compound [102].

The autoignition temperature of N,N-dimethylformamide is 445 C (833 F) [043,062,371,451].

LEL: 2.2% @ 100 C [043,058,102,45UEL: 15.2% @ 100 C [043,058,102,451]

REACTIVITY: N,N-Dimethylformamide may react with oxidizing materials, bromine, potassium permanganate, triethylaluminum and heat, chlorine, sodium hydroborate and heat, chromium trioxide, diisocyanatomethane, carbon tetrachloride and iron, 1,2,3,4,5,6-hexachlorocyclohexane and iron, magnesium nitrate, sodium and heat, sodium hydride and heat, sulfinyl chloride and traces of iron or zinc, 2,4,6-trichloro-1,3,5-triazine (with gas evolution), halogenated hydrocarbons, inorganic and organic nitrates and methylene diisocyanate. Forms explosive mixtures with lithium azide (shock sensitive above 200 C) and uranium perchlorate [043]. It is also incompatible with alkylaluminums [346]. It attacks some forms of plastics, rubber and coatings [102]. It is incompatible with acid chlorides, chloroformates and reducing agents [269]. It reacts violently with hexachlorobenzene, phosphorus trioxide and 2,5-dimethylpyrrole and phosphorous oxychloride mixture [451].

STABILITY: N,N-Dimethylformamide may be sensitive to prolonged exposure to air. Solutions of it in water, DMSO, 95% ethanol or acetone should be stable for 24 hours under normal lab conditions [700].

USES: N,N-Dimethylformamide is used as a solvent and reagent in organic synthesis, resin and polymer solvent, spinning solvent for acrylic fibers, booster solvent in coating, printing and adhesive formulations, catalyst and reaction medium in chemical manufacturing, selective absorption and extraction solvent for recovery, purification, absorption, separation and desulfurization of non-paraffinics from paraffin hydrocarbons, removal of vinyl films, epoxy coatings and varnish finishes, solvent and carrier for gases, purified acetylene, pharmaceuticals, dyes, petroleum products and other organic chemicals. It is also used in the manufacturing of pesticides (chlordimeform), paint stripper and butadiene.

COMMENTS: None

NIOSH Registry Number: LQ2100000

ACUTE/CHRONIC HAZARDS: N,N-Dimethylformamide is an irritant [102] and it may be harmful if swallowed, inhaled or absorbed through the skin. When heated to decomposition it may emit toxic fumes of carbon dioxide, carbon monoxide and nitrogen oxides [269].

SYMPTOMS: Symptoms of exposure may include irritation to the eyes, mucous membranes and upper respiratory tract, dizziness, headache, liver and kidney damage and intolerance to alcohol up to four days after exposure [269]. It may cause abdominal distress, colicky abdominal pain, loss of appetite, nausea, vomiting, constipation, diarrhea, facial flushing (especially after drinking alcohol), agitation, increased blood pressure, skin irritation and skin rash [102]. It may also cause weakness and abnormal liver function [058]. Exposure may cause corneal damage and central nervous system effects [301]. Other symptoms include dermatitis, anorexia and hepatomegaly [346]. It may cause incoordination, hallucinations and digestive disturbances [295]. It may burn the skin and eyes [371] and it may also cause epigastric cramps [421].

1,1-Dimethylhydrazine

NOTE: The form of this name in the Clean Air Act Amendments Section 112(b) list is 1,1-Dimethyl hydrazine. The correct notation for the compound is 1,1-Dimethylhydrazine.

CAS NUMBER: 57-14-7

SYNONYMS: Asymmetric dimethylhydrazine Dimethylhydrazine unsymmetrical 1,1-Dimethylhydrazine

CHEMICAL FORMULA: C2H8N2

MOLECULAR WEIGHT: 60.10

WLN: ZN1&1

PHYSICAL DESCRIPTION: Clear, colorless liquid

SPECIFIC GRAVITY: 0.791 @ 22/4 C [058,205,275,430]

DENSITY: 0.7914 g/mL @ 22 C [016]

MELTING POINT: -58 C [043,062,205,421]

BOILING POINT: 63 C [025,395,421,430]

SOLUBILITY: Water: Decomposes [700] DMSO: >=100 mg/mL @ 24 C [700] 95% Ethanol: >=100 mg/mL @ 24 C [700] Acetone: >=100 mg/mL @ 24 C [700]

OTHER SOLVENTS: Dimethyl formamide: Miscible [031,395,421] Hydrocarbons: Miscible [031,395,421] Petroleum ether: Miscible [205] Alcohol: Soluble [016] Ether: Miscible [013,205,395,421]

OTHER PHYSICAL DATA: Ammonia-like, fishy odor [043,058,371,430] Refractive index: 1.4075 @ 22.3 C [025,395,430] 1.4075 @ 20 C [205,269,275] Saturation concentration: 505 g/m3 @ 25 C [055] Strongly alkaline [099] Critical pressure: 865 psia [371] Liquid surface tension: 28 dynes/cm [371] Latent heat of vaporization: 145 cal/g [371] Heat of combustion: -7870 cal/g [371] Heat of solution: -10 cal/g [371] Vapor pressure: 10 mm Hg @ -22 C [051] 103 mm Hg @ 20 C [421] Specific gravity:0.782 @ 25/4 C [043] 0.8274 @ 20/4 C [031] Boiling point: 62.5 C @ 717 mm Hg[051,055] 63.9 C @ 760 mm Hg [058,205] Fumes in air and gradually turns yellow [031,051,058,395]

HAP WEIGHTING FACTOR: 1 [713]

VOLATILITY:

Vapor pressure: 157 mm Hg @ 25 C [043,051,055] 100 mm Hg @ 16 C [051,395]

Vapor density: 1.94 [043,058,430]

FIRE HAZARD: 1,1-Dimethylhydrazine has a flash point of 1 C (34 F) [051,058,275,430] and it is flammable. Fires involving this material may be controlled with a dry chemical, carbon dioxide or Halon extinguisher.

The autoignition temperature of 1,1-dimethylhydrazine is 249 C (480 F) [043,062,430,451].

LEL: 2% [043,058,371,430] UEL: 95% [043,058,371,430]

REACTIVITY: 1,1-Dimethylhydrazine ignites violently with oxidants such as dinitrogen tetroxide, nitric acid, and hydrogen peroxide [036,043]. It may react with copper and its alloys, brass, iron and iron salts [058,269]. It may also react with halogens, metal oxides and metallic mercury [051]. It may be explosive on contact with dicyanofurazan or its N-oxide [066]. It dissolves, swells and disintegrates many plastics [051,058]. Avoid contact with organic materials like sawdust, excelsior and cotton waste [051]. Evolution of gas occurs when mixed with water [700]. It is a powerful reducing agent [043].

STABILITY: 1,1-Dimethylhydrazine is hygroscopic [031,043,269,395]. It is also sensitive to air [269].

USES: 1,1-Dimethylhydrazine is used as a chemical intermediate in the manufacture of plant growth inhibitors and aminimides, as a protecting group for carbonyl compounds, as a reagent for conversion of aldehydes to nitriles, as a component of jet and rocket fuels, in chemical synthesis, as a stabilizer for organic peroxide fuel additives, as an absorbent for acid gases and in photography.

COMMENTS: None

NIOSH Registry Number: MV2450000

ACUTE/CHRONIC HAZARDS: 1,1-Dimethylhydrazine is highly corrosive [025,051,099] and harmful if swallowed or absorbed through the skin [051,058,269,371]. When heated to decomposition it emits very toxic fumes of nitrogen oxides [043].

SYMPTOMS: Symptoms of exposure may include burns, damage to liver and kidneys, blood effects and gastrointestinal disturbances [269]. It may cause irritation to the nose and throat, respiratory distress, nausea, vomiting, headache, facial numbness, twitching, pulmonary edema, seizures, coma, corrosion, pneumonia and death [058]. It may also cause corneal damage [301]. Other symptoms may include choking, chest pains, dyspnea, lethargy, convulsions and anoxia [346]. The vapors can cause mild irritation to the skin and eyes [051,099]. It may also cause fatty liver and a rise in SGPT levels [395].

Dimethyl phthalate

CAS NUMBER: 131-11-3

SYNONYMS: Phthalic acid, dimethyl ester Dimethyl 1,2-benzendicarboxylate 1,2-Benzenedicarboxylic acid, dimethyl ester

CHEMICAL FORMULA: C10H10O2

MOLECULAR WEIGHT: 194.18

WLN: 1OVR BVO1

PHYSICAL DESCRIPTION: Clear, colorless, oily liquid

SPECIFIC GRAVITY: 1.190 @ 20/20 C

DENSITY: Not available

MELTING POINT: 0.0 C

BOILING POINT: 283.7 C @ 760 mm Hg

SOLUBILITY: Water: <1 mg/mL @ 20 C [700] DMSO: >=100 mg/mL @ 20 C [700] 95% Ethanol: >=100 mg/mL @ 20 C [700] Acetone: >=100 mg/mL @ 20 C [700]

OTHER SOLVENTS: Mineral oil: Soluble Ether: Soluble Benzene: Soluble Petroleum ether: Insoluble Paraffin hydrocarbons: Insoluble Chloroform: Soluble

OTHER PHYSICAL DATA: Volatility index: 1.5150 @ 20 C Evaporation Rate (Butyl Acetate = 1): <1

HAP WEIGHTING FACTOR: 1 [713]

VOLATILITY:

Vapor pressure: 1 mm Hg @ 100.3 C; 5 mm Hg @ 131.8 C; 10 mm Hg @ 147.6 C

Vapor density: 6.69

FIRE HAZARD: Dimethyl phthalate has a flash point of 146 C (295 F). It is combustible. Fires involving this material should be controlled using a dry chemical, carbon dioxide or Halon extinguisher.

The autoignition temperature of dimethyl phthalate is 556 C (1032 F).

LEL: Not available UEL: Not available

REACTIVITY: Dimethyl phthalate can react with oxidizing materials, nitrates, strong alkalis and strong acids.

STABILITY: Dimethyl phthalate is sensitive to prolonged exposure to light. Solutions of it in water, DMSO, 95% ethanol or acetone should be stable for 24 hours under normal lab conditions [700].

USES: Dimethyl phthalate is used as a solvent and plasticizer for cellulose acetate and cellulose acetate butyrate compositions, in film, and as an insect repellant.

COMMENTS: None

NIOSH Registry Number: TI1575000

ACUTE/CHRONIC HAZARDS: Dimethyl phthalate is an irritant and when heated to decomposition it, emits toxic fumes of carbon monoxide and carbon dioxide. It is an experimental carcinogen.

SYMPTOMS: Symptoms of exposure may include irritation of the skin, lips, tongue, mouth, respiratory tract, eyes, mucous membranes and gastrointestinal tract, central nervous system depression, vomiting, diarrhea, coma, cough, paralysis and conjunctivitis.

Dimethyl sulfate

CAS NUMBER: 77-78-1

SYNONYMS: Methyl sulfate (DOT) Sulfuric acid, dimethyl ester

CHEMICAL FORMULA: C2H6O4S

MOLECULAR WEIGHT: 126.14 [703]

PHYSICAL DESCRIPTION: Colorless, odorless liquid [703]

DENSITY: 1.3322 @ 20C/4C [703]

MELTING POINT: -31.8 C [703]

BOILING POINT: 188 C [703] decomposes [702]

SOLUBILITY: Water: 2.8 g/L [702]

OTHER PHYSICAL DATA: Vapor Density: 4.35 [703] Melting Point: -27 C [702] Heat of Combustion: 667.38 kcal/mole [702]

HAP WEIGHTING FACTOR: 1 [713]

VOLATILITY:

Vapor Pressure: 0.1 mm [704]

Vapor Density: 4.35 [703]

FIRE HAZARD: Flash Point: 182 F (open cup) [703]

Flammable when exposed to heat, flame or oxidizers. To fight fire use water, foam, CO2, dry chemical. [703] Class IIIA combustible liquid [704]. Combustion ranking: 59 [702].

Autoignition temperature: 370 F [703]

LEL: Not available UEL: Not available

REACTIVITY: Can react with oxidizing materials. Violent reaction with NH4OH, NaN3. [703, 704] Ammonia: A violent reaction occurred which shattered the flask when liter quantities of dimethyl sulfate and concentrated aqueous ammonia were accidentally mixed. Use dilute ammonia in small quantities to destroy dimethyl sulfate. [710] Decomposes in water to sulfuric acid [704].

CORROSIVITY:

Corrosive to metals [704].

STABILITY: When heated to decomposition it emits toxic fumes of SOx [703].

USES: Methylating agent in the manufacture of many organic chemicals. War gas. [715]

COMMENTS: Caution: Extremely hazardous. No warning characteristics (e.g. odor, irritation). Delayed appearance of symptoms may permit unnoticed exposure to lethal quantities. [715]

NIOSH Registry Number: WS8225000

ACUTE/CHRONIC HAZARDS: Dimethyl sulfate is a poison by inhalation. It is an experimental poison by ingestion, inhalation, intravenous and subcutaneous routes. It is also an experimental carcinogen, tumorigen and teratogen; some human mutagenic data are reported by Sax to exist. Dimethyl sulfate is a corrosive irritant to skin, eyes and mucous membranes. No odor or initial irritation gives warning of exposure. Exposure to dimethyl sulfate may be fatal; in patients surviving severe exposures, there may be serious injury of liver and kidneys. When heated to decomposition, emits toxic fumes of SOx. [703]

SYMPTOMS: Symptoms of exposure may include irritation of skin, eyes and mucous membranes. Liquid dimethyl sulfate causes ulceration and local necrosis if spilled on skin. If exposure is brief and mild it may result in conjunctivitis, catarrhal inflammation of mucous membranes of nose, throat, larynx and trachea and possibly reddening of skin developing after a latent period. With longer, heavier exposures, corneal clouding, increased nasopharynx irritation occurs. Pulmonary edema may also develop with long exposure times and liver and kidneys can be damaged. [703]

4,6-Dinitro-o-cresol, and salts

CAS NUMBER: 534-52-1

NOTE: The Clean Air Act lists 4,6-Dinitro-o-cresol and salts. Data is given here only for 4,6-Dinitro-o-cresol.

SYNONYMS: 2-Methyl-4,6-dinitrophenol 3,5-Dinitro-2-hydroxytoluene 3,5-Dinitro-o-cresol

CHEMICAL FORMULA: C7H6N2O5

MOLECULAR WEIGHT: 198.13

WLN: WNR BQ C ENW

PHYSICAL DESCRIPTION: Yellow prismatic crystals

SPECIFIC GRAVITY: Not available

DENSITY: 6.82

MELTING POINT: 86.5 C

BOILING POINT: Not available

SOLUBILITY: Acetone: Soluble

OTHER SOLVENTS: Water: Slightly soluble Alcohol: Soluble Petroleum ether: Slightly soluble Ether: Soluble

OTHER PHYSICAL DATA: Not available

HAP WEIGHTING FACTOR: 1 [713]

VOLATILITY:

Vapor density: 6.82

FIRE HAZARD: Not available; it may be flammable or combustible.

LEL: Not available UEL: Not available

REACTIVITY: Not available

STABILITY: Not available; it may not be stable to heat or flame and may explode under these conditions.

USES: 4,6-Dinitro-o-cresol is used as a dormant ovicidal spray for fruit trees (although it is highly phytotoxic and cannot be used successfully on actively growing plants), an herbicide and an insecticide.

COMMENTS: None

NIOSH Registry Number: GO9625000

ACUTE/CHRONIC HAZARDS: 4,6-Dinitro-o-cresol is a severe irritant, corrosive to tissues, and very toxic by ingestion or by absorption through the skin.

SYMPTOMS: Symptoms of exposure may include yellowish skin tissue, headache, nausea, vomiting, fever, dyspnea (difficult breathing), restrostermal pain, sweating, mental disorder, rapid pulse, fatigue, jaundice, circulatory collapse, and cataract formation.

2,4-Dinitrophenol

CAS NUMBER: 51-28-5

SYNONYMS: Phenol, 2,4-dinitro- 1-Hydroxy-2,4-dinitrobenzene

CHEMICAL FORMULA: C6H4N2O5

MOLECULAR WEIGHT: 184.11

WLN: WNR BQ ENW

PHYSICAL DESCRIPTION: Yellow crystals or powder

SPECIFIC GRAVITY: 1.683 [033,051,062,205]

DENSITY: 1.683 g/mL @ 24 C [016,043,047,055]

MELTING POINT: 112-114 C [033,205]

BOILING POINT: Sublimes (when carefully heated) [016,029,033,051]

SOLUBILITY: Water: <1 mg/mL @ 19.5 C [700] DMSO: 50-100 mg/mL @ 22 C [700] 95% Ethanol: 10-50 mg/mL @ 22 C [700] Acetone: 50-100 mg/mL @ 22 C [700] Toluene: 6.36 g/100 g @ 15 C [033]

OTHER SOLVENTS: Chloroform: 5.39 g/100 g @ 15 C [033,205] Ethyl acetate: 15.55 g/100 g @ 15 C [033] Pyridine: 20.08 g/100 g @ 15 C [033,205] Carbon tetrachloride: 0.423 g/100 g @ 15 C [033] Ether: Soluble [016,062,295] Benzene: Soluble [016,033,062,205] Aqueous alkaline solutions: Soluble [033,173] Alcohol: Soluble [016,033,205,295]

OTHER PHYSICAL DATA: Melting point also reported as 106-108 C [275] Sweet, musty odor [371] pKa: 4.00 @ 25 C [029] pH range: 2.6 colorless, 4.4 yellow [033] Density: 1.68 g/mL @ 20 C [371]

HAP WEIGHTING FACTOR: 1 [713]

VOLATILITY:

Vapor density: 6.35 [043,051]

FIRE HAZARD: 2,4-Dinitrophenol is flammable [051,275]. Its dust can form explosive mixtures in air [058]. Fires involving this material may be controlled with a dry chemical, carbon dioxide or Halon extinguisher. A water spray may also be used [058,269,371].

LEL: Not available UEL: Not available

REACTIVITY: 2,4-Dinitrophenol forms explosive salts with alkalis and ammonia [043,066]. It is incompatible with heavy metals and their compounds [058,346]. It is also incompatible with strong oxidizers, strong bases and reducing agents [269]. It reacts with combustibles [371].

STABILITY: 2,4-Dinitrophenol may explode if subjected to heat or flame [058,371]. It may explode if allowed to dry out [058]. Solutions of in water, DMSO, 95% ethanol or acetone should be stable for 24 hours under normal lab conditions [700].

USES: 2,4-Dinitrophenol is used in the manufacture of dyes, diaminophenol, wood preservatives, insecticides, explosives, herbicides, photographic developers, picric acid and picramic acid. It is used as a reagent for the detection of potassium and ammonium ions and in chemical synthesis. It is also a useful tool in biochemical research. This compound was formerly used as a metabolic stimulator to aid in weight reduction.

COMMENTS: None

NIOSH Registry Number: SL2800000

ACUTE/CHRONIC HAZARDS: 2,4-Dinitrophenol may be toxic by inhalation, ingestion and skin contact [036, 058,295,346]. There is a danger of cumulative effects [036,151,295]. It is a severe skin and eye irritant [043,058] and is readily absorbed through the skin [033,058]. Vapors are absorbed through the respiratory tract [033]. It is also corrosive [151] and, when heated to decomposition, it emits toxic fumes of carbon monoxide, carbon dioxide and nitrogen oxides [043,058,269].

SYMPTOMS: Symptoms of exposure may include profuse sweating, fever, shortness of breath and yellow coloration of the skin of the hands and feet [036]. Other symptoms may include a marked increase in metabolism, nausea, vomiting, collapse, death, cataracts, weight loss, granulocytopenia, polyneuropathy and exfoliative dermatitis [033]. Skin irritation may also occur [043]. Exposure may also cause disruption of oxidative phosphorylation, increased oxygen consumption, fatigue, intense thirst, sweating, oppression of the chest, anorexia, weakness, dizziness, vertigo, headache, sensitivity of the liver to pressure, jaundice and hypersensitivity, neutropenia, skin rashes and peripheral neuritis [346]. Other symptoms may include degenerative changes in the heart, liver and kidneys, cyanosis, lividity, tremors and coma [301]. Contact may also cause irritation of the eyes and mucous membranes [269]. Secondary glaucoma may also result [099]. Exposure may cause restlessness, flushed skin, tachycardia and fatal hyperthermia [406]. Allergic reactions may also occur [051]. Other symptoms may include hyperpyrexia, acidosis, dehydration, muscle rigor, pulmonary edema, abdominal pain and diarrhea [151]. Anxiety and delirium may also result [430]. Other symptoms may include liver damage, dilation of the pupils, smarting of the skin and skin burns [371].

2,4-Dinitrotoluene

CAS NUMBER: 121-14-2

SYNONYMS: 1-Methyl-2,4-dinitrobenzene

CHEMICAL FORMULA: C7H6N2O4

MOLECULAR WEIGHT: 182.14

WLN: WNR CNW D

PHYSICAL DESCRIPTION: Light yellow powder

SPECIFIC GRAVITY: 1.379 @ 20 C

DENSITY: 1.321 g/mL @ 71 C

MELTING POINT: 67-70 C

BOILING POINT: 300 C (decomposes)

SOLUBILITY: Water: <0.1 mg/mL @ 17 C [700] DMSO: >=100 mg/mL @ 23.5 C [700] 95% Ethanol: 10-50 mg/mL @ 23.5 C [700] Acetone: >=100 mg/mL @ 23.5 C [700]

OTHER SOLVENTS: Chloroform: Very soluble Pyrimidine: Very soluble Carbon disulfide: Very soluble Toluene: Very soluble Pyridine: Soluble Ether: Soluble Benzene: Soluble

OTHER PHYSICAL DATA: Refractive index: 1.442 pKa1: 17.12 (ethylene diamine aq.) pKa2: -12.6 (oleum) Solid and liquid sink in water Density: 1.521 g/mL @ 15 C; 1.3208 g/mL @ 71 C

HAP WEIGHTING FACTOR: 1 [713]

VOLATILITY:

Vapor pressure: 1 mm Hg @ 20 C

Vapor density: 6.27

FIRE HAZARD: 2,4-Dinitrotoluene has a flash point of 207 C (405 F). It is combustible. Fires involving this material may be controlled with a dry chemical, carbon dioxide or Halon extinguisher. A water spray may also be used.

LEL: Not available UEL: Not available

REACTIVITY: 2,4-Dinitrotoluene reacts with oxidizers, caustics, nitric acid, reducing materials and metals, such as zinc or tin. It may react violently in the presence of a base or when heated to the boiling point. It attacks some forms of plastics, rubbers and coatings.

STABILITY: 2,4-Dinitrotoluene is stable under normal laboratory conditions. Solutions of it in water, DMSO, 95% ethanol or acetone should be stable for 24 hours under normal lab conditions [700].

USES: 2,4-Dinitrotoluene is used in organic synthesis for dyes, toluidines and it is also used in explosives.

COMMENTS: Derivation: nitration of nitrotoluene with mixed acid.

NIOSH Registry Number: XT1575000

ACUTE/CHRONIC HAZARDS: 2,4-Dinitrotoluene is an irritant and is readily absorbed through the skin. When heated to decomposition it emits toxic fumes.

SYMPTOMS: Symptoms of exposure may include methemoglobinemia, anemia, leukopenia, liver necrosis, headache, irritability, dizziness, weakness, nausea, vomiting, dyspnea, drowsiness, unconsciousness, cyanosis, vertigo, fatigue, shortness of breath, anorexia, palpitations, arthralgia, insomnia, tremors, paralysis and irritation of exposed tissues.

1,4-Dioxane

CAS NUMBER: 123-91-1

SYNONYMS: 1,4-Diethyleneoxide Dioxane p-Dioxane Diethylene oxide Diethylene dioxide Tetrahydro-p-dioxin

CHEMICAL FORMULA: $C_4H_8O_2$

MOLECULAR WEIGHT: 88.11

WLN: T60 DOTJ

PHYSICAL DESCRIPTION: Clear colorless liquid

SPECIFIC GRAVITY: 1.0329 @ 20/4 C [033,205,395,421]

DENSITY: 1.03 g/mL [295]

MELTING POINT: 11.8 C [016,033,275,395]

BOILING POINT: 101 C [036,055,058,451]

SOLUBILITY: Water: >=100 mg/mL @ 20 C [700] DMSO: >=100 mg/mL @ 20 C [700] 95% Ethanol: >=100 mg/mL @ 20 C [700] Acetone: >=100 mg/mL @ 20 C [700]

OTHER SOLVENTS: Alcohol: Miscible [205,295] Ether: Miscible [205,295] Benzene: Miscible [205] Chloroform: Miscible [205] Petroleum ether: Miscible [205] Oils: Miscible [395] Aromatic hydrocarbons: Miscible [395] Most organic solvents: Miscible [062,395,421,430]

OTHER PHYSICAL DATA: Specific gravity: 1.0356 @ 20/20 C [062] 1.035 @ 25/4 C [430] Boiling point: 81.8 C @ 400 mm Hg; 62.3 C @ 200 mm Hg [033] Boiling point: 45.1 C @ 100 mm Hg; 33.8 C @ 60 mm Hg; 25.2 C @ 40 mm Hg [033] Boiling point: 12 C @ 20 mm Hg [033] Vapor pressure: 20 mm Hg @ 12 C [038,051] 40 mm Hg @ 25.2 C [038,043,051] Vapor pressure: 50 mm Hg @ 30 C [055] 60 mm Hg @ 33.8 C [038] Vapor pressure: 100 mm Hg @ 45.1 C [038,051] 200 mm Hg @ 62.3 C [038] Vapor pressure: 400 mm Hg @ 81.8 C; 760 mm Hg @ 101.1 C [038] Flash point (open cup): 18.3 C (65 F) [062,421,430] Ethereal odor [058,062,295,371] Odor threshold: 170 ppm [102,421,430] Refractive index: 1.4175 @ 20 C [029,033,395] 1.422 @ 25 C [430] Heat of combustion: -6440 cal/g [371] Heat of fusion: 34.85 cal/g [371] Specific heat: 0.0370 kcal/mol/C @ 20 C [033] Viscosity: 0.0120 poise @ 25 C [033] Dipole moment: 0 [033] Critical temperature: 312 C [029,033] Critical pressure: 50.7 atmospheres [029,033] Cryoscopic constant: 4.83 [033] Trouton constant: 21.90 [033] Evaporation rate (butyl acetate = 1): 2.7 [102] Forms an azeotrope with water containing 18.4% water (boiling point of 87.8 C) [029,033,430] Forms an azeotrope with ethanol containing 9.3% of this compound (boiling point of 78.1 C) [033,430] Heat of decomposition: 0.19 kJ/g [066] Lambda max (in water): 400-295 nm, 275 nm, 250 nm, 235 nm, 225 nm (A = 0.01, 0.05, 0.20, 0.50, 1.0) [275] Latent heat of vaporization: 98.6 cal/g [371]

HAP WEIGHTING FACTOR: 1 [713]

VOLATILITY:

Vapor pressure: 29 mm Hg @ 20 C [102,421] 37 mm Hg @ 25 C [055,395,430]

Vapor density: 3.03 [043,055,058]

FIRE HAZARD: 1,4-Dioxane has a flash point of 12 C (54 F) [036,043,275,451] and it is flammable. Fires involving this material can be controlled with a dry chemical, carbon dioxide or Halon extinguisher.

The autoignition temperature of this compound is 180 C (356 F) [036,043,371,451].

LEL: 2% [036,043,058,451] UEL: 22% [036,058,066,451]

REACTIVITY: 1,4-Dioxane may form explosive peroxides [029,062,346,451]. It can react vigorously with oxidizers [043,058,269,346]. It reacts with oxygen to form peroxides [269,395]. It reacts almost explosively with Raney nickel catalyst above 210 C [036,058,066,451]. The addition complex with sulfur trioxide decomposes violently on storage [036,043,066]. An explosion has occurred during distillation from $LiAlH_4$ [036]. An explosion has also occurred during reaction with (perchloric acid + nitric acid) [036,043,066]. 1,4-Dioxane forms impact-sensitive mixtures with decaborane. Mixtures with triethynylaluminum are sensitive to heating or drying [043,066]. Violent reactions may occur with silver perchlorate [043,058,451]. 1,4-Dioxane is incompatible with halogens and reducing agents [269] and it can explode when redistilled [058].

STABILITY: 1,4-Dioxane may form peroxides on exposure to air [036,043,066,395]. Exposure to sunlight accelerates this formation [036,058]. This compound is hygroscopic [102,295,430]. It is also sensitive to heat [102,269]. It is dangerous to distill or evaporate this compound unless precautions have been taken to remove explosive peroxides [295]. Solutions of it in water, DMSO, 95% ethanol or acetone should be stable for 24 hours under normal lab conditions [700].

USES: 1,4-Dioxane is used as a solvent for cellulose acetate, ethyl cellulose, benzyl cellulose, resins, oils, waxes, dyes, many organic as well as some inorganic compounds, lacquers, paints, varnishes, paint and varnish removers, cleaning and detergent preparations, cements, cosmetics, deodorants, fumigants, fats, greases, mineral oil, polyvinyl polymers, plastics, adhesives, sealants, pharmaceuticals, rubber chemicals, surface coatings and electrical, agricultural and biochemical intermediates. It is also used in the preparation of histological slides, as a wetting and dispersing agent in textile processing, in dye baths, in stain and printing compositions, in emulsions, in polishing compounds, as a stabilizer for chlorinated solvents and as a scintillation counter.

COMMENTS: 1,4-Dioxane has poor warning properties [371]. Because it is water-miscible, treatment by shaking with aqueous reductants (such as iron (II) sulfate, sodium sulfide, etc,) is impractical. Peroxides may be removed under anhydrous conditions by passing down a column of activated alumina [066].

NIOSH Registry Number: JG8225000

ACUTE/CHRONIC HAZARDS: 1,4-Dioxane is toxic by ingestion, inhalation and skin absorption [029,036,058,269]. It is readily absorbed through the skin [062,151,395, 451]. It is an irritant and when heated to decomposition it emits acrid smoke, irritating fumes and toxic fumes of carbon monoxide and carbon dioxide [043,058,102,269].

SYMPTOMS: Symptoms of exposure may include irritation of the nose and eyes, headache, nausea, vomiting, drowsiness, dizziness, liver damage and kidney damage [102,295,346]. Central nervous system depression may also occur [033,058]. Other symptoms include lacrimation, convulsions, high blood pressure and unspecified respiratory system and gastrointestinal system effects [015,043]. Irritation of the throat and respiratory system may also occur [102,395]. Exposure may cause anesthesia, gastric distress, dyspnea, vertigo, tenderness in the lumbar and abdominal regions, and death due to acute renal failure [151]. It may also cause stomach pain and loss of appetite. Prolonged skin exposure may cause drying and cracking [102, 346]. It may also cause irritation and eczema [430]. Other symptoms from prolonged skin exposure include rash or burns [371] and dermatitis [058,102]. Eye contact may cause pain [430] and overexposure may lead to corneal injury [058,371]. Other symptoms of exposure include coughing, anorexia, uremia and coma [395]. Damage to the lungs and brain have been reported [102] and narcosis may also occur [058].

1,2-Diphenylhydrazine

CAS NUMBER: 122-66-7

SYNONYMS: Hydrazobenzene N,N'-Bianiline

CHEMICAL FORMULA: C12H12N2

MOLECULAR WEIGHT: 184.26 [703]

PHYSICAL DESCRIPTION: Light or yellow crystals from ethanol [703]

DENSITY: 1.58 [703]

MELTING POINT: 131C [703]

BOILING POINT: Decomposes [703]

SOLUBILITY: Water: Very slightly soluble [703] Acetylene: Insoluble [703]

OTHER PHYSICAL DATA: Heat of Combustion: 1591.00 kcal/mole

HAP WEIGHTING FACTOR: 1 [713]

VOLATILITY: Not available

FIRE HAZARD: Combustion Ranking: 11 [702]

LEL: Not available UEL: Not available

REACTIVITY: Not available

STABILITY: When heated to decomposition it emits toxic fumes of NOx [703].

UDRI Thermal Stability Class: 5 [702] UDRI Thermal Stability Ranking: 251 [702]

USES: Not available

COMMENTS: None

NIOSH Registry Number: MW2625000

ACUTE/CHRONIC HAZARDS: 1,2-Diphenylhydrazine is a poison by ingestion. It is an experimental carcinogen and tumorigen. Mutagenic data exists. When heated to decomposition it emits toxic fumes of NOx. [703]

SYMPTOMS: Not available

Epichlorohydrin

CAS NUMBER: 106-89-8

SYNONYMS: 1-Chloro-2,3-epoxypropane 3-Chloro-1,2-epoxypropane (Chloromethyl)ethylene oxide 2-(Chloromethyl)oxirane

CHEMICAL FORMULA: C_3H_5ClO

MOLECULAR WEIGHT: 92.53

WLN: T3OTJ B1G

PHYSICAL DESCRIPTION: Clear colorless liquid

SPECIFIC GRAVITY: 1.181 @ 20/4 C [026,043]

DENSITY: 1.182 g/cm3 @ 18.5 C

MELTING POINT: -25.6 C [031,038,043,395]

BOILING POINT: 117.9 C @ 760 mm Hg [031,043,051,421]

SOLUBILITY: Water: 50-100 mg/mL @ 22 C DMSO: >=100 mg/mL @ 20.5 C 95% Ethanol: >=100 mg/mL @ 20.5 C Acetone: >=100 mg/mL @ 20.5 C

OTHER SOLVENTS: Chloroform: Miscible [031,205] Carbon tetrachloride: Miscible [031,205,430] Trichloroethylene: Miscible [031] Alcohol: Miscible [031,205,430] Ether: Miscible [031,430] Benzene: Miscible [430] Petroleum hydrocarbons: Immiscible [031,395] Chlorinated aliphatic hydrocarbons: Miscible [395] Most organic solvents: Miscible [421]

OTHER PHYSICAL DATA: Melting point also reported as -57.2 C [058,205,430] Irritating, chloroform-like odor [036,043,102,430] Refractive index: 1.44195 @ 11.6 C; 1.43969 @ 16 C; 1.43585 @ 25 C [031] Refractive index: 1.4361 @ 20 C [017] Specific gravity: 1.1761 @ 20/20 C [043,051] 1.750 @ 25/4 C [395] Specific gravity: 1.1436 @ 50/4 C; 1.1101 @ 75/4 C [031] Boiling point: 79.3 C @ 200 mm Hg, 62.0 C @ 100 mm Hg, 42.0 C @ 40 mm Hg [031] Boiling point: 16.6 C @ 10 mm Hg, 16.5 C @ 1 mm Hg [031] Vapor pressure: 13 mm Hg @ 20 C [058,102,421,430] 1 mm Hg @ -16.5 C [038] Vapor pressure: 5 mm Hg @ 5.6 C; 20 mm Hg @ 29.0 C; 60 mm Hg @ 50.6 C [038] Liquid surface tension: 37.0 dynes/cm [371] Latent heat of vaporization: 97.9 cal/g [371] Heat of combustion: -4524 cal/g [371] Evaporation rate (butyl acetate = 1): 1.35 [102]

HAP WEIGHTING FACTOR: 1 [713]

VOLATILITY:

Vapor pressure: 10 mm Hg @ 16.6 C [038,043,051] 40 mm Hg @ 42.0 C [038]

Vapor density: 3.29 [031,043]

FIRE HAZARD: Epichlorohydrin has a flash point of 33 C (91 F) [205,275] and it is flammable. Fires involving this material may be controlled with a dry chemical, carbon dioxide or Halon extinguisher. A water spray may also be used 036,058,371,421].

The autoignition temperature of epichlorohydrin is 411 C (772 F) [451].

LEL: 3.8% [058,102,451] UEL: 21.0% [058,102,451]

REACTIVITY: Epichlorohydrin is incompatible with strong oxidizers, strong acids and caustics [102,269]. It is also incompatible with zinc, aluminum and chlorides of iron and aluminum [043,102]. It reacts violently with aniline, isopropylamine and trichloroethylene [036,043]. It also reacts violently with sulfuric acid, nitric acid, 2-aminoethanol, chlorosulfonic acid, ethylenediamine, ethyleneimine, oleum and potassium-tert-butoxide [051,451]. It will attack some forms of plastics, rubber and coatings [102]. The wet product will pit steel [051].

STABILITY: Epichlorohydrin polymerizes at temperatures >325 C [102]. Hydrolysis is slow, but is accelerated by heat or traces of acids [395]. Solutions of it in water, DMSO, 95% ethanol and acetone should be stable for at least 24 hours under normal lab conditions [700].

USES: Epichlorohydrin is used in the manufacture of epoxy resins, glycerol and various other intermediates. It is a solvent for natural and synthetic resins, gums, cellulose esters and ethers, paints, varnishes, nail enamels and lacquers, and cements for celluloid. It is used in surface-active agents, pharmaceuticals, insecticides, agricultural chemicals, textile chemicals, coatings, adhesives, ion-exchange resins, plasticizers, glycidyl esters, ethymyl-ethylenic alcohol and fatty acid derivatives. It is a stabilizer in chlorine-containing materials and an intermediate in the preparation of condensates with polyfunctional substances.

COMMENTS: None

NIOSH Registry Number: TX4900000

ACUTE/CHRONIC HAZARDS: Epichlorohydrin is corrosive [269] and is a strong skin irritant and sensitizer. It may be toxic by inhalation, ingestion and skin contact [036]. When heated to decomposition it emits toxic fumes of carbon monoxide, carbon dioxide, phosgene and hydrogen chloride gas [058,102].

SYMPTOMS: Symptoms of exposure may include respiratory paralysis, conjunctivitis, skin blistering, severe pain, sensitization, dermatitis, chronic weariness and stomach upset [036]. It may also cause irritation, nausea, vomiting, coughing, difficult breathing, blue skin, blistering, skin burns, liver injury, lung injury and kidney injury [102]. Other symptoms may include changes in lymphocytes, central nervous system depression, unconsciousness [058] and convulsions [051]. It may also cause headaches and collapse [371]. Exposure may also cause irritation to the eyes and throat, chronic asthmatic bronchitis and fatty infiltration of the liver [395]. It may cause painful irritation of subcutaneous tissues, lung edema, kidney lesions, dyspnea, bronchitis and enlarged liver [421]. It may also cause sterility [055].

1,2-Epoxybutane

CAS NUMBER: 106-88-7

SYNONYMS: 1-Butene oxide 1,2-Butene oxide Butylene oxide Ethylethylene oxide

CHEMICAL FORMULA: C4H8O

MOLECULAR WEIGHT: 72.11

WLN: T3OTJ B2

PHYSICAL DESCRIPTION: Clear colorless liquid

SPECIFIC GRAVITY: 0.8312 @ 20/20 C [042,062]

DENSITY: 0.8297 g/mL @ 20 C [900]

MELTING POINT: 60 C [055]

BOILING POINT: 63 C [025,042,062,275]

SOLUBILITY: Water: >=100 mg/mL @ 17 C DMSO: >=100 mg/mL @ 17 C 95% Ethanol: >=100 mg/mL @ 17 C Acetone: >=100 mg/mL @ 17 C

OTHER SOLVENTS: Ether: Soluble [017,047] Most organic solvents: Miscible [042,062]

OTHER PHYSICAL DATA: Sets to a glass below -150 C [062] Pleasant, sweet alcohol odor [055] Refractive index: 1.3851 @ 20 C [017,047] Specific gravity: 0.837 @ 17/4 C [017,047] 0.82271 @ 25/22 C [052] Vapor pressure: 284 mm Hg @ 29.4 C [371]

HAP WEIGHTING FACTOR: 1 [713]

VOLATILITY:

Vapor pressure: 160 mm Hg @ 12.8 C; 215 mm Hg @ 21.1 C [371]

Vapor density: 2.49 [055,371]

FIRE HAZARD: 1,2-Epoxybutane has a flash point of -12 C (10 F) [269,275] and it is flammable. Fires involving this material may be controlled with a dry chemical, carbon dioxide or Halon extinguisher.

The autoignition temperature is 515 C (959 F) [371].

LEL: 1.5% [042,371] UEL: 18.3% [042,371]

REACTIVITY: 1,2-Epoxybutane reacts with strong oxidizers [042]. It also reacts with acids and bases [269,371]. Contact with anhydrous metal halides; amino, hydroxyl and carboxyl functions; inorganic acids and charcoal may cause polymerization [052].

STABILITY: 1,2-Epoxybutane is sensitive to heat, moisture and light [269]. Solutions of it in water, DMSO, 95% ethanol or acetone should be stable for 24 hours when protected from light [700].

USES: 1,2-Epoxybutane is used as an intermediate, especially for various polymers. It is also used as a stabilizer for chlorinated solvents.

COMMENTS: None

NIOSH Registry Number: EK3675000

ACUTE/CHRONIC HAZARDS: Toxic concentrations of 1,2-epoxybutane vapors can occur at room temperature [062]. It is an irritant and, when heated to decomposition, it emits toxic fumes of carbon monoxide and carbon dioxide [269].

SYMPTOMS: Symptoms of exposure via inhalation may include irritation and possible injury of the respiratory tract. Ingestion may cause irritation of the mouth and stomach. Eye contact with either the liquid or vapor may cause burns of the eyes. The liquid produces a frostbite-type of skin burn if free to evaporate. If confined to the skin, the burn may cause sensitization [371]. Other symptoms via inhalation may include depression of the central nervous system and possible blood and bone changes [052]. Swallowing large quantities may cause gastrointestinal irritation, nausea and vomiting [058].

Ethyl acrylate

CAS NUMBER: 140-88-5

SYNONYMS: Ethyl acrylate Acrylic acid, ethyl ester Propenoic acid, ethyl ester Ethyl Propenoate Ethyoxycarbonylethylene Ethyl 2-propenoate 2-Propenoic acid, ethyl ester

CHEMICAL FORMULA: C5H8O2

MOLECULAR WEIGHT: 100.12

WLN: 2OV1U1

PHYSICAL DESCRIPTION: Colorless liquid

SPECIFIC GRAVITY: 0.9234 @ 20/4 C [017,047,395]

DENSITY: Not available

MELTING POINT: -71.2 C [047,205,395,421]

BOILING POINT: 99.8 C [017,043,051,395]

SOLUBILITY: Water: 10-50 mg/mL @ 21 C DMSO: >=100 mg/mL @ 21 C 95% Ethanol: >=100 mg/mL @ 21 C Acetone: >=100 mg/mL @ 21 C

OTHER SOLVENTS: Chloroform: Soluble [017,047,395] Ether: Soluble [017,031,047,062] Alcohol: Soluble [017,031,062,205]

OTHER PHYSICAL DATA: This compound is inhibited. Odor threshold: 0.47 ppb [430] Evaporation rate (butyl acetate=1): 3.3 [058,102] Powerful, unpleasant, acrid, ester-like odor [058] Refractive index: 1.4068 @ 20 C [017,047,205,430] Ultraviolet maximum (epsilon = 692): 208 nm [395] Specific gravity: 0.9230 @ 20/20 C [062,371] Boiling point: 20 C @ 39.2 mm Hg [031] 95 C @ 746.3 mm Hg [052] Specific heat: 0.442 cal/g/C @ -60 C [031] Bulk density: 7.6 lb/gal @ 20 C [062] Heat of vaporization: 8.27 kcal/mol [031] Heat of combustion: 655.49 kcal/mol [031] Forms an azeotrope with water containing 45% water (boiling point of 81 C) [031] Forms an azeotrope with ethanol containing 56.8% ethanol (boiling point of 76 C) [031] Liquid surface tension: 25 dynes/cm [371] Liquid water interfacial tension: 40 dynes/cm [371] Specific heat of vapor (gas): 1.080 [371] Heat of polymerization: -186 cal/g [371] Reid vapor pressure: 1.4 psia [371]

HAP WEIGHTING FACTOR: 1 [713]

VOLATILITY:

Vapor pressure: 29.3 mm Hg @ 20 C [043,051,421,430] 40 mm Hg @ 26 C [038]

Vapor density: 3.45 [031,043,421,430]

FIRE HAZARD: Ethyl acrylate has a flash point of 9 C (48 F) [102] and it is flammable. Fires involving this material may be controlled with a dry chemical, carbon dioxide or Halon extinguisher.

The autoignition temperature of ethyl acrylate is 383 C (721 F) [102,421].

LEL: 1.8% [043,051,058,430] UEL: 12.1% [058]

REACTIVITY: Ethyl acrylate is incompatible with oxidizers, peroxides, strong alkalis and polymerization initiators [058,346,395]. It is also incompatible with acids [269]. It will react violently with chlorosulfonic acid [043,051]. It readily polymerizes without the presence of an inhibitor (polymerization is speeded up by exposure to light) [031,395].

STABILITY: Ethyl acrylate is sensitive to exposure to moisture, light and heat [058]. It may polymerize when exposed to light [031,269]. High temperatures can negate the effects of inhibitors. It is subject to slow hydrolysis [051]. Inhibitors do not function in the absence of air [102]. Proton NMR stability screening indicates that solutions of this material in DMSO are stable for at least 24 hours under normal lab conditions [700].

USES: Ethyl acrylate is used in the manufacturing of acrylic resins for use in paint formulations, industrial coatings and latexes, in the manufacturing of plastics such as ethylene ethyl acrylate and in the manufacturing of polyacrylate elastomers and acrylic rubber. It is also used in the forming of denture materials, water emulsion vehicle for paints, textiles and paper coatings, leather finishes, resins or adhesives and lends flexibility to hard films.

COMMENTS: Ethyl acrylate is often inhibited with 15 ppm hydroquinone monomethyl ether.

NIOSH Registry Number: AT0700000

ACUTE/CHRONIC HAZARDS: Ethyl acrylate may be toxic by ingestion, inhalation or skin absorption. It can be absorbed through the skin and is a lachrymator and an irritant. It also is corrosive [058]. When heated to decomposition it emits smoke and acrid fumes of carbon dioxide and carbon monoxide [043,269].

SYMPTOMS: Symptoms of exposure may include irritation of the skin, eyes, mucous membranes, lungs and upper respiratory tract [269]. It may also cause irritation of the mouth, stomach and throat. Other symptoms may include possible burns of the eyes (damage irreversible), injury to the skin including reddening and swelling, skin sensitization (allergic reaction), dizziness, labored breathing and nervousness [058]. Central nervous system effects and kidney, liver and heart damage have been reported [301]. Other symptoms may include conjunctivitis, lacrimation and chronic lung disease [107]. Inhalation of high concentrations may cause rapid breathing, headache, nausea, pulmonary edema, lethargy, nose irritation, convulsions and death [151]. It may also cause chest pain [269]. Loss of consciousness has been reported [371]. It may also cause shortness of breath [051]. Other symptoms may include coughing, abdominal pain, vomiting, dermatitis and blurred vision. Persons with pre-existing skin disorders, eye problems, or impaired liver, kidney or respiratory function may be more susceptible to the effects of this chemical [058]. It may also cause gastrointestinal tract irritation [421,430].

Ethylbenzene

NOTE: The form of this name in the Clean Air Act Amendments Section 112(b) list is Ethyl benzene. The correct notation for the compound is Ethylbenzene.

CAS NUMBER: 100-41-4

SYNONYMS: Ethylbenzol Phenylethane Ethylbenzene

CHEMICAL FORMULA: C8H10

MOLECULAR WEIGHT: 106.16

WLN: 2R

PHYSICAL DESCRIPTION: Clear, colorless liquid

SPECIFIC GRAVITY: 0.867 @ 20/4 C

DENSITY: 0.866 g/cm3 @ 21 C

MELTING POINT: -95 C

BOILING POINT: 136.2 C

SOLUBILITY: Water: <1 mg/mL @ 23 C DMSO: >=100 mg/mL @ 23 C 95% Ethanol: 1-10 mg/mL @ 23 C Acetone: >=100 mg/mL @ 23 C

OTHER SOLVENTS: Most organic solvents: Soluble Carbon tetrachloride: Soluble Sulfur dioxide: Soluble Ammonia: Insoluble Ether: Soluble Benzene: Soluble

OTHER PHYSICAL DATA: Aromatic odor Refractive index: 1.49594 @ 20 C; 1.4932 @ 25 C Specific heat: 0.41 cal/gal/degrees C Viscosity: 0.64 centipoise @ 25 C Specific gravity: 0.866 @ 25/25 C Boiling point: 25.8 C @ 10 mm Hg

HAP WEIGHTING FACTOR: 1 [713]

VOLATILITY:

Vapor pressure: 10 mm Hg @ 29.5 C; 5 mm Hg @ 13.9 C; 760 mm Hg @ 136.2 C

Vapor density: 3.66

FIRE HAZARD: Ethylbenzene has a flash point of 15 C (59 F) and it is flammable. Fires involving this compound can be controlled with a dry chemical, carbon dioxide or Halon extinguisher.

The autoignition temperature is 432 C (810 F).

LEL: 1.2% UEL: 6.8

%REACTIVITY: Ethylbenzene can react vigorously with strong oxidizing materials.

STABILITY: Ethylbenzene is stable under normal laboratory conditions. Solutions of it should be stable for 24 hours under normal lab conditions [700].

USES: Ethylbenzene is used as an intermediate in production of styrene, in organic synthesis, as a solvent, an antiknock agent, in the manufacture of acetophenone, and as a constituent of asphalt and naphtha.

COMMENTS: None

NIOSH Registry Number: DA0700000

ACUTE/CHRONIC HAZARDS: Ethylbenzene is an irritant, a lachrymator, and it may be toxic by ingestion, inhalation and skin absorption. In high concentrations it also can be narcotic.

SYMPTOMS: Symptoms of exposure may include irritation, redness and inflammation of the skin; irritation of the nose, throat, and eyes; lacrimation, conjunctivitis, corneal erosion, dermatitis, dizziness, narcosis and a sensation of constriction of the chest.

Ethyl carbamate

CAS NUMBER: 51-79-6

SYNONYMS: Ethyl urethane Carbamic acid, ethyl ester Ethyl aminoformate Urethane

CHEMICAL FORMULA: C3H7NO2

MOLECULAR WEIGHT: 89.09

WLN: ZVO2

PHYSICAL DESCRIPTION: Colorless crystals or white granular powder

SPECIFIC GRAVITY: 0.9862 @ 21/4 C [016,047]

DENSITY: 1.1 g/mL [025,031]

MELTING POINT: 48-50 C [016,031,295,395]

BOILING POINT: 182-184 C [031,058,205,269]

SOLUBILITY: Water: >=100 mg/mL @ 22 C [700] DMSO: >=100 mg/mL @ 22 C [700] 95% Ethanol: >=100 mg/mL @ 22 C [700] Acetone: >=100 mg/mL @ 22 C [700]

OTHER SOLVENTS: Benzene: Soluble [016,025] Ether: 1 g/1.5 mL [031,395] Chloroform: 1 g/0.9 mL [031,395] Glycerol: 1 g/2.5 mL [031,395] Ligroin: Sparingly soluble [025] Olive oil: 1 g/32 mL [031,395] Alcohol: 1 g/0.8 mL [031]

OTHER PHYSICAL DATA: Sublimes readily @ 103 C and 54 mm Hg [031,395] Refractive index: 1.4144 @ 51 C [016,047] Cooling, saline taste [025,031,295] Aqueous solutions are neutral to litmus [031,062,295,395] Odorless [043,058,062,346] Vapor pressure: 40 mm Hg @ 105.6 C; 760 mm Hg @ 184.0 C [038] A 2.93% solution is iso-osmotic with serum [295]

VOLATILITY:

Vapor pressure: 5 mm Hg @ 65.8 C; 10 mm Hg @ 77.8 C; 20 mm Hg @ 91.0 C [038]

Vapor density: 3.07 [043,058]

FIRE HAZARD: Ethyl carbamate has a flash point of 92 C (198 F) [269,275]. It is combustible. Fires involving this material can be controlled with a dry chemical, carbon dioxide or Halon extinguisher. A water spray may also be used [058].

LEL: Not available UEL: Not available

REACTIVITY: Ethyl carbamate is incompatible with alkalies, acids, chloral hydrate, camphor, menthol and thymol [033,058,269,295]. It is also incompatible with antipyrine and salol [033,058,269]. It may react with strong oxidizers [058, 269]. It liquifies with benzoic acid, resorcinol and salicylic acid [295]. It reacts with phosphorus pentachloride to form an explosive product [043].

STABILITY

Ethyl carbamate is stable under normal laboratory conditions. Solutions of this chemical in water, DMSO, 95% ethanol or acetone should be stable for 24 hours under normal lab conditions [700].

USES: Ethyl carbamate is a chemical intermediate for pesticides, fungicides and pharmaceuticals. It is used in biochemical research, as an antineoplastic agent, as an intermediate in organic synthesis, in preparation and modifica- tion of amino resins, as a solubilizer and cosolvent for pesticides, in fumigants and cosmetics, as a fish narcotic and in cross-linking agents. It is also used in veterinary medicine as an anesthetic. Molten urethane is used as a solvent for various organic materials. It was formerly used as a hypnotic, as an adjunct to sulfonamide therapy, as a topical bactericide, as a component (with quinine) of a sclerosing solution for varicose veins and in the treatment of chronic leukemia and multiple myeloma.

COMMENTS: None

NIOSH Registry Number: FA8400000

ACUTE/CHRONIC HAZARDS: Ethyl carbamate may be toxic by ingestion [062], inhalation or skin absorption [058,269]. It may cause irritation [269]. It is also an irritant and, when heated to decomposition it emits toxic fumes of carbon monoxide, carbon dioxide and nitrogen oxides [043,058,269].

SYMPTOMS: Symptoms of exposure may include burning in the throat, watery to bloody diarrhea, abdominal pain, oliguria, fall in blood pressure, anuria, cardiovascular collapse, delirium, convulsions, muscular weakness with respiratory failure, coma, hemorrhages and kidney and liver injury [301]. It also may cause nausea, vomiting and bone marrow depression [043,058,269,295]. Other symptoms may include central nervous system depression and focal degeneration in the brain [043,058,269]. Exposure may lead to gastroenteric hemorrhages and drowsiness [346]. It may also cause irritation [058,269]. Anorexia and dizziness have been reported [295]. Large doses make debilitated persons more prone to hepatitis or fatal hepatic necrosis [346].

Ethyl chloride

CAS NUMBER: 75-00-3

SYNONYMS: Chloroethane Chloryl

CHEMICAL FORMULA: C2H5Cl

MOLECULAR WEIGHT: 64.52

WLN: G2

PHYSICAL DESCRIPTION: Gas @ room temperature

SPECIFIC GRAVITY: 0.8978 @ 20/4 C (liquid) [017,047]

DENSITY: Not available

MELTING POINT: -139 C [038,102,430]

BOILING POINT: 12.3 C [017,031,205,421]

SOLUBILITY: Water: 0.574 g/100 mL @ 20 C [031,102,421,430] DMSO: Not available 95% Ethanol: 48 g/100 mL @ 21 C [421,430] Acetone: Not available

OTHER SOLVENTS: Ether: Soluble [017,047,421,430] Most organic solvents: Miscible [051,062]

OTHER PHYSICAL DATA: @ low temperatures or under increased pressure, this compound is a very volatile liquid [031,062,430]. The liquid vaporizes immediately when re- leased @ ordinary room temperature [031] Ether-like odor, burning taste Specific gravity: 0.9214 @ 0/4 C [031,051,062,421] 0.917 @ 6/6 C [029] Refractive index: 1.3676 @ 20 C; 1.3742 @ 10 C Burns with a smokey, green flame Critical temperature: 187.2 C Critical pressurc: 52.0 atm Evaporation rate: much greater than 1 Boiling point: 32.5 C @ 1520 mm Hg; 92.6 C @ 7600 mm Hg [031] Gas density: 0.00266 g/cm3 [058]

HAP WEIGHTING FACTOR: 1 [713]

VOLATILITY:

Vapor pressure: 1000 mm Hg @ 20 C [042,051,055,062] 457 mm Hg @ 0 C [055]

Vapor density: 2.22

FIRE HAZARD: The flash point for ethyl chloride is -50 C (-58 F) [029,036,062,421] and it is flammable. Fires involving this compound may be controlled using a dry chemical, carbon dioxide or Halon extinguisher. A water fog may also be used [371].

The autoignition temperature of ethyl chloride is 519 C (966 F) [036, 058,102,371].

LEL: 3.8% [042,058,102,430] UEL: 15.4% [031,042,051,062]

REACTIVITY: Ethyl chloride will hydrolyze in the presence of alkalis and water [051,052, 062]. It reacts with water or steam to produce toxic and corrosive fumes [042,051]. It can also react vigorously with oxidizing materials [042,051, 052]. The vapor forms highly flammable mixtures with air [051,058]. A mixture of this compound with potassium is shock-sensitive [036,042,066]. Contact with chemically active metals such as Na, K, Ca, powdered Al, Z and Mg may result in violent reactions [102,346].

STABILITY: Ethyl chloride is heat sensitive [102].

USES: Ethyl chloride is used in the manufacture of tetraethyl lead, ethyl cellulose, sulfanol and trional; as an anesthetic; in organic synthesis; as an alkylating agent; in refrigeration; as an analytical reagent; as a solvent for phosphorous, sulfur, fats, oils, resins and waxes; in insecticides; and as an ethylating agent in the manufacture of dyes and drugs.

COMMENTS: Ethyl chloride is flammable at concentrations necessary for anesthesia, rendering it obsolete [430].

NIOSH Registry Number: KH7525000

ACUTE/CHRONIC HAZARDS: Ethyl chloride is an irritant at even slight concentrations [051,346]. High concentrations are narcotic [031,036,051,102]. It may be toxic via the oral and inhalation routes [051]. It may also be readily absorbed through the skin [058,102]. When heated to decomposition, this compound may emit toxic fumes of hydrogen chloride [058,102,346], carbon monoxide [102], and phosgene [042,051,058,346]

SYMPTOMS: Exposure can cause slight irritation of the eyes [062,099, 102,346]. It may also be irritating to the nose, throat and respiratory tract [346,371]. At concentrations of approximately 2% (molar), it can cause an anesthetic or narcotic effect [058]. This can result in headache and nausea [058,346]. It may also result in dizziness [058,346,371,430]. At higher concentrations, exposure can cause unconsciousness [031,058,102,346]. It may also produce central nervous system depression [151,301]. This depression is usually brief and reversible [151]. Prolonged exposure to high concentrations have resulted in death [058,102]. Due to its rapid rate of evaporation, it can cause tissue freezing or frostbite on dermal contact [058, 102,346,430]. Other symptoms may include drowsiness, noisy talkativeness and sensitizing effects on the myocardium [102]. It may also cause irregular heartbeat [102,301]. It may cause liver and kidney damage, incoordination and abdominal cramps [102,346,421,430]. It may cause slight symptoms of inebriation [102,371,430]. Exposure may also cause lung irritation, damage to internal organs, excitement and paralysis of respiration [301]. It may cause lung damage [371,421]. This compound may irritate the kidneys and cause fat accumulation in the kidneys, cardiac muscles and liver [051]. It may be absorbed readily through the lungs, and rapidly given off through the lungs [058,430]. Other symptoms reported may include cardiac arrhythmias at high concentrations; rare allergic eczematous eruptions when sprayed on skin; and hyperemia, edema, and hemorrhages in the internal organs, brain and lungs [430]. It may also cause weak analgesia [102,421,430] and stupor. The most serious problem from severe acute exposure, other than the anesthetic effect, is the possibility of the potentiation of adrenalin, and the resultant cardiac problems [421,430]. Exposure to this compound may cause death due to respiratory or cardiac arrest [102,346]. Deaths are anesthetic in nature [430].

Ethylene dibromide

CAS NUMBER: 106-93-4

SYNONYMS: Dibromoethane 1,2-Dibromoethane Ethylene bromide Glycol bromide

CHEMICAL FORMULA: C2H4Br2

MOLECULAR WEIGHT: 187.88

WLN: E2E

PHYSICAL DESCRIPTION: Colorless, heavy liquid

SPECIFIC GRAVITY: 2.172 @ 25/25 C [031,043,072,171]

DENSITY: 2.177 g/cc @ 21 C

MELTING POINT: 9.97 C [055,395,430]

BOILING POINT: 131-132 C [031,058,173,269]

SOLUBILITY: Water: <1 mg/mL @ 21 C DMSO: >=100 mg/mL @ 21 C 95% Ethanol: >=100 mg/mL @ 21 C Acetone: >=100 mg/mL @ 21 C

OTHER SOLVENTS: Cyclohexane: >=100 mg/mL @ 21 C Most organic solvents: Soluble [172,295,395,421] Most thinners: Miscible [051,062] Ether: Miscible [017,031,047,295] Benzene: Soluble [017,047] Alcohol: Miscible [031,173,205] Absolute ethanol: Soluble [052] Non-alkaline organic liquids: Miscible [168]

OTHER PHYSICAL DATA: Specific gravity: 2.1792 @ 20/4 C [017,047] 2.1707 @ 25/4 C [043,051] Density: 2.17-2.18 g/mL @ 20 C [062] Emulsifiable [051,062] Refractive index: 1.5387 @ 20 C [017,047] 1.5337 @ 25 C [062] Sweet, chloroform-like odor [031,036,430,451] Odor threshold: 25 ppm [151] Boiling point: 29.1 C @ 10 mm Hg [017,047] 34 C @ 14 mm Hg [025] Vapor pressure: 5 mm Hg @ 4.7 C; 10 mm Hg @ 18.6 C; 20 mm Hg @ 32.7 C [038] Solidifies @ 9 C [031,173] Bulk density: 18.1 lb/gal [062] Saturated concentration: 113 g/m3 @ 20 C; 168 g/m3 @ 30 C [055]

HAP WEIGHTING FACTOR: 10 [713]

VOLATILITY:

Vapor pressure: 11 mm Hg @ 20 C [055,102,421] 17.4 mm Hg @ 30 C [043,062]

Vapor density: 6.48 [043,051]

FIRE HAZARD: Ethylene dibromide is nonflammable [058, 062, 205, 269].

LEL: Not available UEL: Not available

REACTIVITY: Ethylene dibromide may be corrosive to iron and other metals and may decompose upon contact with alkalis [168]. It is incompatible with oxidizing agents and it may react with chemically active metals, sodium, potassium, calcium, powdered aluminum, zinc, magnesium and liquid ammonia

[058,102,346,451]. It may attack some forms of plastics, rubber and coatings [102]. It reacts as an alkylating agent [395].

STABILITY: Ethylene dibromide slowly decomposes in the presence of light and heat [058]. It turns brown upon exposure to light [395]. It may hydrolyze slowly with moisture [058]. Solutions of it in water, DMSO, 95% ethanol or acetone should be stable for 24 hours under normal laboratory conditions [700].

USES: Ethylene dibromide is used as a scavenger for lead in gasoline, grain, vegetable, fruit and tree crop fumigant, general solvent, waterproofing preparations, organic synthesis, insecticide, medicine, entrainment reagent for conversion of unreactive halides to Grignard reagents, control of nematodes, termites and pine bark beetles, production of fire extinguishing agents and gauge fluids during the manufacture of measuring instruments, organic synthesis in production of dyes, pharmaceuticals and ethylene oxide, and specialty solvent for resins, gums and waxes.

COMMENTS: Ethylene dibromide may poison platinum catalysts. Fatal acute intoxications are rare since fume concentrations great enough to cause serious illness in short exposures have a definite and sickening odor.

NIOSH Registry Number: KH9275000

ACUTE/CHRONIC HAZARDS: Ethylene dibromide may be toxic by inhalation, ingestion or skin absorption [036, 062,269]. It is an irritant [058,269,301], and it may also cause narcosis [036,102,171]. It may be absorbed through the skin [058,102,346,430]. When heated to decomposition it emits toxic fumes of carbon monoxide, carbon dioxide and hydrogen bromide gas [269]. It may also emit ionic and oxidized halogen [058].

SYMPTOMS: This compound may cause skin, eye, mucous membrane and respiratory tract irritation [058,269,301]. It may also cause damage to the lungs, liver and kidneys [102,269,295,371]. Inhalation may result in delayed pulmonary lesions [031,151]. The vapor may have a narcotic action [036,102,171]. Prolonged skin contact with the liquid may cause erythema, blistering and skin ulcers (may be delayed 24-48 hours). Dermal sensitization to the liquid may develop. Inhalation may cause severe acute respiratory injury, central nervous system depression and severe vomiting [186,346]. Other symptoms may include abdominal pain, nausea, diarrhea, dark and scant urine, anuria, mild icterus, marked agitation, jaundice, moderate anemia, systemic effects and chemical burns [171]. Pulmonary congestion may result after exposure by inhalation, and ingestion may cause gastrointestinal distress and pulmonary edema [406]. After short exposure to very high concentrations, drowsiness occurs. Other symptoms may include tachypnea, diffuse inflammation of the gastric and intestinal mucosa, massive central lobular necrosis of the liver and patchy necrosis of the tubular epithelium [151]. It may cause vesiculation [102]. It may also cause skin burns. Inhalation may be fatal as a result of spasm, inflammation and edema of the larynx and bronchi. Other symptoms may include chemical pneumonitis and eye damage. High concentrations are extremely destructive to tissues of the mucous membranes, upper respiratory tract, skin and eyes [269]. Anesthesia, conjunctivitis, pharyngeal and bronchial irritation, severe loss of appetite, headache and depression have also been reported [421]. Eye contact may cause a temporary loss of vision [058]. Death may result from respiratory or circulatory failure [151]. Lung injury may result in pneumonia and eventually may cause death [430]. Persons with pre-existing skin disorders or eye problems, or impaired liver, kidney or respiratory function may be more susceptible to the effects of this material [058].

Ethylene dichloride

CAS NUMBER: 107-06-2

SYNONYMS: 1,2-Dichloroethane

CHEMICAL FORMULA: C2H4Cl2

MOLECULAR WEIGHT: 98.96

WLN: G2G

PHYSICAL DESCRIPTION: Clear, colorless, oily liquid

SPECIFIC GRAVITY: 1.2569 @ 20/4 C [031,173]

DENSITY: 1.253 g/mL @ 20 C [395]

MELTING POINT: -35.3 C [017,430]

BOILING POINT: 83.5 C [017,043,395,430]

SOLUBILITY: Water: 5-10 mg/mL @ 19 C [700] DMSO: >=100 mg/mL @ 19 C [700] 95% Ethanol: >=100 mg/mL @ 19 C [700] Acetone: >=100 mg/mL @ 19 C [700]

OTHER SOLVENTS: Chloroform: Soluble [031,173,205] Ether: Soluble [031,173,205] Benzene: Soluble [017] Alcohol: Soluble [017,031,173,205]

OTHER PHYSICAL DATA: Refractive index: 1.4448 @ 20 C [047,205] Sweet taste [043,062,173,346] Chloroform-like odor [036,062,173,295] Burns with a smoky flame [031] Odor threshold: 6-40 ppm [151] Specific gravity: 1.25 @ 25/25 C [058]

HAP WEIGHTING FACTOR: 1 [713]

VOLATILITY:

Vapor pressure: 60 mm Hg @ 20 C [071] 100 mm Hg @ 29.4 C [043]

Vapor density: 3.4 [395,451]

FIRE HAZARD: Ethylene dichloride has a flash point of 13 C (56 F) [031,043,430,451] and it is flammable. Fires involving this material can be controlled with a dry chemical, carbon dioxide or Halon extinguisher. A water spray may also be used [036,371].

The autoignition temperature of ethylene dichloride is 413 C (775 F) [036,451].

LEL: 6.2% [043,062,371,451] UEL: 15.9% [043,062,430,451]

REACTIVITY: Ethylene dichloride is incompatible with strong alkalis, strong caustics, oxidizing materials, active metals such as aluminum, magnesium, sodium, or potassium. It reacts violently with nitrogen tetraoxide, dimethylaminopropylamine, or liquid ammonia. A vigorous reaction also occurs when a mixture of this compound, propylene dichloride, and o-dichlorobenzene comes in contact with aluminum. It can corrode iron, zinc and aluminum in the presence of moisture [071]. Mixtures with HNO3 easily detonate [036].

STABILITY: Ethylene dichloride will darken in color in the presence of air, moisture and light [053,395]. Solutions of it in water, DMSO, 95% ethanol or acetone should be stable for 24 hours under normal lab conditions [700].

USES: Ethylene dichloride is used as a chemical intermediate in the manufacturing of methyl chloroform, perchloroethylene and ethylene amines, polyvinyl chloride, sulfide compounds, vinyl chloride, acetyl cellulose, trichloroethylene, vinylidene chloride and trichloroethane. It is also used as an antiknock additive in leaded fuels, in coatings, pharmaceutical products, color film, pesticides, in extraction of oil from seeds, in the processing of animal fats, cleaning agent for textiles, formulation ingredient in grain fumigants, in the extraction of copper from copper ores, solvent for fats, oils, waxes, gums, resins, particularly rubber, tobacco extract, fumigant, paint, varnish and finish removers, metal degreasing, soaps, scouring compounds, wetting and penetrating agents, organic synthesis, ore flotation, fumigant, ingredient in fingernail polish, used in extracting spices, solvent in printing inks, adhesives, asphalt, bitumen, rubber, cellulose, acetate, cellulose ester, degreaser in engineering, textile, petroleum, extracting agent for soybean oil and caffeine, dry cleaning agent, photography, xerography, water softening and cosmetics.

COMMENTS: None

NIOSH Registry Number: KI0525000

ACUTE/CHRONIC HAZARDS: Ethylene dichloride is toxic by ingestion, inhalation, and skin contact [071]. It can be absorbed through the skin [058,071,269]. It is an irritant of the skin, eyes and respiratory tract [031,269,346]. When heated to decomposition it emits toxic fumes of carbon monoxide, carbon dioxide, hydrogen chloride gas and phosgene [269].

SYMPTOMS: Symptoms of exposure may include irritation of the skin, eyes and respiratory tract, corneal clouding and dermatitis [058]. It may cause conjunctivitis, corneal ulceration, headache, mental confusion, depression, fatigue, albuminuria, central nervous system depression, convulsions, diarrhea, hepatomegaly, hypoglycemia, jaundice, narcosis and pulmonary edema [071]. It may also cause flaccid paralysis without anesthesia, somnolence, cough, nausea, vomiting, hypermotility, ulceration, fatty liver degeneration, change in cardiac rate, cyanosis, coma, edema of the lungs and toxic effects on the kidneys [043]. It may cause feeling of drunkenness, drowsiness, unconsciousness and death from respiratory and cardiac failure, defatting of the skin, swelling of the skin and chemical pneumonia [058]. Other symptoms may include mental confusion, abdominal pains and liver and kidney damage [036]. It can also cause watery stool, weak and rapid pulse and internal bleeding [151]. It may cause corneal opacity [099]. Other symptoms may include edema of the brain, vascular congestion in the lungs, heart and spleen, weight loss and oliguria [301]. It may also cause dizziness, narcosis, intestinal hemorrhages, weakness, trembling and severe shock [173]. Chronic exposure may result in loss of appetite, epigastric distress, tremors, nystagmus, leukocytosis and low blood sugar levels [071].

Ethylene glycol

CAS NUMBER: 107-21-1

SYNONYMS: Ethane-1,2-diol 1,2-Dihydroxyethane 1,2-Ethanediol Ethylene alcohol Ethylene dihydrate

CHEMICAL FORMULA: C2H6O2

MOLECULAR WEIGHT: 62.07

WLN: Q2Q

PHYSICAL DESCRIPTION: Clear, colorless, viscous liquid

SPECIFIC GRAVITY: 1.1135 @ 20/4 C [033]

DENSITY: 1.1155 g/mL @ 20 C [062]

MELTING POINT: -13 C [033,043,205,269]

BOILING POINT: 197.6 C [033,071,205]

SOLUBILITY: Water: >=100 mg/mL @ 17.5 C DMSO: >=100 mg/mL @ 17.5 C 95% Ethanol: >=100 mg/mL @ 17.5 C Acetone: >=100 mg/mL @ 17.5 C Methanol: Soluble [029]

OTHER SOLVENTS: Alcohol: Soluble [017,062,205] Lower aliphatic alcohols: Miscible [033,051,430] Glycerol: Miscible [033,051,205] Acetic acid: Miscible [029,033,051,205] Aldehydes: Miscible [033,051,430] Pyridine and similar coal tar bases: Miscible [029,033,051,205] Chlorinated hydrocarbons: Practically insoluble [033] Petroleum ether: Practically insoluble [033] Carbon tetrachloride: Insoluble [029] Chloroform: Insoluble [029,205] Ether: 1 in 200 [033] Benzene and its homologs: Practically insoluble [033] Oils: Practically insoluble [033] Ketones: Miscible [033,051,430] Carbon disulfide: Insoluble [029]

OTHER PHYSICAL DATA: Specific gravity: 1.113 @ 25/25 C [043,051,269] 1.1274 @ 0/4 C [033] Specific gravity: 1.1065 @ 30/4 C; 1.1204 @ 10/4 C [033] 1.11 @ 25/4 C [430] Boiling point: 93 C @ 13 mm Hg [017] 140 C @ 97 mm Hg; 100 C @ 18 mm Hg [033] Boiling point: 70 C @ 3.0 mm Hg; 20 C @ 0.06 mm Hg [033] Vapor pressure: 5 mm Hg @ 79.7 C; 10 mm Hg @ 92.1 C; 20 mm Hg @ 105.8 C [038] Vapor pressure: 40 mm Hg @ 120 C; 760 mm Hg @ 197.3 C [038] Odorless [062,346,430] Viscosity: 26 centipoise @ 15 C; 21 centipoise @ 20 C [033] Viscosity: 17.3 centipoise @ 25 C [033] Refractive index: 1.43063 @ 25 C; 1.43312 @ 15 C [033] Refractive index: 1.4318 @ 20 C [017,047,205] Sweet taste [029,033,036,062] Dielectric constant: 38.66 esu @ 20 C and 150 m wavelength [033] Dipole moment: 2.20 [033] Specific heat: 0.561 cal/g/C @ 20 C [033] Heat of formation: -108.1 kcal/mol [033] Heat of fusion: 44.7 cal/g [033] Heat of vaporization: 191 cal/g [033] Heat of solution: -6.5 cal/g @ 17 C (when 37 parts are mixed with 63 parts water (w/w)) [033] Surface tension: 48.4 dynes/cm @ 20 C [033] Evaporation rate (butyl acetate=1): <0.01 [058] Absorbs twice its weight of water @ 100% relative humidity [033]

HAP WEIGHTING FACTOR: 1 [713]

VOLATILITY:

Vapor pressure: 0.06 mm Hg @ 20 C [058,430] 1 mm Hg @ 53.0 C [038]

Vapor density: 2.14 [043,051,055]

FIRE HAZARD: Ethylene glycol has a flash point of 111 C (232 F) [043,058,371,451]. It is combustible. Fires involving this material may be controlled with a dry chemical, carbon dioxide or Halon extinguisher. A water spray may also be used [043,058,451].

The autoignition temperature of ethylene glycol is 400 C (752 F) [043,051].

LEL: 3.2% [047,051,058,371] UEL: Not available

REACTIVITY: Ethylene glycol is incompatible with oxidizers, chromium trioxide, oleum, chlorosulfonic acid, sulfuric acid, HClO4, phosphorous pentasulfide, potassium permanganate and sodium peroxide. Mixtures with ammonium dichromate, silver chlorate, sodium chlorite and uranyl nitrate ignite when heated to 100 C. Aqueous solutions may ignite silvered copper wires which have applied D.C. voltage [043].

STABILITY: Ethylene glycol is hygroscopic [033,043,062,275].

USES: Ethylene glycol is used in antifreeze, in hydraulic brake fluids, as an industrial humectant, as an ingredient of electrolytic condensers, as a solvent in the paint and plastics industries, in the formulation of printers inks, stamp pad inks and ball point pen inks, as a softening agent for cellophane and as a stabilizer for soybean foam used to extinguish oil and gasoline fires. It is used in the synthesis of safety explosives, glyoxal, unsaturated ester type alkyd resins, plasticizers, elastomers, synthetic fibers and synthetic waxes. It is also used in asphalt emulsion, as a heat transfer agent and as an ingredient for deicing airport runways.

COMMENTS: None

NIOSH Registry Number: KW2975000

ACUTE/CHRONIC HAZARDS: Ethylene glycol may be toxic by ingestion [033] and it may be readily absorbed through the skin [058,269]. When heated to decomposition it emits acrid smoke, irritating fumes and toxic fumes of carbon monoxide, carbon dioxide and unidentified organic compounds [043,058].

SYMPTOMS: Symptoms of exposure may include restlessness, unsteady gait, drowsiness, coma, transient stimulation of central nervous system followed by depression, vomiting, renal damage, anuria and uremia [036]. Other symptoms may include lacrimation, general anesthesia, headache, cough, respiratory stimulation, nausea and pulmonary and liver damage [043]. It may cause congestion, edema and damage to the brain, acidosis, focal hemorrhagic necrosis of the renal cortex, hydropic degeneration of the liver and kidneys, calcium oxalate crystals in the brain, spinal cord and kidneys; narcosis, cyanosis, tachypnea, pulmonary edema, muscle tenderness, stupor, prostration, unconsciousness with convulsions, hypoglycemia, death from respiratory failure, hypocalcemic tetany, intravascular hemolysis, oliguria, nystagmus, lymphocytosis, reduced blood pH or glucose, methemoglobinemia and hyperkalemia [301]. It may also cause low-grade fever, depressed reflexes, generalized or focal seizures, tetanic contractions, myoclonic jerks, ophthalmoplegia, papilledema, optic atrophy, tachycardia, mild hypotension, broncho pneumonia, cardiac enlargement and congestive failure [430]. Exposure may lead to anorexia, hematopoietic dysfunction and depression followed by respiratory and cardiac failure [346]. It may also lead to blood or central nervous system damage, eye irritation, dizziness, abdominal pain and discomfort, malaise, lumbar pain, loss of appetite and neural dysfunction [058].

Ethyleneimine

NOTE: The form of this name in the Clean Air Act Amendments Section 112(b) list is Ethylene imine. The correct notation for the compound is Ethyleneimine.

CAS NUMBER: 151-56-4

SYNONYMS: Aziridine Hydrazobenzene

CHEMICAL FORMULA: C2H5N

MOLECULAR WEIGHT: 43.08 [703]

PHYSICAL DESCRIPTION: Oily, water-white liquid with pungent ammoniacal odor [703]

DENSITY: 0.832 @ 20C/4C [703]

MELTING POINT: Not available

BOILING POINT: 55-56 C [703]

SOLUBILITY: Water: Miscible [703]

OTHER PHYSICAL DATA: Freezing point: -71.5 C [703] Vapor pressure: 160 mm @ 20 C [703] Vapor density: 1.48 [703]

HAP WEIGHTING FACTOR: 1 [713]

VOLATILITY:

Vapor Pressure: 160 mm @ 20 C [703]

Vapor Density: 1.48 [703]

FIRE HAZARD: Flash Point: 12 F [703, 704] -24 C [702]

A very dangerous fire and explosion hazard when exposed to heat, flame or oxidizers. To fight fire, use alcohol foam, CO_2, dry chemical. [703] Class IB flammable liquid [704]. Combustion ranking: 86 [702].

Autoignition temperature: 608 F [703]

LEL: 3.6% UEL: 46

REACTIVITY: Very dangerous fire and explosion hazard when exposed to oxidizers. Heat and/or catalytically active metals or chloride ions can cause a violent exothermic reaction. [703]

Acids: Reacts violently with acids; aluminum chloride + substituted anilines; acetic acid; acetic anhydride; acrolein; acrylic acid; allyl chloride; CS_2; Cl_2; chlorosulfonic acid; epichlorohydrin; glyoxal; HCl; HF; HNO_3; oleum; beta-propiolactone; Ag; NaOCl; H_2SO_4; vinyl acetate. [703] Very reactive chemically and subject to aqueous auto-catalyzed exothermic polymerization, which may be violent if uncontrolled by dilution, slow addition or cooling. It is normally stored over solid caustic alkali, to minimize polymerization catalyzed by presence of carbon dioxide. [710]

Aluminum chloride: In the preparation of a series of substituted phenylethylenediamines, it is essential to add the reagents to an aromatic solvent at 30-80 C in the order aniline, then aluminum chloride, then ethyleneimine to prevent uncontrollable exothermic reaction. [710]

Chlorinating agents: It gives the explosive 1-chloroaziridine on treatment with, e.g. sodium hypochlorite solution. [710] Reacts with chlorinating agents (e.g., sodium hypochlorite solution) to form the explosive silver derivatives. [703]

Silver: Explosive silver derivatives may be formed in contact with silver or its alloys, including silver solder, which is therefore unsuitable in handling equipment. [710, 704]

STABILITY: When heated to decomposition it emits acrid smoke and irritating fumes [703].

UDRI Thermal Stability Class: 5 [702] UDRI Thermal Stability Ranking: 235 [702]

USES: Not available

COMMENTS: DOT Classification: Flammable Liquid; Label: Flammable Liquid and Poison. [703]

NIOSH Registry Number: KX5075000

ACUTE/CHRONIC HAZARDS: Poison by ingestion, skin contact, inhalation, intraperitoneal and possibly other routes. An experimental carcinogen, neoplastigen, tumorigen and teratogen; human mutagenic data exist. A skin, mucous membrane and severe eye irritant; an allergic sensitizer of skin; causes opaque cornea, keratoconus and necrosis of cornea (experimentally); has been known to cause severe human eye injury. Drinking carbonated beverages is recommended as an antidote to this material in stomach. [703] Ethyleneimine is a suspected human carcinogen. [723]

SYMPTOMS: Symptoms of exposure may include nausea, vomiting; headache, dizziness; pulmonary edema; liver, kidney damage; eye burns; skin sensitivity; irritation of the nose and throat. [704] Corneal damage may occur; skin contact may cause painless, slowly healing, necrotic burns, or dermatitis with vesiculations. Dermal sensitization may occur. Respiratory tract irritation with coughing and possible delayed pulmonary edema may develop. Nausea and vomiting may occur, irritation or burns could develop following ingestion. Bone marrow depression might develop. Dermal irritation or burns may be seen. [723] Prolonged keratitis and conjunctivitis, which respond to therapy with extreme difficulty, have been observed and ulcerations have been found on the septum and vocal cords. Headache, dizziness, pain in the vicinity of the temples, and dullness are other symptoms. Other ailments caused by inhalation of ethyleneimine are nasal secretion, laryngeal edema, pronounced diphtherial-like changes of the trachea and bronchi, bronchitis, shortness of breath, edema of the lungs and secondary bronchial pneumonia. [723]

Ethylene oxide

CAS NUMBER: 75-21-8

SYNONYMS: Dimethylene oxide 1,2-Epoxyethane Oxacyclopropane

CHEMICAL FORMULA: C_2H_4O

MOLECULAR WEIGHT: 44.05

WLN: T3OTJ

PHYSICAL DESCRIPTION: Colorless gas at room temperature

SPECIFIC GRAVITY: 0.8824 @ 10/10 C [016,047]

DENSITY: 0.869 g/mL @ 20 C [371]

MELTING POINT: -111 C [016,033,055,395]

BOILING POINT: 10.7 C [033,043,395,451]

SOLUBILITY: Water: Miscible [043,062,205,395] 95% Ethanol: Miscible [395] Acetone: Miscible [430] Methanol: Miscible [430]

OTHER SOLVENTS: Alcohol: Soluble [016,033,205,295] Benzene: Miscible [430] Ether: Miscible [395,430] Most organic solvents: Miscible [395] Carbon tetrachloride: Miscible [430]

OTHER PHYSICAL DATA: Liquid below 12 C [033,062,099,451] Specific gravity: 0.8711 @ 20/20 C [043,430] 0.891 @ 0/4 C [205,395] Specific gravity: 0.891 @ 4/4 C; 0.887 @ 7/7 C [033] 0.869 @ 20/4 C [395] Vapor pressure: 1 mm Hg @ -89.7 C; 5 mm Hg @ -73.8 C; 10 mm Hg @ -65.7 C [038] Vapor pressure: 20 mm Hg @ -56.6 C; 40 mm Hg @ -46.9 C [038] Vapor pressure: 60 mm Hg @ -40.7 C; 100 mm Hg @ -32.1 C [038] Vapor pressure: 200 mm Hg @ -19.5 C; 400 mm Hg @ -4.9 C [038] Vapor pressure: 760 mm Hg @ 10.7 C [038] 494 mm Hg @ 0 C [395] Viscosity: 0.32 centipoise @ 0 C [062,395] Ethereal odor [071,327,451] Odor threshold: 50 ppm [371] Burning rate: 3.5 mm/minute [371] Critical temperature: 196 C [371] Critical pressure: 71.0 atmospheres [371] Liquid surface tension: 24.3 dynes/cm [371] Latent heat of vaporization: 138.5 cal/g [371] Heat of fusion: 28.07 cal/g [371] Heat of combustion: -6380 cal/g [371] Heat of solution: -34 cal/g [371] Refractive index: 1.3597 @ 7 C [016,029,033,205] 1.3614 @ 4 C [430]

HAP WEIGHTING FACTOR: 10 [713]

VOLATILITY:

Vapor pressure: 1095 mm Hg @ 20 C [043,055,327,395]

Vapor density: 1.52 [043,055,071]

FIRE HAZARD: Ethylene oxide has a flash point of -20 C (-4 F) [043,066,430,451] and it is flammable. Fires involving this material may be controlled with a dry chemical, carbon dioxide or Halon extinguisher.

The autoignition temperature of ethylene oxide is 429 C (804 F) [043,062,371,451]. Mixtures of this compound with oxygen or air are explosive [295].

LEL: 3.0% [043,066,371,451] UEL: 100% [043,066,371,451]

REACTIVITY: Violent polymerization occurs on contact with ammonia, alkali hydroxides, amines, metallic potassium, acids and covalent halides (e.g. aluminum chloride, iron (III) chloride, tin (IV) chloride, aluminum oxide, iron oxide and rust). Explosive reactions occur with glycerol at 200 C. Rapid compression of the Vapor with air causes explosions [043,066]. This chemical is incompatible with bases, alcohols, copper and mercaptans [043,071]. It is also incompatible with m-nitroaniline, trimethylamine, magnesium perchlorate and contaminants [043, 066]. It may react with alkane thiols or bromoethane [043]. Violent decomposition occurs with polyhydric alcohol, propylene oxide and sucroglyceride [066]. Iron blue pigment reacts with this compound to give a pyrophoric solid [036,066]. Ethylene oxide reacts with water [151]. It is incompatible with oxidizers [346]. It is also incompatible with sulfhydryl compound, inorganic chloride and 4-(4'-nitrobenzyl)pyridine [395]. It polymerizes violently on contact with highly catalytic surfaces [451].

STABILITY: The vapor of ethylene oxide is readily initiated into explosive decomposition [036,066]. It decomposes violently at temperatures greater than 427 C [327].

USES: Ethylene oxide is used as an intermediate in the production of ethylene glycol, as an intermediate for polyethylene terephthalate polyester fiber and film production, and in the manufacture of non-ionic surface-active agents, diethylene glycol, triethylene glycol, ethanolamines, choline and choline chloride, and other organic chemicals. It is also used as a fungicide for treatment by fumigation of books; dental, pharmaceutical, medical and scientific equipment and supplies (glass, metals, plastics, rubber or textiles), drugs, leather, motor oil, paper, soil, bedding for experimental animals, clothing, furs, furniture and transportation vehicles (jet aircraft, buses and railroad passenger cars). It is also used to sterilize foodstuffs such as spices, cocoa, flour, dried egg powder, desiccated coconut, dried fruits and dehydrated vegetables. It is used to accelerate the maturing of tobacco leaves. It is also used as a rocket propellant, to sterilize surgical instruments, as a starting material for the manufacture of acrylonitrile, as a petroleum demulsifier and as an industrial sterilant (e.g. medical plastic tubing).

COMMENTS: None

NIOSH Registry Number: KX2450000

ACUTE/CHRONIC HAZARDS: Ethylene oxide is an irritant and may be very toxic by inhalation [036]. When heated to decomposition it emits acrid smoke, irritating fumes and toxic fumes of carbon monoxide and carbon dioxide [043,327].

SYMPTOMS: Symptoms of exposure may include irritation of the skin, eyes, mucous membranes and upper respiratory tract [033,451]. Other symptoms may include convulsions, nausea, vomiting, olfactory and pulmonary changes, and in high concentrations, pulmonary edema [043]. It may cause bronchitis, coma, conjunctivitis, corneal damage and delayed burns and blistering [036]. It may also cause headache, toxic dermatitis with large bullae, and in high concentrations, unconsciousness and seizures [099]. Anesthesia may occur [071]. Drowsiness may also occur [151,186,327,346]. Exposure may cause erythema, marked desquamation, formation of residual pigment, urticarial wheal, weakness, dullness of head, stupor, coughing and bradycardia [173]. It may also cause loss of taste and smell, incoordination, dyspnea, cyanosis, hemolysis, sensitization, anaphylaxis, kidney damage and death. Chronic exposure may lead to lymphocytosis, peripheral neuropathy, chromosomal damage to lymphocytes and leukemia [151]. Liver damage may occur [295,301,327]. Severe dermatitis may also occur [346]. Other symptoms may include diarrhea, vertigo and central nervous system depression [295]. This compound may cause frostbite, gastric irritation, lung injury, shortness of breath, reproductive effects and neurotoxicity [327]. It may also cause redness, edema and ulceration of the skin, and encephalopathy (rare) [395]. Necrosis of the skin has been reported [430]. Difficult breathing may occur [371]. Exposure may also cause emphysema [058].

Ethylene thiourea

CAS NUMBER: 96-45-7

SYNONYMS: 2-Imidazolidinethione 4,5-Dihydroimidazole-2(3H)-thione 2-Mercapto-4,5-dihydroimidazole

CHEMICAL FORMULA: C3H6N2S

MOLECULAR WEIGHT: 102.16

WLN: T5MYMTJ BUS

PHYSICAL DESCRIPTION: White to pale green crystals

SPECIFIC GRAVITY: Not available

DENSITY: Not available

MELTING POINT: 203-204 C [026,031,205,395]

BOILING POINT: Not available

SOLUBILITY: Water: 1-5 mg/mL @ 18 C DMSO: >=100 mg/mL @ 18 C 95% Ethanol: 1-5 mg/mL @ 18 C Acetone: 5-10 mg/mL @ 18 C Methanol: Moderately soluble [031,395]

OTHER SOLVENTS: Alcohol: Soluble [016] Ethanol: Soluble [031,205] Ethylene glycol: Moderately soluble [031,053,395] Acetic acid: Slightly soluble [053,062] Naphtha: Slightly soluble [062] Chloroform: Insoluble [026,031,205,395] Ligroin: Insoluble [031] Pyridine: Moderately soluble [031,053,395] Ether: Insoluble [026,031,205,395] Benzene: Insoluble [026,031,205,395] Hot water: Very soluble [053,062] Naphthalene: Slightly soluble [053]

OTHER PHYSICAL DATA: The pure compound is odorless, but the technical product may have an amine odor [051]

HAP WEIGHTING FACTOR: 1 [713]

VOLATILITY: Vapor pressure: <1 mm Hg @ 20 C [053,058]

FIRE HAZARD: Ethylene thiourea has a flash point of 252 C (486 F) [051,053,900]. It is combustible. Fires involving this material may be controlled with a dry chemical, carbon dioxide or Halon extinguisher.

LEL: Not available UEL: Not available

REACTIVITY: Ethylene thiourea is incompatible with strong oxidizing agents [053,058,269].

STABILITY: Ethylene thiourea may be sensitive to prolonged exposure to light. Solutions of it in water, DMSO or 95% ethanol should be stable for 24 hours under normal lab conditions [700]. Solutions in acetone may be decomposed by ultraviolet radiation [053,900].

USES: Ethylene thiourea is used as a polymer vulcanizing and curing agent. It is extensively used as an accelerator in the curing of polychloroprene (neoprene) and other elastomers. It is also used in electroplating baths, as an intermediate for anti-oxidants, in insecticides, dyes, pharmaceuticals and synthetic resins.

COMMENTS: Ethylene thiourea may be present as a contaminant in the ethylene bis- dithiocarbamate fungicides and can also be found when food containing the fungicides is cooked [055]. Its industrial use is subject to severe restriction [029].

NIOSH Registry Number: NI9625000

ACUTE/CHRONIC HAZARDS: Ethylene thiourea is an irritant [269] and may be toxic by ingestion and skin absorption [053,151]. When heated to decomposition it emits very toxic fumes of carbon monoxide, carbon dioxide, nitrogen oxides and sulfur oxides [043,053,269].

SYMPTOMS: Exposure may cause myxedema, the symptoms of which may include thickening and drying of the skin with slowing of mental and physical activity, goiter and other effects of decreased thyroid hormone output [346]. Specific symptoms of myxedema also may include an unusual puffy swelling of subcutaneous tissues (non-pitting edema), a yellowish or ivory pallor to the complexion, enlarged tongue, husky voice, dry and brittle hair and a decrease in all metabolic rates. Talking, breathing and the pulse rate may be slowed [900]. It may also cause irritation [269].

---

bis(2-Ethylhexyl) phthalate

CAS NUMBER: 117-81-7

SYNONYMS: DEHP Di(2-ethylhexyl) phthalate Dioctyl phthalate 1,2-Benzenedicarboxylic acid, bis(2-ethylhexyl) ester

CHEMICAL FORMULA: C24H38O4

MOLECULAR WEIGHT: 390.54

WLN: 4Y2&1OVR BVO1Y4&2

PHYSICAL DESCRIPTION: Colorless oily liquid

SPECIFIC GRAVITY: 0.9861 @ 20/20 C [051,062,395,421]

DENSITY: 0.9732 g/mL @ 24 C [052]

MELTING POINT: -50 C [205,269,275,430]

BOILING POINT: 384 C [047,205,269,275]

SOLUBILITY: Water: <0.1 mg/mL @ 22 C DMSO: 10-50 mg/mL @ 22 C 95% Ethanol: >=100 mg/mL @ 22 C Acetone: >=100 mg/mL @ 22 C

OTHER SOLVENTS: Mineral oil: Miscible [051,062,395,421] Hexane: Miscible [421]

OTHER PHYSICAL DATA: Refractive index: 1.4853 @ 20 C [047,205,269,275] Pour point: -46 C [062,395] Mild odor [371] Boiling point: 222-230 C @ 4 mm Hg [047] 231 C @ 5 mm Hg [062,395,421] Specific gravity: 0.9732 @ 24/21 C [052] 0.981 @ 20/4 C [269] 0.981 @ 25/25 C [205] Viscosity: 81.4 centipoise @ 20 C [062,395,421] Lambda max: 281(shoulder) nm, 275 nm, 225 nm (epsilon = 1130,1260,873) [052] Liquid surface tension (estimated): 15 dynes/cm [371] Liquid water interfacial tension (estimated): 30 dynes/cm [371] Heat of combustion: -8410 cal/g [371]

HAP WEIGHTING FACTOR: 1 [713]

VOLATILITY:

Vapor pressure: 1.32 mm Hg @ 200 C [395,421]

Vapor density: 13.45 [055]

FIRE HAZARD: bis(2-Ethylhexyl) phthalate has a flash point of 207 C (405 F) [205,269,275]. It is combustible. Fires involving this material may be controlled with a dry chemical, carbon dioxide or Halon extinguisher.

The autoignition temperature of bis(2-ethylhexyl) phthalate is 390 C (735 F) [451].

LEL: 0.3% @ 245 C [451] UEL: Not available

REACTIVITY: bis(2-Ethylhexyl) phthalate is incompatible with oxidizing materials [051,058,269,346]. It is also incompatible with nitrates, strong acids and strong alkalis [346].

STABILITY: bis(2-Ethylhexyl) phthalate is stable under normal laboratory conditions [051,058,371]. UV spectrophotometric stability screening indicates that solutions of it in 95% ethanol are stable for at least 24 hours [700].

USES: bis(2-Ethylhexyl) phthalate is used in vacuum pumps. It is also used as a plasticizer for polyvinyl chloride, especially in the manufacture of medical devices, and as a plasticizer for resins and elastomers. It is a solvent in erasable ink and dielectric fluid. It is also used as an acaricide for use in orchards, an inert ingredient in pesticides, a detector for leaks in respirators, testing of air filtration systems and component in cosmetic products.

COMMENTS: None

NIOSH Registry Number: TI0350000

ACUTE/CHRONIC HAZARDS: bis(2-Ethylhexyl) phthalate is an irritant and may be harmful if ingested or inhaled [269]. It may be poorly absorbed through the skin and has low toxicity by all routes of exposure [421]. When heated to decomposition it emits acrid smoke and toxic fumes of carbon monoxide, carbon dioxide and traces of incompletely burned carbon products [043,058,269].

SYMPTOMS: Symptoms of exposure may include irritation of the eyes, mucous membranes [346], and skin [043,058,395]. It may also cause nausea, diarrhea [346], and mild gastric disturbances [151].

Ethylidene dichloride

CAS NUMBER: 75-34-3

SYNONYMS: 1,1-Dichloroethane

CHEMICAL FORMULA: $C_2H_4Cl_2$

MOLECULAR WEIGHT: 98.96

WLN: GYG1

PHYSICAL DESCRIPTION: Clear, colorless, oily liquid

SPECIFIC GRAVITY: 1.1757 @ 20/4 C [017,031,047,205]

DENSITY: 1.256 g/cc @ 21 C [700]

MELTING POINT: -97 C [017,051,205,269]

BOILING POINT: 57.3 C [017,031,421,430]

SOLUBILITY: Water: <1 mg/mL @ 20 C [700] DMSO: >=100 mg/mL @ 20 C [700] 95% Ethanol: >=100 mg/mL @ 20 C [700] Acetone: >=100 mg/mL @ 20 C [700]

OTHER SOLVENTS: Ether: Soluble [017,051,062,430] Benzene: Soluble [017,421] Alcohol: Soluble [017,051,062,071] Most organic solvents: Miscible [051,071]

OTHER PHYSICAL DATA: Refractive index: 1.4164 @ 20 C [017,047,205] Saccharine taste [062] Ethereal odor [062] Saturation concentration: 986 g/cu m @ 20 C; 1,406 g/cu m @ 30 C [055] Evaporation rate (butyl acetate = 1): 11.6 [102] Critical temperature: 261.5 C [371] Critical pressure: 734.8 psia [371] Liquid surface tension: 24.75 dynes/cm [371] Ratio of specific heat of vapor (gas): 1.136 @ 20 C [371] Vapor (gas) specific gravity: 3.42 [371] Latent heat of vaporization: 73.1 cal/g [371] Heat of combustion: -2,652 cal/g [371] Reid vapor pressure: 7.35 psia [371] Percent in saturated air: 30.8 @ 25 C [430]

HAP WEIGHTING FACTOR: 1 [713]

VOLATILITY:

Vapor pressure: 234 mm Hg @ 25 C [051,055] 182 mm Hg @ 20 C [051,071,102]

Vapor density: 3.44 [051,071]

FIRE HAZARD: Ethylidene dichloride has a flash point of -5 C (22 F) [205,269] and it is flammable. Fires involving this material may be controlled with a dry chemical, carbon dioxide or Halon extinguisher. A water spray may also be used [051,071, 371].

The autoignition temperature of ethylidene dichloride is 458 C (856 F) [051,071,371]

LEL: 5.6% [043,071,371,430] UEL: 11.4% [043,071,371,430]

REACTIVITY: Ethylidene dichloride can react vigorously with oxidizing materials [051,071,102, 346]. It is incompatible with strong bases [269]. Contact with strong caustics will cause formation of flammable and toxic gas. It will attack some forms of plastics, rubber and coatings [102].

STABILITY: Ethylidene dichloride is heat sensitive [051,071,102]. Solutions of it in water, DMSO, 95% ethanol or acetone should be stable for 24 hours under normal lab conditions [700].

USES: Ethylidene dichloride is used in vinyl chloride manufacturing, chlorinated solvent intermediate, coupling agent in antiknock gasoline, metal degreasing agent, organic synthesis, ore flotation, manufacturing of methyl chloroform, chemical intermediate, an industrial solvent, in fumigant formulations, extraction solvent, as a dewaxer of mineral oils, extractant for heat-sensitive substances, manufacturing of high vacuum rubber and silicon grease and as a paint, varnish and finish remover. It was formerly used as an anesthetic.

COMMENTS: None

NIOSH Registry Number: KI0175000

ACUTE/CHRONIC HAZARDS: Ethylidene dichloride may be harmful by inhalation, ingestion or skin absorption. Vapor or mist is irritating to the eyes, mucous membranes, skin and upper respiratory tract. When heated to decomposition it emits toxic fumes of carbon monoxide, carbon dioxide, hydrogen chloride gas and phosgene [269]. It is narcotic in high concentrations [031,071,151]. It may also have anesthetic effects at high concentrations [051,071,430]and it is a lachrymator [371].

SYMPTOMS: Symptoms of exposure may include liver and kidney damage, skin and eye irritation, dermatitis and skin burns [051,071]. It may cause unconsciousness, central nervous system depression and drowsiness [346]. It may also cause nausea, vomiting, faintness, irritation of the respiratory tract, salivation, sneezing, coughing, dizziness, lacrimation, reddening of the conjunctiva, cyanosis, circulatory failure and slight smarting of the eyes and respiratory system [371]. It produces narcotic effects in high concentrations [031,051, 071,151].

Formaldehyde

CAS NUMBER: 50-00-0

SYNONYMS: 1) Gaseous form: Methanal Methyl aldehyde Methylene oxide 2) Aqueous solution: Formalin (30 to 50% formaldehyde by weight, which usually contains 6 to 12% methanol) [704]

CHEMICAL FORMULA: CH2O

MOLECULAR WEIGHT: 30.03

WLN: VHH

PHYSICAL DESCRIPTION: Gas: nearly colorless gas with a pungent, suffocating odor [704] Solution: clear, colorless liquid.

SPECIFIC GRAVITY: 1.081-1.085 @ 25/25 C (37% solution) [031]

DENSITY: 0.815 @ 20C/4C [706]

MELTING POINT: -92 C [706]

BOILING POINT: 96 C (37% solution) [031,036] -21 C (gas) [706]

SOLUBILITY: Water: >=100 mg/mL @ 20.5 C [700] DMSO: >=100 mg/mL @ 20.5 C [700] 95% Ethanol: >=100 mg/mL @ 20.5 C [700] Acetone: >=100 mg/mL @ 20.5 C [700]

OTHER SOLVENTS: Petroleum ether: Insoluble [025] Ether: Soluble [017,172] Benzene: Soluble [017] Most organic solvents: Soluble [025] Chloroform: Immiscible [295]

OTHER PHYSICAL DATA: Refractive index: 1.3746 @ 20 C [031] Pungent, suffocating odor [025] pH: 2.8-4.0 [031] Burning taste [455] Specific gravity: 0.815 @ 20/4 C [172]

HAP WEIGHTING FACTOR: 1 [713]

VOLATILITY:

Vapor pressure: 93.60 mm Hg @ 38 C [700]

Vapor density: 1.0 [451]

FIRE HAZARD: Formaldehyde has a flash point of 85 C (185 F) (058). It is combustible. Fires involving this material may be controlled with a dry chemical, carbon dioxide or Halon extinguisher. A water spray may also be used [036,058].

The autoignition temperature of formaldehyde is 430 C (806 F) [043].

LEL: 7.0% [043,051,451] UEL: 73.0% [043,051,451]

REACTIVITY: Formaldehyde is a strong reducing agent, especially in the presence of alkalis. It is incompatible with ammonia, alkalis, tannin, bisulfides, iron preparations, copper salts, iron salts, silver salts, iodine and potassium permanganate. It combines directly with albumin, casein, gelatin, agar and starch to form insoluble compounds [031]. It reacts violently with nitrous oxides at about 180 C, ($HClO_4$ + aniline), performic acid, nitromethane, manganese carbonate and hydrogen peroxide [043]. It reacts with strong oxidizers and acids [346]. It is also incompatible with phenols [295].

STABILITY: Formaldehyde may become cloudy upon standing, especially at cool temperatures. It slowly oxidizes in air. It is sensitive to exposure to light [169,295]. It is polymerized in aqueous solutions if unstabilized [169]. Solutions of it in water, DMSO, 95% ethanol or acetone should be stable for 24 hours under normal lab conditions [700].

USES: Formaldehyde is used as a preservative, disinfectant and antiseptic, in embalming solutions and in the manufacture of phenolic resins, artificial silk, cellulose esters, dyes, urea, thiourea, melamine resins, organic chemicals, glass mirrors and explosives. It is also used in improving fastness of dyes on fabrics, in tanning and preserving hides, in mordanting and waterproofing fabrics, as a germicide and fungicide for vegetables and other plants, in destroying flies and other insects, in preserving and coagulating rubber latex and to prevent mildew and spelt in wheat and rot in oats. It is used to render casein, albumin, and gelatin insoluble, in chemical analysis, as a tissue fixative, as a component of particle board and plywood and in the manufacture of pentaerythritol, hexamethylenetetramine and 1,4-butanediol. It is also used in ceiling and wall insulation, in resins used to wrinkle-proof fabrics, in photography for hardening gelatin plates and papers, for toning gelatin-chloride papers and for chrome printing and developing. It is an intermediate in drug manufacture and a pesticide intermediate.

COMMENTS: Commercial formulations generally consist of a 37% by weight solution of formaldehyde in water with 10-15% methanol added as a stabilizer.

NIOSH Registry Number: LP8925000

ACUTE/CHRONIC HAZARDS: Formaldehyde and its vapors are irritants to the skin, eyes and mucous membranes [036,151,301,406]. It is also an irritant to all parts of the respiratory system [036,301,406]. It can be absorbed through the skin [169]. It can cause lacrimation [455]. Thermal decomposition products may include carbon monoxide and carbon dioxide [058].

SYMPTOMS: Inhalation of formaldehyde may cause irritation of the eyes, mucous membranes and upper respiratory tract [036,151,301,406]. It may also cause irritation of the nose [151]. Higher concentrations may cause bronchitis, pneumonia or laryngitis [036,151]. Exposure may also cause headache, dizziness, difficult breathing and pulmonary edema [215]. Coughing or dysphagia may also result. Contact with the vapor or solution causes skin to become white, rough, hard and anesthetic due to superficial coagulation necrosis. With long exposure, dermatitis and hypersensitivity frequently result [151]. Prolonged exposure may also cause cracking of skin and ulceration, especially around the fingernails and may also cause conjunctivitis [036]. Ingestion may cause immediate intense pain in the mouth and pharynx [151]. It may also cause abdominal pains with nausea, vomiting and possible loss of consciousness [036,151,301]. Other symptoms following ingestion include proteinuria, acidosis, hematemesis, hematuria, anuria, vertigo, coma and even death due to respiratory failure [031]. Occasional diarrhea (possibly bloody), pale, clammy skin and other signs of shock, difficult micturition, convulsions and stupor may also occur. Ingestion also leads to inflammation, ulceration and/or coagulation necrosis of the gastrointestinal mucosa [151]. Corrosive damage in the stomach and esophageal strictures sometimes occur and tissue destruction may extend as far as the jejunum. Circulatory collapse and kidney damage may also occur soon after ingestion. Severe lung changes may result from aspiration of the ingested compound in combination with stomach acid [151]. Degenerative changes may be found in the liver, kidneys, heart and brain. Primary points of attack for this compound include the respiratory system, lungs, eyes and skin [301]. Lacrimation may occur [455].

Glycol ethers

NOTE: The Clean Air Act specifies glycol ethers as including mono- and di- ethers of ethylene glycol, diethylene glycol, and triethylene glycol R-(OCH2CH2)n-OR' where: n = 1, 2, or 3 R = alkyl or aryl groups R' = R, H, or groups which, when removed, yield glycol ethers with the structure: R-(OCH2CH2)n-OH. Polymers are excluded from the glycol category. [708] The following compounds are intended as selected examples of glycol ethers and do not comprise a comprehensive list.

Ethylene glycol diethyl ether

CAS NUMBER: 629-14-1

SYNONYMS: 1,2-Diethoxyethane Diethyl cellosolve

CHEMICAL FORMULA: C6H14O2

MOLECULAR WEIGHT: 118.20

WLN: Not available

PHYSICAL DESCRIPTION: Colorless liquid

SPECIFIC GRAVITY: 0.8417 @ 20/20 C [042,053,062]

DENSITY: 0.8484 g/mL @ 20 C [017,047,053,371]

MELTING POINT: -74 C [026,042,269,275]

BOILING POINT: 121.4 C [042,062,205,430]

SOLUBILITY: Water: >=100 mg/mL @ 19 C DMSO: >=100 mg/mL @ 19 C 95% Ethanol: >=100 mg/mL @ 19 C Acetone: >=100 mg/mL @ 19 C

OTHER SOLVENTS: Benzene: >10% [047] Ether: >10% [047]

OTHER PHYSICAL DATA: Boiling point: 49 C @ 56 mm Hg [026] Vapor pressure: 60 mm Hg @ 39.0 C; 100 mm Hg @ 51.8 C [038] Vapor pressure: 200 mm Hg @ 71.8 C; 400 mm Hg @ 94.1 C [038] Slight ethereal odor [042,053] Refractive index: 1.3923 @ 20 C [269,275] Burning rate: 4.1 mm/min [371] Liquid surface tension (estimated): 26 dynes/cm @ 20 C [371] Latent heat of vaporization: 107 cal/g [371] Heat of combustion: -8100 cal/g [371]

HAP WEIGHTING FACTOR: 1 [713]

VOLATILITY:

Vapor pressure: 20 mm Hg @ 14.7 C; 40 mm Hg @ 29.7 C [038]

Vapor density: 4.1 [053,371,430]

FIRE HAZARD: Ethylene glycol diethyl ether has a flash point of 35 C (95 F) [042,062,430,451] and it is flammable. Fires involving this material may be controlled with a dry chemical, carbon dioxide or Halon extinguisher.

The autoignition temperature for ethylene glycol diethyl ether is 208 C (406 F) [042, 053,371,451].

LEL: Not available UEL: Not available

REACTIVITY: Ethylene glycol diethyl ether can react with oxidizers [042,053,269]. It is incompatible with strong acids [269].

STABILITY: Ethylene glycol diethyl ether may form explosive peroxides on prolonged storage [269]. Solutions of it in water, DMSO, 95% ethanol or acetone should be stable for 24 hours under normal lab conditions [700].

USES: Ethylene glycol diethyl ether is used as a solvent for ester gum, shellac, some resins and some oils. It is also used as a solvent and diluent for detergents and in organic synthesis as a reaction medium.

COMMENTS: None

NIOSH Registry Number: KI1225000

ACUTE/CHRONIC HAZARDS: Ethylene glycol diethyl ether may be toxic by inhalation [430], ingestion or skin absorption [269]. It may be readily absorbed through the skin [053,151] and it is an irritant [269,275,371,430]. When heated to decomposition it emits toxic fumes of carbon monoxide and carbon dioxide [269].

SYMPTOMS: Symptoms of exposure may include irritation of the skin, eyes and mucous membranes [053,371,430]. It may also irritate the nose and throat [371]. Other symptoms may include vomiting, pulmonary edema, stupor, muscle tenderness, anuria, oliguria, convulsions, metabolic acidosis, cyanosis, kidney damage, coma and hypocalcemic tetany [151,301]. Exposure may cause central nervous system depression, hematuria, transient exhilaration, drunkenness, ataxia, vertigo, nausea, abdominal pain, lumbar pain, dehydration, weakness, hyperpnea, albuminuria, acute renal failure, peripheral edema, ascites, drowsiness and death from cardiovascular collapse [151]. Exposure may also cause alcohol intoxication-like symptoms, headache, tachycardia, tachypnea, hypotension, prostration, hypoglycemia, narcosis, uremia, nystagmus, intravascular hemolysis, lymphocytosis, unconsciousness and brain damage. Death may occur within a few hours from respiratory failure or within the first 24 hours from pulmonary edema [301]. Ingestion may cause irritation of the mouth and stomach [371]. Corneal injury may also occur [053,430].

Ethylene glycol dimethyl ether

CAS NUMBER: 110-71-4

SYNONYMS: 1,2-Dimethoxyethane alpha,beta-Dimethoxyethane Dimethylcellosolve Ethylene dimethyl ether Monoethylene glycol dimethyl ether

CHEMICAL FORMULA: C4H10O2

MOLECULAR WEIGHT: 90.12

WLN: 1O2O1

PHYSICAL DESCRIPTION: Clear colorless liquid

SPECIFIC GRAVITY: 0.86285 @ 20/4 C [031]

DENSITY: 0.8683 g/mL @ 20 C [062]

MELTING POINT: -58 C [017,031,058,275]

BOILING POINT: 85 C [036,058,269,275]

SOLUBILITY: Water: >=100 mg/mL @ 22 C DMSO: >=100 mg/mL @ 22 C 95% Ethanol: >=100 mg/mL @ 22 C Acetone: >=100 mg/mL @ 22 C

OTHER SOLVENTS: Benzene: Soluble [017] Ether: Soluble [017,047] Hydrocarbons: Soluble [025,031,062] Petroleum ether: Soluble [205] Alcohol: Miscible [043]

OTHER PHYSICAL DATA: Specific gravity: 0.86877 @ 15/4 C; 0.8602 @ 33/4 C; 0.8692 @ 20/20 C [031] Melting point also reported as -71 C [025,031] Boiling point: 20 C @ 61.2 mm Hg; 16 C @ 50 mm Hg; -14 C @ 10 mm Hg [031] Vapor pressure: 61.2 mm Hg @ 20 C [058] 100 mm Hg @ 31.8 C, 200 mm Hg @ 50 C, 400 mm Hg @ 70.8 C [038] Refractive index: 1.3796 @ 20 C [017,205] 1.3739 @ 24 C [031,043] Sharp ethereal odor [025,031,036,043] pH: 8.2 [062] 100% Volatile by volume [058] UV max (in water): 400-350 nm, 300 nm, 250 nm, 220 nm (A = 0.01, 0.03, 0.20, 1.0) [275] Burning rate: 4.9 mm/min. [371] Critical temperature: 263 C [371] Critical pressure: 38.2 atmospheres [371] Latent heat of vaporization: 74.6 cal/g [371] Heat of combustion: -6680 cal/g [371] Evaporation rate (butyl acetate = 1): 4.99 [058]

HAP WEIGHTING FACTOR: 1 [713]

VOLATILITY:

Vapor pressure: 40 mm Hg @ 10.7 C; 60 mm Hg @ 19.7 C [038]

Vapor density: 3.1 [058,371]

FIRE HAZARD: Ethylene glycol dimethyl ether has a flash point of 0 C (32 F) [269,275] and it is flammable. Fires involving this material may be controlled with a dry chemical, carbon dioxide or Halon extinguisher. A water spray may also be used [036,269].

The autoignition temperature of ethylene glycol dimethyl ether is 202 C (395 F) [371,451].

LEL: 1.6% [058] UEL: 10.4% [058]

REACTIVITY: Ethylene glycol dimethyl ether is incompatible with strong oxidizers [058,269]. It is also incompatible with strong acids [269]. Explosive decomposition has occurred with lithium tetrahydroaluminate [043,066]. It can form peroxides from contact with oxygen [058].

STABILITY: Ethylene glycol dimethyl ether is liable to form explosive peroxides on exposure to air and light [036]. NMR stability screening indicates that solutions of it in DMSO are stable for 24 hours [700].

USES: Ethylene glycol dimethyl ether is used as a solvent. It is also used to facilitate the formation of alkali metal-hydrocarbon adducts and is used in the Reformatsky reaction with methyl gamma-bromocrotonate.

COMMENTS: None

NIOSH Registry Number: KI1451000

ACUTE/CHRONIC HAZARDS: Ethylene glycol dimethyl ether may be harmful by inhalation [036,058,269], ingestion or skin absorption [058,269] and it is an irritant [025,058,269]. When heated to decomposition it emits toxic fumes of carbon monoxide and carbon dioxide [058,269].

SYMPTOMS: Symptoms of exposure via ingestion may include nausea and vomiting [058,371]. Other symptoms by ingestion may include cramps, weakness, unconsciousness and coma [371]. Abdominal pain and central nervous system depression may also result. Ingestion of large amounts may lead to damage to the blood, liver and kidneys [058]. Inhalation of this compound may cause dizziness or difficult breathing [371]. Other symptoms via inhalation include mild irritation, coughing and sneezing. Skin contact may result in local inflammation. Prolonged or large area contact may cause central nervous system effects. Irritating effects from eye contact may cause pain, redness, blurred vision or tearing. Prolonged exposure may lead to hemolysis and liver or kidney damage. It may also cause adverse reproductive effects [058].

Ingestion of this class of compounds may cause symptoms of alcohol intoxication soon progressing to cyanosis, headache, tachypnea, tachycardia, hypotension, pulmonary edema, muscle tenderness, stupor, prostration, unconsciousness with convulsions and possible hypoglycemia. Death may occur within a few hours from respiratory failure or within the first 24 hours from pulmonary edema. Persons who have prolonged coma or convulsions may have irreversible brain damage. Hypocalcemic tetany has also been reported. Overdosage may cause intravascular hemolysis. Repeated doses may cause oliguria progressing to anuria and uremia [301].

Ethylene glycol monobutyl ether

CAS NUMBER: 111-76-2

SYNONYMS: Bytoxyethanol n-Butoxyethanol 2-Butoxyethanol 2-Butoxy-1-ethanol Butyl Cellosolve

CHEMICAL FORMULA: $C_6H_{14}O_2$

MOLECULAR WEIGHT: 118.18

WLN: Q204

PHYSICAL DESCRIPTION: Clear, colorless, mobile liquid

SPECIFIC GRAVITY: 0.9012 @ 20/20 C

DENSITY: Not available

MELTING POINT: -70 C

BOILING POINT: 171 C @ 743 mm Hg

SOLUBILITY: Water: >=100 mg/mL @ 22 C DMSO: >=100 mg/mL @ 22 C 95% Ethanol: >=100 mg/mL @ 22 C Acetone: >=100 mg/mL @ 22 C

OTHER SOLVENTS: Most organic solvents: Soluble Mineral oil: Soluble Ether: Soluble

OTHER PHYSICAL DATA: Specific gravity: 0.9015 @ 20/4 C Boiling point: 50 C @ 4 mm Hg Pleasant odor; sour taste Oily liquid Refractive index: 1.4198 @ 20 C Weight/gallon: 7.51 lb @ 20 C Evaporation rate: 0.1

HAP WEIGHTING FACTOR: 1 [713]

VOLATILITY:

Vapor pressure: 0.76 mm Hg @ 20 C; 0.88 mm Hg @ 25 C; 300 mm Hg @ 140 C

Vapor density: 4.07

FIRE HAZARD: Ethylene glycol monobutyl ether has a flash point of 60 C (141 F). It is combustible. Fires involving this material can be controlled with a dry chemical, carbon dioxide or Halon extinguisher.

The autoignition temperature for ethylene glycol monobutyl ether is 244 C (472 F).

LEL: 1.1% UEL: 10.6

%REACTIVITY: Ethylene glycol monobutyl ether may react with bases, aluminum and oxidizing materials. It is liable to form peroxides on exposure to air and light. It attacks some forms of plastics, rubber and coatings.

STABILITY: Ethylene glycol monobutyl ether is sensitive to air and light. Solutions of it in DMSO, 95% ethanol or acetone should be stable for 24 hours under normal lab conditions [700].

USES: Ethylene glycol monobutyl ether is used as a solvent for nitrocellulose, resins, grease, oil and albumin; dry cleaning; spray lacquers; quick-drying lacquers; varnishes; enamels; varnish removers; textiles (preventing spotting in printing or dyeing); and as an emulsifier for petroleum.

COMMENTS: None

NIOSH Registry Number: KJ8575000

ACUTE/CHRONIC HAZARDS: Ethylene glycol monobutyl ether is an irritant, may be toxic by skin absorption, and may be narcotic if ingested. When heated to decomposition this compound emits acrid smoke and irritating fumes.

SYMPTOMS: Symptoms of exposure may include irritation of the eyes and respiratory tract; headache, hepatic hemoglobinemia, albuminuria, nausea, vomiting, dizziness, drowsiness, unconsciousness, central nervous system effects, narcotic effects, bone marrow damage, kidney and liver damage; dark red urine and hemolysis.

Ethylene glycol monophenyl ether

CAS NUMBER: 122-99-6

SYNONYMS: Ethylene glycol phenyl ether 2-Phenoxyethanol Phenyl cellosolve

CHEMICAL FORMULA: C8H10O2

MOLECULAR WEIGHT: 138.17

WLN: Not available

PHYSICAL DESCRIPTION: Clear, colorless, viscous liquid

SPECIFIC GRAVITY: 1.1094 @ 20/20 C [031,062]

DENSITY: 1.105-1.110 g/mL @ 20 C [455]

MELTING POINT: 14 C [031,042,062]

BOILING POINT: 245.2 C [026,031]

SOLUBILITY: Water: 10-50 mg/mL @ 20 C DMSO: >=100 mg/mL @ 20 C 95% Ethanol: >=100 mg/mL @ 20 C Acetone: >=100 mg/mL @ 20 C

OTHER SOLVENTS: Ether: Soluble [017,031,047] Sodium hydroxide solutions: Freely soluble [031] Arachis oil: 1 in 50 @ 20 C [295,455] Olive oil: 1 in 50 @ 20 C [295,455] Glycerol: Miscible [295,455]

OTHER PHYSICAL DATA: Boiling point: 116 C @ 12.5 mm Hg [026] 134-135 C @ 18 mm Hg [017,047] Boiling point: 137 C @ 25 mm Hg; 165 C @ 80 mm Hg [031] Specific gravity: 1.1059 @ 24/21 C [052] 1.102 @ 20/4 C [031] Specific gravity: 1.11 @ 20/20 C [026] Density: 1.102 g/mL @ 22 C [017,047] Vapor pressure: 5 mm Hg @ 106.6 C; 10 mm Hg @ 121.2 C [038] Burning taste [031,107] Refractive index: 1.5340 @ 20 C [031,047] Faint aromatic odor [031,062,107] Lambda max (in methanol): 278 nm, 271 nm, 266 nm (shoulder), 220 nm (epsilon= 1383, 1670, 1243, 7710) [052]

HAP WEIGHTING FACTOR: 1 [713]

VOLATILITY:

Vapor pressure: <0.01 mm Hg @ 20 C [062] 1 mm Hg @ 78.0 C [038,107]

Vapor density: 4.77 [107]

FIRE HAZARD: Ethylene glycol monophenyl ether has a flash point of 121 C (250 F) [031,042,062,107]. It is combustible. Fires involving this material may be controlled with a dry chemical, carbon dioxide or Halon extinguisher. Water or foam may cause frothing [451].

LEL: Not available UEL: Not available

REACTIVITY: Ethylene glycol monophenyl ether is incompatible with strong oxidizers [107].

STABILITY: Ethylene glycol monophenyl ether is stable for 2 weeks at temperatures up to 60 C [052]. It is stable in the presence of acids and alkalis [062]. Solutions of it in water, DMSO, 95% ethanol or acetone should be stable for 24 hours under normal lab conditions [700].

USES: Ethylene glycol monophenyl ether is used as a fixative for perfumes, a bactericide (in conjunction with quaternary ammonium compounds), an insect repellent, a topical antiseptic, a solvent for cellulose acetate, dyes, inks and resins, in organic synthesis of plasticizers, in germicides, in pharmaceuticals, and in cosmetics and in preservatives.

COMMENTS: None

NIOSH Registry Number: KM0350000

ACUTE/CHRONIC HAZARDS: Ethylene glycol monophenyl ether is an irritant and, when heated to decomposition, it emits acrid smoke and fumes [042].

SYMPTOMS: Symptoms of exposure may include irritation of the skin, eyes, mucous membranes and respiratory tract; coughing, headache, abdominal pain and nausea [107]. Exposure also may cause severe eye damage in rabbits [430].

Heptachlor

CAS NUMBER: 76-44-8

SYNONYMS: 1,4,5,6,7,8,8-Heptachloro-3a,4,7,7a-tetrahydro-4,7-methano-1H-indene

CHEMICAL FORMULA: C10H5Cl7

MOLECULAR WEIGHT: 373.32

WLN: L C555 A DU IUTJ AG AG BG FG HG IG JG

PHYSICAL DESCRIPTION: White crystals

SPECIFIC GRAVITY: 1.57-1.59 [062,430]

DENSITY: 1.57-1.59 g/mL [051,062,072,173]

MELTING POINT: 95-96 C [016,025,031,062]

BOILING POINT: Decomposes [051,058,072]

SOLUBILITY: Water: <0.1 mg/mL @ 18 C [700] DMSO: >=100 mg/mL @ 24 C [700] 95% Ethanol: 10-50 mg/mL @ 24 C [700] Acetone: >=100 mg/mL @ 24 C [700]

OTHER SOLVENTS: Ligroin: Soluble [016,047] Xylene: 102 g/100 mL @ 27 C [051,072,173,395] Carbon tetrachloride: 112 g/100 mL @ 27 C [051,072,173,395] Cyclohexane: 119 g/100 mL @ 27 C [051,072,173,395] Cyclohexanone: 119 g/100 mL [169] Kerosene: 189 g/100 mL @ 27 C [173] Hexane: Slightly soluble [421]

Ether: Soluble [016,047] Benzene: 106 g/100 mL @ 27 C [051,072,173,395] Most organic solvents: Soluble [043,169] Petroleum distillates: Soluble [151] Ketones: Soluble [151] Alcohol: 4.5 g/100 mL @ 27 C [051,072,173] Deodorized kerosene: 263 g/L [172]

OTHER PHYSICAL DATA: Density: 1.66 g/mL @ 20 C [051,072] Physical description (technical grade): Light tan waxy solid [169,172] Melting point (technical grade): 46-74 C [058] Boiling point also reported as 135-145 C [169,395,421] Camphor-like odor [051,058,072,173]

HAP WEIGHTING FACTOR: 10 [713]

VOLATILITY:

Vapor pressure: 0.0003 mm Hg @ 25 C [033,055,395,421]

FIRE HAZARD: Heptachlor is not combustible [058, 421]. Fires involving this material may be controlled with a dry chemical, carbon dioxide or Halon extinguisher.

LEL: Not available UEL: Not available

REACTIVITY: Heptachlor is incompatible with strong alkali [169,173]. It is corrosive to metals [169]. It can react with iron and rust to form toxic gases [058]. It can react vigorously with oxidizing materials [051]. It is susceptible to epoxidation [172].

STABILITY: Heptachlor is stable under normal lab conditions [169,173,186,295]. Solutions of it in water, DMSO, 95% ethanol or acetone should be stable for 24 hours under normal lab conditions [700].

USES: Heptachlor is an insecticide used for control of the cotton boll weevil, termites, ants, grasshoppers, cutworms, maggots, thrips, wireworms, flies, mosquitoes, soil insects, household insects and field insects. It has some fumigant action, and is applied as a soil treatment, a seed treatment or directly to foliage.

COMMENTS: The EPA has cancelled registration of pesticides containing Heptachlor with the exception of its use through subsurface ground insertion for termite control and dipping of roots or tops of nonfood plants [051,072,421].

NIOSH Registry Number: PC0700000

ACUTE/CHRONIC HAZARDS: Heptachlor may be toxic by ingestion, inhalation and skin absorption [033, 062]. It may be readily absorbed by the skin as well as by the lungs and gastrointestinal tract [173,430]. When heated to decomposition it emits smoke, acrid fumes and toxic fumes of carbon monoxide [051]. Decomposition may also produce hydrogen chloride gas [058].

SYMPTOMS: Symptoms of exposure may include irritation of the skin, central nervous system stimulation, ataxia, renal damage, tremors, convulsions, respiratory collapse, kidney damage and death [051,072]. It can also cause congestion, edema, scattered petechial hemorrhages in the lungs, kidneys and brain, hyperexcitability, central nervous system depression and anuria [301]. Other symptoms may include paresthesia, excitement, giddiness, fatigue and coma [295]. If ingested, it can cause nausea, vomiting, diarrhea and irritation of the gastrointestinal tract [051,072]. It may also cause liver necrosis and blood dyscrasias [033].

Hexachlorobenzene

CAS NUMBER: 118-74-1

SYNONYMS: Perchlorobenzene Phenyl perchloryl

CHEMICAL FORMULA: $C6Cl6$

MOLECULAR WEIGHT: 284.76

WLN: GR BG CG DG EG FG

PHYSICAL DESCRIPTION: White needles

SPECIFIC GRAVITY: 2.044 @ 24 C

DENSITY: Not available

MELTING POINT: 229 C

BOILING POINT: 322 C (sublimes)

SOLUBILITY: Water: <1 mg/mL @ 20 C [700] DMSO: <1 mg/mL @ 20 C [700] 95% Ethanol: <1 mg/mL @ 20 C [700] Acetone: 1-5 mg/mL @ 23 C [700]

OTHER SOLVENTS: Chloroform: Soluble Carbon disulfide: Soluble Carbon tetrachloride: Sparingly soluble Ether: Soluble Benzene: Soluble

OTHER PHYSICAL DATA: Easily sublimable

HAP WEIGHTING FACTOR: 10 [713]

VOLATILITY:

Vapor pressure: 1 mm Hg @ 114.4 C

Vapor density: 9.8

FIRE HAZARD: Hexachlorobenzene has a flash point of 242 C (468 F). It is combustible. Fires involving this material can be controlled with a dry chemical, carbon dioxide or Halon extinguisher.

LEL: Not available UEL: Not available

REACTIVITY: Hexachlorobenzene reacts violently with dimethylformamide.

STABILITY: Hexachlorobenzene is sensitive to moisture. Solutions of it in water, DMSO, 95% ethanol or acetone should be stable for 24 hours under normal lab conditions [700].

USES: Hexachlorobenzene is used in organic synthesis, as a fungicide for seeds, as a wood preservative, in the manufacture of pentachlorophenol, in the production of aromatic fluorocarbons and in the impregnation of paper.

COMMENTS: None

NIOSH Registry Number: DA2975000

ACUTE/CHRONIC HAZARDS: Hexachlorobenzene is a irritant and, when heated to decomposition, it emits toxic fumes of chlorides, carbon monoxide and carbon dioxide.

SYMPTOMS: Symptoms of exposure may include irritation of the eyes, skin, mucous membranes and upper respiratory tract, corneal opacity, focal alopecia, atrophic hands, hypertrichosis, hepatomegaly, porphyria, anorexia, weight loss, enlargement of thyroid and lymph nodes, skin photosensitization and abnormal growth of hair.

Hexachlorobutadiene

CAS NUMBER: 87-68-3

SYNONYMS: 1,1,2,3,4,4-Hexachloro-1,3-butadiene Perchlorobutadiene Hexachloro-1,3-butadiene

CHEMICAL FORMULA: C4Cl6

MOLECULAR WEIGHT: 260.76

WLN: GYGUYGYGUYGG

PHYSICAL DESCRIPTION: Heavy, clear, colorless liquid

SPECIFIC GRAVITY: 1.675 @ 15.5/15.5 C [051,055,062,395]

DENSITY: 1.68 g/mL @ 20 C [029]

MELTING POINT: -21 C [017,029,047,421]

BOILING POINT: 215 C [017,047,051,346]

SOLUBILITY: Water: <0.1 mg/mL @ 22 C [700] DMSO: >=100 mg/mL @ 19 C [700] 95% Ethanol: >=100 mg/mL @ 19 C [700] Acetone: >=100 mg/mL @ 19 C [700]

OTHER SOLVENTS: Ether: Soluble [017,062,395,421]

OTHER PHYSICAL DATA: Specific gravity: 1.675 @ 25/22 C [052] 1.68 @ 25/4 C [430] Boiling point: 101 C @ 20 mm Hg [017,047] 64-65 C @ 0.5 mm Hg [029] Vapor pressure: 22 mm Hg @ 100 C; 500 mm Hg @ 200 C [051,055,062] Viscosity: 2.447 centipoise, 1.479 centistokes (@ 37.7 C) [062] Viscosity: 1.131 centipoise, 0.731 centistokes (@ 98.9 C) [062] Refractive index: 1.5550 @ 20 C [205,275] 1.5535 @ 25 C [430] Faint turpentine odor [051] Odor threshold: 0.006 ppm [051] Lambda max (in methanol): 249 nm, 220 nm (epsilon=4300, 17700) [052]

HAP WEIGHTING FACTOR: 1 [713]

VOLATILITY:

Vapor pressure: 0.15 mm Hg @ 20 C [051] ~0.3 mm Hg @ 25 C [430]

Vapor density: 8.99 [043,051,053,451]

FIRE HAZARD: Hexachloro-1,3-butadiene is nonflammable [062,275, 395,430].

The autoignition temperature is 610 C (1130 F) [043, 051,053].

LEL: Not available UEL: Not available

REACTIVITY: Hexachloro-1,3-butadiene rapidly decomposes rubber on contact [051]. It can react vigorously with oxidizing materials and it reacts to form an explosive product with bromine perchlorate [043].

STABILITY: Hexachloro-1,3-butadiene is stable for 2 weeks at temperatures up to 60 C when protected from light [052]. Solutions of it in water, DMSO, 95% ethanol or acetone should be stable for 24 hours under normal lab conditions [700].

USES: Hexachloro-1,3-butadiene is used as a solvent for elastomers; heat transfer liquid; transformer and hydraulic fluid; to wash liquor for removing C4 and higher hydrocarbons; solvent for natural rubber, synthetic rubber and other polymers; pesticide; insecticide; herbicide; algicide; chemical intermediate; recovery of chlorine fluid for gyroscopes; recovery of chlorine from "sniff" gas in chlorine plants; and as an intermediate in the manufacture of rubber compounds and lubricants.

COMMENTS: None

NIOSH Registry Number: EJ0700000

ACUTE/CHRONIC HAZARDS: Hexachlorobutadiene may be toxic by ingestion or inhalation [062]. It is an irritant [301] and also is corrosive [275]. When heated to decomposition it emits toxic fumes of chlorine [043].

SYMPTOMS: Symptoms of exposure may include irritation of the skin, eyes, mucous membranes and lungs; liver and kidney damage; and central nervous system effects [301]. It may cause corrosion of exposed tissues [275]. Other symptoms may include hypotension, cardiac disease, chronic bronchitis, disturbance of nervous function and chronic hepatitis [395].

Hexachlorocyclopentadiene

CAS NUMBER: 77-47-4

SYNONYMS: Perchlorocyclopentadiene 1,2,3,4,5,5-Hexachloro-1,3-cyclopentadiene

CHEMICAL FORMULA: C5Cl6

MOLECULAR WEIGHT: 272.75

WLN: L5 AHJ AG AG BG CG DG EG

PHYSICAL DESCRIPTION: Clear pale yellow liquid

SPECIFIC GRAVITY: 1.7019 @ 25/4 C [017,047]

DENSITY: Not available

MELTING POINT: -9 C [017,047,421]

BOILING POINT: 239 C @ 753 mm Hg [269,275]

SOLUBILITY: Water: <0.1 mg/mL @ 21.5 C [700] DMSO: >=100 mg/mL @ 21 C [700] 95% Ethanol: >=100 mg/mL @ 21 C [700] Acetone: >=100 mg/mL @ 21 C [700] Methanol: Soluble [421]

OTHER SOLVENTS: Carbon tetrachloride: Soluble [421] Hexane: Soluble [421]

OTHER PHYSICAL DATA: Refractive index: 1.5658 @ 20 C [017,047] Odor threshold: 0.0014-0.0016 mg/L [055] Pungent odor [042,058,062,421] Viscosity: 7.8 centipoise @ 25 C [058] UV lambda max: 323 nm [052] Surface tension: 47 dynes/cm [051] Latent heat of vaporization: 42 cal/g [371] Specific gravity: 1.70-1.71 @ 15.5/15.5 C [058], 1.702 @ 20/4 C [269,275] Boiling point: 48-49 C @ 0.3 mm Hg [017,047], 68-70 C @ 1-1.3 mm Hg [025]

HAP WEIGHTING FACTOR: 1 [713]

VOLATILITY:

Vapor pressure: 0.08 mm Hg @ 25 C [055,421,430] 1 mm Hg @ 60 C [430]

Vapor density: 9.42 [042,051,071]

FIRE HAZARD: Hexachlorocyclopentadiene is nonflammable [062, 071, 371, 421].

LEL: Not available UEL: Not available

REACTIVITY: Hexachlorocyclopentadiene is explosive when mixed with sodium [053,066]. It is highly reactive with olefins and polynuclear hydrocarbons [051]. It reacts with common metals [058]. In the presence of moisture, it will corrode iron and other metals [371]. It is incompatible with strong oxidizing agents [269].

STABILITY: Hexachlorocyclopentadiene is light sensitive [051]. Solutions of it in DMSO, 95% ethanol or acetone should be stable for 24 hours under normal lab conditions [700].

USES: Hexachlorocyclopentadiene is used as an intermediate for resins, dyes, pesticides, fungicides, and pharmaceuticals. It is also used in the synthesis of cyclodiene pesticides which includes chlordane, heptachlor, aldrin, isodrin, dieldrin, endrin, endosulfan, alodan, beomadan and telodrin. It is also used to make kepone, mirex and chlorinated anhydride for flame retardants in plastics and polymers. It is also used to make nonflammable resins and shock proof plastics, acids, esters, ketones, and fluorocarbons. It is also used to produce chlorendic acid which is used as a flame retardant in resins.

COMMENTS: Hexachlorocyclopentadiene may be shipped as a corrosive under the DOT regulations; however, it must be noted that it is a "poison inhalation hazard".

NIOSH Registry Number: GY1225000

ACUTE/CHRONIC HAZARDS: Hexachlorocyclopentadiene may be readily absorbed through the skin [052,062,071,269]. It may be toxic through absorption, inhalation and ingestion [051,053,062]. It is also extremely irritating to eyes, skin, and mucous membranes [053, 062, 269,371] and to the upper respiratory tract [058,269,301]. This compound is corrosive to tissues [051,071] and it may cause lacrimation [053,371]. When heated to decomposition it emits toxic fumes of hydrogen chloride gas, phosgene, carbon monoxide and carbon dioxide [051,058,269,371].

SYMPTOMS: Symptoms of exposure may include headache, dizziness, irritation of the eyes, throat, nose and skin, dermatitis, corrosion to tissues, nausea, vomiting, hematemesis, abdominal cramps, diarrhea, nervousness, dyspnea, cyanosis, oliguria, proteinuria, hematuria, jaundice, hepatomegaly, optic neuritis, unconsciousness, coma, ventricular fibrillation [051, 071]. It may cause lacrimation [053,151,346,371]. It may also cause respiratory failure [071]. Other symptoms may include severe burns to the eyes and skin, blistering and burning of the skin and depression. Inhalation may cause coughing, difficult breathing, sneezing and salivation [371]. Inhalation may be fatal as a result of spasm, inflammation, edema of the larynx and bronchi, chemical pneumonitis and pulmonary edema. Exposure may also lead to a burning sensation, wheezing, laryngitis and shortness of breath [269]. Degenerative changes of the brain, heart, liver, adrenals, kidneys and severe lung damage may result from exposure to this compound [346]. Lethargy, mucous membrane irritation, gasping respiration, tremors and skin inflammation may also occur [053].

Hexachloroethane

CAS NUMBER: 67-72-1

SYNONYMS: Carbon hexachloride

CHEMICAL FORMULA: C2Cl6

MOLECULAR WEIGHT: 236.74

WLN: GXGGXGGG

PHYSICAL DESCRIPTION: Colorless crystals

SPECIFIC GRAVITY: 2.091 @ 20/4 C [047,061,269]

DENSITY: Not available

MELTING POINT: Sublimes [031]

BOILING POINT: 186.8 C (sublimes) [031]

SOLUBILITY: Water: <1 mg/mL @ 21 C [700] DMSO: >=100 mg/mL @ 22 C [700] 95% Ethanol: 50-100 mg/mL @ 22 C [700] Acetone: >=100 mg/mL @ 22 C [700]

OTHER SOLVENTS: Chloroform: Soluble [031,042] Oils: Soluble [031,042] Ether: Soluble [031,042] Benzene: Soluble [031,042]

OTHER PHYSICAL DATA: Camphor like odor Triple point: 186.8 C [031] Sublimes readily without melting [042]

HAP WEIGHTING FACTOR: 1 [713]

VOLATILITY:

Vapor pressure: 0.4 mm Hg @ 20 C; 0.8 mm Hg @ 30 C [055]

Vapor density: 8.16 [055]

FIRE HAZARD: Hexachloroethane nonflammable [269,396].

LEL: Not available UEL: Not available

REACTIVITY: Hexachloroethane can react with hot iron, zinc and aluminum [346]. Dehalogenation of this material by reaction with alkalis and metals will produce unstable chloroacetylenes [042]. It can also react with strong oxidizing agents [269].

STABILITY: Hexachloroethane is stable under normal laboratory conditions. Solutions of it in water, DMSO, 95% ethanol or acetone should be stable for 24 hours under normal lab conditions [RAD].

USES: Hexachloroethane is used in organic synthesis, as a retarding agent in fermentation, as a camphor substitute in nitrocellulose, in pyrotechnics and smoke devices, as a solvent, in explosives, as a rubber vulcanizing accelerator, as an anthelmintic (vet), as a degassing agent for magnesium, as a component of extreme pressure lubricants, as an ignition suppressant in combustible liquids, and as a moth repellant.

COMMENTS: None

NIOSH Registry Number: KI4025000

ACUTE/CHRONIC HAZARDS: Hexachloroethane is an irritant and it can be absorbed through the skin [061]. When heated to decomposition it emits toxic fumes of carbon monoxide, carbon dioxide, hydrogen chloride, and phosgene [269]. It is a positive animal carcinogen [015].

SYMPTOMS: Symptoms of exposure may include skin, eye, mucous membrane and upper respiratory tract irritation. Chronic effects may include liver damage and nervous system disturbances [269]. High concentrations may cause narcosis [346].

Hexamethylene-1,6-diisocyanate

CAS NUMBER: 822-06-0

SYNONYMS: 1,6-Diisocyanatohexane 1,6-Hexanediol diisoocyanate Isocyanic acid, hexamethylene ester

CHEMICAL FORMULA: C8H12N2O2

MOLECULAR WEIGHT: 168.22 [703]

SPECIFIC GRAVITY: Not available

MELTING POINT: 140 C [702]

BOILING POINT: 255 C [702]

SOLUBILITY: Water: Not Available

OTHER PHYSICAL DATA: Not available

HAP WEIGHTING FACTOR: 1 [713]

VOLATILITY: Not available

FIRE HAZARD: Flash Point: 140 C [702]

Autoignition temperature: Not available

LEL: Not available UEL: Not available

REACTIVITY: Compound reacts with water during sampling [702]. Potentially explosive reaction with alcohols + base [703].

STABILITY: When heated to decomposition it emits toxic fumes of NOx [703].

USES: Not available

COMMENTS: Compound must be derivatizated during sampling. Method is under development. [702]

NIOSH Registry Number: MO1740000

ACUTE/CHRONIC HAZARDS: Hexamethylene-1,6-diisocyanate is a poison by inhalation and intravenous routes. It is moderately toxic by ingestion and skin contact. When heated to decomposition it emits toxic fumes of NOx. [703]

SYMPTOMS: Not available

Hexamethylphosphoramide

CAS NUMBER: 680-31-9

SYNONYMS: Phosphoric tris(dimethylamide) Phosphoryl hexamethyltriamide tris(Dimethylamino)phosphine oxide tris(Dimethylamino)phosphorus oxide

CHEMICAL FORMULA: $C_6H_{18}N_3OP$

MOLECULAR WEIGHT: 179.2

WLN: 1N1&PO&N1&1&N1&1

PHYSICAL DESCRIPTION: Clear, colorless liquid

SPECIFIC GRAVITY: 1.024 @ 25/25 C

DENSITY: 1.03 g/mL

MELTING POINT: 5-7 C

BOILING POINT: 233 C

SOLUBILITY: Water: >=100 mg/mL @ 18 C [700] DMSO: >=100 mg/mL @ 18 C [700] 95% Ethanol: >=100 mg/mL @ 18 C [700] Acetone: >=100 mg/mL @ 18 C [700]

OTHER SOLVENTS: Most organic liquids except high-boiling saturated hydrocarbons: Soluble Ether: Soluble

OTHER PHYSICAL DATA: Refractive index: 1.4582 @ 20 C Spicy odor Boiling point: 78 C @ 2.5 mm Hg; 115 C @ 15 mm Hg

HAP WEIGHTING FACTOR: 1 [713]

VOLATILITY:

Vapor pressure: 0.07 mm Hg @ 25 C

Vapor density: 6.18

FIRE HAZARD: The flash point for hexamethylphosphoramide is 105 C (222 F). It is combustible. Fires involving this compound should be controlled using a dry chemical, carbon dioxide or Halon extinguisher.

LEL: Not available UEL: Not available

REACTIVITY: Hexamethylphosphoramide may react with strong oxidizing agents and strong acids.

STABILITY: Hexamethylphosphoramide is stable under normal laboratory conditions. Solutions of it should be stable for 24 hours under normal lab conditions [700].

USES: Hexamethylphosphoramide is a solvent widely used in synthesis, a jet fuel additive, a chemosterilant for insect pests, a chemical mutagen, an ultraviolet inhibitor in polyvinyl chloride, an anti-static additive, and a flame retardant.

COMMENTS: None

NIOSH Registry Number: TD0875000

ACUTE/CHRONIC HAZARDS: Hexamethylphosphoramide may be toxic and an irritant. When heated to decomposition it emits toxic fumes.

SYMPTOMS: Symptoms of exposure may include eye, skin, mucous membrane and upper respiratory tract irritation; kidney and liver damage. In laboratory animals this compound has caused weight loss, altered gastrointestinal function; nervous system dysfunction, kidney disease, testicular atrophy and aspermia; nasal tumors, dilatation of bronchi and bronchopneumonia.

Hexane

CAS NUMBER: 110-54-3

SYNONYMS: Dipropyl n-Hexane

CHEMICAL FORMULA: C6H14

MOLECULAR WEIGHT: 86.18

WLN: 6H

PHYSICAL DESCRIPTION: Clear colorless liquid

SPECIFIC GRAVITY: 0.6603 @ 20/4 C [017,042,055]

DENSITY: 0.6548 g/mL @ 25 C [047]

MELTING POINT: -95 C [017,062,102,274]

BOILING POINT: 69 C [017,031,036,042]

SOLUBILITY: Water: <1 mg/mL @ 16.5 C [700] DMSO: <1 mg/mL @ 16.5 C [700] 95% Ethanol: >=100 mg/mL @ 16.5 C [700] Acetone: >=100 mg/mL @ 16.5 C [700]

OTHER SOLVENTS: Chloroform: Miscible [031,042] Ether: Miscible [031,042] Most organic solvents: Miscible [421]

OTHER PHYSICAL DATA: Specific gravity: 0.655 @ 25/4 C [025] Vapor pressure: 100 mm Hg @ 15.8 C [038,042] 200 mm Hg @ 31.6 C [038] Refractive index: 1.3749 @ 20 C; 1.3723 @ 25 C Faint, gasoline-like odor Solidifies between -95 C and -100 C [031,421] Critical temperature: 234.2 C Critical pressure: 29.7 atm Very volatile Pour point: <-18 C [058] Viscosity: 0.31 centipoise @ 25 C

HAP WEIGHTING FACTOR: 1 [713]

VOLATILITY:

Vapor pressure: 120 mm Hg @ 20 C [055] 180 mm Hg @ 25 C [058]

Vapor density: 2.97

FIRE HAZARD: n-Hexane has a flash point of -23 C (-9.4 F) [025,036,042] and it is flammable. Fires involving this material may be controlled with a dry chemical, carbon dioxide or Halon extinguisher.

The autoignition temperature of n-hexane is 225 C (437 F) [042,102,371].

LEL: 1.2% [036,042,371,421] UEL: 7.5% [036,042,058,102]

REACTIVITY: n-Hexane can react vigorously with oxidizing materials [042,058,102,346]. This would include compounds such as liquid chlorine, concentrated O2, sodium hypochlorite and calcium hypochlorite [058]. It is also incompatible with dinitrogen tetraoxide [042,066]. It will attack some forms of plastics, rubber and coatings [102].

STABILITY: n-Hexane is stable under typical ambient storage conditions but it may be sensitive to light [052]. It may also be sensitive to prolonged exposure to heat [102]. Solutions of it in water, DMSO, 95% ethanol or acetone are stable for 24 hours under normal laboratory conditions [700].

USES: n-Hexane is a widely used solvent, especially for vegetable oils. It is also used in low temperature thermometers, as a paint diluent, an alcohol denaturant, a liquid for determining refractive index of minerals, and a common laboratory reagent.

COMMENTS: n-Hexane is a common constituent of petroleum.

NIOSH Registry Number: MN9275000

ACUTE/CHRONIC HAZARDS: Hexane is an irritant [058,102, 371,430] and it is narcotic in high concentrations [036,102]. When heated to decomposition it emits smoke and toxic fumes of carbon monoxide and carbon dioxide [058,102,269]. It may also emit fumes of aldehydes [058].

SYMPTOMS: Symptoms of exposure may include irritation of the eyes, skin, and mucous membranes of the upper respiratory tract [058,102,371,421]. It may also cause nausea, headache, dizziness and dermatitis [215,346,371,430]. It may cause asphyxia leading to brain damage or cardiac arrest [151,215,346]. Other symptoms may include an anesthetic affect, vomiting and unconsciousness [058,151,102,371]. It may also cause polyneuropathy [058,099,151,102,346]. Exposure may result in irritation of the respiratory system and loss of sensation of hands and feet [036,099]. It may also cause narcosis [036,102,215, 346], throat irritation, and giddiness [102,421]. Myocardial sensitization to epinephrine has been reported [346]. Other symptoms may include mild depression, cardiac arrhythmias and swelling of the abdomen. Aspiration of this compound may cause severe lung irritation, coughing, and excitement followed by depression [371]. Pulmonary edema and hemorrhage, and chemical pneumonitis may also be caused by aspiration [058]. Other symptoms may include vertigo, bronchial irritation, intestinal irritation, central nervous system effects, euphoria and numbness of the limbs [430]. Damage to the eyes may include blurred vision, constriction of visual field, optic nerve atrophy and retrobulbar neuritis [099]. Other symptoms may include lightheadedness, nose irritation, convulsions, respiratory arrest, drying of skin and irreversible nerve damage [102]. Chest pain and edema have also been reported [269]. Persons with previous chronic respiratory disease, or impaired renal or liver function may be more susceptible to harmful effects from exposure to this chemical. Heavy exposure can result in death [102].

Hydrazine

CAS NUMBER: 302-01-2

SYNONYMS: Diamine

CHEMICAL FORMULA: H4N2

MOLECULAR WEIGHT: 32.05

WLN: ZZ

PHYSICAL DESCRIPTION: Not available

SPECIFIC GRAVITY: Not available

DENSITY: Not available

MELTING POINT: 1.4 C

BOILING POINT: 113.5 C

SOLUBILITY: Water: Miscible DMSO: Not available 95% Ethanol: Not available Acetone: Not available

OTHER SOLVENTS: Alcohol: Miscible Chloroform: Insoluble Ether: Insoluble

OTHER PHYSICAL DATA: Burns with a violet flame Explodes during distillation if traces of air are present Affected by uv and metal ion catalysis Powerful reducing agent Weak base

HAP WEIGHTING FACTOR: 100 [713]

VOLATILITY: Not available; however it is a highly volatile flammable gas.

FIRE HAZARD: Hydrazine has a flash point of 38 C and it is flammable. Fires involving this chemical should be extinguished with alcohol foam, carbon dioxide, and/or dry chemical extinguishers.

LEL: Not available UEL: Not available

REACTIVITY: Hydrazine is very reactive with most oxidizing agents.

STABILITY: Hydrazine is unstable to heat and oxidizing agents.

USES: Hydrazine is used as a rocket fuel and jet fuel, as a cleaning agent, and in the synthesis of drugs and agricultural chemicals.

COMMENTS: None

NIOSH Registry Number: MV7180000

ACUTE/CHRONIC HAZARDS: Hydrazine is very volatile and may be toxic by inhalation. When heated to decomposition it emits toxic fumes of nitrogen compounds.

SYMPTOMS: Not available

Hydrochloric acid

CAS NUMBER: 7647-01-0

SYNONYMS: Hydrogen chloride Muriatic acid (DOT) [703]

CHEMICAL FORMULA: HCl

MOLECULAR WEIGHT: 36.46

PHYSICAL DESCRIPTION: Colorless, fuming gas or colorless, fuming liquid; strongly corrosive [703] pungent, irritating odor [704]

DENSITY: 1.639 g/L (gas) @ 0 C [703] 1.194 @ -26 C (liquid) [703]

MELTING POINT: -114.8 C [706]

BOILING POINT: -84.9 C [706]

SOLUBILITY: Water: 67% @ 86 F [704]

OTHER PHYSICAL DATA: Vapor Pressure: 4.0 atm @ 17.8 C [703] Freezing Point: -174 F [704]

HAP WEIGHTING FACTOR: 1 [713]

VOLATILITY:

Vapor Pressure: 4.0 atm @ 17.8 C [703]

FIRE HAZARD: Flash Point: Not applicable

Nonflammable gas [703,704].

Autoignition temperature: Not available

LEL: Not applicable UEL: Not applicable

REACTIVITY: Explosive reaction with potassium permanganate; sodium; tetraselenium tetranitride. Ignition on contact with fluorine; hexalithium disilicide; metal acetylides or carbides (e.g., cesium acetylide; rubidium acetylide). Violent reactions with acetic anhydride; 2-amino ethanol; NH4OH; Ca3P2; chlorosulfonic acid; 1,1-difluoroethylene; ethylene diamine; ethylene imine; oleum; HClO4; beta- propiolactone; propylene oxide; (AgClO4 + CCl4); NaOH; H2SO4; U3P4; vinyl acetate; CaC2; CsC2H; Cs2C2; Mg3B2;

CsC_2H; Cs_2C_2; Mg_3B_2; $HgSO_4$; RbC_2H; Rb_2C_2; Na. Vigorous reaction with aluminum. [703] Hydrochloric acid is highly corrosive to most metals [704].

Alcohols, Hydrogen cyanide: Preparation of alkyliminioformate chlorides (imidoester hydrochlorides) by passing HCl rapidly into alcoholic HCN proceeds explosively (probably owing to a rapid exotherm), even with strong cooling. Alternative procedures involving very slow addition of HCl into a well-stirred mixture kept cooled to ambient temperature, or rapid addition of cold alcoholic HCN to cold alcoholic Hcl, are free of this hazard. [703, 710]

Chlorine, Dinitroanilines: The previously reported cleavage of dinitroanilines to chloronitrodiazonium salts by Hcl is apparently catalyzed by chlorine or other oxidants. The reaction takes place vigorously with copious gas evolution, but at low temperatures (40-25 C or lower) there can be a very long induction period before onset of the vigorous reaction. This could be hazardous on attempting to scale-up laboratory processes. The reactivity of the isomeric dinitroanilines varies, 2,3- being most reactive and 3,5- being unreactive. The cleavage reaction is specific for concentrated Hcl, and has not been observed with hydrobromic acid or 30% sulfuric acid. [710, 703]

Sulfuric acid: Accidental addition of 6500 l of concentrated HCl to a bulk sulfuric acid storage tank released sufficient HCl gas by dehydration to cause the tank to burst violently. Complete dehydration of HCl solution releases some 250 volumes of gas. [710] Potentially dangerous reaction with sulfuric acid releases HCl gas. [703]

STABILITY: When heated to decomposition it emits toxic fumes of Cl- [703].

USES: Hydrogen chloride is used as a general purpose food additive. It is a common air contaminant and is heavily used in industry. [703]

COMMENTS: In general, hydrochloric acid causes little trouble in industry other than from accidental splashes and burns [703].

NIOSH Registry Number: MW4025000

ACUTE/CHRONIC HAZARDS: Human poison by an unspecified route; mildly toxic to humans by inhalation, moderately toxic experimentally by ingestion. A corrosive irritant to the skin, eyes, and mucous membranes. Mutagenic data exists; an experimental teratogen. A concentration of 35 ppm causes irritation of the throat after short exposure. In general, HCl causes little trouble in industry other than from accidental splashes and burns. When heated to decomposition it emits toxic fumes of Cl-. [703]

SYMPTOMS: Symptoms of exposure may include inflammation of the nose, throat, and larynx; coughing, burning of the throat, choking; burning of the eyes, skin; dermatitis; in animals: laryngeal spasm; pulmonary edema. [704]

Hydrogen fluoride

CAS NUMBER: 7664-39-3

SYNONYMS: Hydrofluoric acid

CHEMICAL FORMULA: HF

MOLECULAR WEIGHT: 20.01 [703]

PHYSICAL DESCRIPTION: Clear, colorless, fuming corrosive liquid or gas [703]

SPECIFIC GRAVITY: 1.00 (Liquid @ 67 F) [703]

MELTING POINT: -83.1 C [703]

BOILING POINT: 19.54 C [703]

SOLUBILITY: Water: Very soluble [702]

OTHER PHYSICAL DATA: Density: 0.901 g/L (gas) [703] 0.699 @ 2 C (liquid) [703] Vapor Pressure: 400 mm @ 2.5 C [703]

HAP WEIGHTING FACTOR: 1 [713]

VOLATILITY:

Vapor Pressure: 400 mm @ 2.5 C [703]

FIRE HAZARD: Flash Point: Not applicable

Nonflammable gas [704]

Autoignition temperature: Not available

REACTIVITY: Explosive reaction with cyanogen fluoride; glycerol nitric acid; sodium (with aqueous acid). Violent reaction with As2O3; P2O5; acetic anhydride; 2-amino ethanol; NH4OH; HBiO3; CaO; chlorosulfonic acid; ethylene diamine; ethylene imine; F2; mercury (II) oxide + organic materials above 0 C; n-phenylazopiperidine; oleum; beta-propiolactone; propylene oxide; Na; NaOH; H2SO4; vinyl acetate; HgO; sodium tetrafluorosilicate; n-phenyl azo piperidine. Reacts with water or steam to produce toxic and corrosive fumes. [703] Violent reaction of 40% HF with bismuthic acid; reaction with mercury(II) oxide suspension above 0 C; explosive with methanesulfonic acid; unstable mixed with nitric acid and either lactic acid, propylene glycol, or glycerol; causes arsenic trioxide and calcium oxide to incandesce; violent reaction with potassium tetrafluorosilicate(2-); violent exotherm with light emission with potassium permanganate. [710]

CORROSIVITY:

Corrosive to metals. Will attack glass and concrete [704]. Dissolves silica, silicic acid, glass. Should be stored in steel cylinders. [715]

STABILITY: Fumes in air [715]. When heated to decomposition it emits highly corrosive fumes of F- [703].

USES: Catalyst, especially in the petroleum industry (paraffin alkylation); in fluorination processes, especially in the aluminum industry; in the manufacture of fluorides; for separating uranium isotopes; in making fluorine containing plastics; in dye chemistry [715].

COMMENTS: Highly irritating, corrosive and poisonous. Extremely corrosive to skin and eyes. Causes severe burns which may not be painful or visible for several hours. [715]

NIOSH Registry Number: MW7875000

ACUTE/CHRONIC HAZARDS: Human poison by inhalation; poison experimentally by subcutaneous and intraperitoneal routes. Corrosive irritant to skin, eyes (@ 0.05 mg/L), and mucous membranes. Experimental teratogenic effects and reproductive effects; mutagenic data exists. Dangerous at 50-250 ppm even for brief exposures. Produces severe skin burns. A common air contaminant. [703]

SYMPTOMS: Irritation of the eyes, nose, and throat; pulmonary edema; skin and eye burns; nasal congestion; and bronchitis. [704] Inhalation of the vapor may cause ulcers of the upper respiratory tract. Hydrogen fluoride produces severe skin burns which are slow in healing. The subcutaneous tissues may by affected, becoming blanched and bloodless. Gangrene of the affected areas may follow. [703]

Hydroquinone

CAS NUMBER: 123-31-9

SYNONYMS: p-Benzenediol

CHEMICAL FORMULA: C6H6O2

MOLECULAR WEIGHT: 110.11

WLN: QR DQ

PHYSICAL DESCRIPTION: Fine white crystals or crystalline powder

SPECIFIC GRAVITY: 1.358 @ 20/4 C [043]

DENSITY: 1.332 g/mL @ 15 C [033,205,421,430]

MELTING POINT: 170 C [036,062,371]

BOILING POINT: 285 C [062,269,275,371]

SOLUBILITY: Water: 10-50 mg/mL @ 20 [700] DMSO: >=100 mg/mL @ 18 C [700] 95% Ethanol: >=100 mg/mL @ 18 C [700] Acetone: 10-50 mg/mL @ 18 C [700]

OTHER SOLVENTS: Ether: Soluble [017,047,062,395] Benzene: Slightly soluble [033,043,205,395] Carbon tetrachloride: Very soluble [395] Glycerol: 1 in 1 [295] Chloroform: 1 in 51 [295] Alcohol: Very soluble [033,043]

OTHER PHYSICAL DATA: Vapor pressure: 4 mm Hg @ 150 C [058,430] 10 mm Hg @ 163.5 C [038] Vapor pressure: 20 mm Hg @ 174.6 C [038] pKa1: 9.91 (@ 30 C, in water) [029] pKa2: 12.04 (@ 30 C, in water) [029] Odorless [058]

HAP WEIGHTING FACTOR: 1 [713]

VOLATILITY:

Vapor pressure: 0.001 mm Hg @ 20 C [421] 1 mm Hg @ 132.4 C [038,043]

Vapor density: 3.81 [043,058,421,430]

FIRE HAZARD: Hydroquinone has a flash point of 165 C (329 F) [043,058,421,451]. It is combustible. Fires involving this material may be controlled with a dry chemical, carbon dioxide or Halon extinguisher. A water spray may also be used [058].

The autoignition temperature for hydroquinone is 516 C (960 F) [043, 058, 062, 371].

LEL: Not available UEL: Not available

REACTIVITY: Hydroquinone is incompatible with strong oxidizing agents [058,269, 295,346]. It is also incompatible with bases [269,295]. It reacts with oxygen and sodium hydroxide [036,043,066]. It also reacts with ferric salts [295].

STABILITY: Hydroquinone darkens on exposure to air and light [295]. Solutions of it become brown in air due to oxidation [033,058,099]. Oxidation is rapid in the presence of alkali [033]. Hydroquinone is a slight explosion hazard when exposed to heat [058]. UV spectrophotometric stability screening indicates that solutions of it in water and in 95% ethanol are stable for less than two hours [700].

USES: Hydroquinone is used as a photographic reducer and developer (except in color film), as a dye intermediate inhibitor, as a stabilizer in paints, varnishes, motor fuels and oils, as an antioxidant for fats and oils, as an inhibitor of polymerization, as a reagent in the determination of small quantities of phosphate and as a depigmentor.

COMMENTS: None

NIOSH Registry Number: MX3500000

ACUTE/CHRONIC HAZARDS: Hydroquinone may be fatal by ingestion or inhalation and it is a severe irritant [058].

SYMPTOMS: Symptoms of exposure may include tinnitus, nausea, vomiting, sense of suffocation, shortness of breath, cyanosis, convulsions, delirium, collapse, death, irritation of the intestinal tract, dermatitis and staining or opacification of the cornea [033]. Other symptoms may include skin depigmentation, slurred speech, tremors, muscular twitching, headache, dyspnea, methemoglobinemia, coma, greenish or brownish green urine, severe irritation, photophobia, lacrimation, injury of the corneal epithelium and frank corneal ulceration [295]. It may cause increased motor activity, hypersensitivity to external stimuli, hyperactive reflexes, hypothermia, paralysis and loss of reflexes [430]. It may also cause stupor, fall in blood pressure, hyperpnea, abdominal pain, diarrhea, intense thirst, sweating, dizziness, increased rate of respiration, pallor, jaundice, oliguria, anuria, lesions of the skin and eyes, irritation of the eyes, discoloration of the conjunctiva, keratitis, bronchopneumonia and liver, kidney and neurological disorders [058]. It can cause an increased pulse rate [043].

Isophorone

CAS NUMBER: 78-59-1

SYNONYMS: 3,5,5-Trimethyl-2-cylohexene-1-one 3,5,5-Trimethyl-5-cyclohexene-1-one 1,1,3-Trimethyl-3-cyclohexene-5-one Isoacetophorone

CHEMICAL FORMULA: $C_9H_{14}O$

MOLECULAR WEIGHT: 138.21

WLN: L6V BUTJ C1 E1 E1

PHYSICAL DESCRIPTION: Clear colorless liquid

SPECIFIC GRAVITY: 0.925 @ 20/20 C [205]

DENSITY: 0.9229 g/mL @ 20 C [017,047]

MELTING POINT: -8 C [051,062]

BOILING POINT: 215.2 C [043,062,205,421]

SOLUBILITY: Water: 0.1-1 mg/mL @ 18 C [700] DMSO: >=100 mg/mL @ 20 C [700] 95% Ethanol: >=100 mg/mL @ 20 C [700] Acetone: >=100 mg/mL @ 20 C [700]

OTHER SOLVENTS: Acid: Soluble [047] Most organic solvents: Miscible [421] Ether: Soluble [017,047] Absolute ethanol: Soluble [052]

OTHER PHYSICAL DATA: Sharp, peppermint or camphor-like odor [055,151,430] Viscosity: 2.62 centipoise @ 20 C [062] Refractive index: 1.4759 @ 20 C [047] Specific gravity: 0.9199 @ 22/22 C [052] 0.923 @ 20/4 C [269,275] Boiling point: 214 C @ 754 mm Hg; 99 C @ 18 mm Hg [017,047] 89 C @ 10 mm Hg [025] Vapor pressure: 5 mm Hg @ 66.7 C; 10 mm Hg @ 81.2 C; 760 mm Hg @ 215.2 C [038] Evaporation rate (butyl acetate = 1): 0.02 [058]

HAP WEIGHTING FACTOR: 1 [713]

VOLATILITY:

Vapor pressure: 0.2 mm Hg @ 20 C [062,421] 1 mm Hg @ 38 C [038,043,051]

Vapor density: 4.77 [043,051,053,055]

FIRE HAZARD: Isophorone has a flash point of 84 C (184 F) [043,275,371,421]. It is combustible. Fires involving this material may be controlled with a dry chemical, carbon dioxide or Halon extinguisher. A water spray may also be used [058]. The autoignition temperature is 462 C (864 F) [043,062,371].

LEL: 0.8% [043,051,058,421] UEL: 3.8% [043,058,371,421]

REACTIVITY: Isophorone can react with oxidizing materials [043,052,058,346]. It is incompatible with strong acids and bases [269].

STABILITY: Isophorone is sensitive to heat. It is stable at temperatures up to 25 C for 2 weeks. Some decomposition occurs between 25 C and 60 C [052]. Solutions of it in water, DMSO, 95% ethanol or acetone should be stable for 24 hours under normal lab conditions [700].

USES: Isophorone is used in solvent mixtures for finishes. It is also a high solvent power for vinyl resins, cellulose esters and similar substances, solvent for lacquers, plastics and printing inks and a pesticide ingredient.

COMMENTS: Isophorone is a priority toxic pollutant (EPA) [346].

NIOSH Registry Number: GW7700000

ACUTE/CHRONIC HAZARDS: Isophorone is a lachrymator [269,275], an irritant [036,062,346,371], is absorbed through the skin, and may be moderately toxic by ingestion [058]. When heated to decomposition it emits toxic fumes of carbon monoxide, carbon dioxide and unidentified organic compounds in black smoke [058].

SYMPTOMS: Symptoms of exposure may include eye, nose and throat irritation [055,346,421,430]. It may also cause skin irritation [036,062, 371,430]. It may cause nausea, headache, dizziness, faintness and narcosis [058,269,346,430]. Other effects may include burning of the eyes, reddening of the skin, vomiting, unconsciousness, fatigue and malaise [058,430]. Further results of exposure may include burning sensation, coughing, wheezing, shortness of breath and laryngitis [269]. It may cause an anesthetic effect, eye tissue damage, cracking of the skin, and mouth and stomach irritation [371]. It may have harmful central nervous system effects [301,371,430]. Other symptoms may include olfactory fatigue, severe ocular damage, inebriation and a feeling of suffocation [430]. Skin contact may result in dermatitis or a degreasing action on the skin [036]. Eye contact may cause corneal damage or conjunctiva irritation [036,099]. Other symptoms may include respiratory tract irritation, and liver and kidney damage [036,269,301]. Lung damage may also result from exposure [430].

Lead compounds

NOTE: For all listings which contain the word "compounds," the following applies: Unless otherwise specified, these listings are defined as including any unique chemical substance that contains the named chemical as part of that chemical's infrastructure. [708] Information for the metallic form of lead is given below. Other compounds described are intended as selected examples of lead compounds and do not comprise a comprehensive list.

Lead

CAS NUMBER: 7439-92-1

ATOMIC FORMULA: Pb

ATOMIC WEIGHT: 207.19

WLN: Not available

PHYSICAL DESCRIPTION: Silver-bluish white soft solid

SPECIFIC GRAVITY: Not available

DENSITY: 11.3437 g/mL @ 16 C

MELTING POINT: 327.50 C

BOILING POINT: 1740 C

SOLUBILITY: Water: Insoluble DMSO: Insoluble 95% Ethanol: Insoluble Acetone: Insoluble

OTHER SOLVENTS: Not available

OTHER PHYSICAL DATA: Not available

VOLATILITY:

Vapor Pressure: 1 mm @ 973 C [703]

FIRE HAZARD: Flash Point: Not available

Flammable in the form of dust when exposed to heat or flame [703]. Metal is a noncombustible solid in bulk form [704].

Autoignition temperature: Not available

LEL: Not available UEL: Not available

REACTIVITY: Mixtures of hydrogen peroxide + trioxane explode on contact with lead. Rubber gloves containing lead may ignite in nitric acid. Violent reaction on ignition with chlorine trifluoride; concentrated hydrogen peroxide; ammonium nitrate (below 200 C with powdered lead); sodium acetylide (with powdered lead). Incompatible with NaN3; Zr; disodium acetylide; oxidants. Can react vigorously with oxidizing materials. [703]

STABILITY: Moderately explosive in the form of dust when exposed to heat or flame. When heated to decomposition it emits highly toxic fumes of Pb. [703]

USES: Lead is very resistant to corrosion. In fact, lead pipes used as drains for baths in ancientRome are still in service. Lead is also used for construction of containers designed to hold corrosive liquids (such as sulfuric acid). Lead containers, which are relatively soft and weak, can be made stronger by the addition of a small percentage of antimony or other metals. Natural lead is a mixture of four stable isotopes: 204Pb (1.48%), 206Pb (23.6%), 207Pb (22.6%), and 208Pb (52.3%). Alloys of lead include solder, type metal, and various antifriction metals. Great quantities of lead, both as the metal and as lead dioxide, are used in storage batteries. A lot of lead is also used in cable coverings, plumbing, ammunition, and in the manufacture of lead tetraethyl. Metalic lead is very effective as a sound absorber, and it is often used as a radiation shield around X-ray equipment and nuclear reactors. White lead, the basic carbonate, sublimed white lead, chrome yellow, red lead, and other lead compounds are used extensively in paints, although in recent years the use of lead in paints has been drastically curtailed to eliminate or reduce health hazards. Lead oxide is used in producing fine "crystal glass" and "flint glass" of a high index of refraction for achromatic lenses. Lead nitrate and lead acetate are soluble salts. Lead salts such as lead arsenate have been used as insecticides, but their use in recent years has been mostly eliminated in favor of less toxic organic compounds. Environmental concern with lead poisoning has resulted in a national program to eliminate the lead in gasoline and also in paints; the concern with lead in paints has been from children ingesting old flaking paint in old houses.. [706]

COMMENTS: Care must be used in handling lead as it is a cumulative poison [706].

NIOSH Registry Number: OF7525000

ACUTE/CHRONIC HAZARDS: Lead powder may be toxic and the effects of exposure are cumulative.

SYMPTOMS: Symptoms of acute exposure may include abdominal pain, diarrhea, shock, muscular weakness and pain, headache, kidney damage, and coma.

Symptoms of chronic exposure may include lead encephalopathy (especially in children), headache, vomiting, delirium/hallucinations, convulsions, coma, death from exhaustion and respiratory failure. Children typically show weight loss, weakness, anemia and exhibit GI and CNS complaints.

Lead acetate

CAS NUMBER: 301-04-2

SYNONYMS: Lead(II)salt acetic acid

CHEMICAL FORMULA: C4H6O4.Pb

MOLECULAR WEIGHT: 325.28

WLN: QV1 &-PB-

PHYSICAL DESCRIPTION: Colorless or white crystals

SPECIFIC GRAVITY: 3.25 @ 20 C [395,430]

DENSITY: 2.55 [031,043,055]

MELTING POINT: 280 C [395,430]

BOILING POINT: Not available

SOLUBILITY: Water: 10-50 mg/mL @ 20 C [700] DMSO: 10-50 mg/mL @ 20 C [700] 95% Ethanol: <1 mg/mL @ 20 C [700] Acetone: <1 mg/mL @ 20 C [700]

OTHER SOLVENTS: Glycerol: Freely soluble [031,043,062] Glycerin: Soluble [395,430]

OTHER PHYSICAL DATA: pH in 5% aqueous solution @ 25 C is 5.5-6.5 [031] Sweetish taste [062] Slight acetic odor [031,043,058]

VOLATILITY: Not available

FIRE HAZARD: Lead acetate is nonflammable [058,371].

LEL: Not available UEL: Not available

REACTIVITY: Lead acetate is incompatible with acids, soluble sulfates, citrates, tartrates, chlorides, carbonates, alkalis, tannin, phosphates, resorcinol, salicylic acid, phenol, chloral hydrate, sulfites, vegetable infusions,and tinctures [031,043,051]. It is reactive at high temperatures and pressures [051]. It explodes when in contact with $KBrO_3$ [451]. Lead acetate-Lead bromate may be formed during the manufacture of this compound, and is explosive and sensitive to friction [066]. It takes up CO_2 from the air and becomes incompletely soluble [031,062].

STABILITY: Lead acetate is stable under ambient conditions [058,371]

USES: Lead acetate is used for dyeing and printing cottons, weighting silks, in the manufacture of lead salts and chrome yellow, in analytical procedures for detection of sulfide, and as a reagent for the determination of chromium trioxide (CrO_3) and molybdinum trioxide (MoO_3). It is also used as an astringent and as a local sedative for bruises and superficial inflammation. It is used in waterproofing, as a component in some varnishes, in the gold cyanization process, as an inecticide, a component in antifouling paints, as a hair dye, in chemicals production, and lead coating of metals. It is also used in pharmaceutical reagent solutions.

COMMENTS: None

NIOSH Registry Number: AI5250000

ACUTE/CHRONIC HAZARDS: Lead acetate may be highly toxic by all routes [051] and it is a cumulative poison. It is absorbed at about 1.5 times the rate of other lead compounds [346]. Hazardous decomposition products at high temperatures include acetic acid, carbon monoxide, and lead oxide [058].

SYMPTOMS: Symptoms of exposure in early stages of lead poisoning may include fatigue, disturbance of sleep, and constipation [058]. Symptoms of a more severe exposure may include abdominal pain, nausea, headache, loss of appetite, metallic taste, muscle and joint pain, dizziness, and hypertension [058,406]. Other symptoms of mild poisoning are lethargy, moroseness, and flatulence [051]. Prolonged overexposure may lead to severe damage of red blood cell formation, of the central nervous system, the peripheral nervous system, kidneys, and liver. The resulting high levels of this compound in the blood lead to convulsions, coma, and death [058]. It may also impair male and female reproductive systems and damage unborn fetuses [058]. Superficial injury of the eyes has been reported, use as an astringent may cause incrustations which could cause long lasting irritation to the eyes [099]. Abdominal pain is a sensitive index of intoxication with this compound [151,406].

This compound may cause encephalopathy (especially in children), symptoms of which are neurologic defects, retarded mental development, chronic hyperactivity, appearance of the brain is hyperemic or pale, exudates containing plasma proteins, lesions throughout the brain, remarkable facial deformity (one case), calcification of the cerebrum, and edema and hemorrhage of the brain [151,406]. It may also cause paresthesias and muscle weakness which may proceed to paralysis [371].

Lead dioxide

CAS NUMBER: 1309-60-0

CHEMICAL FORMULA: O2Pb

MOLECULAR WEIGHT: 239.21

WLN: .PB..O2

PHYSICAL DESCRIPTION: Dark brown crystals

SPECIFIC GRAVITY: Not available

DENSITY: 9.375 g/mL @ 19 C

MELTING POINT: 290 C (decomposes)

BOILING POINT: Not available

SOLUBILITY: Water: Insoluble DMSO: Insoluble 95% Ethanol: Insoluble Acetone: Not available

OTHER SOLVENTS: Glacial acetic acid: Soluble

OTHER PHYSICAL DATA: Not available

VOLATILITY: Lead dioxide is not volatile.

FIRE HAZARD: Lead dioxide is not flammable.

LEL: Not available UEL: Not available

REACTIVITY: Lead dioxide is incompatible with aluminum carbide, barium sulfide, hydroxylamine, molybdenum, phenylhydrazine and phosphorus.

STABILITY: Lead dioxide is stable under normal laboratory conditions.

USES: Lead dioxide is used as an oxidizing agent, in electrodes, in lead-acid storage batteries, as a curing agent for polysulfide elastomers, in textiles, in matches, in explosives, and as an analytical reagent.

COMMENTS: None

NIOSH Registry Number: OG0700000

ACUTE/CHRONIC HAZARDS: Lead dioxide may be toxic.

SYMPTOMS: Symptoms of exposure may include abdominal pains, diarrhea, constipation, loss of appetite, muscular weakness, headache, pyorrhea, vomiting, aching bones and muscles and a blue "lead line" on gums.

Lead subacetate

CAS NUMBER: 1335-32-6

SYNONYMS: bis(Aceto)tetrahydroxytrilead bis(Aceto)dihydroxytrilead Monobasic lead acetate

CHEMICAL FORMULA: C4H10O8Pb3

MOLECULAR WEIGHT: 807.71

WLN: OV1 &OV1 &.PB3.Q4

PHYSICAL DESCRIPTION: White, heavy powder

SPECIFIC GRAVITY: Not available

DENSITY: Not available

MELTING POINT: Not available

BOILING POINT: Not available

SOLUBILITY: Water: Soluble DMSO: Not available 95% Ethanol: Not available Acetone: Not available

OTHER SOLVENTS: Not available

OTHER PHYSICAL DATA: Not available

VOLATILITY: Not available

FIRE HAZARD: Not available

LEL: Not available UEL: Not available

REACTIVITY: Not available

STABILITY: Not available

USES: Lead subacetate is used in sugar analysis to remove coloring matters and for clarifying and decolorizing other solutions of organic substances.

COMMENTS: None

NIOSH Registry Number: OF8750000

ACUTE/CHRONIC HAZARDS: Lead subacetate may be toxic.

SYMPTOMS: Not available

Lindane (all isomers)

NOTE: There is only one isomer of lindane. However, lindanc is thc gamma isomer of 1,2,3,4,5,6-hexachlorocyclohexane and it is that specific compound for which data is provided in this database.

CAS NUMBER: 58-89-9

SYNONYMS: gamma-Benzene hexachloride gamma-BHC 1,2,3,4,5,6-Hexachlorocyclohexane, gamma-isomer

CHEMICAL FORMULA: C6H6Cl6

MOLECULAR WEIGHT: 290.83

WLN: L6TJ AG BG CG DG EG FG GAMMA

PHYSICAL DESCRIPTION: Colorless or white crystalline solid

SPECIFIC GRAVITY: 1.87 @ 20/4 C [055]

DENSITY: 1.87 g/mL @ 20 C [055,205]

MELTING POINT: 112-113 C [016,169,173]

BOILING POINT: 323.4 C [016,395]

SOLUBILITY: Water: <1 mg/mL @ 24 C [700] DMSO: 50-100 mg/mL @ 23 C [700] 95% Ethanol: 1-5 mg/mL @ 23 C [700] Acetone: >=100 mg/mL @ 23 C [700] Methanol: 7.4% [173,169] Toluene: >50 g/L [172]

OTHER SOLVENTS: Petroleum ether: 2.9% [169] Chloroform: 24% [033,169,173] Chlorinated hydrocarbons: Moderately soluble [173] Ethyl acetate: >50 g/L [172] Ether: >50 g/L [172] Benzene: >50 g/L [172] Acetic acid: 12.8% [169] Carbon tetrachloride: 6.7% [169] Cyclohexane: 36.7% [169,173] Dioxane: 31.4% [169] Kerosene: 2.0-3.2% [173] Acid: Soluble [047] Xylene: 24.7% [169,173] Rat fat: 12.55% [173]

OTHER PHYSICAL DATA: Vapor pressure is also reported as 0.0000094 mm Hg @ 20 C [033,173] Boiling point: 176.2 C @ 10 mm Hg [017] 288 C [051] Slight, musty odor [033,421] Refractive index: 1.644 @ 20 C [033] Sublimes slowly [058] Bitter taste [051] Odor threshold: 12.0 mg/kg [055]

HAP WEIGHTING FACTOR: 1 [713]

VOLATILITY:

Vapor pressure: 0.03 mm Hg @ 20 C [058,395]

FIRE HAZARD: Lindane is nonflammable [102,371].

LEL: Not available UEL: Not available

REACTIVITY: Lindane is incompatible with strong bases [058]. It is incompatible with powdered metals such as iron, zinc and aluminum [058,395]. It is also incompatible with oxidizing agents [269]. It can undergo oxidation when in contact with ozone [186]. When exposed to alkalis, lindane undergoes dehydrochlorination [172,173,395].

STABILITY: Lindane is stable under normal laboratory conditions. Solutions of it in water, DMSO, 95% ethanol or acetone should be stable for 24 hours under normal lab conditions [700]. Lindane is extremely stable to light, air and temperatures up to 180 C [169]. It decomposes at temperatures above 177 C [102].

USES: Lindane is used as an insecticide, pediculicide, scabicide, ectoparasiticide and a pesticide. It is also used as a foliar spray and soil application for insecticidal control of a broad spectrum of phytophagous and soil dwelling insects, animal ectoparasites and public health pests. It is used on ornamentals, fruit trees, nut trees, vegetables, tobacco and timber. Lindane is found in baits and seed treatments for rodent control. It acts as a stomach and contact poison and has some fumigant action. It has been of value in the control of malaria and other vector-borne diseases and in the control of grasshoppers, cotton insects, rice insects, wireworms and other soil pests. It is used in pet shampoo to maintain the natural luster of the coat and aid in the prevention of ticks, lice and sarcoptic mange mites. It is also used on patients in the treatment of head and crab lice and their ova.

COMMENTS: None

NIOSH Registry Number: GV4900000

ACUTE/CHRONIC HAZARDS: Lindane is an irritant and may be toxic by ingestion, inhalation or skin absorption [033, 062,269]. Oils may enhance skin absorption [158]. When heated to decomposition it emits toxic fumes of chlorine, hydrochloric acid and phosgene [043].

SYMPTOMS: Symptoms of exposure may include epileptic convulsions and serious EEG disturbances [421]. Other symptoms may include gastrointestinal disturbances, severe central nervous system involvement-cerebellar derangement, muscle spasms, blindness from optic nerve atrophy and diminution of vision [058].

It may cause malaise, faintness, dizziness followed by collapse and convulsions sometimes preceded by screaming and accompanied by foaming at the mouth and biting of the tongue, unconsciousness, retrograde amnesia, moderate rise in temperature, facial pallor, slight circumoral cyanosis, severe cyanosis of the face and extremities, slightly enlarged liver, depression and death from ingestion [173].

Effects of acute overexposure may be central nervous system stimulation, dyspnea, headache, nausea and irritation of the respiratory tract.

Effects of chronic overexposure may include irreversible renal changes, conjunctivitis, ecchymosis, staggering, fever, vomiting, mental confusion, pulmonary edema, dilation of the heart, extensive necrosis of blood vessels in the lungs, liver and kidney, fatty degeneration of the liver and kidneys and some cases of hypoplastic anemia.

Animal symptoms that have been observed are increased respiration, restlessness accompanied by frequency of micturition, intermittent muscular spasms of the whole body, salivation, grinding of the teeth and consequent bleeding from the mouth, backward movement with loss of balance and somersaulting, retraction of the head, convulsions, gasping and biting, collapse and death usually within a day. It may cause degenerative changes in the kidneys, pancreas, testes, nasal mucous membranes and liver (in extremely high doses). It may also cause immunosuppression [173]. It may cause respiratory failure [102].

Maleic anhydride

CAS NUMBER: 108-31-6

SYNONYMS: cis-Butenedioic anhydride 2,5-Furandione

CHEMICAL FORMULA: C4H2O3

MOLECULAR WEIGHT: 98.06

WLN: T5VOVJ

PHYSICAL DESCRIPTION: Colorless needles

SPECIFIC GRAVITY: 0.934 @ 20/4 C

DENSITY: 1.314 g/mL @ 60 C

MELTING POINT: 52.8 C

BOILING POINT: 197-9 C

SOLUBILITY: Water: Soluble; decomposes in hot water DMSO: Soluble 95% Ethanol: Soluble Acetone: Soluble

OTHER SOLVENTS: Chloroform: Soluble Ligroin: Slightly soluble Toluene: Soluble Dioxane: Soluble Ether: Soluble Benzene: Soluble

OTHER PHYSICAL DATA: Not available

HAP WEIGHTING FACTOR: 1 [713]

VOLATILITY: Not available

FIRE HAZARD: The flash point of maleic anhydride is 103 C (218 F). It is combustible. Fires involving this material should be controlled using a dry chemical, carbon dioxide or Halon extinguisher.

LEL: Not available UEL: Not available

REACTIVITY: Maleic anhydride reacts with strong oxidizers, alkali metals, caustics and amines at >65.5 C (>150 F).

STABILITY: Maleic anhydride is stable under normal laboratory conditions.

USES: Maleic anhydride is used in the manufacture of unsaturated polyester resins, in the manufacture of fumaric acid, pesticides, and pharmaceuticals, and in Diels-Alder synthesis.

COMMENTS: None

NIOSH Registry Number: ON3675000

ACUTE/CHRONIC HAZARDS: Maleic anhydride is an irritant and may be toxic if ingested.

SYMPTOMS: Symptoms of exposure may include irritation and burns of the skin, eyes and mucous membranes, headaches, nosebleed, nausea, temporary impairment of vision and conjunctivitis.

Manganese compounds

NOTE: For all listings which contain the word "compounds," the following applies: Unless otherwise specified, these listings are defined as including any unique chemical substance that contains the named chemical as part of that chemical's infrastructure. [708] Information for the metallic form of manganese is given below. Other compounds described are intended as selected examples of manganese compounds and do not comprise a comprehensive list.

Manganese

CAS NUMBER: 7439-96-5

ATOMIC FORMULA: Mn

ATOMIC WEIGHT: 54.94 [703]

PHYSICAL DESCRIPTION: Reddish-gray or silvery, brittle, metallic element [703]

SPECIFIC GRAVITY: 7.20 @ 20C/4C [704]

MELTING POINT: 1260 C [703]

BOILING POINT: 1900 C [703]

SOLUBILITY: Water: Insoluble [704]

OTHER PHYSICAL DATA: Vapor Pressure: 1 mm @ 1292 C [703] Melting Point: 1244 C [702] Boiling Point: 1962 C [702]

HAP WEIGHTING FACTOR: 1 [713]

VOLATILITY:

Vapor Pressure: 1 mm @ 1292 C [703]

FIRE HAZARD: Flash Point: Not available

Flammable and moderately explosive in the form of dust or powder when exposed to flame. To fight fire, use special dry chemical. The dust may be pyrophoric in air. [703]

Autoignition temperature: Not available

LEL: Not available UEL: Not available

REACTIVITY: Dust may be pyrophoric in air and explode when heated in CO_2, and mixed with Al dust may explode in air. Mixtures with ammonium nitrate may explode when heated. Powdered metal ignites on contact with fluorine; chlorine + heat; hydrogen peroxide; BrF_5; sulfur dioxide + heat. Violent reaction with NO_2 and oxidants. Incandescent reaction with phosphorus; nitryl fluoride, nitric acid. Can react with oxidizing materials. [703]

STABILITY: Moderately explosive in the form of dust or powder when exposed to flame. Will react with water or steam to produce hydrogen. [703]

USES: Many alloys of manganese are important in industry. For example, manganese improves the rolling and forging qualities, strength, toughness, stiffness, wear resistance, hardness, and hardenability of steel. Manganese also forms feromagnitic alloys with aluminum and antimony, especially when it is mixed with small amounts of copper.

Pure manganese metal exists in four allotropic forms. Two common forms include the alpha form (which is stable at ambient temperatures) and the gamma form (which changes to the alpha form at ordinary temperatures and which is reported to be flexible, soft, easily cut, and capable of being bent). Manganese dioxide (pyrolusite) is used as a depolarizer in dry cells. Manganese dioxide also is used to "decolorize" glass colored green by impurities of iron and it also gives glass an amethyst color. In fact, manganese dioxide provides the color of true amethyst. Manganese dioxide is also used in the preparation of oxygen and chlorine. The permanganate salt of manganese is a powerful oxidizing agent commonly used in quantitative analysis and in medicine. Manganese is an important trace element for health and has been reported to be essential for utilization of vitamin B1. [706]

COMMENTS: None

NIOSH Registry Number: OO9275000

ACUTE/CHRONIC HAZARDS: Manganese is an experimental tumorigen. Human systemic effects by inhalation include degenerative brain changes, change in motor activity, and muscle weakness. It is a skin and eye irritant. Mutagenic data exists. [703] Target organs include the respiratory system, central nervous system, blood and kidneys. [704]

SYMPTOMS: Symptoms of exposure may include Parkinson's disease; asthenia, insomnia, mental confusion; metal fume fever; dry throat, cough, tight chest, dyspnea, flu-like fever; low-back pain; vomiting; malaise; fatigue. [704] Human systemic effects by inhalation include degenerative brain changes, change in motor activity, and muscle weakness. It is a skin and eye irritant. [703]

Manganese (II) sulfate monohydrate

CAS NUMBER: 10034-96-5

SYNONYMS: Manganous(II)sulfate, hydrate

CHEMICAL FORMULA: MnSO4*H2O

MOLECULAR WEIGHT: 169.01

WLN: MN S-O4

PHYSICAL DESCRIPTION: Pale red, slightly efflorescent crystals

SPECIFIC GRAVITY: 2.950 @ 20/4 C

DENSITY: Not available

MELTING POINT: > 400 C

BOILING POINT: Not available

SOLUBILITY: Water: 50-100 mg/mL @ 21 C [700] DMSO: <1 mg/mL @ 18 C [700] 95% Ethanol: <1 mg/mL @ 18 C [700] Acetone: <1 mg/mL @ 21 C [700]

OTHER SOLVENTS: Not available

OTHER PHYSICAL DATA: pH of 5% solution @ 25 C: 2.0-4.0 Metal corrosivity: Al 0.002 in/yr

VOLATILITY: Not available

FIRE HAZARD: Manganese (II) sulfate monohydrate is nonflammable.

LEL: Not available UEL: Not available

REACTIVITY: Manganese (II) sulfate monohydrate is incompatible with aluminum and magnesium [058].

STABILITY: Manganese (II) sulfate monohydrate is hygroscopic [058,275]. Solutions of it in water, DMSO, 95% ethanol or acetone should be stable for 24 hours under normal lab conditions [700].

USES: Manganese (II) sulfate monohydrate is used as a porcelain glaze, as a fertilizer additive and as a catalyst. It is used in paints, varnishes, textile dyes, fungicides, medicines and ceramics. In foods, it is used as a nutrient and dietary supplement. It is also used in ore flotation, as a catalyst in viscose process and in synthetic manganese dioxide. In veterinary medicine, it is used as a nutritional factor (essential trace element in all animals) and in the prevention of perosis in poultry.

COMMENTS: None

NIOSH Registry Number: OP0893500

ACUTE/CHRONIC HAZARDS: Manganese (II) sulfate monohydrate is an irritant and may be toxic by inhalation or ingestion. When heated to decomposition this compound emits toxic fumes of sulfur oxides [058].

SYMPTOMS: Symptoms of exposure may include mild irritation of the eyes and mucous membranes of the respiratory tract [053,053,346]. Symptoms of exposure to this type of compound may include apathy, anorexia, headache, recurring leg cramps, loss of balance, clumsiness, pneumonia and associated pulmonary problems [052]. Other symptoms may include central nervous system damage, pulmonary system damage, upper respiratory infections, languor, sleepiness, weakness in the legs, a stolid, mask-like face, muscular twitchings, varying from a fine tremor of the hands to coarse, rhythmical movements of the arms, legs and trunk; slight increase in tendon reflexes, ankle and patellar clonus, typical Parkinsonian slapping gait and minute handwriting (affected by micrographia) [043]. It can cause spastic gait, insomnia, dystonia, fatiguability, asthenia and an inability to concentrate [151]. It can also cause impaired mentation, ataxia, difficulty in walking and decreased movement of the eyelids and eyes [099]. Other symptoms may include lethargy, edema, extrapyramidal effects, sleep disturbances, dermatitis, irritability, liver enlargement, increased muscle tone, muscular cramps, mental deterioration, excessive salivation and perspiration, sexual disturbances, blood changes and, very rarely, hypothyroidism [295]. Exposure can cause spasms, arthralgias and speech disturbances such as slurred speech, slow and monotonous voice tone, inability to speak above a whisper, difficult articulation and incoherence, and even complete muteness. Psychosis may occur with unaccountable laughter, euphoria, impulsive acts, absentmindedness, mental confusion, aggressiveness and hallucinations. Propulsion, retropropulsion and lateropropulsion are affected with no movement for protection when falling. Absolute detachment may occur, broken by sporadic or spasmodic laughter [346]. Inhalation may cause acute bronchitis, nasopharyngitis and itching. Numbness of the extremities and impairment of libido may occur [301]. Other effects may include weakness, delusions, compulsions, rigidity, bradykinesia, sudden crying, stuttering, hoarse voice, nervousness, inability to walk backward, hyporeflexia, Romberg's sign, adiadochokinesia, forgetfulness, general malaise, drowsiness, stiffness of the arms or legs, urinary difficulty, somnolence, sexual excitement followed by impotence, "hen's gait" and frequent falling [430]. Fever may occur, with chills, upset stomach (nausea), vomiting, dryness of the throat, cough and body aches. Other symptoms include trouble with memory and judgement, unstable emotions, incoordination, chest pain, restlessness, double vision, impaired hearing, anemia, lassitude and low back pain [102]. Symptoms of exposure may simulate progressive bulbar paralysis, post encephalitic Parkinsonism, multiple sclerosis, amyotrophic lateral sclerosis and progressive lenticular degeneration (Wilson's Disease) [043].

Mercury compounds

NOTE: For all listings which contain the word "compounds," the following applies: Unless otherwise specified, these listings are defined as including any unique chemical substance that contains the named chemical as part of that chemical's infrastructure. [708] Information for the metallic form of mercury is given below. Other compounds described are intended as selected examples of mercury compounds and do not comprise a comprehensive list.

Mercury

CAS NUMBER: 7439-97-6

ATOMIC FORMULA: Hg

ATOMIC WEIGHT: 200.59 [702]

PHYSICAL DESCRIPTION: Silver-white, heavy, odorless liquid [704] a liquid metallic element [703] Solid: tin-white, ductile, malleable mass which can be cut with a knife [703]

SPECIFIC GRAVITY: 13.6 @ 20C/4C [704]

MELTING POINT: -38.8 C [702]

BOILING POINT: 357 C [702]

SOLUBILITY: Water: 0.28 umoles/L [702]

OTHER PHYSICAL DATA: Freezing Point: -38 F [704] Vapor Pressure: 0.0012 mm [704] Solubility, water: Insoluble [704]

VOLATILITY:

Vapor Pressure: 0.0012 mm [704]

FIRE HAZARD: Flash Point: Not applicable

Mercury is a noncombustible liquid [704].

Autoignition temperature: Not available

LEL: Not applicable UEL: Not applicable

REACTIVITY: It may explode on contact with 3-bromopropyne; alkynes + silver perchlorate; ethylene oxide; lithium; methylsilane + oxygen (when shaken); peroxyformic acid; chlorine dioxide; tetracarbonylnickel + oxygen. With methyl azide explodes with shock or spark. The vapor ignites on contact with boron diiodophosphide. It also reacts violently with acetylenic compounds, chlorine gas, chlorine dioxide, methyl azide, nitromethane, and is also often incompatible with other oxidants. [703]

Ammonia: May react with ammonia to form explosive product. [703] The use of mercury manometers with ammonia should be avoided as intrinsically unsafe. Although pure dry ammonia and mercury do not react even under pressure at 340 kbar and 200 C, the presence of traces of water leads to the formation of an explosive compound, which may explode during depressurization of the system. Explosions in mercury-ammonia systems had been reported previously. [710]

Metals: Reacts violently with metals (e.g., aluminum, calcium, potassium, sodium, rubidium; exothermic formation of amalgams). [703] The high mobility and tendency to dispersion exhibited by mercury, and the ease with which it forms alloys (amalgams) with many laboratory and electrical contact metals, can cause severe corrosion problems in laboratories. A filter-cyclone trap is suggested to contain completely mercury ejected accidentally by over- pressuring of mercury manometers and similar items. [710]

CORROSIVITY:

The high mobility and tendency to dispersion exhibited by mercury, and the ease with which it forms alloys (amalgams) with many laboratory and electrical contact metals, can cause severe corrosion problems in laboratories [710].

STABILITY: When heated to decomposition it emits toxic fumes of Hg. [703]

USES: Mercury easily forms alloys with many metals, such as gold, silver, and tin, which are called amalgams. Mercury is used to recover gold from gold ores by using its amalgamating properties. Metalic mercury is also widely used to make laboratory equipment such as thermometers, barometers, diffusion pumps, and many other instruments. It is also used to manufacture commercial products such as mercury-vapor lamps, advertising signs, mercury switches, and many other types of electrical apparatus. Mercury also finds other uses in mercury cells, for caustic soda and chlorine production, dental preparations, antifouling paint, batteries, and catalysts. It used to be used in the manufacture of some pesticides but this use has been mostly discontinued because of environmental concerns. The most important salts are mercuric chloride, mercurous chloride, mercury fulminate, and mercuric sulfide. [706]

COMMENTS: Mercury is a virulent poison and it is readily absorbed through the respiratory tract, the gastrointestinal tract, and even through unbroken skin. It acts as a cumulative poison and dangerous levels are easily accumulated in air because of the relative volatility (high vapor pressure) of mercury at ambient temperatures. Containers of mercury should be securely covered and spillage should be immediately cleaned up. When it is necessary to heat mercury or mercury compounds, this should be done in a well-ventilated hood. Methyl mercury is a dangerous environmental pollutant that can be formed as a metabolic product and it can sometimes be found fairly frequently in water and streams. [706]

NIOSH Registry Number: OV4550000

ACUTE/CHRONIC HAZARDS: Poison by inhalation. Corrosive to skin, eyes, and mucous membranes. Human systemic effects by inhalation: wakefulness, muscle weakness, anorexia, headache, tinnitus, hypermotility, diarrhea, liver changes, dermatitis, fever. An experimental tumorigen and teratogen. Experimental reproductive effects and human mutagenic data exist. When heated to decomposition, emits toxic fumes of Hg. [703] Elemental mercury (metallic) is usually non-toxic when ingested unless GI fistula or other GI inflammatory disease is present or the mercury is retained for a prolonged period in the GI tract. Mercury vapor will cause severe pulmonary damage if inhaled, as well as nephrotoxicity and gingivitis. Elemental mercury will readily vaporize if heated and form extremely toxic oxide fumes. The most consistent and pronounced effects of chronic exposure to elemental mercury vapor are on the CNS. Effects are neurological and psychiatric. Either acute or chronic exposure may produce permanent changes to affected organs and organ systems. Acute poisoning due to mercury vapors affects the lungs primarily, in the form of acute interstitial pneumonitis, bronchitis, and bronchiolitis. In general, chronic exposure produces four classical signs: gingivitis, sialorrhea, increased irritability, and muscular tremors. Rarely are all four seen together in an individual case. [723] It is still not known to what degree renal damage may occur in connection with chronic exposure to mercury vapor. Severe nephrotic changes have not been described in patients exposed only to mercury vapor. In patients exposed to a combination of mercury dust and vapor, such changes have been reported. The brain is the critical organ in humans for chronic mercury vapor exposure; in severe cases, spongeous degeneration of brain cortex can occur as a late sequela to past exposure. Renal proteinuria has been described following exposure to mercury vapor. Other reported effects from elemental mercury are contact dermatitis from mercury amalgam fillings and mercury sensitivity occurring among dental students. [723]

SYMPTOMS: Symptoms of exposure to mercury vapor may include coughing, chest pain, dyspnea, bronchial pneuitis; tremor, insomnia; irritability, indecision; headache, fatigue, weakness; stomatitis, salivation; gastrointestinal disturbance, anorexia, weight loss; proteinuria; irritation of eyes and skin. [704] Human systemic effects by inhalation include wakefulness, muscle weakness, anorexia, headache, tinnitus, hypermotility, diarrhea, liver changes, dermatitis and fever. [703] Symptoms of chronic exposure to elemental mercury vapor may include depression, irritability, exaggerated response to stimulation (erethism), excessive shyness, insomnia, emotional instability, forgetfulness, confusion, and vasomotor disturbances such as excessive perspiration and uncontrolled blushing. Tremors are also common; these are exaggerated when task is required but minimal when patient is at rest or asleep. A fine trembling of fingers, eyelids, lips and tongue may be interrupted intermittently by coarse shaking movements. Erethism and tremors are reversible. [723] Acute intoxication from inhaling mercury vapor in high concentrations used to be common among those who extracted mercury from its ores. The condition is characterized by metallic taste, nausea, abdominal pain, vomiting, diarrhea, headache, and sometimes albuminuria. After a few days, salivary glands swell, stomatitis and gingivitis develop, and a dark line of mercury sulfide forms on inflamed gums. Teeth may loosen, and ulcers may form on lips and cheeks. In milder cases, recovery occurs within 10-14 days; but in others, poisoning of the chronic type may ensue. [723]

Mercuric chloride

CAS NUMBER: 7487-94-7

SYNONYMS: Dichloromercury Mercury (II) chloride Mercury bichloride Mercury perchloride

CHEMICAL FORMULA: Cl2Hg

MOLECULAR WEIGHT: 271.50

WLN: .HG..G2

PHYSICAL DESCRIPTION: White solid

SPECIFIC GRAVITY: Not available

DENSITY: 5.44 g/mL @ 25 C

MELTING POINT: 276 C

BOILING POINT: 302 C

SOLUBILITY: Water: 5-10 mg/mL @ 22 C [700] DMSO: >=100 mg/mL @ 22 C [700] 95% Ethanol: >=100 mg/mL @ 22 C [700] Acetone: >=100 mg/mL @ 22 C [700]

OTHER SOLVENTS: Methanol: Soluble Pyridine: Soluble Carbon disulfide: Slightly soluble Acetic acid: 1g/40 mL Glycerol: 1g/12 mL Ether: 0.88-1.0 g/100 mL Benzene: 1 g/200 mL

OTHER PHYSICAL DATA: Density (liquid): 4.44 g/mL @ 280 C

VOLATILITY:

Vapor pressure: 0.00014 mm Hg @ 35 C, 1 mm Hg @ 136.2 C

FIRE HAZARD: Mercuric chloride is nonflammable.

LEL: Not available UEL: Not available

REACTIVITY: Mercuric chloride may react with a wide variety of materials, including formates sulfides, albumin, gelatin, alkalis, alkaloid salts, ammonia, lime water, antimony, arsenic, bromides, borax, carbonates and reduced iron; iron, copper, lead and silver salts; infusions of cinchoma, oak bark or senna, tannic acids and vegetable astringents.

STABILITY: Mercuric chloride is decomposed by sunlight.

USES: Mercuric chloride is used for preserving wood and anatomical specimens, for embalming, as a disinfectant, for browning and etching steel and iron, as an intensifier in photography, a white reserve in fabric printing, for tanning, for electroplating aluminum, in dry batteries, and as a fungicide and insecticide.

COMMENTS: None

NIOSH Registry Number: OV9100000

ACUTE/CHRONIC HAZARDS: Mercuric chloride is an irritant and may be toxic by ingestion or absorption through the skin. It is also corrosive to tissues and may cause ulceration of the conjunctiva and cornea. When heated to decomposition, toxic fumes are evolved.

SYMPTOMS: Symptoms of exposure may include burning of the mouth and throat, corrosion of tissues and mucous membranes, severe nausea, vomiting, hematemesis, necrosis, abdominal and/or chest pain, melena, corneal ulceration, dermatitis, convulsions, thirst, spitting of blood, diarrhea, kidney dysfunction, exhaustion, shock, cardiac arrhythmias, reduction of neurocyte alkali phosphatase activity and death.

Mercury((O-carboxyphenyl)thio)ethyl, sodium salt

CAS NUMBER: 54-64-8

SYNONYMS: o-(Ethylmercurithio)benzoic acid, sodium salt

CHEMICAL FORMULA: C9H9HgO2S.Na

MOLECULAR WEIGHT: 404.82

WLN: QVR BS-HG-2 &-NA-

PHYSICAL DESCRIPTION: Light cream-colored, crystalline powder

SPECIFIC GRAVITY: Not available

DENSITY: Not available

MELTING POINT: 232-233 C (decomposes) [269,275]

BOILING POINT: Not available

SOLUBILITY: Water: >=100 mg/mL @ 19 C [700] DMSO: >=100 mg/mL @ 19 C [700] 95% Ethanol: 10-50 mg/mL @ 19 C [700] Acetone: <1 mg/mL @ 19 C [700] Methanol: Soluble [295]

OTHER SOLVENTS: Ether: Practically insoluble [033,062,233,295] Benzene: Practically insoluble [033,062] Alcohol: 1 g/8 mL [033,295]

OTHER PHYSICAL DATA: pH of 1% aqueous solution: 6.7 [062] Slight odor [062,295]

VOLATILITY: Not available

FIRE HAZARD: Flash point data for mercury((O-carboxyphenyl)thio)ethyl, sodium salt are not available. It is probably combustible. Fires involving this material may be controlled with a dry chemical, carbon dioxide or Halon extinguisher. A water spray may also be used [269].

LEL: Not available UEL: Not available

REACTIVITY: Mercury((O-carboxyphenyl)thio)ethyl, sodium salt is incompatible with acids, iodine, heavy metal salts and many alkaloids [295,455]. It can be absorbed by rubber caps and it is incompatible with strong oxidizing agents and strong bases. The rate of oxidation for this compound is greatly increased by traces of copper ions [295].

STABILITY: Mercury((O-carboxyphenyl)thio)ethyl, sodium salt is stable in air, but not sunlight [033,233]. It may discolor on exposure to light [269]. Dilute aqueous solutions of this chemical are fairly stable to heat but labile to light. Solutions are less stable to heat when acidic than when alkaline. Solutions are most stable to light at pH 5 to 7. Solutions are unstable to heat but not to light in the presence of copper, iron or zinc ions but not in the presence of calcium or magnesium ions [295].

Solutions of it in water, DMSO, 95% ethanol or acetone should be stable for 24 hours when protected from light [700].

USES: Mercury((O-carboxyphenyl)thio)ethyl, sodium salt is used as an ophthalmic preservative, a topical antiinfective and a topical veterinary antibacterial and antifungal agent. It is also used as a bacteriostat and a fungistat. It is an FDA over-the-counter drug.

COMMENTS: None

NIOSH Registry Number: OV8400000

ACUTE/CHRONIC HAZARDS: Mercury((O-carboxyphenyl)thio)ethyl, sodium salt is an irritant and may be toxic by ingestion and inhalation [062]. When heated to decomposition it emits toxic fumes of mercury, sodium oxide and sulfur oxides [043].

SYMPTOMS: Symptoms of exposure to this class of compounds may include aphthous stomatitis, catarrhal gingivitis, nausea, liquid stools, pain, liver disorder, injury to the cardiovascular system and hematopoietic system, deafness and ataxia. Exposure may be fatal. Headache, paresthesia of the tongue, lips, fingers and toes, other non-specific dysfunctions, metallic taste, slight gastrointestinal disturbances, excessive flatus and diarrhea may occur. Acute poisoning may cause gastrointestinal irritation and renal failure. Early signs of severe poisoning may include fine tremors of extended hands, loss of side vision, slight loss of coordination in the eyes, speech, writing and gait, inability to stand or carry out voluntary movements, occasional muscle atrophy and flexure contractures, generalized myoclonic movements, difficulty understanding ordinary speech, irritability and bad temper progressing to mania, stupor, coma, mental retardation in children, skin irritation, blisters and dermatitis [173]. Other symptoms may include chorea, athetosis, tremors, convulsions, pain and numbness in the extremities, nephritis, salivation, loosening of the teeth, blue line on the gums, anxiety, mental depression, insomnia, hallucinations and central nervous system effects [301]. Exposure may also cause irritation of the eyes, mucous membranes and upper respiratory tract [269].

Methyl mercury (II) chloride

CAS NUMBER: 115-09-3

SYNONYMS: Methylmercuric chloride

CHEMICAL FORMULA: CH3ClHg

MOLECULAR WEIGHT: 251.08

WLN: G-HG-1

PHYSICAL DESCRIPTION: White microcrystals

SPECIFIC GRAVITY: Not available

DENSITY: 4.06 g/mL @ 25 C [205]

MELTING POINT: 170 C [205,233]

BOILING POINT: Not available

SOLUBILITY: Water: <0.1 mg/mL @ 21 C [700] DMSO: >=100 mg/mL @ 27 C [700] 95% Ethanol: 10-50 mg/mL @ 27 C [700] Acetone: >=100 mg/mL @ 27 C [700]

OTHER SOLVENTS: Most organic solvents: Soluble [900]

OTHER PHYSICAL DATA: Dipole moment: 2.99D [233] Lambda max: 206 nm (log epsilon=3.17) [233]

VOLATILITY:

Vapor pressure: 0.0085 mm Hg [900]

FIRE HAZARD: Flash point data for methyl mercury (II) chloride are not available. It is probably nonflammable.

LEL: Not availablc UEL: Not available

REACTIVITY: Not available

STABILITY: Methyl mercury (II) chloride may be sensitive to light [052]. Aqueous solutions containing this compound at a concentration of 0.25 mg/mL are stable for 3 weeks in the dark at room temperature. Aqueous 0.0001 M solutions show no degradation after 17.1 hours of midday sunlight. High intensity UV irradiation of solutions of this compound causes decomposition [900]. NMR stability screening indicates that solutions of it in DMSO are stable for at least 24 hours [700].

USES: Methyl mercury (II) chloride is used as a fungicide.

COMMENTS: None

NIOSH Rcgistry Number: OW1225000

ACUTE/CHRONIC HAZARDS: Methyl mercury (II) chloride may be toxic by all route of exposure and it may be absorbed through the skin. When heated to decomposition it emits toxic fumes of chlorine and mercury [043].

SYMPTOMS: Symptoms of exposure may include ataxia, paralysis, incoordination and death [099]. Other symptoms may include nausea, abdominal pain, vomiting, diarrhea, shock, kidney damage, nervous disturbances, tremors of the hands, insomnia, loss of memory, irritability, depression, loosening of teeth and excessive salivation [036]. Paresthesias, neurasthenia, hearing loss, dysarthria and mental deterioration has also occurred [406]. Chorea, athetosis and convulsions have also developed [301].

Methyl mercury(II) hydroxide

CAS NUMBER: 1184-57-2

SYNONYMS: Methylmercuric hydroxide

CHEMICAL FORMULA: CH4HgO

MOLECULAR WEIGHT: 232.64

WLN: Q-HG-1

PHYSICAL DESCRIPTION: Pale green solid

SPECIFIC GRAVITY: Not available

DENSITY: Not available

MELTING POINT: 137 C [025,233]

BOILING POINT: Not available

SOLUBILITY: Water: 1-10 mg/mL @ 21 C [700] DMSO: 1-10 mg/mL @ 21 C [700] 95% Ethanol: 50-100 mg/mL @ 21 C [700] Acetone: <1 mg/mL @ 21 C [700]

OTHER SOLVENTS: Not available

OTHER PHYSICAL DATA: Unpleasant odor This compound is manufactured and produced as a 1 M aqueous solution; but when crystallized, it is not soluble

VOLATILITY: Not available

FIRE HAZARD: Flash point data for methyl mercury (II) hydroxide are not available. It is probably combustible. Fires involving this material may be controlled with a dry chemical, carbon dioxide or Halon extinguisher.

LEL: Not available UEL: Not available

REACTIVITY: Methyl mercury (II) hydroxide is incompatible with oxidizing materials, acid and acid fumes [900].

STABILITY: Methyl mercury (II) hydroxide is sensitive to air and light [900]. Solutions of it in water and methanol are stable; therefore, it should be stable in 95% ethanol [700].

USES: Methyl mercury (II) hydroxide is used as a defoliant, an epoxidation catalyst and as a constituent of seed disinfectants for fungicides.

COMMENTS: The actual composition of this compound is not known. It is believed to be a mixture of CH3HgOH and oxonium salts.

NIOSH Registry Number: OW4900000

ACUTE/CHRONIC HAZARDS: Methyl mercury (II) hydroxide may be toxic by ingestion, inhalation or skin absorption. It may cause irritation to the eyes, skin and mucous membranes [070,900]. When heated to decomposition it emits toxic fumes of mercury [042,900].

SYMPTOMS: Symptoms associated with compounds of this class may occur weeks to months after an acute exposure to toxic concentrations. Acute and chronic symptoms are similar and may include numbness and tingling of the lips, hands and feet; dysarthria, concentric constriction of the visual fields, hearing impairment and ataxia [070,099]. Clumsiness, speech impairment, slight ptosis, irregular nystagmus on extremes of gaze, "jerky pursuit" and abnormal saccadic movements are common [099]. It can cause chorea, athetosis, tremor and convulsions [301]. Other symptoms may include irritation of the eyes, skin and mucous membranes, and emotional disturbances. Severe intoxication can produce periods of spasticity and jerking movements of the limbs, head or shoulders; bouts of groaning, moaning, shouting or crying; and incontinence. Less frequent symptoms may include dizziness, hypersalivation, lacrimation, nausea, vomiting, diarrhea, constipation, kidney damage, dermatitis and skin burns [070]. Damage tends to be irreversible [070,301]. This type of compound passes through the blood brain barrier and placenta very rapidly [421].

Methanol

CAS NUMBER: 67-56-1

SYNONYMS: Carbinol Methyl alcohol Methyl hydroxide Methylol

CHEMICAL FORMULA: CH4O

MOLECULAR WEIGHT: 32.04

WLN: Q1

PHYSICAL DESCRIPTION: Colorless liquid

SPECIFIC GRAVITY: 0.7913 @ 20/4 C

DENSITY: 0.796 g/mL @ 15 C

MELTING POINT: -98 C

BOILING POINT: 64.6 C

SOLUBILITY: Water: >=100 mg/mL @ 21 C [700] DMSO: >=100 mg/mL @ 21 C [700] 95% Ethanol: >=100 mg/mL @ 21 C [700] Acetone: >=100 mg/mL @ 21 C [700]

OTHER SOLVENTS: Chloroform: Miscible Ketones: Miscible Most organic solvents: Soluble Ether: Miscible Benzene: Miscible

OTHER PHYSICAL DATA: Refractive index: 1.3292 @ 20 C Odor threshold: 4.3 ppm Highly polar Slight alcoholic odor

HAP WEIGHTING FACTOR: 1 [713]

VOLATILITY:

Vapor pressure: 100 mm Hg @ 21.2 C; 237.87 mm Hg @ 38 C [700]

Vapor density: 1.11

FIRE HAZARD: The flash point of methanol is 11 C (52 F) and it is flammable. Fires involving this material can be controlled with a dry chemical, carbon dioxide or Halon extinguisher.

The autoignition temperature of methanol is 463.89 C (867 F).

LEL: 6.0%   UEL: 36.5

%REACTIVITY: Methanol may react violently with acids, acid chlorides, acid anhydrides, oxidizing agents, reducing agents and alkali metals.

STABILITY: Methanol is stable under normal laboratory conditions. Solutions of it in water, DMSO, 95% ethanol and acetone should be stable for 24 hours under normal lab conditions [700].

USES: Methanol is used as an industrial solvent, and as a starting material for organic synthesis of formaldehydes, methyl esters of organic and inorganic acids, methacrylates, methylamines, methyl anhydrides, ethylene glycol and pesticides. It is also used as an antifreeze for automotive radiators and air brakes and it is an ingredient of gasoline and diesel oil antifreezes. Other uses include: an octane booster in gasoline, as fuel for picnic stoves and soldering torches, an extractant for animal and vegetable oils, for denaturing ethanol, a softening agent for pyroxylin plastics, a solvent adjuvant for polymers, and a solvent in thc manufacturing of cholesterol, streptomycin, vitamins, hormones and other pharmaceuticals. Methanol is also an ingredient in paint, varnish removers, cleaning and dewaxing preparations, spirit duplicating fluids and embalming fluids. It is also used in the manufacture of photographic film, celluloid, textile soap, wood stains, coated fabrics, shatter-proof glass, paper coating, waterproofing formulations, artificial leather, synthetic indigo and other dyes.

COMMENTS: None

NIOSH Registry Number: PC1400000

ACUTE/CHRONIC HAZARDS: Methanol may be toxic and an irritant. It can be absorbed through the skin. It may also cause narcosis.

SYMPTOMS: Symptoms of exposure may include headache, fatigue, nausea, dizziness, stupor, cramps, dermatitis, visual impairment or complete blindness (may be permanent); acidosis, convulsions, mydriasis, circulatory collapse, respiratory failure, death, irritation of mucous membranes, damage to the central nervous system (especially the optic nerve), injury to the kidneys, liver, heart and other organs; peripheral neuritis, gastrointestinal disturbances, photophobia and conjunctivitis, followed by definite eye lesions; narcosis, unconsciousness, shallow breathing, cyanosis, coma, fall in blood pressure, hyperemia of the optic disk with blurring of the margin; burning scnsation, coughing, wheezing, laryngitis, vomiting, delirium, pain in the eyes, giddiness, vertigo, severe abdominal pain, back pain, dyspnea, motor restlessness, cold clammy extremities and diarrhea.

Methoxychlor

CAS NUMBER: 72-43-5

SYNONYMS: 2,2-bis(p-Methoxyphenyl)-1,1,1-trichloroethane

CHEMICAL FORMULA: C16H15Cl3O2

MOLECULAR WEIGHT: 345.66

WLN: GXGGYR DO1&R DO1

PHYSICAL DESCRIPTION: White to tan powder

SPECIFIC GRAVITY: 1.41 @ 25 C [055,169,173,371]

DENSITY: Not available

MELTING POINT: 86-88 C [025,031,395,421]

BOILING POINT: Decomposes [051,102,371,421]

SOLUBILITY: Water: <1 mg/mL @ 23 C [700] DMSO: >=100 mg/mL @ 23 C [700] 95% Ethanol: 10-50 mg/mL @ 23 C [700] Acetone: >=100 mg/mL @ 23 C [700] Methanol: 50 g/kg @ 20 C [169,172]

OTHER SOLVENTS: Aromatic solvents: Readily soluble [169,173,395] Alcohol: Soluble [031,062,173,395] Petroleum oils: Moderately soluble [173,395] Xylene: 440 g/kg @ 20 C [169,172] Chlorinated solvents: Readily soluble [169] Vegetable oils: Readily soluble [169] Chloroform: 440 g/kg @ 20 C [169,172] Ketonic solvents: Readily soluble [169]

OTHER PHYSICAL DATA: This compound exists as dimorphic crystals [025,031,047,395] Gray flaky powder (technical grade) [165,169,172,395] Melting point also reported as 78-78.2 C [025,031,395,421] Threshold odor concentration (in water): 4.7 mg/kg [055] Slightly fruity odor [051,346,371,421] Spectroscopy data: lambda max (in benzene): 275 nm, 270 nm, 238 nm, 230 nm (E = 183, 241, 458, 575) [395]

HAP WEIGHTING FACTOR: 1 [713]

VOLATILITY:

Vapor pressure: Very low [102,169]

Vapor density: 12 [043]

FIRE HAZARD: Methoxychlor is combustible. It burns only at high temperatures [051,371]. Fires involving this material may be controlled with a dry chemical, carbon dioxide or Halon extinguisher. A water spray may also be used [051,102,371].

LEL: Not available UEL: Not available

REACTIVITY: Methoxychlor is incompatible with alkaline materials, especially in the presence of catalytically-active metals [169,173,186,395]. It is slightly corrosive to iron and aluminum [169]. It is decomposed by refluxing with sodium in isopropyl alcohol [186]. It is also incompatible with strong oxidizers [051,102,346]. It will attack some forms of plastics, rubber and coatings [102].

STABILITY: Methoxychlor turns pink or tan on exposure to light [169]. It is described as resistant to ultraviolet light, but other studies have shown it to break down rapidly under UV in hexane solution [186]. Solutions of it in water, DMSO, 95% ethanol or acetone should be stable for 24 hours under normal lab conditions [700].

USES: Methoxychlor is an insecticide used to control a wide range of insect pests (particularly chewing insects) in field crops, forage crops, fruit, vines, flowers, vegetables and in forestry. It is also used for the control of insect pests in animal houses and dairies, and in household and industrial premises. It is used in veterinary medicine as an ectoparasiticide.

COMMENTS: None

NIOSH Registry Number: KJ3675000

ACUTE/CHRONIC HAZARDS: Methoxychlor may be toxic by ingestion and inhalation [051,371,421] and an irritant [033,051,371]. When heated to decomposition it emits toxic fumes of chlorides, hydrogen chloride gas and carbon monoxide [043,051,102,371].

SYMPTOMS: Symptoms of exposure may include vomiting, tremors, convulsions and liver damage [301]. Somnolence may also occur [043]. Other symptoms may include generalized depression, headache, staggering, nausea and lethargy [051]. It may also cause diarrhea, numbness and partial paralysis [058]. Exposure over a prolonged period may cause kidney damage [033,058].

Symptoms of exposure to this type of compound may include central nervous system stimulation, paresthesia, excitement, giddiness, fatigue, pulmonary edema, myocardial toxicity, hypothermia and coma. Respiration is initially accelerated and later depressed. Chronic exposure to this type of compound may cause loss of appetite, muscular weakness, apprehensive mental state and enhancement of hepatic microsomal enzyme activity [295].

Symptoms of exposure to a related compound may include gastric irritation, delayed emesis, malaise, sore throat, weakness, ataxia, confusion, paresis, irregular and abnormally slow pulse and death due to respiratory failure [151].

Methyl bromide

CAS NUMBER: 74-83-9

SYNONYMS: Bromomethane

CHEMICAL FORMULA: CH3Br

MOLECULAR WEIGHT: 94.94

WLN: E1

PHYSICAL DESCRIPTION: Colorless liquid or gas

SPECIFIC GRAVITY: 1.6755 @ 20/4 C [017,395]

DENSITY: 1.732 g/mL @ 0 C [062,169]

MELTING POINT: -93 C [043,055,058,169]

BOILING POINT: 3.56 C [043,047,205,421]

SOLUBILITY: Water: Slightly soluble [025,058,295,421] DMSO: Reacts [043] 95% Ethanol: Soluble [395] Acetone: Not available

OTHER SOLVENTS: Benzene: Soluble [053,395] Alcohol: Soluble [017,047,053,205] Chloroform: Soluble [047,053,205,395] Carbon disulfide: Soluble [053,169,395] Ether: Soluble [017,169,205,395] Most organic solvents: Soluble [052,169,172,421] Carbon tetrachloride: Soluble [053,395] Lower alcohols: Soluble [169] Esters: Soluble [169] Ketones: Soluble [169] Aromatic hydrocarbons: Soluble [169] Halogenated hydrocarbons: Soluble [169]

OTHER PHYSICAL DATA: Burning taste [043,062,173,295] Specific gravity: 1.732 @ 0/0 C [043,205] 1.730 @ 0/4 C [031,047] Forms a hydrate @ 0-5 C [025] Vapor pressure: 1 mm Hg @ -96.3 C; 40 mm Hg @ -54.2 C [038] Will cling to the ground but will not dissolve to any great extent [051,071] Forms a voluminous crystalline hydrate with cold water [062] Viscosity: 0.397 centipoise @ 0 C [031] Specific heat: 0.165 cal/g/C @ -96.6 C; 1.97 cal/g/C @ -13.0 C; 0.107 cal/g/C @ 25 C [031] Critical temperature: 194 C [031] Evaporation rate (butyl acetate = 1): >1 [102] Liquid density @ boiling point (4.6 C): 1722 kg/m3 [058] Liquid surface tension: 24.5 dynes/cm @ 15 C [371] Vapor (gas) specific gravity: 3.3 [371] Ratio of specific heats of vapor (gas): 1.247 [371] Latent heat of vaporization: 59.7 cal/g [371] Heat of combustion: -1771 cal/g [371] Heat of fusion: 15.05 cal/g [371] Reid vapor pressure: 45 psia [371] Odorless in low concentrations [052,172,295,421] Sweetish, chloroform-like odor @ high concentrations [051,058,172,346] Refractive index: 1.4432 @ -20 C [031]

HAP WEIGHTING FACTOR: 1 [713]

VOLATILITY:

Vapor pressure: 1250 mm Hg @ 20 C [053,051,062,071]

Vapor density: 3.3 [053,102,395,451]

FIRE HAZARD: Methyl bromide forms flammable mixtures between 10 and 15% (13.5-14.5% but also 8.6-20% have been noted), the limits in oxygen, or under pressure being wider [066]. Fires involving this material may be controlled with a dry chemical, carbon dioxide or Halon extinguisher. A water spray may also be used [043,058, 451,610].

The autoignition temperature of methyl bromide is 537 C (999 F) [058,102, 371,451].

LEL: 13.5% [043,053,058,102] UEL: 14.5% [043,053,102]

REACTIVITY: Methyl bromide's explosive sensitivity of mixtures with air may be increased by the presence of aluminum, magnesium, zinc or their alloys [043, 071]. It is incompatible with metals, dimethyl sulfoxide and ethylene oxide [043]. It forms a crystalline hydrate with cold water and penetrates rubber [036]. It is incompatible with strong oxidizers [052,053,102,346]. It is also incompatible with trimethylsulfoxonium bromide [066]. It may attack some forms of plastics, rubber and coatings [053,102]. In the presence of moisture, it is corrosive to aluminum, tin, magnesium, zinc and their alloys [169].

STABILITY: Methyl bromide is sensitive to sunlight [071]. It burns in oxygen [031, 062,173,395]. It is hydrolyzed slowly in water and more rapidly in alkaline media [169].

USES: Methyl bromide is used in ionization chambers, refrigerants, fire extinguishing agents, organic synthesis as an methylating agent, preparation of quarternary ammonium compounds, organo-tin derivatives and antipyrine. It is also used as a soil and space fumigant, in the disinfestation of potatoes, tomatoes and other crops, acaricidal fumigant, industrial solvent for extraction of plant oils, fungicide, herbicide, nematocide, rodenticide, insecticide and in degreasing wool, nuts, seeds and flowers. Other uses include food sterilization for pest control in fruits, vegetables, dairy products, nuts and grains, selective solvent in aniline dyes, laboratory procedures and as an intermediate in the manufacturing of many drugs.

COMMENTS: None

NIOSH Registry Number: PA4900000

ACUTE/CHRONIC HAZARDS: Methyl bromide may be toxic by inhalation, ingestion or through skin absorption [051,062,071,451]. It may also be readily absorbed through the skin [051, 071, 102, 301]. It is an irritant , a lachrymator [395,102], and it is narcotic at high concentrations [043,102]. When heated to decomposition it may emit toxic fumes of carbon dioxide [058] and hydrogen bromide [053,058,102,151].

SYMPTOMS: Symptoms of exposure may include anorexia, nausea, vomiting, corrosion to the skin, severe skin burns, enlarged liver, kidney damage with development of albuminuria and in fatal cases, cloudy swelling and tubular degeneration. Central nervous system effects may include, blurred vision, mental confusion, numbness and tremors. Death following acute poisoning is usually caused by its irritant effect on the lungs. In chronic poisoning, death is due to injury to the central nervous system [043]. Direct skin contact may cause prickling, itching, cold sensation, erythema, vesication, damage to peripheral nerve tissue and delayed dermatitis. It may also cause double vision, dizziness, headache, convulsions, muscular tremors, fatal pulmonary edema and neurological and gastrointestinal disturbances [051,071]. It may cause skin irritation [062]. Other symptoms may include mental excitement, acute mania, bronchitis, pneumonia and severe eye damage [036]. It may cause abdominal pain and death from respiratory or circulatory collapse [031,151]. It may also cause unconsciousness leading to a prompt "anesthetic" death, malaise, ataxia, myoclonus, exaggerated (or absent) deep reflexes, positive Romberg's sign, paroxysmal abnormalities of the EEG, great agitation, change of personality, coma and mild euphoria [173]. Exposure to this compound may also cause muscle weakness, loss of coordination and gait, hyperthermia, hepato and nephrotoxicity, behavioral changes, paralysis of extremities, delirium, epileptiform attacks and skin lesions [421]. It may cause respiratory tract irritation, organic brain syndrome, psychological depression, seizures, prominent cerebellar and parkinsonian signs, renal failure due to tubular necrosis, jaundice, elevations of liver enzyme activity in serum and abnormal liver function, lassitude, slurring of speech, staggering gait, diplopia, sometimes strabismus, epileptiform convulsions, perhaps with a Jacksonian-type of progression, rapid respirations, cyanosis, pallor and collapse, areflexia, impaired superficial sensation, absent or hypoactive distal-tendon reflexes and bronchopneumonia after severe pulmonary lesions [151]. It may also cause severe mucous membrane burns [451]. Other symptoms may include DNA mutations and large chromosome alterations [051]. It may cause congestion with coughing, chest pain, shortness of breath, confusion, shaking and unconsciousness, severe pulmonary irritation and neurotoxicity, narcosis at high concentrations, vertigo, tremor of the hands, dyspnea, hallucinations, anxiety, inability to concentrate, conjunctivitis and dry, scaling, itching dermatitis [102]. It may also cause lacrimation from irritation of the eyes, transient dimming of vision and blindness for 12 hours, nystagmus on lateral gaze, delirium, apathy, aphasia, edema of the papilla and punctiform hemorrhages may be found and optic atrophy [099]. In addition, it causes sleepiness, digestive problems, loss of stability, lack of motor coordination, sensorial problems and impaired hearing [058]. Somnolence, permanent brain damage, lethargy and sensory disturbances can occur [346]. It may cause anuria, hyperactivity, blood pressure fall, papilledema, fainting attacks and bronchospasm [301]. It may also cause retinal and submucous hemorrhages, stomach congestion and congestion of the brain with multiple hemorrhages associated with degenerative changes, such as necrosis [395].

Methyl tert-butyl ether

CAS NUMBER: 1634-04-4

SYNONYMS: Methyl 1,1-dimethylethyl ether Propane, 2-methoxy-2-methyl Methyl t-butyl ether Methyl tert butyl ether

CHEMICAL FORMULA: C5H120

MOLECULAR WEIGHT: 88.15 [702]

PHYSICAL DESCRIPTION: Not available

SPECIFIC GRAVITY: Not available

MELTING POINT: -109 C [702]

BOILING POINT: 55.2 C [702]

SOLUBILITY: Water: Soluble [702]

OTHER PHYSICAL DATA: Not available

HAP WEIGHTING FACTOR: 1 [713]

VOLATILITY:

Vapor Pressure: 245 mm Hg @ 25 C [715]

FIRE HAZARD: Flash Point: -28 C [715]

Flammable when exposed to heat or flame. DOT Classification: Flammable Liquid. [703]

Ignition temperature: 224 C [715]

LEL: Not available UEL: Not available

REACTIVITY: Unstable in acid solution [715].

STABILITY: When heated to decomposition it emits toxic fumes of NOx [703].

USES: Octane booster in gasoline. Chromatographic eluent especially in HPLC. [715]

COMMENTS: None

NIOSH Registry Number: KN5250000

ACUTE/CHRONIC HAZARDS: When heated to decomposition, methyl tert-butyl ether emits acrid smoke and irritating fumes [703].

SYMPTOMS: Oral ingestion of methyl tert-butyl ether (MTBE) has not been reported in humans. Based on studies of gallbladder instillation of MTBE in humans and oral ingestion studies in animals, nausea, vomiting, and sedation are likely, followed by general anesthesia (CNS and respiratory depression). Liver function test abnormalities have been seen in animals following intraductal administration. Ingestion of gasoline-MTBE mixtures may result in aspiration pneumonitis. Inhalation of MTBE produces anesthesia in animals. Chronic inhalation causes tracheal and nasal inflammation. Respiratory depression may be noted. Conjunctivitis lasting 72 hours and corneal opacities and chemosis have been seen in rabbits following instillation into unwashed eyes. Labored and irregular breathing were associated with anesthetic effects in rats. Aspiration pneumonitis is possible following ingestion of gasoline-MTBE mixtures. Somnolence was seen after intraductal instillation of as little as 8 mL, with pronounced sedation after 140 mL. Since approximately 30% is retrieved by aspiration during this procedure, this represents absorption of approximately 6 mL and 100 mL, respectively. Nausea and vomiting have been seen in animals and humans. [723]

Methyl chloride

CAS NUMBER: 74-87-3

SYNONYMS: Chloromethane Monochloromethane

CHEMICAL FORMULA: CH3Cl

MOLECULAR WEIGHT: 50.49

WLN: G1

PHYSICAL DESCRIPTION: Colorless gas

SPECIFIC GRAVITY: 0.9159 @ 20/4 C [016,395]

DENSITY: 0.92 g/mL @ 20 C [062,205]

MELTING POINT: -97 C [025,031,043,047]

BOILING POINT: -24.2 C @ 760 mm Hg [102,371,421]

SOLUBILITY: Water: Slightly soluble [031,062,421,451] DMSO: Not available 95% Ethanol: 7.8 g/100 mL @ 20 C [430] Acetone: Soluble [016,430]

OTHER SOLVENTS: Alcohol: Soluble [016,031,062,421] Chloroform: Miscible [031,051,205,421] Carbon tetrachloride: 3756 mL/100 mL @ 20 C [395] Glacial acetic acid: Miscible [031,051,205,421] Ether: Miscible [031,051,205,421] Benzene: Soluble [016,062]

OTHER PHYSICAL DATA: Sweet taste [031,051,295,395] Ethereal odor [031,058,062,395] Refractive index: 1.3389 @ 20 C [016] 1.3712 @ -24 C [031,205] Burns with a greenish smoky flame [031] Critical pressure: 65.9 atm [371] Critical temperature: 143.6 C [371] Liquid surface tension: 0.0162 N/m @ 20 C [371] Latent heat of vaporization: 101.3 cal/g [371] Heat of combustion: -2939 cal/g [371] Liquid viscosity: .332 centipoise @ -34 C, .320 centipoise @ -29 C [371] pH of solution: neutral [058] Density: 0.991 g/mL @ -25 C [055] 0.997 g/mL @ -24 C [371]

HAP WEIGHTING FACTOR: 1 [713]

VOLATILITY:

Vapor pressure: 760 mm Hg @ -24 C [038] 3672 mm Hg @ 20 C [058]

Vapor density: 1.8 [055,102,395,451]

FIRE HAZARD: Methyl chloride has a flash point of -45.5 C (-49.9 F) [058,395] and it is flammable. Fires involving this material may be controlled with a dry chemical, carbon dioxide or Halon extinguisher. A water spray may also be used [043].

The autoignition temperature of methyl chloride is 632 C (1170 F) [036, 058, 371, 451].

LEL: 8.1% [031,058,371,451] UEL: 17.2% [031,051,058,371]

REACTIVITY: Methyl chloride is a gas that is a dangerous fire hazard when exposed to heat, flame or powerful oxidizers. It explodes on contact with interhalogens (bromine trifluoride, bromine pentafluoride), magnesium and its alloys, potassium and its alloys, sodium and its alloys and zinc. It may react with (aluminum chloride + ethylene), aluminum chloride or powdered aluminum [043]. This gas reacts with sodium-potassium alloy [051,451].

STABILITY: Methyl chloride reacts with moisture in air [058]. Solutions of it in water, DMSO, 95% ethanol or acetone should be stable for 24 hours under normal lab conditions [700].

USES: Methyl chloride is used as a methylating and chlorinating agent in organic chemistry, as an extractant for greases, oils and resins, as a solvent in the synthetic rubber industry, as a refrigerant, propellant and blowing agent in polystyrene foam production, as a local anesthetic, as an intermediate in drug manufacturing, as a fumigant, as a food additive, as a catalyst carrier in low temperature polymerization, as a fluid for thermometric and thermostatic equipment, as a herbicide, in the manufacture of silicones, tetraethyllead and methyl cellulose, as a low temperature solvent, as a dye intermediate and in the production of methyl silicon polymers and resins.

COMMENTS: None

NIOSH Registry Number: PA6300000

ACUTE/CHRONIC HAZARDS: Methyl chloride is a gas that can be absorbed through the skin [421]. It may be toxic and an irritant and, when heated to decomposition, it emits very toxic fumes of chloride, chlorine, phosgene, hydrogen chloride and carbon monoxide [043, 051,058,102].

SYMPTOMS: Symptoms of exposure may include nausea, dizziness, blurred vision, vomiting, convulsive seizures and renal and hepatic damage [295,346]. Other symptoms may include frostbite, eye damage, staggering gait, difficult speech, headache, coma, central nervous system depression and depression of bone marrow activity [346]. Weakness, confusion, drowsiness, abdominal pain, paresthesia, mental changes, muscle dysfunction and pulmonary edema may also occur [295]. Additional symptoms may include nystagmus, tremors, spasticity, ataxia, epileptiform convulsions, tachycardia, dyspnea, nervousness, insomnia, intention tremor, memory disturbances and anorexia [151]. Other symptoms may include stomach pains, visual disturbances, mental confusion, unconsciousness and death [036]. Exposure to this compound may lead to slight irritation, narcotic effects, incoordination, hiccups, diplopia and dimness of vision followed by delirium, convulsions and coma, degenerative changes in the central nervous system, acute nephritis, anemia, degenerative changes in the heart, damage to the cardiovascular system, anesthesia and hemorrhages into the lungs, intestinal tract and dura [043,051]. Other symptoms may include fatigue, loss of appetite, muscular weakness, depression and psychic disturbances [051]. Additional symptoms may include breathing difficulties, diarrhea, blood damage, slurred speech, double vision, sleepiness, paralysis, toxic encephalopathy, memory lapses and cyanosis [102]. Somnolence, vertigo and hyperemia and edema of the brain may also occur [099]. Exposure to this compound may also cause irritation and burns of the skin and eyes, neurological damage, increased pulse rate, respiration rate and temperature, appearance of drunkenness, trembling, degenerative changes in the lungs and brain and personality changes [058]. Emotional disturbances may also occur [371]. Other symptoms may include giddiness, respiratory failure and incapacitation [421]. Loss of balance, congestion of the lungs and kidneys and fatty degeneration of the liver may also occur [395]. Exposure to this compound may also lead to optical difficulties [451].

Methyl chloroform

CAS NUMBER: 71-55-6

SYNONYMS: Methyltrichloromethane alpha-Trichloroethane Chloroethene Chlorothene 1,1,1-TCE 1,1,1-Trichloroethane

CHEMICAL FORMULA: C2H3Cl3

MOLECULAR WEIGHT: 133.40

WLN: GXGG1

PHYSICAL DESCRIPTION: Clear colorless liquid

SPECIFIC GRAVITY: 1.3376 @ 20/4 C [031,043,205]

DENSITY: 1.3376 g/mL @ 20 C [421]

MELTING POINT: -32.5 C [025,031,051]

BOILING POINT: 74.1 C [031,043,395,421]

SOLUBILITY: Water: <1 mg/mL @ 20 C [700] DMSO: >=100 mg/mL @ 20 C [700] 95% Ethanol: >=100 mg/mL @ 20 C [700] Acetone: Reacts [051,269,451] Methanol: Soluble [031,043,051,395]

OTHER SOLVENTS: Benzene: Soluble [031,043,205,395] Ether: Soluble [017,031,043,395] Chloroform: Soluble [017,047] Carbon tetrachloride: Soluble [031,043,051,395] Carbon disulfide: Soluble [051,395] Most organic solvents: Miscible [295,421]

OTHER PHYSICAL DATA: Specific gravity: 1.3249 @ 26/4 C [430] 1.3366 @ 25/4 C [051] Density: 1.336 g/mL @ 25 C [395] Vapor pressure: 60 mm Hg @ 9.5 C; 200 mm Hg @ 36.2 C [038] Vapor pressure: 155 mm Hg @ 30 C [055] Refractive index: 1.4379 @ 20 C [017,047,205] 1.43765 @ 21 C [430] Chloroform-like odor [051,346,371,451] Odor threshold: 100 ppm [371] Burning rate: 2.9 mm/min [371] Liquid surface tension: 25.4 dynes/cm @ 20 C [371] Latent heat of vaporization: 58 cal/g [371] Heat of combustion: 2600 cal/g [371] Evaporation rate (butyl acetate=1): 6.0 (at 1 atmosphere and 25 C) [058]

HAP WEIGHTING FACTOR: 1 [713]

VOLATILITY:

Vapor pressure: 100 mm Hg @ 20 C [038,043,055,058] 125 mm Hg @ 25 C [051]

Vapor density: 4.6 [051,058,371]

FIRE HAZARD: Methyl chloroform (1,1,1-Trichloroethane) has a flash point of >93.3 C (>200 F) [700]. It burns only in excess oxygen or air if a strong source of ignition is present [421]. Fires involving this material can be controlled with a dry chemical, carbon dioxide or Halon extinguisher.

The autoignition temperature is 537 C (998 F) [051, 451].

LEL: 7.5% [451]   UEL: 12.5% [451]

REACTIVITY: Methyl chloroform decomposes in the presence of chemically active metals [102,346,395]. This includes aluminum, magnesium and their alloys [036,066, 269,421]. It will react violently with dinitrogen tetraoxide, oxygen, liquid oxygen, sodium and sodium-potassium alloys [043,051,066,451]. It will also react violently with acetone, zinc and nitrates [269]. It can react with sodium hydroxide [043,051,451]. It is incompatible with strong oxidizers and strong bases [058,102,269,346]. Mixtures with potassium or its alloys are shock-sensitive and may explode on light impact [036,066]. This chemical can react with an aqueous suspension of calcium hydroxide, and with chlorine in sunlight [395]. It will attack some forms of plastics, rubber and coatings [102]. Upon contact with hot metal or on exposure to ultraviolet radiation, it will decompose to form irritant gases [346]. A cobalt/molybdenum-alumina catalyst will generate a substantial exotherm on contact with its vapor at ambient temperatures [066]. Hazardous reactions also occur with (aluminum oxide + heavy metals) [043].

STABILITY: Methyl chloroform is hygroscopic [031,051,102,275]. It is oxidized by atmospheric oxygen at high temperatures [051,395]. It is reactive to sunlight at high altitudes [051]. Solutions of this chemical in anhydrous DMSO or ethanol should be stable for 24 hours under normal lab conditions [700].

USES: Methyl chloroform is used in cold-type metal cleaning, in plastic cleaning, in vapor degreasing, as a chemical intermediate for vinylidene chloride, in aerosols (as a vapor pressure depressant, solvent and carrier), in adhesives (as a resin solvent) and as a lubricant carrier to inject graphite, grease and other lubricants. It is used alone and in cutting oil formulations as a coolant and lubricant for drilling and tapping alloy and stainless steels. It is also used to develop printed circuit boards, in motion picture film cleaning, in stain repellants for upholstery fabrics, in wig cleaning, in textile processing and finishing and as a solvent in drain cleaners, shoe polishes, spot cleaners, insecticide formulations and printing inks. It is also used as a solvent for cleaning precision instruments.

COMMENTS: Inhibitors are generally added to this compound to increase its stability [421].

NIOSH Registry Number: KJ2975000

ACUTE/CHRONIC HAZARDS: Methyl chloroform may be harmful by inhalation, ingestion and skin absorption [036,269]. It can be absorbed through the lungs, gastrointestinal tract and skin [395]. It is an irritant [031,036,269,371] and it may be narcotic in high concentrations [031,036,051,058]. It may also cause slight lacrimation [371]. When heated to decomposition it emits irritating gases and toxic fumes of carbon monoxide, carbon dioxide, hydrogen chloride gas, chlorine and phosgene [058,102,269,451].

SYMPTOMS: Symptoms of exposure may include irritation of the eyes, dizziness, incoordination, unconsciousness and death [102,346,371]. Irritation of the mucous membranes and respiratory tract may occur [269]. Fainting may also occur [295]. Other symptoms may include decreased reaction time, impaired manual dexterity, ataxia, lightheadedness, positive Romberg test, diarrhea, respiratory arrest and nausea [151]. It causes a proarrhythmic activity which sensitizes the heart to epinephrine, resulting in cardiac arrhythmias [043,151,301,406]. This sometimes will cause cardiac arrest, particularly when massive amounts are inhaled [043,051,151]. Inhalation can cause euphoria [043,051,151]. High concentrations may cause narcosis [031,036,058,406]. Exposure can cause headache, drowsiness, burning sensation on the eyes and skin, irritation of the throat, cardiac sensitization, aspiration of vomitus during anesthesia, blood pressure depression, chemical pneumonitis and pulmonary edema with hemorrhage [058]. It can also cause anesthesia, cardiac fibrillations, slight reddening of the skin, central nervous system impairment, helplessness, loss in equilibrium and mild eye and nasal discomfort [430]. Other symptoms are hallucinations, distorted perceptions, motor activity changes, irritability, aggression, hypermotility and other gastrointestinal changes [043]. Impaired judgement has been reported [451]. Increased reaction time has also been reported [346]. Repeated skin contact may result in a dry, scaly and fissured dermatitis due to its defatting properties [346,371,421]. Prolonged skin contact may result in considerable pain and irritation [430]. Other symptoms may include difficult breathing, asphyxiation, slight lacrimation and slight smarting of the eyes and respiratory system [371]. It may cause impaired psychophysiological functions [051,395]. It may also cause irregular heart beat, lassitude and coma [102]. High concentrations cause central nervous system depression [051,151, 301,346]. Hemorrhage in the brain may also result from exposure to high concentrations [301]. Eye contact may lead to superficial and transient injury to the eyes [058,099]. It may also lead to mild conjunctivitis [346]. Chronic exposure may result in liver and kidney damage [051,151,269,295]. Exposure to and/or consumption of alcohol may increase its toxic effects [151,269].

Methyl ethyl ketone

CAS NUMBER: 78-93-3

SYNONYMS: 2-Butanone Butanone Ethyl methyl ketone Methyl acetone MEK

CHEMICAL FORMULA: C4H8O

MOLECULAR WEIGHT: 72.11

WLN: 2V1

PHYSICAL DESCRIPTION: Clear colorless liquid

SPECIFIC GRAVITY: 0.805 @ 20/4 C [031,055,062,421]

DENSITY: 0.806 g/mL @ 20 C [371]

MELTING POINT: -86.3 C [017,047,371]

BOILING POINT: 79.6 C [017,031,055,205,371]

SOLUBILITY: Water: >=100 mg/mL @ 19 C [700] DMSO: >=100 mg/mL @ 19 C [700] 95% Ethanol: >=100 mg/mL @ 19 C [700] Acetone: >=100 mg/mL @ 19 C [700]

OTHER SOLVENTS: Benzene: Miscible [031,205] Ether: Miscible [031,205] Most organic solvents: Soluble [421] Oils: Miscible [062]

OTHER PHYSICAL DATA: Specific gravity: 0.7997 @ 25/4 C; 0.8255 @ 20/4 C [062] Specific gravity: 0.80615 @ 20/20 C [043,051] Boiling point: 30 C @ 119 mm Hg [017] Refractive index: 1.3788 @ 20 C [047] 1.3814 @ 15 C [029] Fragrant, mint-like, moderately sharp odor [346] Odor threshold: 2 ppm [055,430] Specific heat: 0.549 cal/g [062] Viscosity: 0.40 centipoise @ 25 C [062] Constant boiling mixture with water (88.7%) with a boiling point of 73.4 C [[031] Lambda max: 400-350 nm, 340 nm, 330 nm (A=0.01, 0.10, 1.0) [275]

IIAP WEIGHTING FACTOR: 1 [713]

VOLATILITY:

Vapor pressure: 77.5 mm Hg @ 20 C [055,421,430] 100 mm Hg @ 25.0 C [038]

Vapor density: 2.42 [043]

FIRE HAZARD: Methyl ethyl ketone has a flash point of -3 C (26 F) [205,269,275 and it is flammable. Fires involving this material can be controlled with a dry chemical, carbon dioxide or Halon extinguisher. A water spray may also be used [036,269,451].

The autoignition temperature of methyl ethyl ketone is 515 C (960 F) [036,043,051,062,421].

LEL: 1.8% [043,058,066,430] UEL: 10% [036,051,062]

REACTIVITY: Methyl ethyl ketone is incompatible with strong oxidizers [058,269,346]. It is also incompatible with chlorosulfonic acid, oleum, hydrogen peroxide, and nitric acid [043]. It can react with potassium tert-butoxide and with chloroform [043,066]. It can also react with bases and strong reducing agents [269]. Vigorous reactions occur with chloroform in the presence of bases, and explosive peroxides are formed with (hydrogen peroxide + nitric acid) [036]. It is also incompatible with 2-propanol [066].

STABILITY: Methyl ethyl ketone is stable under normal laboratory conditions. Solutions of it in water, DMSO, 95% ethanol or acetone should be stable for 24 hours under normal lab conditions [700].

USES: Methyl ethyl ketone is used as a solvent in nitrocellulose coating and vinyl film manufacture, in smokeless powder manufacture, in cements and adhesives, dewaxing of lubricating oils, "Glyptal" resins, paint removers, organic synthesis, cleaning fluids, acrylic coatings; intermediate in drug manufacture; manufacture of colorless synthetic resins; swelling agent of resins; intermediate in the manufacture of ketones and amines; and printing catalyst and carrier.

COMMENTS: None

NIOSH Registry Number: EL6475000

ACUTE/CHRONIC HAZARDS: Methyl ethyl ketone may be harmful by inhalation, ingestion or skin absorption. It is a severe irritant and when heated to decomposition, it emits toxic fumes of carbon monoxide and carbon dioxide [269].

SYMPTOMS: Symptoms of exposure may include severe irritation of the eyes, irritation of the skin and upper respiratory tract, headache, dizziness and nausea [058,269]. Other symptoms may include weakness, fatigue, possible respiratory arrest, diarrhea, vomiting, stomach and/or intestinal irritation and unconsciousness [058]. It may cause irritation of the nose and throat, diminished vision, mild vertigo and narcosis [430]. It may also cause central nervous system effects and neuropathy [301]. It may irritate the mucous membranes [151]. Prolonged exposure may produce central nervous system depression [430]. Repeated exposure is reported to cause permanent brain and nervous system damage [058]. Eye contact may result in slight conjunctival hyperemia developing to severe anterior uveitis [099]. It may also result in corneal injury. Prolonged skin contact may defat the skin and produce dermatitis [430]. A drop in rabbit eyes has produced moderate reversible injury [099].

Methylhydrazine

NOTE: The form of this name in the Clean Air Act Amendments Section 112(b) list is Methyl hydrazine. The correct notation for the compound is Methylhydrazine.

CAS NUMBER: 60-34-4

SYNONYMS: Hydrazomethane

CHEMICAL FORMULA: CH6N2

MOLECULAR WEIGHT: 46.09

WLN: ZM1

PHYSICAL DESCRIPTION: Clear, colorless liquid

SPECIFIC GRAVITY: 0.874 @ 20/4 C

DENSITY: 0.87 g/mL

MELTING POINT: -20.9 C

BOILING POINT: 87.8 C

SOLUBILITY: Water: Insoluble (<1 mg/mL @ 24 C) [700] DMSO. Soluble (>=10 mg/mL @ 24 C) [700] 95% Ethanol: Soluble (>=10 mg/mL @ 24 C) [700] Acetone: Not available

OTHER SOLVENTS: Hydrazine: Miscible Low molecular wt. monohydric alcohols: Miscible Hydrocarbons: Soluble Ether: Soluble

OTHER PHYSICAL DATA: Ammonia-like odor Heat capacity (25 C): 32.25 cal/mol/degree C Odor threshold: 1-3 ppm Refractive index: 1.4325 @ 20 C

HAP WEIGHTING FACTOR: 1 [713]

VOLATILITY:

Vapor pressure: 49.6 mm Hg @ 25 C

Vapor density: 1.6

FIRE HAZARD: Methylhydrazine has a flash point of 16.7 C (62) and it is flammable. Fires involving this chemical can be controlled with a dry chemical, carbon dioxide, foam or Halon extinguisher. Vapors can travel along surfaces to an ignition source and flash back.

The autoignition temperature of methyl hydrazine is 196 C (385 F).

LEL: 2.5% UEL: 97

%REACTIVITY: Methylhydrazine is a strong reducing agent. It ignites spontaneously on contact with strong oxidizing agents such as fluorine, chlorine trifluoride, nitrogen tetroxide and fuming nitric acid.

STABILITY: Methylhydrazine is hygroscopic. It is reported to be stable if not in contact with copper, iron or their alloys.

USES: Methylhydrazine is used as a rocket fuel and as an intermediate in chemical synthesis.

COMMENTS: None

NIOSH Registry Number: MV5600000

ACUTE/CHRONIC HAZARDS: Methyl hydrazine may be toxic by inhalation and is a severe irritant.

SYMPTOMS: Symptoms of exposure may include tremors, convulsions, irritation of the skin, eyes, mouth, respiratory tract and gastrointestinal tract, respiratory distress, systemic effects, skin and eye burns, hemolytic anemia, vomiting, diarrhea, cyanosis and ataxia.

Methyl iodide

CAS NUMBER: 74-88-4

SYNONYMS: Iodomethane

CHEMICAL FORMULA: CH3I

MOLECULAR WEIGHT: 141.94

WLN: I1

PHYSICAL DESCRIPTION: Colorless, transparent liquid

SPECIFIC GRAVITY: 2.279 @ 20/4 C

DENSITY: Not available

MELTING POINT: -66.45 C

BOILING POINT: 42.4 C

SOLUBILITY: Water: 10-50 mg/mL @ 18 C [700] DMSO: >=100 mg/mL @ 18 C [700] 95% Ethanol: >=100 mg/mL @ 18 C [700] Acetone: >=100 mg/mL @ 18 C [700]

OTHER SOLVENTS: Water: Slightly soluble Alcohol: Miscible Ether: Miscible Benzene: Soluble

OTHER PHYSICAL DATA: Not available

HAP WEIGHTING FACTOR: 1 [713]

VOLATILITY: Not available

FIRE HAZARD: Not available

LEL: Not available UEL: Not available

REACTIVITY: Not available

STABILITY: Methyl iodide is unstable to heat or light; it turns brown on exposure to light.

USES: Methyl iodide is used in methylations; in microscopy because of its high refractive index; as embedding material for examining diatoms; and also in testing for pyridine.

COMMENTS: None.

NIOSH Registry Number: PA9450000

ACUTE/CHRONIC HAZARDS: Methyl iodide is an irritant and may be toxic by inhalation or ingestion.

SYMPTOMS: Symptoms of exposure may include severe narcosis, lung irritation from acute exposure. Prolonged contact with skin can cause vesicant burns.

Methyl isobutyl ketone

CAS NUMBER: 108-10-1

SYNONYMS: Hexone 4-Methyl-2-pentanone Isobutyl methyl ketone Isopropylacetone MIBK

CHEMICAL FORMULA: C6H12O

MOLECULAR WEIGHT: 100.16

WLN: 1Y1&1V1

PHYSICAL DESCRIPTION: Clear, colorless, mobile liquid

SPECIFIC GRAVITY: 0.8017 @ 20/4 C [055,421]

DENSITY: 0.7978 g/mL @ 20 C [017]

MELTING POINT: -84.7 C [017,031,038,421]

BOILING POINT: 117-118 C [031,269,275,346]

SOLUBILITY: Water: 1-5 mg/mL @ 21 C [700] DMSO: >=100 mg/mL @ 21 C [700] 95% Ethanol: >=100 mg/mL @ 21 C [700] Acetone: >=100 mg/mL @ 21 C [700]

OTHER SOLVENTS: Alcohol: Miscible [031,051,295] Chloroform: Soluble [017] Ether: Miscible [031,051,205,295] Benzene: Miscible [031,051,205] Most organic solvents: Miscible [062,421]

OTHER PHYSICAL DATA: Boiling point: 35-40 C @ 16 mm Hg [017] Refractive index: 1.3960 @ 20 C [031,269,275] 1.3969 @ 17 C [029] Odor threshold: 0.10 ppm [430] Faint, ketonic and camphor-like odor [036,051,295] Vapor pressure: 1 mm Hg @ -1.4 C; 20 mm Hg @ 40.8 C; 40 mm Hg @ 52.8 C [038] Vapor pressure: 60 mm Hg @ 60.4 C; 100 mm Hg @ 70.4 C [038] Flash point also reported as 13 C (55.4 F) [205,275] UV max (in water): 400 nm, 380 nm, 360 nm, 340 nm, 335 nm (A = 0.01, 0.02, 0.15, 0.50, 1.0) [275] Heat of combustion: -5800 cal/g [371] Heat of solution: -5 cal/g [371] Critical temperature: 298.3 C [371] Critical pressure: 32.3 atmospheres

[371] Liquid surface tension: 23.6 dynes/cm [371] Liquid-water interfacial tension: 15.7 dynes/cm [371] Latent heat of vaporization: 82.5 cal/g [371] Specific gravity: 0.801 @ 20/4 C [029,031] 0.8020 @ 20/20 C [058] Vapor pressure: 200 mm Hg @ 85.6 C; 400 mm Hg @ 102.0 C [038] Evaporation rate (butyl acetate = 1): 1.6 [058]

HAP WEIGHTING FACTOR: 1 [713]

VOLATILITY:

Vapor pressure: 5 mm Hg @ 19.7 C; 10 mm Hg @ 30.0 C [038]

Vapor density: 3.45 [043,051,055]

FIRE HAZARD: Methyl isobutyl ketone has a flash point of 23 C (73 F) [031,051,062,37 and it is flammable. Fires involving this material can be controlled with a dry chemical, carbon dioxide or Halon extinguisher.

The autoignition temperature for methyl isobutyl ketone is 448 C (854 F) [371,451].

LEL: 1.4% [043,062,371,421] UEL: 7.5% [043,062,371]

REACTIVITY: Methyl isobutyl ketone is incompatible with caustic soda and other strong alkalis, hydrochloric acid, sulfuric acid and other strong inorganic acids, amines and oxidizing agents such as hydrogen peroxide, nitric acid, perchloric acid and chromium trioxide [058]. It reacts violently with potassium tert-butoxide [043,051,066]. It reacts vigorously with reducing materials [043].

STABILITY: Methyl isobutyl ketone is sensitive to air (may form explosive peroxides) [043, 295]. It is also sensitive to heat [043,058]. Solutions of it in water, DMSO, 95% ethanol or acetone should be stable for 24 hours under normal lab conditions [700].

USES: Methyl isobutyl ketone is used as a solvent for paints, varnishes, nitrocellulose, lacquers, fats, oils, waxes, natural and synthetic gums, resins, cellulose esters and other coating systems. It is also used in adhesives, as an alcohol denaturant, in the manufacture of methyl amyl alcohol, in extraction processes including extraction of uranium from fission products and in organic synthesis.

COMMENTS: None

NIOSH Registry Number: SA9275000

ACUTE/CHRONIC HAZARDS: Methyl isobutyl ketone is an irritant and it is narcotic in high concentrations [043]. It may be readily absorbed by the skin [062].

When heated to decomposition it emits toxic fumes of carbon monoxide and carbon dioxide [058, 269].

SYMPTOMS: Symptoms of exposure may include irritation of the skin, eyes and mucous membranes [043]. Other symptoms may include irritation of the nasal passages and throat and mental sluggishness [058]. Irritation of the respiratory tract may occur [036,430]. Gastroenteritis may also occur [151]. Exposure may cause dizziness and unconsciousness [371]. It may also cause weakness, headache, nausea and vomiting [421,430]. Lightheadedness, narcosis and incoordination have been reported [421]. Loss of appetite and diarrhea have also been reported [430]. High concentrations may cause central nervous system depression [151,295]. Prolonged skin contact may cause drying of the skin. Eye injury may also occur [058].

Methyl isocyanate

CAS NUMBER: 624-83-9

SYNONYMS: Isocyanic acid, methyl ester Iso-cyanatomethane Isocyanatomethane

CHEMICAL FORMULA: C2H3NO

MOLECULAR WEIGHT: 57.06

WLN: OCN1

PHYSICAL DESCRIPTION: Colorless, volatile liquid

SPECIFIC GRAVITY: 0.9230 @ 27/4 C

DENSITY: Not available

MELTING POINT: -80 C

BOILING POINT: 37-39 C

SOLUBILITY: Water: Decomposes DMSO: >=100 mg/mL @ 24 C [700] 95% Ethanol: >=100 mg/mL @ 24 C [700] Acetone: >=100 mg/mL @ 24 C [700] Methanol: Soluble

OTHER SOLVENTS: Not available

OTHER PHYSICAL DATA: Refractive index: 1.3419 @ 18 C; 1.3695 @ 20 C Specific gravity: 0.96 @ 20/4 C; 0.9599 @ 20/20 C Powerful odor (causes tears)

HAP WEIGHTING FACTOR: 10 [713]

VOLATILITY:

Vapor pressure: 348 mm Hg @ 20 C

Vapor density: 2.0

FIRE HAZARD: The flash point of methyl isocyanate is <-18 C (<0 F) and it is flammable. Fires involving this material should be controlled using a dry chemical, carbon dioxide or Halon extinguisher.

The autoignition temperature for methyl isocyanate is 535 C (995 F).

LEL: 5.3% UEL: 26

%REACTIVITY: Methyl isocyanate reacts with water; rapid reaction takes place in the presence of acids, alkalis and amines. Contact with iron, tin, copper (or salts of these elements) and with certain other catalysts (such as triphenylarsenic oxide, triethyl phosphine and tributyltin oxide) may cause violent polymerization. It can also polymerize at elevated temperatures. It will attack some plastics, rubber and coatings.

STABILITY: Gas chromatography stability screening indicates that solutions of methyl isocyanate in acetone are stable for at least 24 hours [700]. The neat compound is sensitive to moisture.

USES: Methyl isocyanate is used for the manufacture of synthetic rubbers and adhesives, as an insecticide and herbicide intermediate, and as a reagent for conversion of aldoximes to nitriles.

COMMENTS: None

NIOSH Registry Number: NQ9450000

ACUTE/CHRONIC HAZARDS: Methyl isocyanate is an irritant, a sensitizer, and may be toxic by inhalation and skin absorption. Also, it is extremely volatile and is a lachrymator. When heated to decomposition it emits highly toxic fumes.

SYMPTOMS: Symptoms of exposure may include irritation of the eyes, nose, throat, skin and mucous membranes, pulmonary edema, coughing with secretions, chest pain, dyspnea, asthma, skin and eye injury, pulmonary sensitization, lung injury, shortness of breath, erythema skin sensitization and a change in the olfactory nerve.

Methyl methacrylate

CAS NUMBER: 80-62-6

SYNONYMS: Methacrylic acid, methyl ester 2-Methylacrylic acid methyl ester Methyl methylacrylate 2-Methyl-2-propenoic acid methyl ester

CHEMICAL FORMULA: C5H8O2

MOLECULAR WEIGHT: 100.12

WLN: 1UY1&V01

PHYSICAL DESCRIPTION: Clear colorless liquid

SPECIFIC GRAVITY: 0.9440 @ 20/4 C [016,395]

DENSITY: 0.9433 g/mL @ 20 C [205]

MELTING POINT: -48 C [016,275,395,430]

BOILING POINT: 101.0 C [036,058,371,430]

SOLUBILITY: Water: 1-10 mg/mL @ 17.5 C [700] DMSO: >=100 mg/mL @ 23 C [700] 95% Ethanol: >=100 mg/mL @ 23 C [700] Acetone: >=100 mg/mL @ 23 C [700]

OTHER SOLVENTS: Alcohol: Soluble [016,430] Ether: Soluble [016,430] Ethylene glycol: Sparingly soluble [395] Methyl ethyl ketone: Soluble [031,295] Tetrahydrofuran: Soluble [031] Carbon tetrachloride: Soluble [205] Aromatic and chlorinated hydrocarbons: Soluble [031,295] Esters: Soluble [031,205] Ketones: Soluble [205] Most organic solvents: Miscible [421]

OTHER PHYSICAL DATA: Specific gravity: 0.9380 @ 25/22 C [052] 0.940 @ 25/25 C [062] Boiling point: 24 C @ 32 mm Hg [395] Vapor pressure: 1 mm Hg @ -30.5 C, 5 mm Hg @ -10.0 C, 10 mm Hg @ 1.0 C, 20 mm Hg @ 11.0 C, 60 mm Hg @ 34.5 C, 100 mm Hg @ 47.0 C, 200 mm Hg @ 63.0C, 400 mm Hg @ 82.0 C, 760 mm Hg @ 101.0 C [038] 17 mm Hg @ 7.5 C [051,072] 28 mm Hg @ 20 C, 49 mm Hg @ 30 C [055] Refractive index: 1.4142 @ 20 C [016,395] Sulfuric, sweet, sharp odor [055] Odor threshold: 0.21 ppm [102] Lambda max: 231 nm (E=10) [395] Evaporation rate (butyl acetate=1): 3.1 [058,102] Critical temperature: 294 C [371]

Critical pressure: 33 atmospheres [371] Liquid surface tension: 28 dynes/cm @ 20 C [371] Latent heat of vaporization: 77 cal/g [371] Heat of combustion: -6310 cal/g [371] Viscosity: 0.53 centipoise @ 25 C [058] Heat of polymerization: -138 cal/g [371]

HAP WEIGHTING FACTOR: 1 [713]

VOLATILITY:

Vapor pressure: 40 mm Hg @ 25.5 C [038,043,051,395]

Vapor density: 3.45 [043,055,058,430]

FIRE HAZARD: Methyl methacrylate has a flash point of 10 C (50 F) [036,062,205,451] and it is flammable. Fires involving this material may be controlled with a dry chemical, carbon dioxide or Halon extinguisher.

The autoignition temperature of methyl methacrylate is 421 C (790 F) [036,062,102,371].

LEL: 2.1% [036,062,066,371] UEL: 12.5% [036,066,371,421]

REACTIVITY: Methyl methacrylate is incompatible with nitrates, oxidizers, polymerizers and strong alkalis [102,346]. It polymerizes easily, especially when heated or in the presence of hydrochloric acid [430]. It is incompatible with reducing agents, polymerization catalysts such as peroxides and persulfates, nitric acid, ammonia, amines, halogens and halogen compounds [058]. Methyl methacrylate is also incompatible with benzoyl peroxide [043,451]. A potentially violent reaction may occur with the polymerization initiators azoisobutyronitrile, di-tert-butyl peroxide and propionaldehyde [043]. It is incompatible with acids and bases [269]. It is readily polymerized by ionizing radiation [052, 062].

STABILITY: Methyl methacrylate polymerizes on exposure to light or heat in the presence of oxygen [025,395]. It may form explosive peroxides [058]. It is incompatible with moisture [346]. Inert atmospheres may contribute to hazardous polymerization [058]. Ultraviolet light may cause polymerization [371]. Methyl methacrylate is reactive at high temperatures and pressures [072]. By analogy with a similar compound, solutions of it in DMSO should be stable for at least 24 hours under normal lab conditions [700].

USES: Methyl methacrylate is used in acrylic bone cements used in orthopedic surgery, in the production of acrylic polymers, polymethylmethacrylate and copolymers used in acrylic surface coatings, in the manufacture of emulsion polymers, in the modification of unsaturated polyester resins, in the production of higher methacrylates, acrylic fibers, acrylic film, inks, radiation-polymerized impregnants for wood, solvent-based adhesives and binders and as an impact modifier of polyvinyl chloride. It is also used in medicinal spray adhesives, in nonirritant bandage solvents, in dental technology, as a ceramic filler or cement, to coat corneal contact lenses, in intraocular lenses, as a monomer for polymethacrylate resins and in the impregnation of concrete.

COMMENTS: None

NIOSH Registry Number: OZ5075000

ACUTE/CHRONIC HAZARDS: Methyl methacrylate is a lachrymator [269], an irritant, an allergin, and a sensitizer. When heated to decomposition it emits acrid smoke, irritating fumes and toxic fumes of carbon monoxide and carbon dioxide [043, 058, 102,269].

SYMPTOMS: Symptoms of exposure may include hypotension, dizziness, dyspnea, nausea and vomiting [295]. Other symptoms may include headache, pain in the extremities, excessive fatigue, sleep disturbances, memory loss, irritability and allergic responses on skin contact [395]. It may also cause central nervous system effects, contact dermatitis and sensitization of the skin and oral mucosa [151]. Irritation of the eyes, nose and throat may also occur [102,346,371]. Other symptoms may include irritation of the skin and respiratory tract, respiratory depression and systemic effects [430]. Narcosis, corneal damage, unconsciousness, gastrointestinal disturbances, drowsiness, anorexia and smarting of the eyes may also result [058]. Exposure may also cause difficult breathing and skin and eye burns [371]. Irritation of mucous membranes may also occur [269,421]. Skin rashes may develop as a result of prolonged exposure [102]. Eye damage may also occur [269]. Exposure to this compound may lead to alimentary system irritation [036].

4,4'-Methylenebis(2-chloroaniline)

NOTE: The form of this name in the Clean Air Act Amendments Section 112(b) list is 4,4-Methylene bis(2-chloroaniline). The correct notation for the compound is 4,4'-Methylenebis(2-chloroaniline).

CAS NUMBER: 101-14-4

SYNONYMS: 4,4'-Methylenebis(2-chlorobenzenamine) 3,3'-Dichloro-4,4'-diaminodiphenylmethane 4,4'-Methylenebis(2-chloroaniline)

CHEMICAL FORMULA: $C_{13}H_{12}Cl_2N_2$

MOLECULAR WEIGHT: 267.17

WLN: ZR BG D1R DZ CG

PHYSICAL DESCRIPTION: Tan-colored pellets

SPECIFIC GRAVITY: 1.44 [055,062,430]

DENSITY: 1.44 g/mL @ 4 C [421]

MELTING POINT: 99-107 C [055,062,430]

BOILING POINT: Not available

SOLUBILITY: Water: <1 mg/mL @ 25 C [700] DMSO: >=100 mg/mL @ 23 C [700] 95% Ethanol: >=100 mg/mL @ 23 C [700] Acetone: >=100 mg/mL @ 23 C [700]

OTHER SOLVENTS: Ether: Soluble [031,051,395] Methyl ethyl ketone (hot): Soluble [051,062,421,430] Esters: Soluble [051,062,421,430] Aromatic hydrocarbons: Soluble [051,062,421,430] Dilute acids: Soluble [031] Dimethylformamide: Soluble [421] Most organic solvents: Soluble [051,395] Lipids: Soluble [051,395]

OTHER PHYSICAL DATA: Slight odor

HAP WEIGHTING FACTOR: 1 [713]

VOLATILITY:

Vapor pressure: 0.0013 mm Hg @ 60 C [421]

FIRE HAZARD: Flash point data for 4,4'-methylenebis(2-chloroaniline) are not available. It is probably combustible. Fires involving this material may be controlled with a dry chemical, carbon dioxide or Halon extinguisher.

LEL: Not available UEL: Not available

REACTIVITY: 4,4'-Methylenebis(2-chloroaniline) is a weak base.

STABILITY: When exposed to heat above 200 C, 4,4'-methylenebis(2-chloroaniline) undergoes an exothermic and self-sustaining decomposition reaction. In a closed container, the pressure buildup can be rapid enough to cause an explosion [051]. UV spectrophotometric stability screenings indicate that solutions of it in 95% ethanol are stable for at least 24 hours [700].

USES: 4,4'-Methylenebis(2-chloroaniline) is used in the production of solid elastomeric parts, as a curing agent for epoxy resins and in the manufacture of crosslinked urethane foams used in automobile seats and safety padded dashboards, and in the manufacture of gun mounts, jet engine turbine blades, radar systems, and components in home appliances.

COMMENTS: None

NIOSH Registry Number: CY1050000

ACUTE/CHRONIC HAZARDS: 4,4'-Methylenebis(2-chloroaniline) is an irritant and can be absorbed through the skin [051, 062, 346]. It is a positive animal carcinogen [015,346,395]. When heated to decomposition it emits toxic fumes of hydrogen chloride and nitrogen oxides [042].

SYMPTOMS: Symptoms of exposure may include cyanosis [051,058,421]. Other symptoms may include methemoglobinemia and hematuria [051,421]. It can cause kidney irritation [421]. It can also cause a bluish-gray color of the skin [051]. Mild cystitis has been reported from chronic exposure [346].

Methylene chloride

CAS NUMBER: 75-09-2

SYNONYMS: Methane dichloride Dichloromethane Methylene bichloride

CHEMICAL FORMULA: CH2Cl2

MOLECULAR WEIGHT: 84.93

WLN: G1G

PHYSICAL DESCRIPTION: Clear colorless liquid

SPECIFIC GRAVITY: 1.326 @ 20/4 C [043,051]

DENSITY: 1.3255 g/mL @ 20 C [173,421]

MELTING POINT: -96.7 C [038,205,421,430]

BOILING POINT: 39.8 C @ 760 mm Hg [051,058,102,371]

SOLUBILITY: Water: 10-50 mg/mL @ 21 C [700] DMSO: >=100 mg/mL @ 21 C [700] 95% Ethanol: >=100 mg/mL @ 21 C [700] Acetone: >=100 mg/mL @ 21 C [700]

OTHER SOLVENTS: Alcohol: Miscible [031,173,205,295] Ether: Miscible [031,173,205,295] Dimethylformamide: Miscible [031,173,205] Most organic solvents: Completely miscible [421]

OTHER PHYSICAL DATA: Specific gravity: 1.36174 @ 0/4 C; 1.33479 @ 15/4 C; 1.30777 @ 30/4 C [031] Density: 1.321 g/mL @ 25 C [047] Vapor pressure: 20 mm Hg @ -33.4 C; 40 mm Hg @ -22.3 C [038] Vapor pressure: 60 mm Hg @ -15.7 C; 100 mm Hg @ -6.3 C [038] Vapor pressure: 200 mm Hg @ 8.0 C; 400 mm Hg @ 24.1 C [038] Vapor pressure: 500 mm Hg @ 30 C [055] 349 mm Hg @ 20 C [055,058] Vapor pressure: 380 mm Hg @ 22 C [043,051] 760 mm Hg @ 40.7 C [038] Not unpleasant, sweetish odor above 300 ppm; odor becomes unpleasant ~1000 ppm [430] Odor threshold: 205-307 ppm [051,371] UV max: 400-340 nm, 260 nm, 250 nm, 240 nm, 235 nm [275] Evaporation rate (butyl acetate = 1): 27.5 [102] Evaporation rate (ethyl ether = 1): 0.71 [058] Critical temperature: 245 C [051,371] Critical pressure: 60.9 atm [051,371] Latent heat of vaporization: 78.7 cal/g [051,371] Heat of fusion: 16.89 cal/g [051,371] pH of solutions: neutral [058] Viscosity: 0.430 centipoise @ 20 C [062] Refractive index: 1.4244 @ 20 C [031,052,062]

HAP WEIGHTING FACTOR: 1 [713]

VOLATILITY:

Vapor pressure: 440 mm Hg @ 25 C [301,421,430]

Vapor density: 2.93 [043,055,058,395]

FIRE HAZARD: Methylene chloride is nonflammable under normal conditions [062,173,371,421]. However, it is flammable from 12%-19% in air with high ignition energy [036,043,066]. Methylene chloride is not explosive when mixed with air but may form explosive mixtures in atmospheres with higher oxygen content [043,051,395]. It forms flammable vapor-air mixtures at >=100 C [051,451]. Fires involving this material may be controlled with a dry chemical, carbon dioxide or Halon extinguisher. A water spray may also be used [058].

The autoignition temperature of methylene chloride is 556 C (1033 F) [102, 451].

LEL: 12% [058,102,371,451] UEL: 19% [058,102,371,451]

REACTIVITY: Methylene chloride reacts vigorously with active metals such as lithium, sodium and potassium, and with strong bases such as potassium tert-butoxide [051,395,451]. It is incompatible with strong oxidizers, strong caustics and chemically active metals such as aluminum or magnesium powders [051, 058,102,346]. The liquid will attack some forms of plastic, rubber and coatings [102]. Methylene chloride reacts with sodium-potassium alloy, (potassium hydrogen + N-methyl-N-nitrosourea), nitrogen tetroxide and liquid oxygen. It also reacts with titanium. On contact with water it corrodes iron, some stainless steels, copper and nickel [051]. It is incompatible with alkali metals [058,269]. It is incompatible with amines, zinc and alloys of aluminum, magnesium and zinc [058]. Methylene chloride is liable to explode when mixed with dinitrogen pentoxide or nitric acid [036]. Mixtures of it in air with methanol vapor are flammable [043].

STABILITY: Methylene chloride is sensitive to heat [043,269]. It is also sensitive to exposure to moisture [058]. It is subject to slow hydrolysis which is accelerated by light [051]. Solutions of it in water, DMSO, 95% ethanol or acetone should be stable for 24 hours under normal lab conditions [700].

USES: Methylene chloride is used in rubber adhesives and other rubber solutions, in the pharmaceutical industry, as a paint and varnish remover, in solvent degreasing, in aerosol formulations, in food and drug processing, in the plastic industry, in hairsprays, insecticides and spray paints, as a cosolvent or vapor pressure depressant, as a blowing agent for flexible polyurethane foams, as a cleaning solvent for circuit boards, as a stripper solvent for photoresists, as a solvent for cellulose acetate fiber, in plastic film, in protective coatings, in chemical processing, as a carrier solvent for herbicides and insecticides, to extract heat-sensitive, naturally occurring substances such as cocoa, edible fats, spices and beer hops, for decaffeinating coffee, as a refrigerant, in oil dewaxing, as a dye and perfume intermediate, in the textile industry, as a post-harvest fumigant for strawberries, as a grain fumigant, for degreening citrus fruits, as an industrial solvent, in low temperature extraction, as a solvent for oil, fats, bitumen and esters, in coating photographic films, as a solvent for resins and rubber, as a food additive, in synthetic fibers and leather coatings, as a spotting agent and in organic synthesis.

COMMENTS: None

NIOSH Registry Number: PA8050000

ACUTE/CHRONIC HAZARDS: Methylene chloride may be toxic by all routes [051]. It is an irritant [102,451] and may be readily absorbed through the skin [051,058,151,395]. When heated to decomposition it emits highly toxic fumes of chlorine, hydrogen chloride gas, carbon monoxide, carbon dioxide and phosgene [043,058,102].

SYMPTOMS: Symptoms of exposure may include headache, elevated blood concentrations of carboxyhemoglobin, nausea and irritation of the skin and eyes [058,102,295]. Central nervous system depression, pulmonary edema, hemolysis, chronic intoxication and paresthesia may also occur [295]. Other symptoms may include narcosis, temporary neurobehavioral effects, increase in serum bilirubin, increased urinary formic acid concentrations and increased risk of spontaneous abortion [395]. In addition, intravascular hemolysis, unconsciousness, lack of response to painful stimuli, rapid followed by slowed respiration, erythema, blistering, toxic encephalopathy, painful joints, swelling of the extremities, mental impairment, diabetes, skin rash, aspiration pneumonia, gross hematuria, reduction of blood pH, gastrointestinal injury and narrowing of the intestinal lumen may also occur [301]. Other symptoms may include upper respiratory tract irritation, giddiness, stupor, irritability, numbness, tingling in the limbs and hallucinations [346,421]. A dry, scaly and fissured dermatitis, skin burns, coma and death may also result [346]. Other symptoms may include dizziness, sense of fullness in the head, sense of heat, dullness, lethargy and drunkenness [058,430]. In addition, mental confusion, lightheadedness, vomiting, weakness, somnolence, lassitude, anorexia, depression, fatigue, vertigo, liver damage, nose and throat irritation, anesthetic effects, smarting and reddening of the skin, blood dyscrasias, acceleration of the pulse and congestion in the head may result [051]. Staggering may also occur [102]. Other symptoms of exposure may include neurasthenic disorders, digestive disturbances and acoustical and optical delusions [421]. Arrhythmias produced by catecholamines may also result [406]. Additional symptoms may include edema, faintness, loss of appetite and apathy [173]. Hyporeflexia, gross hemoglobinuria, epiglottal edema, metabolic acidosis, gastrointestinal hemorrhage, ulceration of the duodenojejunal junction and diverticula may also occur [151]. Other symptoms may include kidney damage, lung damage, corneal injury, abdominal pain and an increase in salivary gland tumors [058]. Cyanosis may also occur [036]. Exposure may also cause altered sleep time, convulsions, euphoria and a change in cardiac rate [043].

Methylenediphenyl diisocyanate

NOTE: The form of this name in the Clean Air Act Amendments Section 112(b) list is Methylene diphenyl diisocyanate. The correct notation for the compound is Methylenediphenyl diisocyanate.

CAS NUMBER: 101-68-8

SYNONYMS: 4,4'-Diphenylmethane diisocyanate 4,4'-Methylene diphenyl diisocyanate MDI

CHEMICAL FORMULA: $C_{15}H_{10}N_2O_2$

MOLECULAR WEIGHT: 250.27

WLN: OCNR D1R DNCO

PHYSICAL DESCRIPTION: Yellow crystals or fused solid

SPECIFIC GRAVITY: 1.2

DENSITY: 1.2 g/mL @ 20 C

MELTING POINT: 37.2 C

BOILING POINT: 194-199 C @ 5 mm Hg

SOLUBILITY: Water: Insoluble DMSO: Soluble 95% Ethanol: Reacts violently Acetone: Soluble

OTHER SOLVENTS: Kerosene: Soluble Nitrobenzene: Soluble Benzene: Soluble

OTHER PHYSICAL DATA: Not available

HAP WEIGHTING FACTOR: 10 [713]

VOLATILITY: Not available

FIRE HAZARD: Methylenediphenyl diisocyanate has a flash point of 218 C (425 F). It is combustible. Fires involving this material should be controlled using a dry chemical, carbon dioxide or Halon extinguisher.

LEL: Not available UEL: Not available

REACTIVITY: Methylenediphenyl diisocyanate reacts violently with alcohols.

STABILITY: Methylenediphenyl diisocyanate is stable under normal laboratory conditions.

USES: Methylenediphenyl diisocyanate is used for the preparation of polyurethane resin and spandex fibers, bonding rubber to rayon.

COMMENTS: None

NIOSH Registry Number: NQ9350000

ACUTE/CHRONIC HAZARDS: Methlyenediphenyl diisocyanate is an irritant and may be toxic if ingested. It generates toxic irritating vapors when heated to decomposition.

SYMPTOMS: Symptoms of exposure may include local irritation of the eyes and skin, breathlessness, chest discomfort, reduced pulmonary function, and allergic sensitization in susceptible individuals.

4,4'-Methylenedianiline

CAS NUMBER: 101-77-9

SYNONYMS: 4,4'-Diaminodiphenylmethane bis(p-Aminophenyl)methane bis(4-Aminophenyl)methane 4,4'Methylenebisbenzenamine

CHEMICAL FORMULA: C13H14N2

MOLECULAR WEIGHT: 198.27

WLN: ZR D1R DZ

PHYSICAL DESCRIPTION: Tan flakes

SPECIFIC GRAVITY: Not available

DENSITY: 1.15 g/mL @ 25 C [058]

MELTING POINT: 91.5-92 C [031,346,395,421]

BOILING POINT: 398-399 C @ 768 mm Hg [017,031,047,395]

SOLUBILITY: Water: <1 mg/mL @ 19 C [700] DMSO: >=100 mg/mL @ 19 C [700] 95% Ethanol: >=100 mg/mL @ 19 C [700] Acetone: >=100 mg/mL @ 19 C [700]

OTHER SOLVENTS: Most organic solvents: Soluble [395] Ether: Very soluble [031,062,205,421] Benzene: Very soluble [031,062,205,395] Alcohol: Very soluble [031,062,205,395]

OTHER PHYSICAL DATA: Faint amine odor [043,058,421] Vapor pressure: 0.1 mm Hg @ 152 C [421] 1 mm Hg @ 197 C [395] Boiling point: 257 C @ 18 mm Hg [017,031,047] 249-253 C @ 15 mm Hg [031] Boiling point: 232 C @ 9 mm Hg [025,031]

HAP WEIGHTING FACTOR: 1 [713]

VOLATILITY: Not available

FIRE HAZARD: The flash point of 4,4'-Methylenedianiline is 221 C (430 F) [058,205,269,275]. It is combustible. Fires involving this material may be controlled using a dry chemical, carbon dioxide or Halon extinguisher. A water spray may also be used [058,451,610].

LEL: Not available UEL: Not available

REACTIVITY: 4,4'-Methylenedianiline is incompatible with strong oxidizing agents [058,269]. It is also incompatible with acids. It will catalyze isocyanate-alcohol and epoxy reactions [058].

STABILITY: 4,4'-Methylenedianiline oxidizes in air [395]. It is also light sensitive especially in solution [052,058,395]. Slow air oxidation is catalyzed by light (visibly darkens). 4,4'-Methylenedianiline may also be somewhat hygroscopic [052,058]. It will polymerize if heated above 125 C [058]. Solutions of it in water, DMSO, 95% ethanol or acetone should be stable for 24 hours when protected from air and light [700].

USES: 4,4'-Methylenedianiline compound is used as a curing agent for epoxy resins and urethane elastomers, as a intermediate in the preparation of polyurethanes and Spandex fibers, in the determination of tungsten and sulfates and in the preparation of azo dyes. It is also used as a corrosion inhibitor, cross-linking agent for epoxy resins, in the preparation of isocyanates and polyisocyanates, in the rubber industry, as a curative for neoprene, as a anti-frosting agent (anti-oxidant) in footwear and raw material in preparation of poly(amide-imide) resins (used in magnet wire enamels).

COMMENTS: The technical grade of 4,4'-methylenedianiline may contain 2,4'-Methyleneaniline as an impurity. [055].

NIOSH Registry Number: BY5425000

ACUTE/CHRONIC HAZARDS: 4,4'-Methylenedianiline is an irritant and may be harmful by inhalation, ingestion or skin contact [036,058,269] and it may be absorbed through the skin [058, 269, 346,421]. When heated to decomposition it emits toxic fumes of carbon monoxide, carbon dioxide, nitrogen oxides, hydrogen cyanide and aniline [043,058,269].

SYMPTOMS: Exposure may cause toxic hepatitis [099,275,346,395]. It is also the causative agent in "Epping Jaundice", a condition which includes severe right-upper-quadrant pain, and high fever and chills, with subsequent jaundice [031,421,430]. This condition may also result in portal inflammation, eosinophil infiltration, cholangitis and cholestasis [395]. Other symptoms of exposure may include weakness, abdominal pain, nausea and/or vomiting and anorexia [031,058].

Inhalation and/or ingestion of the dust leads to liver damage [036]. Eye contact may cause burning, itching and tearing [058]. It is painfully irritating to the eyes at 4 ppm [421]. Skin contact can result in reddening and swelling, as well as skin sensitization causing allergic dermatitis with rash, itching, hives and swelling of the arms and legs [058]. Myocardial damage has also been reported [099,395,421]. It is suspect in the development of choroidal melanoma [099]. It may also cause kidney damage [301]. Persons with pre-existing skin disorders or impaired pulmonary function may be more susceptible to the effects of this chemical. Other symptoms may include corneal burns, irritation to the skin, eyes and respiratory tract, blood and spleen damage and conjunctivitis [058].

Mineral fibers (fine)

NOTE: This Clean Air Act category includes mineral fiber emissions from facilities manufacturing or processing glass, rock, or slag fibers (or other mineral derived fibers) of average diameter 1 micrometer or less. [708] No chemical or physical properties are able to be listed as descriptive of this general category.

Naphthalene

CAS NUMBER: 91-20-3

SYNONYMS: Naphthene Naphthaline

CHEMICAL FORMULA: $C_{10}H_8$

MOLECULAR WEIGHT: 128.17

WLN: L66J

PHYSICAL DESCRIPTION: White crystalline solid

SPECIFIC GRAVITY: 1.162 @ 20/4 C [031,205]

DENSITY: 1.162 g/mL @ 20 C [043,421]

MELTING POINT: 80.2 C [031,062,205,371]

BOILING POINT: 218 C [017,102,346,371]

SOLUBILITY: Water: <1 mg/mL @ 22 C [700] DMSO: >=100 mg/mL @ 22 C [700] 95% Ethanol: 10-50 mg/mL @ 22 C [700] Acetone: >=100 mg/mL @ 22 C [700] Methanol: 1 g/13 mL [031,172] Toluene: 1 g/3.5 mL [031,172]

OTHER SOLVENTS: Benzene: 1 g/3.5 mL [031,172] Ether: Very soluble [031,043,173,430] Absolute alcohol: Soluble [043,062] Petroleum ether: Sparingly soluble [025] Chloroform: 1 g/2 mL [031] Hydronaphthalenes: Very soluble [031,043,173] Olive oil: 1 g/8 mL [031] Turpentine: 1 g/8 mL [031] Carbon tetrachloride: 1 g/2 mL [031,172] Carbon disulfide: 1 g/1.2 mL [031] 1,2-Dichloroethane: Very soluble [172,173] Most organic solvents: Soluble [173,421]

OTHER PHYSICAL DATA: Boiling point: 193.2 C @ 400 mm Hg; 145.5 C @ 100 mm Hg [031] Boiling point: 130.2 C @ 60 mm Hg; 101.7 C @ 20 mm Hg; 85.8 C @ 10 mm Hg [031] Vapor pressure: 0.23 mm Hg @ 25 C [051,071] 9.8 mm Hg @ 80 C [301] Refractive index: 1.4003 @ 24 C [017] 1.5821 @ 100 C [205] Refractive index: 1.5898 @ 85 C [017] Sublimes appreciably @ temperatures above the melting point [031,205] Odor of moth balls [031,071,346,430] Odor threshold: 0.003 ppm [051,071] Purple fluorescence in Hg light (petroleum ether solution) [031] UV absorption: Several characteristic bands between 217.5 and 320 nm in hexane [031] Specific gravity: 1.517 @ 15/4 C [172] 0.9628 @ 100/4 C [031,205] Density: 1.15 g/mL @ 25 C [430]

HAP WEIGHTING FACTOR: 1 [713]

VOLATILITY:

Vapor pressure: 0.05 mm Hg @ 20 C [055] 1 mm Hg @ 52.6 C [038,043,071]

Vapor density: 4.42 [043,051,071,430]

FIRE HAZARD: Naphthalene has a flash point of 88 C (190 F) [031,172,371,421]. It is combustible. Fires involving this material may be controlled with a dry chemical, carbon dioxide or Halon extinguisher. A water spray may also be used [043,051,071,371].

The autoignition temperature of naphthalene is 567 C (1053 F) [031, 043,051,071].

LEL: 0.9% [043,102,371,451] UEL: 5.9% [043,102,371,451]

REACTIVITY: Naphthalene is incompatible with oxidizing agents [052,102,269,346]. It reacts violently with dinitrogen pentaoxide [036,043,066]. It also reacts violently with CrO3 and (aluminum chloride + benzoyl chloride) [043]. The molten form of naphthalene (>110 C) in contact with water, may result in violent foaming or the formation of highly reactive mixtures [051,053,371, 430]. When melted, it will attack some forms of plastics, rubber and coatings [053,102].

STABILITY: Naphthalene is sensitive to heat [043]. It volatilizes appreciably at room temperature [031,421]. Solutions of it in water, DMSO, 95% ethanol or acetone should be stable for 24 hours under normal lab conditions [700].

USES: Naphthalene is used in the manufacture of phthalic and anthranilic acids which are used in making indigo, indanthrene and triphenylmethane dyes. It is also used in the manufacture of hydroxyl naphthols, naphthylamines, naphthyl sulfonic acids and similar compounds used in the dye industries. It is used in the manufacture of synthetic resins, celluloid, lampblack, smokeless powder and hydronaphthalenes (tetralin, decalin) which are used as solvents, in lubricants and in motor fuels. Other uses include moth repellents, insecticides, antiseptics (topical and intestinal), anthelmintics and in veterinary medicine in dusting powders, as an insecticide and internally as an intestinal antiseptic and vermicide. It has also been used in fungicides, cutting fluids, synthetic tanning, preservatives, textile chemicals, emulsion breakers, scintillation counters, raw materials and intermediates in the plastics industry, air refreshers and the manufacture of lacquers and varnishes.

COMMENTS: Naphthalene is the most abundant single constituent of coal tar (dry coal tar contains about 11% of it) [031,151,173]. Its use as a moth repellent and insecticide is decreasing due to the introduction of chlorinated compounds [031].

NIOSH Registry Number: QJ0525000

ACUTE/CHRONIC HAZARDS: Naphthalene may be toxic by ingestion or inhalation [031,043,051,052]. It may be harmful if absorbed through the skin [051,071,102,269] and it is an irritant [052,071,269,371]. When heated to decomposition it emits acrid smoke, irritating fumes and toxic fumes of carbon monoxide and carbon dioxide [102,269].

SYMPTOMS: Symptoms of exposure may include irritation of the skin, eyes, mucous membranes and respiratory tract, nausea and headache [051,186,269,346]. Other symptoms may include vomiting, diaphoresis (profuse sweating), hematuria (blood in the urine), fever, hemolytic anemia, hepatic necrosis, convulsions, coma and dermatitis [031,042,051,071]. This dermatitis is characterized by itching, redness, scaling, weeping and crusting of the skin [301]. It can cause erythema, malaise, abdominal pain, irritation of the bladder, confusion, intravascular hemolysis, cataracts and excitement [186,346]. It can also cause allergic skin reactions, tiredness, painful urination, jaundice, diarrhea, conjunctivitis, dark urine and kidney and liver damage [102]. Other symptoms may include urethral pain, pallor, listlessness, anorexia, frequent urination, optic neuritis and azotemia. The liver and/or spleen may be palpable [173]. It can cause visual disturbances and anuria [301]. It can also cause hemoglobinuria, lens opacities and corneal injury [421]. It sometimes causes dizziness [371]. Absorption into the body leads to the formation of methemoglobin, which in sufficient concentration causes cyanosis. Onset may be delayed 2-4 hours or longer [269]. Exposure can result in renal tubular blockage and acute renal shutdown [102,346]. Newborn infants exposed to heavy amounts can develop kernicterus [173].

Nickel compounds

NOTE: For all listings which contain the word "compounds," the following applies: Unless otherwise specified, these listings are defined as including any unique chemical substance that contains the named chemical as part of that chemical's infrastructure. [708] Information for the metallic form of nickel is given below. Other compounds described are intended as selected examples of nickel compounds and do not comprise a comprehensive list.

Nickel

CAS NUMBER: 7440-02-0

CHEMICAL FORMULA: Ni

MOLECULAR WEIGHT: 58.71

WLN: NI

PHYSICAL DESCRIPTION: Silver metal

SPECIFIC GRAVITY: 8.908

DENSITY: 8.9 g/mL @ 25 C

MELTING POINT: 1455 C

BOILING POINT: 2730 C @ 760 mm Hg

SOLUBILITY: Water: Insoluble DMSO: Insoluble 95% Ethanol: Insoluble Acetone: Insoluble

OTHER SOLVENTS: Acids: Soluble Ammonia: Insoluble

OTHER PHYSICAL DATA: Ferromagnetic Malleable Ductile Odorless

VOLATILITY:

Vapor pressure: 1 mm Hg @ 1810 C

FIRE HAZARD: Nickel metal is nonflammable. However nickel dust or powder (such as in nickel catalyst) is a combustible solid which can form explosive mixtures in air. Raney nickel ignites spontaneously in air. Under fire conditions the highly toxic substance, nickel carbonyl may form. NFPA Flammability Code 4. [720]

LEL: Not available UEL: Not available

REACTIVITY: Nickel powder reacts violently with titanium, ammonium nitrate, potassium perchlorate, potassium, performic acid, fluorine, selenium, sulfur and sulfur compounds; hydrazine, ammonia and hydrazoic acid. It is attacked by dilute hydrochloric acid, dilute sulfuric acid and nitric acid. It also reacts with acids, oxidizing agents, aluminum trichloride, ethylene-p-dioxane, hydrogen, methanol, oxidants, wood and other combustibles, magnesium silicate.

STABILITY: Raney nickel ignites spontaneously in air. [720] Nickel metal not in dust or powder form is stable under normal laboratory conditions.

USES: Nickel powder is used for nickel-plating, various alloys, coins, batteries, magnets, lightening rod tips, electrical contacts and electrodes, spark plugs, machinery parts, catalyst for hydrogenation of oils and other organic substances.

COMMENTS: None

NIOSH Registry Number: QR5950000

ACUTE/CHRONIC HAZARDS: This element is a positive animal carcinogen.

SYMPTOMS: Symptoms of exposure to powdered nickel may include wheezing, coughing, shortness of breath or burning in the mouth, throat or chest) may develop.

Nickel cyanide

CAS NUMBER: 557-19-7

CHEMICAL FORMULA: C2N2Ni

MOLECULAR WEIGHT: 110.75

WLN: NI CN2

PHYSICAL DESCRIPTION: Yellow-brown [706] Nickel cyanide, tetrahydrate (NiC2N2H8O4): light-green plates or powder [706]

SPECIFIC GRAVITY: Not available

DENSITY: Not available

MELTING POINT: Nickel cyanide, tetrahydrate: -4H2O, 200 C [706]

BOILING POINT: Nickel cyanide, tetrahydrate: decomposes [706]

SOLUBILITY: Cold Water: Insoluble [706] Hot Water: Insoluble [706] KCN: Soluble [706] DMSO: Not available 95% Ethanol: Not available Acetone: Not available

OTHER PHYSICAL DATA: Solubility, Nickel cyanide, tetrahydrate: KCN: Soluble [706] NH4OH: Soluble [706] Alkali: Soluble [706] Dilute acid: Slightly soluble [706]

VOLATILITY: Not available

FIRE HAZARD: Not available

LEL: Not available UEL: Not available

REACTIVITY: Nickel cyanide reacts violently with magnesium metal.

STABILITY: Not available

USES: It is used for nickel plating.

COMMENTS: None

NIOSH Registry Number: Not available

ACUTE/CHRONIC HAZARDS: Nickel cyanide may be severely toxic.

SYMPTOMS: Not available

Nickelocene

CAS NUMBER: 1271-28-9

SYNONYMS: Nickelocene (8-10 percent in toluene) Nickel, compound with pi-cyclopentadienyl (1:2)

CHEMICAL FORMULA: C10H10.Ni

MOLECULAR WEIGHT: 188.81

WLN: L50J 0-NI-- 0L50J

PHYSICAL DESCRIPTION: Dark brown liquid (8-10% solution)

SPECIFIC GRAVITY: Not available

DENSITY: Not available

MELTING POINT: 171-173 C

BOILING POINT: Not available

SOLUBILITY: Water: Insoluble (<1 mg/mL @ 24 C) [700] DMSO: Insoluble (<1 mg/mL @ 24 C) [700] 95% Ethanol: Decomposes Acetone: Decomposes

OTHER SOLVENTS: Toluene: Soluble Ether: Decomposes Benzene: Soluble

OTHER PHYSICAL DATA: This material is an 8-10% solution in toluene The neat material decomposes rapidly in air

VOLATILITY: Not available

FIRE HAZARD: Nickelocene has a flash point of 4 C (solution) and it is flammable. Fires involving this material can be controlled with a dry chemical, carbon dioxide or Halon extinguisher.

LEL: Not available UEL: Not available

REACTIVITY: Not available

STABILITY: NMR(H) studies indicate that solutions of nickelocene in DMSO are stable for less than 4 hours [700]. Fresh solutions should be prepared prior to each use. The neat material is sensitive to air.

USES: Nickelocene is used as a catalyst and as a complexing agent.

COMMENTS: None

NIOSH Registry Number: QR6500000

ACUTE/CHRONIC HAZARDS: Nickelocene can be absorbed through the skin.

SYMPTOMS: Not available

Nickel oxide

CAS NUMBER: 1313-99-1

SYNONYMS: Nickel monoxide Nickelous oxide Nickel protoxide

CHEMICAL FORMULA: NiO

MOLECULAR WEIGHT: 74.71

WLN: Not available

PHYSICAL DESCRIPTION: Green-black cubic crystals

SPECIFIC GRAVITY: 6.670 [274]

DENSITY: 6.7 g/cm3 [058]

MELTING POINT: 1984 C [017,053]

BOILING POINT: Not available

SOLUBILITY: Water: <1 mg/mL @ 20 C [700] DMSO: <1 mg/mL @ 20 C [700] 95% Ethanol: <1 mg/mL @ 20 C [700] Acetone: <1 mg/mL @ 20 C [700] Methanol: <1 mg/mL @ 17 C [700] Toluene: <1 mg/mL @ 17 C [700]

OTHER SOLVENTS: Acids: Soluble [017,031,042,062] Ammonium hydroxide: Soluble [017,062,430] Potassium cyanide: Soluble [430] Caustic solutions: Insoluble [062] Hot mineral acids: Insoluble [052] Most organic solvents: Practically insoluble [052]

OTHER PHYSICAL DATA: Refractive index: 2.1818 Odorless

VOLATILITY:

Vapor pressure: 0 mm Hg @ 20 C [058]

FIRE HAZARD: Flash point data for nickel oxide are not available. It is probably combustible. Fires involving it may be controlled with a dry chemical, carbon dioxide or Halon extinguisher. A water spray may also be used [058].

LEL: Not available UEL: Not available

REACTIVITY: Nickel oxide reacts violently with iodine, hydrogen sulfide and (BaO + air) [042,053,058,066]. It is incompatible with anilinium perchlorate and hydrogen peroxide. It incandesces in cold fluorine [066].

STABILITY: Nickel oxide may be light-sensitive. It should be thermally stable at temperatures up to 340 C [052].

USES: Nickel oxide is used to prepare other nickel salts, in porcelain painting, in fuel cell electrodes, for production of stainless and alloy steels, in enamel frits and ceramic glazes, and in glass manufacturing.

COMMENTS: Nickel oxide occurs naturally as the mineral bunsenite.

NIOSH Registry Number: QR8400000

ACUTE/CHRONIC HAZARDS: Nickel oxide is an irritant [053,058]. There is evidence that this compound is carcinogenic in animals [015,395]. When heated to decomposition it may emit toxic fumes and metal oxides [058].

SYMPTOMS: Exposure may result in "nickel itch", which includes skin sensitization and itching dermatitis [042,058,099,301,430]. It may cause intestinal disorders [058]. It may also cause irritation to the eyes, skin and upper respiratory tract [058]. It may cause conjunctivitis [053,099]. Other symptoms include asthma, epiphora and pulmonary fibrosis [099,301,421]. Chronic exposure may result in lung and nasal cancer [053,215,395]. It may also cause sinus and laryngeal cancer [215]. This compound has caused convulsions, intestinal disorders and asphyxia in dogs [042].

Nickel subsulfide

CAS NUMBER: 12035-72-2

SYNONYMS: Nickel sulfide (3:2) Nickel tritadisulfphide

CHEMICAL FORMULA: Ni3S2

MOLECULAR WEIGHT: 240.25

WLN: Not available

PHYSICAL DESCRIPTION: Pale yellowish-bronze, metallic, lustrous solid

SPECIFIC GRAVITY: 5.82 [017,053]

DENSITY: Not available

MELTING POINT: 790 C [017,053,395]

BOILING POINT: Not available

SOLUBILITY: Water: <1 mg/mL @ 21.5 C [700] DMSO: <1 mg/mL @ 21.5 C [700] 95% Ethanol: <1 mg/mL @ 21.5 C [700] Acetone: <1 mg/mL @ 21.5 C [700] Methanol: <1 mg/mL @ 17 C [700] Toluene: <1 mg/mL @ 17 C [700]

OTHER SOLVENTS: Nitric acid: Soluble [017,395]

OTHER PHYSICAL DATA: Not available

VOLATILITY: Not available

FIRE HAZARD: Flash point data for nickel subsulfide are not available. It is probably combustible. Fires involving this material may be controlled with a dry chemical, carbon dioxide or Halon extinguisher.

LEL: Not available UEL: Not available

REACTIVITY: Not available

STABILITY: Nickel subsulfide is sensitive to prolonged exposure to air [052]. Solutions of it in water, DMSO, 95% ethanol or acetone should be stable for 24 hours under normal lab conditions [700].

USES: Nickel subsulfide is a major component in the refining of certain nickel ores.

COMMENTS: None

NIOSH Registry Number: QR9800000

ACUTE/CHRONIC HAZARDS: There is evidence that nickel subsulfide is carcinogenic in animals [015,395]. When heated to decomposition it emits toxic fumes of sulfur oxides [042].

SYMPTOMS: Symptoms of exposure may include skin sensitization and dermatitis [301]. This condition is often referred to as "Nickel itch" [042]. Inhalation may lead to asthma, pulmonary fibrosis, pulmonary edema and pneumonitis [053, 421]. It may cause lung or nasal cancer in humans [421]. Ingestion of large doses (1-3 mg/kg) has been shown to cause intestinal disorders, convulsions and asphyxia in dogs [042].

Nickel sulfate

CAS NUMBER: 7786-81-4

SYNONYMS: Sulfuric acid, nickel (2+) salt (1:1) Nickelous sulfate

CHEMICAL FORMULA: NiSO4

MOLECULAR WEIGHT: 154.77

WLN: Not available

PHYSICAL DESCRIPTION: Yellow-green crystalline solid

SPECIFIC GRAVITY: 3.68 @ 20 C [371]

DENSITY: 3.68 g/mL @ 20 C [017,062,205,371]

MELTING POINT: 848 C (decomposes) [051,071,205,395]

BOILING POINT: Decomposes [371]

SOLUBILITY: Water: 27.3 - 27.7 weight % @ 20 C [395] DMSO: Not available 95% Ethanol: Insoluble [017,062] Acetone: Insoluble [017] Methanol: 0.11 weight % @ 35 C [395]

OTHER SOLVENTS: Ether: Insoluble [017,062]

OTHER PHYSICAL DATA: Color in water: Green-blue Odorless

VOLATILITY: Not available

FIRE HAZARD: Nickel sulfate is nonflammable [371].

LEL: Not available UEL: Not available

REACTIVITY: Nickel sulfate may react with strong acids to form flammable/explosive hydrogen gas. Contact with sulfur evolves heat. It may react with other combustible materials [102,346].

STABILITY: Nickel sulfate is stable under normal laboratory conditions.

USES: Nickel sulfate is used in the manufacture of nickel ammonium sulfate; catalysts; plating and as mordant in fabric dyeing and printing. It is also used as a blackening agent for zinc and brass.

COMMENTS: None

NIOSH Registry Number: QR9350000

ACUTE/CHRONIC HAZARDS: When heated to decomposition, nickel sulfate may emit toxic fumes of nickel carbonyl and sulfur oxides [042,102].

SYMPTOMS: Symptoms of exposure may include skin and eye irritation and dermatitis [371]. Ingestion causes nausea, vomiting, diarrhea and giddiness [099, 102, 151,346,430].

Nickel sulfate hexahydrate

CAS NUMBER: 10101-97-0

SYNONYMS: Nickel (II) sulfate hexahydrate Sulfuric acid, nickel (2+) salt, hexahydrate

CHEMICAL FORMULA: NiO4S.6H2O

MOLECULAR WEIGHT: 262.86

WLN: .NI..S-O4 QH6

PHYSICAL DESCRIPTION: Blue or emerald-green crystals

SPECIFIC GRAVITY: 2.070 [017,058,205,275]

DENSITY: 2.031 g/mL [062]

MELTING POINT: Not available

BOILING POINT: Decomposes @ 840 C [058]

SOLUBILITY: Water: >=100 mg/mL @ 20 C [700] DMSO: <1 mg/mL @ 20 C [700] 95% Ethanol: <1 mg/mL @ 20 C [700] Acetone: <1 mg/mL @ 20 C [700] Methanol: 12.5 g/100 cc [017]

OTHER SOLVENTS: Ammonium hydroxide: Very soluble [017,430] Acids: Soluble [058]

OTHER PHYSICAL DATA: Two known phases: alpha-form (blue tetragonal) transition to beta-form (green monoclinic) @ 53.3 C Loses 5 waters @ 100 C [031,058] Loses 6 waters @ 103 C [017,062,102,430] Becomes blue and opaque @ room temperature Refractive index (alpha-form): 1.511 Refractive index (beta-form): 1.487 Odorless Sweet astringent taste Somewhat efflorescent Greenish-yellow anhydrous salt is formed @ 280 C Aqueous solution is acidic (pH ~4.5)

VOLATILITY:

Vapor pressure: Essentially 0 mm Hg @ 20 C [102]

FIRE HAZARD: Nickel sulfate hexahydrate is nonflammable.

LEL: Not available UEL: Not available

REACTIVITY: This type of compound is incompatible with strong acids, sulfur, $Ni(NO_3)_2$, wood and other combustibles [346].

STABILITY: Nickel sulfate hexahydrate is stable under normal laboratory conditions.

USES: Nickel sulfate hexahydrate is used in nickel-plating, as mordant in dyeing and printing textiles, for blackening zinc and brass, for nickel and nickel carbonate catalysts, in coatings, in ceramics and in the manufacture of organic nickel salts (e.g. nickel ammonium sulfate).

COMMENTS: None

NIOSH Registry Number: QR9600000

ACUTE/CHRONIC HAZARDS: Nickel sulfate hexahydrate may be toxic by ingestion and inhalation [036,058,102]. It is an irritant [036,058] and, when heated to decomposition, it emits toxic fumes of sulfur oxides [042,058]

SYMPTOMS: Symptoms of exposure may include irritation of the skin, eyes and respiratory tract; skin sensitization and a dermatitis known as "nickel itch" [058,346]. This begins with a sensation of burning and itching in the hand, followed by erythema and a nodular eruption on the web of the fingers, wrists and forearms. The nodules may become pustules or may ulcerate [430]. In the chronic stage, pigmented or depigmented plaque may be formed [102]. It may also be accompanied by lichenification resembling atopic or neurodermatitis [346]. Other symptoms may include diarrhea, allergic skin rash, giddiness, coughing, sore throat, shortness of breath, irritation of the gastrointestinal tract, abdominal pain, eye damage, nasal or lung damage, and liver and kidney damage [058]. It may cause nausea, vomiting, central nervous system depression and myocardial weakness [058,151]. Other symptoms may include local astringent action, emesis, hyperglycemia, weakness, headache, palpitations and capillary damage (especially in the brain and adrenals) [151]. It may also cause asthma, burning pain in the mouth and throat, tenesmus, retching, hemolysis, hematuria, anuria, jaundice, hypotension, collapse and convulsions [301]. Chronic eczema and fever may occur [430]. Pulmonary fibrosis and pulmonary edema have been reported [421]. Pneumonitis has also been reported [102]. Prolonged exposure to excessive concentrations of dust may cause chronic pulmonary disorders [058]. Chronic repeated application of solutions of this class of compounds to the skin may cause erythematous, papular and granulomatous reactions in susceptible individuals [301]. When added to the feed of sheep this type of compound is said to cause blindness [099]. In animals by subcutaneous or intravenous route it has caused gastroenteritis with nervous symptoms of tremor, chorea-like movements, paralysis and, rarely, convulsions. Death may be due to heart failure [430].

Nitrobenzene

CAS NUMBER: 98-95-3

SYNONYMS: Nitrobenzol

CHEMICAL FORMULA: C6H5NO2

MOLECULAR WEIGHT: 123.12

WLN: WNR

PHYSICAL DESCRIPTION: Colorless or pale yellow to brown, oily liquid

SPECIFIC GRAVITY: 1.203 @ 20/4 C

DENSITY: 1.20 g/mL @ 20 C

MELTING POINT: 5.7 C

BOILING POINT: 210.8 C

SOLUBILITY: Water: Insoluble (<1 mg/mL @ 24 C) [700] DMSO: Soluble (>=10 mg/mL @ 24 C) [700] 95% Ethanol: Soluble (>=10 mg/mL @ 24 C) [700] Acetone: Very soluble

OTHER SOLVENTS: Oils: Very soluble Ether: Very soluble Benzene: Very soluble

OTHER PHYSICAL DATA: Odor of bitter almond oil Refractive index: 1.5529 @ 20 C

HAP WEIGHTING FACTOR: 1 [713]

VOLATILITY:

Vapor pressure: 1 mm Hg @ 44.4 C

Vapor density: 4.25

FIRE HAZARD: The flash point for nitrobenzene is 89 C (188 F). It is combustible. Fires involving this compound should be controlled with a dry chemical, carbon dioxide or Halon extinguisher.

The autoignition temperature of nitrobenzene is 482 C (900 F).

LEL: Not available UEL: Not available

REACTIVITY: Nitrobenzene reacts with reducing agents and, at high temperatures, caustic soda and alkalis. It also may react violently with nitric acid, (aluminum trichloride+phenol), (aniline+ glycerine), dinitrogen tetroxide, silver perchlorate, potassium hydroxide, sodium chlorate and sulfuric acid.

STABILITY: Nitrobenzene is stable under normal laboratory conditions.

USES: Nitrobenzene is used in the manufacture of aniline, benzidine, quinoline, azobenzene, pyroxylin compounds, soaps, shoe and metal polishes It is also used as a solvent for cellulose ester, for modifying esterification of cellulose acetate and for refining lubricating oils.

COMMENTS: Ingestion of alcohol aggravates the toxic effects of nitrobenzene.

NIOSH Registry Number: DA6475000

ACUTE/CHRONIC HAZARDS: Nitrobenzene is rapidly absorbed through the skin and may cause irritation on contact. Vapors from this compound are toxic. Hazardous fumes are evolved when this material is heated to decomposition.

SYMPTOMS: Symptoms of exposure may include irritation of the eyes, skin and mucous membranes, headache, signs of anoxia, including cyanosis of lips, nose, and ear lobes; eye irritation; dermatitis; anemia, nausea, vomiting, cramps, and CNS effects (tremors, twitching). Death may result from the intake of 2 ml of nitrobenzene. Other symptoms may include shock, stupor, coma; initial excitation followed by depression; unsteady gait and vertigo; dyspnea (difficult breathing); respiratory failure; myocardial depression, lowered blood pressure; altered pulse, varied skin temperature and circulatory failure; red blood cell damage, anemia and fatigue; jaundice; dark or brown red urine; black tongue; disturbed vision. Chronic exposure may result in hemolytic anemia, general body fatigue; bladder distress, peripheral neuritis and weight loss.

4-Nitrobiphenyl

SYNONYMS: p-Nitrobiphenyl 4-Nitrodiphenyl 4-phenyl-nitrobenzene

CAS NUMBER: 92-93-3

CHEMICAL FORMULA: C12H9NO2

MOLECULAR WEIGHT: 199.22 [703]

PHYSICAL DESCRIPTION: White to yellow, needle-like crystalline solid with a sweetish odor [704]

SPECIFIC GRAVITY: Not available

MELTING POINT: 113-114 C [703]

BOILING POINT: 340 C [703]

SOLUBILITY: Water: Insoluble [703] Cold alcohol: Slightly soluble [703] Ether: Very soluble [703]

OTHER PHYSICAL DATA: Heat of Combustion: 688.8 kcal/mole [702]

HAP WEIGHTING FACTOR: 1 [713]

VOLATILITY:

Vapor Pressure: Not available

FIRE HAZARD: Flash Point: 290 F [704]

4-Nitrobiphenyl is a combustible solid [704].

Autoignition temperature: Not available

LEL: Not available UEL: Not available

REACTIVITY: Incompatible or reacts with strong reducers [704]. Sampling problem: this compound may form reactions of PAH and NOx/HNO3 in combustion effluent either in the gas phase or on silica or alumina bearing particles. Its occurrence may be the result of reaction on the sampling media. [702]

STABILITY: When heated to decomposition it emits toxic fumes of Nox [703].

USES: Not available

COMMENTS: None

NIOSH Registry Number: DV5600000

ACUTE/CHRONIC HAZARDS: 4-Nitrobiphenyl is a poison by intraperitoneal route. It is moderately toxic by ingestion. It is a human carcinogen as well as an experimental carcinogen, neoplastigen and tumorigen. Mutagenic data exists. When heated to decomposition it emits toxic fumes of Nox. [703] Targeted organs include the bladder and blood. [704]

SYMPTOMS: Symptoms of exposure may include headache, lethargy, and dizziness; dyspnea; ataxia, weakness; methemoglobinemia; urinary burning; acute hemorrhagic cystitis. [704]

4-Nitrophenol

CAS NUMBER: 100-02-7

SYNONYMS: 4-Hydroxynitrobenzene p-Nitrophenol

CHEMICAL FORMULA: C6H6NO3

MOLECULAR WEIGHT: 139.11

WLN: WNR DQ

PHYSICAL DESCRIPTION: Colorless to slightly yellow crystals

SPECIFIC GRAVITY: 1.270 @ 20/4 C

DENSITY: 1.479-1.495 g/mL @ 20 C

MELTING POINT: 113-115 C (sublimes)

BOILING POINT: 279 C (decomposes)

SOLUBILITY: Water: <0.1 mg/mL @ 21 C [700] DMSO: >=100 mg/mL @ 21 C [700] 95% Ethanol: >=100 mg/mL @ 21 C [700] Acetone: >=100 mg/mL @ 21 C [700] Toluene: Soluble

OTHER SOLVENTS: Carbon disulfide: Slightly soluble Pyrimidine: Soluble Chloroform: Very soluble Fixed alkali hydroxides: Soluble Carbonates: Soluble Ether: Very soluble Benzene: Soluble hot; slightly soluble cold

OTHER PHYSICAL DATA: Odorless Sweet then burning taste

HAP WEIGHTING FACTOR: 1 [713]

VOLATILITY:

Vapor pressure: 1 mm Hg @ 20 C; 18.7 mm Hg @ 186 C; 2.2 mm Hg @ 146 C

Vapor density: 1.244 @ 65 C

FIRE HAZARD: The flash point for 4-nitrophenol is 192 C (377 F). It is combustible. Fires involving this compound should be controlled using a dry chemical, carbon dioxide or Halon extinguisher.

LEL: Not available UEL: Not available

REACTIVITY: 4-Nitrophenol is incompatible with oxidizing agents, many organics, combustible substances and reducing agents.

STABILITY: 4-Nitrophenol is stable under normal laboratory conditions. Solutions of it in water, DMSO, 95% ethanol or acetone should be stable for 24 hours under normal lab conditions [700].

USES: 4-Nitrophenol is used as an indicator in 0.1% alcohol solution, as an intermediate in organic synthesis, in the production of parathion, as a fungicide for leather, an indicator in water analysis, and as a bactericide.

COMMENTS: None

NIOSH Registry Number: SM2275000

ACUTE/CHRONIC HAZARDS: 4-Nitrophenol may be toxic by ingestion, inhalation or absorption through the skin. When heated to decomposition it emits toxic fumes. It is also an irritant and is corrosive.

SYMPTOMS: Symptoms following exposure may include irritation of the skin, eyes, nose and throat, headache, loss of consciousness, drowsiness, nausea, cyanosis, liver and kidney damage, methemoglobinemia, central nervous system depression, dyspnea, sweating, dry throat, fever, muscular weakness, fatigue, irritability, abdominal cramps, dermatitis, corneal damage and hypothermia.

2-Nitropropane

CAS NUMBER: 79-46-9

SYNONYMS: Dimethylnitromethanc Isonitropropane Nitroisopropane

CHEMICAL FORMULA: C3H7NO2

MOLECULAR WEIGHT: 89.11

WLN: WNY1&1

PHYSICAL DESCRIPTION: Colorless liquid

SPECIFIC GRAVITY: 0.9876 @ 20/4 C

DENSITY: 0.99 g/mL

MELTING POINT: -93 C

BOILING POINT: 120.3 C @ 760 mm Hg

SOLUBILITY: Water: >=0.1 mg/mL @ 19 C [700] DMSO: Soluble 95% Ethanol: Soluble Acetone: Soluble

OTHER SOLVENTS: Ether: Soluble Benzene: Soluble Chloroform: Very soluble Most organic solvents: Very soluble

OTHER PHYSICAL DATA: Heat of formation: -43.78 kcal/mole @ 25 C Heat of combustion: 477.60 kcal/mole @ 25 C Latent heat of vaporization: 9.88 kcal/mole @ 25 C Refractive index: 1.3941 @ 20 C

HAP WEIGHTING FACTOR: 1 [713]

VOLATILITY:

Vapor pressure: 12.9 mm Hg @ 20 C

Vapor density: 3.06

FIRE HAZARD: 2-Nitropropane has a flash point of 24 C (75 F) and it is flammable. Fires involving this material can be controlled with a dry chemical, carbon dioxide or Halon extinguisher.

The autoignition temperature of 2-nitropropane is 428 C (802 F).

LEL: 1.6% UEL: 11.0

%REACTIVITY: 2-Nitropropane can react with amines, strong acids, strong alkalis, and chlorosulfonic acid.

STABILITY: 2-Nitropropane is sensitive to heat.

USES: 2-Nitropropane is used as a solvent for vinyl and epoxy coatings, cellulose acetate, lacquers, synthetic rubbers, fats, oils, dyes, and other organic materials, in chemical synthesis, as a rocket propellant, and as a gasoline additive.

COMMENTS: None

NIOSH Registry Number: TZ5250000

ACUTE/CHRONIC HAZARDS: 2-Nitropropane is an irritant.

SYMPTOMS: Symptoms of exposure may include irritation of the skin, eyes, mucous membranes and upper respiratory tract; headaches, dizziness, nausea, diarrhea, vomiting, depressed appetite, hepatocellular carcinoma, liver and kidney damage, methemoglobinemia, cyanosis and pulmonary edema.

N-Nitroso-N-methylurea

CAS NUMBER: 684-93-5

SYNONYMS: Methyl nitrosourea N-Nitroso-N-methylcarbamide

CHEMICAL FORMULA: C2H5N3O2

MOLECULAR WEIGHT: 103.08

WLN: ZVN1&NO

PHYSICAL DESCRIPTION: Pale yellow crystals

SPECIFIC GRAVITY: <1.0 [058]

DENSITY: Not available

MELTING POINT: 124 C (decomposes) [025,051,071,430]

BOILING POINT: Not available

SOLUBILITY: Water: <1 mg/mL @ 18 C [700] DMSO: <1 mg/mL @ 18 C [700] 95% Ethanol: <1 mg/mL @ 18 C [700] Acctone: <1 mg/mL @ 18 C [700] Methanol: 10-50 mg/mL @ 18 C [700] Toluene: <1 mg/mL @ 18 C [700]

OTHER SOLVENTS: Benzene: Soluble [017,047] Ether: Soluble [017,047] Polar organic solvents: Soluble [395,430] Non-polar organic solvents: Insoluble [395] Chloroform: Soluble [017]

OTHER PHYSICAL DATA: Lambda max: 231 nm (epsilon=571) in water [395] Half-life @ pH 9.0 is <2 minutes [051,071] Air/water distribution coefficient is very low [071]

HAP WEIGHTING FACTOR: 1 [713]

VOLATILITY:

Vapor pressure: Non-volatile [051,071]

FIRE HAZARD: Flash point data for N-nitroso-N-methylurea are not available. It is probably combustible. Fires involving this material may be controlled with a dry chemical, carbon dioxide or Halon extinguisher. A water spray may also be used [058].

LEL: Not available UEL: Not available

REACTIVITY: N-Nitroso-N-methylurea is highly reactive [051,071,395]. It is incompatible with strong acids and bases [058]. It is also incompatible with water and nucleophilic reagents. Alkaline hydrolysis produces a highly toxic, irritating and explosive gas [051,071,430]. It can detonate with (potassium hydroxide + methylene chloride) [043].

STABILITY: N-Nitroso-N-methylurea slowly decomposes in water [058]. The pure compound is sensitive to humidity and light [395,430]. It can suddenly decompose after storage at ambient temperatures or slightly above (30 C) [066]. It decomposes in alkaline solutions. The stability in aqueous solutions is pH dependent (at 20 C) [395]. Maximum stability occurs at about pH 4 [051,071]. UV spectrophotometric stability screenings indicate that N-nitroso-N-methylurea is stable in 95% ethanol for at least 2 hours and in water for at least 6 hours [700].

USES: No evidence was found that N-nitroso-N-methylurea has been produced in significant commercial quantities. It is available in small quantities for research purposes. Its primary use is for tumor induction and related research in experimental test systems, including tests for mutagenesis. It has been used in clinical trials as an anti-tumor agent. Formerly, this compound was used to produce diazomethane for laboratory use. Its use as a polymerization agent is patented.

COMMENTS: This material is shipped wet with weak acetic acid. It should not be stored dry [058].

NIOSH Registry Number: YT7875000

ACUTE/CHRONIC HAZARDS: N-Nitroso-N-methylurea is and irritant [430] and may be toxic by skin contact or ingestion [051,071]. When heated to decomposition it emits toxic fumes of nitrogen oxides [043,058]. Decomposition products may be explosive [051,071].

SYMPTOMS: Exposure may cause nausea, vomiting, epigastric pain, diarrhea and leukopenia [051,071]. It may also cause skin rashes [051,058, 071]. Symptoms of exposure in laboratory animals may include depression of the hematopoietic system, degeneration of the testes and ovaries, severe damage to hematopoietic, lymphoid and other tissues that have rapid cell turnover rates; pancreatic damage and diabetes [051,071]. Eye contact may cause damage to the retina, degeneration of pigment epithelium and destruction of the rods and outer nuclear layer leading to thinning of the whole retina and migration of pigment [099]. Retinal atrophy and cataracts have also been reported [051, 071,099] and it may cause acetonuria [107].

N-Nitrosodimethylamine

CAS NUMBER: 62-75-9

SYNONYMS: Dimethylnitrosamine N,N-Dimethylnitrosamine Nitrous dimethylamide

CHEMICAL FORMULA: C2H6N2O

MOLECULAR WEIGHT: 74.08

WLN: ONN1&1

PHYSICAL DESCRIPTION: Yellow liquid

SPECIFIC GRAVITY: 1.0059 @ 20/4 C

DENSITY: 1.01 g/mL

MELTING POINT: Not available

BOILING POINT: 151 C

SOLUBILITY: Water: >=100 mg/mL @ 19 C [700] DMSO: >=100 mg/mL @ 19 C [700] 95% Ethanol: >=100 mg/mL @ 19 C [700] Acetone: >=100 mg/mL @ 19 C [700]

OTHER SOLVENTS: Ether: Soluble Benzene: Soluble

OTHER PHYSICAL DATA: Refractive index @ 20 C: 1.4368 Lipophilic

HAP WEIGHTING FACTOR: 1 [713]

VOLATILITY:

Vapor pressure: 40 mm Hg @ 67.1 C [051]

Vapor density: 2.56 [055]

FIRE HAZARD: N-Nitrosodimethylamine has a flash point of 61 C (142 F) [205,269,275]. It is combustible. Fires involving this material may be controlled with a dry chemical, carbon dioxide or Halon extinguisher.

LEL: Not available UEL: Not available

REACTIVITY: N-Nitrosodimethylamine is incompatible with strong oxidizers [107,269,395]. It is also incompatible with strong bases [107,269]. It can be reduced by reducing agents. It is incompatible with hydrogen bromide in acetic acid. It is also photochemically reactive [395].

STABILITY: N-Nitrosodimethylamine is sensitive to exposure to light, especially ultraviolet light. It is stable at room temperature for more than 14 days in aqueous solution at neutral and alkaline pH in the absence of light. It is slightly less stable at strongly acid pH at room temperature [395]. Solutions of it in water, DMSO, 95% ethanol or acetone should be stable for 24 hours under normal lab conditions [700].

USES: N-Nitrosodimethylamine is used in rocket fuels, as a solvent, as a rubber accelerator, in the manufacture of dimethylhydrazine, as a nematocide, as a gasoline and lubricant additive, as an antioxidant, as a pesticide, in the inhibition of nitrification in soil, as a plasticizer for acrylonitrile polymers, in active metal anode electrolyte systems (high energy batteries), in the preparation of thiocarbonyl fluoride polymers and as an additive in condensers to increase the dielectric constant.

COMMENTS: N-Nitrosodimethylamine is emitted during the compounding, forming and curing operations of elastomeric parts by reaction of accelerators/stabilizers used such as tetramethylthiuram disulfide, tetramethylthiuram monosulfide and tetraethylthiuram disulfide [055]. N-Nitrosodimethylamine is often present in fish meal [062]. It may be found in cutting oils [346].

NIOSH Registry Number: IQ0525000

ACUTE/CHRONIC HAZARDS: N-Nitrosodimethylamine may be harmful by eye or skin contact, inhalation or ingestion, and it is an irritant. When heated to decomposition this compound emits toxic fumes of nitrogen oxides.

SYMPTOMS: Symptoms following exposure may include headache, fever, weakness, nausea, liver damage, reproductive effects, vomiting, abdominal pain, diarrhea, gastrointestinal hemorrhage, hepatomegaly, jaundice and ascites.

N-Nitrosomorpholine

CAS NUMBER: 59-89-2

SYNONYMS: 4-Nitrosomorpholine

CHEMICAL FORMULA: C4H8N2O2

MOLECULAR WEIGHT: 116.14

WLN: T6N DOTJ ANO

PHYSICAL DESCRIPTION: Yellow crystals

SPECIFIC GRAVITY: Not available

DENSITY: Not available

MELTING POINT: 29 C [031,107,395]

BOILING POINT: 224-224.5 C @ 747 mm Hg [031,395]

SOLUBILITY: Water: >=100 mg/mL @ 19 C [700] DMSO: >=100 mg/mL @ 19 C [700] 95% Ethanol: >=100 mg/mL @ 19 C [700] Acetone: >=100 mg/mL @ 19 C [700]

OTHER SOLVENTS: Most organic solvents: Soluble [395]

OTHER PHYSICAL DATA: Boiling point: 96 C @ 6 mm Hg [395] 139-140 C @ 25 mm Hg [031] Lambda max (water): 237 nm, 346 nm (epsilon = 685, 7.3) [395]

HAP WEIGHTING FACTOR: 1 [713]

VOLATILITY:

Vapor pressure: 25 mm Hg @ 139-140 C [031]

FIRE HAZARD: Flash point data for N-Nitrosomorpholine are not available. It is probably combustible. Fires involving this material may be controlled with a dry chemical, carbon dioxide or Halon extinguisher.

LEL: Not available UEL: Not available

REACTIVITY: N-Nitrosomorpholine is incompatible with strong oxidizers [107,395].

STABILITY: N-Nitrosomorpholine is sensitive to exposure to light, especially ultraviolet light [107,395]. It is stable for more than 14 days in neutral and alkaline solutions at room temperature in the dark, and slightly less stable in acid solutions [395]. Solutions of it in water, DMSO, 95% ethanol or acetone should be stable for 24 hours under normal lab conditions [700].

USES: N-Nitrosomorpholine is used as a solvent for polyacrylonitrile and as a chemical intermediate in the synthesis of N-aminomorpholine. It is effective against microbial infections.

COMMENTS: None

NIOSH Registry Number: QE7525000

ACUTE/CHRONIC HAZARDS: When heated to decomposition N-nitrosomorpholine emits toxic fumes of nitrogen oxides [043].

SYMPTOMS: Not available

Parathion

CAS NUMBER: 56-38-2

SYNONYMS: Diethyl p-nitrophenyl thiophosphate O,O-Diethyl O-(p-nitrophenyl) phosphorothioate

CHEMICAL FORMULA: C10H14NO5PS

MOLECULAR WEIGHT: 291.27

WLN: WNR DOPS&O2&O2

PHYSICAL DESCRIPTION: Pale yellow to dark brown liquid

SPECIFIC GRAVITY: 1.265 @ 20/4 C [169,172,395,430]

DENSITY: 1.270 g/cm3 @ 18.5 C [700]

MELTING POINT: 6 C [027,031,205,371]

BOILING POINT: 375 C [017,031,205,346]

SOLUBILITY: Water: <1 mg/mL @ 23 C [700] DMSO: >=100 mg/mL @ 22 C [700] 95% Ethanol: >=100 mg/mL @ 22 C [700] Acetone: >=100 mg/mL @ 22 C [700]

OTHER SOLVENTS: Chloroform: Soluble [017,047,169] Alcohol: Soluble [017,031,043,169] Esters: Soluble [031,043,173] Ketones: Soluble [031,043,062,173] Petroleum ether: Insoluble [027,031,173,205] Kerosene: Insoluble [031,043,151,173] Ether: Soluble [017,031,047,205] Benzene: Soluble [169,205] Petroleum oils: Slightly soluble [172,395] Usual spray oils: Insoluble [031,062,173] Most organic solvents: Miscible [058,172] Animal and vegetable oils: Soluble [062] Aromatic hydrocarbons: Soluble [031,043,062,173]

OTHER PHYSICAL DATA: Specific gravity: 1.2704 @ 20/20 C [017,047] 1.266 @ 25/15.6 C [058] Boiling point: 157-162 C @ 0.6 mm Hg [027,031,055,395] Vapor pressure: 0.0000965 mm Hg @ 25 C [058] Refractive index: 1.5420 @ 20 C [027] 1.5370 @ 25 C [017,031,169,205] Surface tension: 39.2 dynes/cm @ 25 C [031] Garlic-like odor [058,102,421] Viscosity: 15.30 centipoise @ 25 C [031] Heat of combustion: -5140 cal/g [371] log P octanol: 3.81 @ 20 C [055]

HAP WEIGHTING FACTOR: 1 [713]

VOLATILITY: Not available

FIRE HAZARD: Parathion has a flash point of >93.3 C (>200 F) [058]. It is combustible [043]. Fires involving this material may be controlled with a dry chemical, carbon dioxide or Halon extinguisher. A water spray may also be used [058].

LEL: Not available UEL: Not available

REACTIVITY: Parathion is incompatible with strong oxidizers [102,346]. It reacts slowly with water and rapidly with bases or caustic solutions [058]. It is incompatible with substances having a pH higher than 7.5 [033]. It will attack some forms of plastics, rubber and coatings [102].

STABILITY: Parathion slowly decomposes in air [062]. It decomposes when exposed to temperatures >100 C [058,102]. It is rapidly hydrolyzed under alkaline conditions [027,151,169,172]. It is stable at a pH below 7.5 [173]. This chemical is stable in distilled water and acid solutions [062,395].

USES: Parathion is a broad spectrum insecticide used in agricultural applications. It is also used as an acaricide, a fumigant and a nematocide.

COMMENTS: Parathion is among the most poisonous materials commonly used for pest control. It is responsible for hundreds of human poisonings, many of which are fatal [395].

NIOSH Registry Number: TF4550000

ACUTE/CHRONIC HAZARDS: Parathion can be fatal by ingestion, inhalation and skin or eye contact [058]. It is a cholinesterase inhibitor [058,102,151,295] and it can be absorbed through the skin [027,052,058,173]. When heated to decomposition it emits toxic fumes of hydrogen sulfide, carbon monoxide, nitrogen oxides, phosphorus oxides and sulfur oxides [043,058,102].

SYMPTOMS: Symptoms of exposure may include nausea, vomiting, diarrhea, excessive salivation, muscle twitching, convulsions, respiratory failure, coma and cholinesterase inhibition [033,058,102,295]. Anorexia, pinpoint pupils and bronchoconstriction may occur [033]. Other symptoms may include tightness of the chest, wheezing, blurred vision, tearing, runny nose, headache, abdominal cramps, dizziness, staggering, slurred speech and irregular or slow heartbeat [058,102,151]. It may cause cyanosis (bluish discoloration of the skin), loss of appetite, sweating and confusion [058,102]. It may also cause aching in and behind the eyes, weakness, loss of muscle coordination and incontinence [151]. Pulmonary edema may also occur [371]. Exposure may result in labored breathing, nervousness, drooling or frothing of the mouth and nose, excessive bronchial secretion, possible muscle paralysis (on exposure to large amounts), loss of reflexes and death [058]. It may also result in general anesthesia and effects of the pulmonary system, kidney, ureter and bladder [043]. This chemical may cause reduced vision, narrowing of peripheral visual fields, congestion or atrophy of optic nerves, difficulty with ocular pursuit movements, abnormality of ERG, miosis and spasm of accommodation for near vision, wide and unreactive pupils, and scotoma [099].

Pentachloronitrobenzene

CAS NUMBER: 82-68-8

SYNONYMS: Nitropentachlorobenzene Quintocene Quintozene Quintobenzene

CHEMICAL FORMULA: C6Cl5NO2

MOLECULAR WEIGHT: 295.34

WLN: WNR BG CG DG EG FG

PHYSICAL DESCRIPTION: Crystalline pale yellow to white solid

SPECIFIC GRAVITY: 1.718 @ 25/4 C [017,031,062,205]

DENSITY: 1.718 g/mL @ 25 C [051,172]

MELTING POINT: 146 C [043,172,346,395]

BOILING POINT: 328 C @ 760 mm Hg (decomposes) [031,051,062,169]

SOLUBILITY: Water: <1 mg/mL @ 22 C [700] DMSO: 50-100 mg/mL @ 20 C [700] 95% Ethanol: 10-50 mg/mL @ 20 C [700] Acetone: >=100 mg/mL @ 20 C [700]

OTHER SOLVENTS: Alcohol: Practically insoluble cold [031] Benzene: Freely soluble [031,051,395] Chloroform: Freely soluble [031,051,395] Carbon disulfide: Freely soluble [031,051,169,395] Ethanol: ~2.0 g/kg @ 25 C [169,172,173] Aromatic solvents: Readily soluble [051,169] Ketones: Readily soluble [051,169] Chlorinated hydrocarbons: Readily soluble [051,169] Alkanols: Slightly soluble [051]

OTHER PHYSICAL DATA: Melting point (technical grade): 142-145 C [058,062] Vapor pressure: 0.00001161 mm Hg @ 10 C; 0.00005 mm Hg @ 20 C [051] pH (saturated solution): 2.5-4.0 [058] Musty odor [051,062] Odor is similar to moth balls [058]

HAP WEIGHTING FACTOR: 1 [713]

VOLATILITY:

Vapor pressure: 0.013 mm Hg @ 25 C [043,173,395]

Vapor density: 10.2 [051]

FIRE HAZARD: Flash point data for pentachloronitrobenzene are not available. It is probably combustible. Fires involving this material may be controlled with a dry chemical, carbon dioxide or Halon extinguisher. A water spray may also be used [058,269].

LEL: Not available UEL: Not available

REACTIVITY: Pentachloronitrobenzene is incompatible with strong oxidizers [058,269]. It is also incompatible with strong bases [269]. It is corrosive to unlined metal containers [058].

STABILITY: Pentachloronitrobenzene is hydrolyzed by alkalis [169]. In one study, ultraviolet radiation has resulted in decomposition [395]. Solutions of it in water, DMSO, 95% ethanol or acetone should be stable for 24 hours under normal lab conditions [700].

USES: Pentachloronitrobenzene is used as a fungicide for seed and soil treatment. It is also used as an herbicide, in slime prevention in industrial waters and to control damping off and other fungal infections.

COMMENTS: None

NIOSH Registry Number: DA6650000

ACUTE/CHRONIC HAZARDS: Pentachloronitrobenzene may be harmful if swallowed, inhaled or absorbed through the skin. It is an irritant [269] and when heated to decomposition it emits highly toxic fumes of chlorine, carbon monoxide, carbon dioxide, nitrogen oxides, hydrogen chloride gas and phosgene [043,058,269].

SYMPTOMS: Symptoms of exposure may include irritation of the skin and eyes [269]. Skin contact may result in erythema, itching, edema and formation of small vesicles [051,173]. Skin sensitization may also occur [151]. Eye contact may result in conjunctivitis and corneal injury [173]. Kidney and liver damage may occur [301]. Vomiting may also occur [058].

Exposure to this type of compound can cause central nervous system stimulation, vomiting, diarrhea, paresthesia, excitement, giddiness, fatigue, tremors, convulsions, coma, pulmonary edema, hypothermia and liver, kidney and myocardial toxicity. Respiration may be initially accelerated and then later depressed. Chronic exposure to this type of compound leads to headache, loss of appetite, muscular weakness, fine tremors and apprehensive mental state [295].

Pentachlorophenol

CAS NUMBER: 87-86-5

CHEMICAL FORMULA: $C6Cl5OH$

MOLECULAR WEIGHT: 266.34

WLN: QR BG CG DG EG FG

PHYSICAL DESCRIPTION: White to light brown crystalline solid

SPECIFIC GRAVITY: 1.978 @ 22/4 C [017,031,062,205]

DENSITY: Not available

MELTING POINT: 191 C [043,169,173,295]

BOILING POINT: 309-310 C (with decomposition) [031,169,421]

SOLUBILITY: Water: <1 mg/mL @ 20 C [700] DMSO: 50-100 mg/mL @ 20 C [700] 95% Ethanol: >=100 mg/mL @ 20 C [700] Acetone: >=100 mg/mL @ 20 C [700] Methanol: Soluble [051,421]

OTHER SOLVENTS: Benzene: Soluble [017,031,062,395] Dilute alkali: Soluble [062] Paraffins: Slightly soluble [169,172] Carbon tetrachloride: Slightly soluble [169,172] Cold petroleum ether: Slightly soluble [031,043,205,395] Alkanes: Slightly soluble [421] Most organic solvents: Soluble [169,172] Alcohol: Very soluble [031,043,205,295] Sodium hydroxide solution: Freely soluble [295] Ether: Soluble [017,043,295,395] Paraffinic petroleum oils: Moderately soluble [173]

OTHER PHYSICAL DATA: Weak acid (pKa 4.71) [172] pKa: 8.2 (in methanol) [025] Melting point (hydrous): 174 C [017,025,047,395] Phenolic (carbolic) odor [036] Pungent taste [421] Molar refraction: 53.5 [051] Log octanol/water partition coefficient: 5.01 [051,055] Odor threshold, medium: 0.857 ppm [051] Taste threshold, upper: 0.857 ppm [051] Vapor pressure: 0.12 mm Hg @ 100 C [051] 20 mm Hg @ 192.2 C [038] Boiling point also given as 293.08 C [051] Sublimes in needles [051] Lambda maximum (in methanol): 326 nm (shoulder), 304 nm, 293 nm (shoulder), 254 nm (shoulder); epsilon 0.0651, 0.2831, 0.2181, 0.4416 [052]

HAP WEIGHTING FACTOR: 1 [713]

VOLATILITY:

Vapor pressure: 40 mm Hg @ 211.2 C [043,058]

Vapor density: 9.2 [055,058]

FIRE HAZARD: Pentachlorophenol is nonflammable [058,172, 371,421].

LEL: Not available UEL: Not available

REACTIVITY: Pentachlorophenol may react with strong oxidizing agents [058,102,269,346]. It is also incompatible with strong bases, acid chlorides and acid anhydrides [269]. It forms salts with alkaline metals [395]. Solutions of pentachlorophenol in oil cause natural rubber to deteriorate, but synthetic rubber may be used in equipment and for protective clothing [172].

STABILITY: Pentachlorophenol is stable under normal laboratory conditions [058,169,395]. Irradiation of a dilute aqueous solution (100 ppm) by sunlight or ultraviolet light results in the formation of photodegradation products [430]. Solutions of it in DMSO, 95% ethanol or acetone should be stable for 24 hours under normal lab conditions [700].

USES: Pentachlorophenol is used an insecticide for termite control, pre-harvest defoliant, general herbicide, wood preservative, synthesis of pentachlorophenyl esters, molluscide, fungicide, bactericide, antimildew agent, slimicide and algicide. The technical material finds extensive use in cooling towers of electric plants, as additives to adhesives based on starch and vegetable and animal protein, in shingles, roof tiles, brick walls, concrete blocks, insulation, pipe sealant compounds, photographic solutions, and textiles and in drilling mud in the petroleum industry.

COMMENTS: None

NIOSH Registry Number: SM6300000

ACUTE/CHRONIC HAZARDS: Pentachlorophenol may be fatal if swallowed, inhaled or absorbed through the skin [036,051]. It is an irritant and may be readily absorbed through the skin [058,102,421]. When heated to decomposition it may emit toxic fumes of carbon monoxide, carbon dioxide [269], hydrogen chloride and chlorinated hydrocarbons [451].

SYMPTOMS: Exposure may cause muscular weakness and unconsciousness [036]. It may cause irritation of the throat, sneezing, coughing, anorexia, weight loss, profuse sweating, headaches, dizziness, nausea, vomiting, dyspnea, and chest pains [346]. It may also cause loss of appetite, shortness of breath and irritation of the respiratory tract. Risk of serious or fatal intoxication is increased in hot weather [058]. Other symptoms may include abdominal pain, intense thirst and death [051]. Ingestion causes an increase then a decrease in respiration, blood pressure and urinary output; fever, increased bowel action, motor weakness, collapse with convulsions and lung, kidney and liver damage [031]. It may cause skin and conjunctiva irritation, exfoliation of the epidermal layer, painful and red skin, impairment of the autonomic function and circulation, visual damage, including an arcuate type of scotoma, dehydration, coma, terminal spasm and chloracne. Children are more susceptible than adults. Symptoms in children may include intermittent delirium, rigors, flushing and excitement. Babies nurse avidly. Tachycardia, hepatomegaly, progressive metabolic acidosis, proteinuria, azotemia, irritability followed by lethargy, pneumonia or bronchitis and aplastic anemia may occur [173]. It may also cause acneform dermatitis, allergic skin response, eye inflammation, permanent corneal injury and irritation to the nose and pharynx [151]. Other symptoms may include inflammation of the conjunctiva, characteristically shaped corneal opacity, corneal numbness, slight mydriasis, tachypnea and hepatic enlargement [421]. It may cause irritation of the eyes, burns to the skin and eyes, irritation of the mucous membranes and anesthesia [371]. Other symptoms may include systemic intoxication, hyperpyrexia, gastrointestinal upset, congestion of the lungs, edema in the brain, cardiac dilatation, centrilobular degeneration in the liver and mild degeneration of the renal tubules [430]. It may also cause profuse diaphoresis and Hodgkin's disease [395].

Phenol

CAS NUMBER: 108-95-2

SYNONYMS: Benzenol Carbolic acid Hydroxybenzene Phenyl hydrate Phenyl hydroxide

CHEMICAL FORMULA: C6H6O

MOLECULAR WEIGHT: 94.11

WLN: QR

PHYSICAL DESCRIPTION: White crystalline solid

SPECIFIC GRAVITY: 1.0576 @ 41/4 C [205]

DENSITY: Not available

MELTING POINT: 43 C [017,025,036,421]

BOILING POINT: 182 C [062,102,421,430]

SOLUBILITY: Water: 50-100 mg/mL @ 19 C [700] DMSO: >=100 mg/mL @ 19 C [700] 95% Ethanol: >=100 mg/mL @ 19 C [700] Acetone: >=100 mg/mL @ 19 C [700] Toluene: Soluble [430]

OTHER SOLVENTS: Cyclohexane: >=100 mg/mL @ 21 C [700] Glycerol: Very soluble [031,421,430] Petrolatum: Very soluble [031,062,421] Carbon disulfide: Very soluble [031,062,421] Aqueous alkali hydroxides: Very soluble [031,421] Ether: Very soluble [031,205,421] Benzene: 1 g/12 mL [031,421] Petroleum ether: Almost insoluble [031,421] Chloroform: Soluble [017,062,430,455] Olive oil: Soluble [430] Alkalies: Soluble [062,205,455] Caustic alkalies: Soluble [295] Alcohol: Very soluble [031,058,205,421] Liquid paraffin: 1 in 100 [295] Castor oil: 1 in ~10 [295] Ethyl acetate: Soluble [430]

OTHER PHYSICAL DATA: Boiling point: 70.9 C @ 10 mm Hg [017] 90.2 C @ 25 mm Hg [025] Vapor pressure: 1 mm Hg @ 40.1 C [038,043,055] 5 mm Hg @ 62.5 C [038] Vapor pressure: 10 mm Hg @ 73.8 C; 20 mm Hg @ 86 C; 40 mm Hg @ 100.1 C [038] Sharp, medicinal, sweet and tarry odor [058] Odor threshold (lower): 0.016 ppm [051] Sharp, burning taste [055,062,430] Sweet, pungent taste (on dilution) [295] Taste threshold (lower): 0.0001 ppm [051] Evaporation rate (butyl acetate = 1): <0.01 [058] log P octanol: 1.46 [055] Refractive index: 1.5425 @ 41 C [031] pKa: 10.0 @ 25 C [031,455] pH (of aqueous solutions): ~6.0 [031,421] Saturation concentration: 0.77 g/m3 @ 20 C; 2.0 g/m3 @ 30 C [055] Burning rate: 3.5 mm/min [371] Critical temperature: 421.1 C [371] Critical pressure: 60.5 atmospheres [371] Liquid surface tension: 36.5 dynes/cm @ 55 C [371] Liquid water interfacial pressure: 20 dynes/cm @ 42 C [371] Latent heat of vaporization: 72 cal/g [371] Heat of combustion: -7445 cal/g [371] Lambda max (in cyclohexane): 259 nm (shoulder), 265 nm, 268 nm (shoulder), 271 nm, 278 nm (epsilon = 790, 1397, 1365, 2027, 1794) [052]

HAP WEIGHTING FACTOR: 1 [713]

VOLATILITY:

Vapor pressure: 0.2 mm Hg @ 20 C [055] 0.35 mm Hg @ 25 C [421]

Vapor density: 3.24 [043,055,102,421]

FIRE HAZARD: Phenol has a flash point of 79 C (175 F) [031,058,275,371]. It is combustible. Fires involving this material may be controlled with a dry chemical, carbon dioxide or Halon extinguisher. A water spray may also be used [051,058,371,430].

The autoignition temperature of phenol is 715 C (1319 F) [043, 062,371,451].

LEL: 1.7% [102,371] UEL: 8.6% [058,102,371]

REACTIVITY: Phenol is incompatible with strong oxidizers [043,058,269,346]. It is also incompatible with calcium hypochlorite, butadiene and nitrobenzene [058]. Violent or explosive reactions occur with aluminum chloride [036,058]. When hot, it will attack aluminum, magnesium, lead and zinc metals [058,102]. It coagulates collodion and acetanilide [031,295,421]. It also coagulates butyl chloral hydrate, camphor, monobrominated camphor, chloral hydrate, lead acetate, diuretin, menthol, naphthalene, naphthol, acetophenetidin, pyrogallol, resorcinol, sodium phosphate, thymol, urethane and terpin hydrate [031,421]. It coagulates with salol, chloralamide, ammonium salt, ammonium phenate and ammonium carbolate. Other incompatibilities include alkaline salts, phenazone, piperazine, quinine salts, phenacetin and iron salts. It coagulates albumin [295]. It is also incompatible with strong acids and bases [269]. Violent or explosive reactions occur with formaldehyde, peroxodisulfuric acid, peroxo- monosulfuric acid, sodium nitrite and (sodium nitrate + trifluoroacetic acid) [036,043,066]. Explosions also occur with (nitrobenzene + aluminum chloride) and (aluminum chloride + carbon monoxide) [066]. It reacts with (aluminum chloride + nitromethane) [043]. This chemical will attack some forms of plastics, rubber and coatings [102]. It also coagulates proteins [421].

STABILITY: Phenol is hygroscopic [062,430]. It is sensitive to exposure to heat [043,058,102]. On exposure to air or light, it turns a pinkish to reddish color [031,421]. Solutions of it in water, DMSO, 95% ethanol or acetone should be stable for 24 hours under normal lab conditions [700].

USES: Phenol is used as an intermediate, as a general disinfectant for toilets, stables, cesspools, floors and drains, and sanitizer. It is used in the manufacture of colorless or light-colored artificial resins, many medical and industrial organic compounds and dyes, phenol-formaldehyde resins, bisphenol A, nylon intermediates, alkylphenols, germicides, antioxidants, preservatives, barn deodorants, explosives, fertilizers, coke, illuminating gas, lampblack, paints, paint removers, rubber, asbestos goods, wood preservatives, textiles, drugs, pharmaceutical preparations, perfumes, bakelite and plastics. It is used as a pharmaceutic aid (antimicrobial agent) and as a reagent in chemical analysis. It is also used topically as an anesthetic in pruritic skin conditions and internally and externally as an antiseptic. It is also used in medicine as a cauterizing agent and in the treatment of severe disability (muscle spasms and paralysis). It is used in caustics, fuel-oil sludge inhibitors, solvents, rubber chemicals, and in the petroleum, leather, paper, soap, toy and agricultural industries.

COMMENTS: This compound is obtained from coal tar [031,430].

NIOSH Registry Number: SJ3325000

ACUTE/CHRONIC HAZARDS: Phenol may be fatal by ingestion, inhalation or skin absorption [269]. It rapidly penetrates the skin [421]. It is a severe irritant and is corrosive [058]. When heated to decomposition it emits toxic fumes of carbon monoxide, carbon dioxide and unidentified organic compounds [058,269].

SYMPTOMS: Symptoms of exposure may include irritation of the skin, eyes, nose and throat, blindness, headache, dizziness, loss of appetite, abdominal pain, skin depigmentation, severe skin burns, vomiting, diarrhea, difficulty in swallowing, muscular weakness, unconsciousness, coma and death [058,102]. Other symptoms may include heart damage, central nervous system depression, respiratory arrest, irritation of the respiratory tract, mouth and stomach, giddiness, jaundice, shortness of breath, bladder damage, cardiac arrest, severe burns of tissues and the eyes, burning pain in the mouth and throat, irregular breathing, corrosion of the skin and eyes and blurred vision [058]. Liver and kidney damage may occur [036,058,301,346]. Nausea may also occur [031,036,058,301]. Exposure may result in circulatory collapse, tachypnea, paralysis, convulsions, greenish to smoky colored urine, necrosis of the mouth and gastrointestinal tract and icterus [031]. It may also result in respiratory alkalosis followed by acidosis, methemoglobinemia, necrosis of the mucous membranes, cerebral edema, bladder necrosis, erythema, pulmonary edema followed by pneumonia, profuse sweating, intense thirst, hyperactivity, stupor, blood pressure fall, hyperpnea, hemolysis, oliguria, anuria, central nervous system damage and muscle contractions [301]. Eye effects may include conjunctiva chemotic, white and hypesthetic cornea, edematous eyelids, iritis and carbolochronosis [099]. Other symptoms may include necrosis of the skin, ulcerative esophagitis, laryngeal edema, white or brownish stains and areas of necrosis about the face, mouth and esophagus; neurolysis of the cervical posterior roots, hypersensitivity, idiosyncrasy, hypothermia, loss of vasoconstrictor tone, cardiac depression, stertorous breathing, mucous rales, frothing of the mouth and nose, ventricular arrhythmias, hyperbilirubinemia in newborns, Heinz body hemolytic anemia, sores and burning in the mouth, pallor, weakness, tinnitus (ringing in the ears), weak and irregular pulse, hypotension, shallow respirations, cyanosis, fleeting excitement, confusion, rhonchi and fever [151]. Exposure may cause antipyresis, tremors, digestive disturbances, ptyalism, anorexia, fainting, vertigo, mental disturbances, ochronosis and gangrene [430]. It may also cause spasm, inflammation and edema of the larynx and bronchi; chemical pneumonitis, burning sensation, coughing, wheezing, laryngitis, dermatitis and central nervous system disturbances [269]. Mucocutaneous and gastrointestinal corrosion may occur [406]. Damage to the pancreas, spleen or lungs may also occur [051]. Skin eruptions and rapid and difficult breathing have been reported [036]. Other reported symptoms may include severe eye damage and skin rash [346]. Exposure may cause irritation of the mucous membranes, weight loss, muscle aches and pain and conjunctival swelling [102].

---

p-Phenylenediamine

CAS NUMBER: 106-50-3

SYNONYMS: p-aminoaniline 4-aminoaniline 1,4-Benzenediamine 1,4-Diaminobenzene

CHEMICAL FORMULA: C6H8N2

MOLECULAR WEIGHT: 108.15

WLN: ZR DZ

PHYSICAL DESCRIPTION: White to light purple crystals

SPECIFIC GRAVITY: Not available

DENSITY: Not available

MELTING POINT: 140 C

BOILING POINT: 267 C

SOLUBILITY: Water: Soluble DMSO: Not available 95% Ethanol: Not available Acetone: Not available

OTHER SOLVENTS: Alcohol: Soluble Chloroform: Soluble Ether: Soluble

OTHER PHYSICAL DATA: Not available

HAP WEIGHTING FACTOR: 1 [713]

VOLATILITY:

Vapor density: 3.72

FIRE HAZARD: The flash point for p-phenylenediamine is 155 C. Fires involving this compound may be controlled with water, dry chemical, foam, carbon dioxide and/or Halon extinguishers.

LEL: Not available UEL: Not available

REACTIVITY: p-Phenylenediamine may react with oxidizing materials.

STABILITY: p-Phenylenediamine oxidizes on exposure to air.

USES: p-Phenylenediamine is used as an azo dye intermediate, a photographic developing agent, in photochemical measurements, as an intermediate in manufacture of antioxidants and accelerators for rubber, a laboratory reagent, and as a dye for hair and fur.

COMMENTS: None

NIOSH Registry Number: SS8050000

ACUTE/CHRONIC HAZARDS: p-Phenylenediamine is an irritant and it may be toxic by ingestion or inhalation.

SYMPTOMS: Exposure may result in keratoconjunctivitis, rhinitis (inflammation of the nose), paroxysmal cough, asthma; papular eczema on face, neck, and arms; macrocytic, and hypochromic anemia.

Phosgene

CAS NUMBER: 75-44-5

SYNONYMS: Carbon oxychloride Carbonyl chloride Chloroformylchloride

CHEMICAL FORMULA: CCl2O

MOLECULAR WEIGHT: 98.91

WLN: Not available

PHYSICAL DESCRIPTION: Colorless gas or volatile liquid

SPECIFIC GRAVITY: Not available

DENSITY: 1.37 g/mL @ 20 C

MELTING POINT: -118 C

BOILING POINT: 83 C

SOLUBILITY: Water: Very slightly soluble DMSO: Not available 95% Ethanol: Not available Acetone: Not available

OTHER SOLVENTS: Acetic acid: Very soluble Toluene: Freely soluble Chloroform: Soluble Most liquid hydrocarbons: Soluble Benzene: Very soluble

OTHER PHYSICAL DATA: Odor of new mown hay or green corn

HAP WEIGHTING FACTOR: 10 [713]

VOLATILITY:

Vapor pressure: 1180 mm Hg @ 20 C

Vapor density: 3.4

FIRE HAZARD: Phosgene is nonflammable.

LEL: Not available UEL: Not available

REACTIVITY: Phosgene is readily hydrolyzed by water to carbonic acid and hydrochloric acid and ammonolysis forms urea. With an excess of amine it forms urea derivates.

STABILITY: Phosgene is sensitive to exposure to moisture.

USES: Phosgene is used in organic synthesis in the manufacture of dyes, pharmaceuticals, herbicides, insecticides, synthetic foams, resins, and polymers and as a chlorinating agent.

COMMENTS: Phosgene is a highly toxic and reactive gas.

NIOSH Registry Number: SY5600000

ACUTE/CHRONIC HAZARDS: Phosgene may be toxic by all routes of exposure. When heated to decomposition, or on contact with water or steam, it will react to produce toxic and corrosive fumes.

SYMPTOMS: Symptoms of exposure may include burning of the eyes and throat, coughing, shortness of breath, burning in the chest, dyspnea, moist rales in the chest, pulmonary edema, pneumonia, and death.

Phosphine

CAS NUMBER: 7803-51-2

SYNONYMS: Hydrogen phosphide Phosphorus trihydride

CHEMICAL FORMULA: PH3

MOLECULAR WEIGHT: 34.00

PHYSICAL DESCRIPTION: Colorless gas or liquid (Weast, 1979) with a disagreeable, fishy odor (Grant, 1974); a colorless gas with a disagreeable garlic-like odor (Student, 1981, p. 405) [701]

SPECIFIC GRAVITY: 0.746 at -90 C (Weast, 1979) [701]

MELTING POINT: -207 F, -133 C (Merck, 1983) [701]

BOILING POINT: -126 F, -87.7 C (Merck, 1983) [701]

SOLUBILITY: Water: Slightly soluble (Merck, 1983); 26cc/100mL at 17 C (Farm Chemicals Handbook, 1984, p. C-47) [701]

OTHER PHYSICAL DATA: Vapor Pressure: 760 mmHg at -87.5 C (Weast, 1979, p. D-201) [701] Vapor Density (AIR=1): 1.17 (NFPA, 1978) [701] Density: 1.529 g/L @ 0 C [703] Freezing Point: -209 F [704] Autoignition temperature: 212 F [703]

HAP WEIGHTING FACTOR: 1 [713]

VOLATILITY:

Phosphine is a highly volatile compound that is a gas under ambient conditions.

FIRE HAZARD: Fires involving phosphine may be controlled with a dry chemical, carbon dioxide or Halon extinguisher.

LEL: Not available UEL: Not available

REACTIVITY: Not available

STABILITY: Not available

USES: Not available

COMMENTS: None

NIOSH Registry Number: Not available

ACUTE/CHRONIC HAZARDS: Phosphine is very toxic.

SYMPTOMS: Not available

Phosphorus

CAS NUMBER: 7723-14-0

CHEMICAL FORMULA: P_4

MOLECULAR WEIGHT: 123.88 [703]

PHYSICAL DESCRIPTION: Cubic crystals; colorless to yellow, wax-like solid [703] white to yellow soft, waxy solid with acrid fumes in air [704]

SPECIFIC GRAVITY: 1.82 @ 20C/4C [704]

MELTING POINT: 44.1 C [703]

BOILING POINT: 280 C [703]

SOLUBILITY: Water: 0.0003% [704]

OTHER PHYSICAL DATA: Vapor Pressure: 1 mm @ 76.6 C [703] Vapor Density: 4.42 [703] Autoignition temperature: 86 F [703]

EDITOR'S NOTE: The following data is for white phosphorus only. However, the data on "USES" refers to several substances in the compound group.

HAP WEIGHTING FACTOR: 1 [713]

VOLATILITY:

Vapor Pressure: 1 mm @ 76.6 [703]

Vapor Density: 4.42 [703]

FIRE HAZARD: Flash Point: Spontaneously flammable in air [702, 703, 704, 706]

Dangerous fire hazard when exposed to heat, flame, or by chemical reaction with oxidizers. Ignites spontaneously in air. To fight fire, use water. [703]

Autoignition temperature: 86 F [703]

LEL: Not available UEL: Not available

REACTIVITY: Phosphorus is a very reactive element. It is incompatible with air, oxygen, alkalis, and reducing agents. White phosphorus takes fire spontaneously in air, burning to the pentoxide. Red phosphorus does not ignite spontaneously and is not as dangerous as white phosphorus, but it does convert to white phosphorus at some temperatures and emits highly toxic fumes of the phosphorus oxides when heated. [706] There is danger of explosion hazards by chemical reaction of phosphorus with many materials such as alkaline hydroxides; ammonium nitrate; beryllium, bromate salts, chlorates of Ba, Ca, Mg, K, Na, Zn, iodates of Ba, Ca, Mg, K, Na, Zn, reactive metals such as cesium and some of its compounds, charcoal in the presence of air; chlorine dioxide; chlorosulfonic acid; chromic acid; copper; iron; lithium and some of its salts, manganese perchlorate; manganese, mercuric oxide and mercurous nitrate, nitrogen dioxide, oxygen, performic acid; platinum, potsssium, potassium hydroxide, potassium permanganate, several selenium salts, silver nitrate, silver oxide, sodium, sodium perchlorate, sodium hydroxide, sulfur, sulfur trioxide, sulfuric acid, zirconium, and many other compounds [703] (the above are not complete but are only provided to show the wide range of dangerous reactivities). Contact with boiling caustic alkalis or hot calcium hydroxide evolves phosphine, which usually ignites in air. [710]

STABILITY: When heated to decomposition it emits highly toxic fumes of POx [703]. See comments below.

USES: Phosphoric acid is an important compound in making super-phosphate fertilizers. Concentrated phosphoric acids, which may contain as much as 70 to 75% P2O5 content, are currently important for farming and agrucultural production activities. World-wide demand for fertilizers has caused record phosphate production. Phosphates are also used in the production of special glasses, such as those used for sodium lamps. Bone-ash, calcium phosphate, is also used to produce fine chinaware and to produce mono-calcium phosphate used in baking powder. Phosphorous is also important in the production of steels, phosphor bronze, and many other products. Trisodium phosphate is important as a cleaning agent, as a water softener, and for preventing boiler scale and corrosion of pipes and boiler tubes. Red phosphorous is used in the manufacture of safety matches, pyrotechnics, pesticides, incendiary shells, smoke bombs, tracer bullets, etc. [706]

COMMENTS: Phosphorous exists in four or more allotropic forms: white (or yellow), red, and black (or violet). White phosphorous has two modifications: alpha and beta with a transition temperature at -3.8 C. Ordinary phosphorous is a waxy white solid and when it is pure it is a colorless and transparent solid. It takes fire spontaneously in air, burning to form phosphorus pentoxide. It is also very poisonous requiring only about 50 mg to constitute a fatal dose. White phosphorous should be stored under water, since it is a very reactive element in air. If phosphorus contacts the skin it can cause severe burns. When exposed to sunlight or when heated in its own vapor to 250 C, white phosphorus is converted to red phosphorus, which does not phosphoresce in air as does the white variety. Red phosphorus does not ignite spontaneously and is not as dangerous as white phosphorous. It should, however, be handled with care as it does convert to the white form at some temperatures and it emits highly toxic fumes of the oxides of phosphorous when heated. [706]

NIOSH Registry Number: TH3500000

ACUTE/CHRONIC HAZARDS: Human poison by ingestion; experimental poison by ingestion and subcutaneous routes. Experimental reproductive effects. Human systemic effects by ingestion: fluid intake, sweating, nausea, diarrhea, cyanosis, cardiomyopathy. Toxic quantities have acute effect on liver and can cause severe eye damage. When heated to decomposition, emits highly toxic fumes of POx. [703] Very poisonous, 50 mg constituting an approximate fatal dose; causes severe skin burns. [706]

SYMPTOMS: The most common symptom of chronic phosphorus poisoning is necrosis of the jaw (phossy-jaw). Anemia, gastrointestinal effects, and brittleness of long bones can occur from chronic inhalation or ingestion of phosphorus. Inhalation of it can cause photophobia myosis, dilation of the pupils, retinal hemorrhage, congestion of blood vessels, and sometimes an optic neuritis. [703] Symptoms of exposure may include irritation of the eyes and respiratory tract; abdominal pain, nausea, jaundice; anemia; cachexia; dental pain, excess salivation, jaw pain, and swelling; burns skin and eyes. [704]

Phthalic anhydride

CAS NUMBER: 85-44-9

SYNONYMS: 1,2-Benzenedicarboxylic acid anhydride 1,3-Dioxophthalan 1,3-Dihydro-1,3-dioxoisobenzofuran 1,3-Isobenzofurandione Phthalandione

CHEMICAL FORMULA: C8H4O3

MOLECULAR WEIGHT: 148.12

WLN: T56 BVOVJ

PHYSICAL DESCRIPTION: White to pale cream powder or flakes

SPECIFIC GRAVITY: 1.527 @ 4 C [043,055,062]

DENSITY: Not available

MELTING POINT: 130.8 C [031,038,055,205]

BOILING POINT: 295 C (sublimes) [017,031,043,047]

SOLUBILITY: Water: Decomposes DMSO: >=100 mg/mL @ 19 C [700] 95% Ethanol: <1 mg/mL @ 19 C [700] Acetone: >=100 mg/mL @ 19 C [700]

OTHER SOLVENTS: Carbon disulfide: Soluble [031,062] Ether: Sparingly soluble [025,031,043,205] Alcohol: Soluble [043]

OTHER PHYSICAL DATA: Choking odor [102,371] Vapor pressure: 1 mm Hg @ 96.5 C [038,043] Saturation concentration: 0.0016 g/m3 @ 20 C; 0.0078 g/m3 @ 30 C [055] Acidic pH [058]

HAP WEIGHTING FACTOR: 1 [713]

VOLATILITY:

Vapor pressure: 0.0002 mm Hg @ 20 C [055] 0.001 mm Hg @ 30 C [055,058]

Vapor density: 5.10 [043,055,058,102]

FIRE HAZARD: Phthalic anhydride has a flash point of 151.6 C (305 F) [043,062,102,275]. It is combustible. Fires involving this material may be controlled with a dry chemical, carbon dioxide or Halon extinguisher. A water spray may also be used [058].

The autoignition temperature is 570 C (1058 F) [043,102,371,451]. It may form explosive mixtures with air [058].

LEL: 1.7% [043,058,102,451] UEL: 10.4% [043,058,102]

REACTIVITY: Phthalic anhydride is incompatible with strong oxidizers [043,058,102,269]. It is also incompatible with strong acids, strong bases and strong reducing agents [269]. It may react violently with copper oxide or sodium nitrite [036,043, 269]. It may also react with nitric acid [036,043]. Nitration with sulfuric acid may also present a danger [036]. It is incompatible with water, alkalis, nitrating mixtures and amines [058].

STABILITY: Phthalic anhydride is sensitive to moisture [269,275]. It may also be sensitive to heat. UV spectrophotometric stability screening indicates that solutions of it in 95% ethanol are stable for less than two hours but solutions in acetone are stable for at least 24 hours [700].

USES: Phthalic anhydride is used in alkyd resins, plasticizers, hardeners for resins, polyesters and synthesis of phenolphthalein and other phthaleins, dyes, chlorinated products, pharmaceutical intermediates, insecticides, diethyl phthalate, dimethyl phthalate and laboratory reagents. It is also used in the manufacture of specialty chemicals, synthetic fibers, pigments, synthetic indigo, artificial resins (glyptal), phthalates and benzoic acid. It is used as a dehydrating agent for alcohols, as a protecting reagent for amino acids, a fire retardant for use in components of polyester resins and in the manufacture of metallic and acid salts.

COMMENTS: None

NIOSH Registry Number: TI3150000

ACUTE/CHRONIC HAZARDS: Phthalic anhydride is a severe irritant [036,058,401] and these irritating effects are worse on moist surfaces [102,275]. When heated to decomposition it emits toxic fumes of carbon monoxide and carbon dioxide [102,269]. It may be readily absorbed through the skin [269].

SYMPTOMS: Symptoms of exposure may include moderate to severe irritation of the eyes and skin [058,102,371,421]. It may also cause irritation of the mucous membranes and upper respiratory tract [036,058,421]. Inhalation of the dust or vapors of this compound may cause coughing or sneezing [102, 371]. It may also cause nosebleeds and asthma attacks in persons who have previously had asthma [102]. It may also cause bronchitis [102,421]. Repeated or prolonged exposure can cause skin burns [058,102,371]. It may also cause skin rash, dermatitis, conjunctivitis and chronic eye irritation. Absorption into the body leads to formation of methemoglobin which in sufficient concentration causes cyanosis. Onset may be delayed 2-4 hours or longer. Exposure may also damage the liver and kidneys [102]. It may possibly cause pulmonary sensitization [421]. Other symptoms may include chronic congestion and ulceration of the nose, eye burns, allergic respiratory reaction, a burning sensation in the nose and throat and gastrointestinal disturbances [058]. Consumption of alcohol may increase toxic effects [269]. It will cause internal irritation if taken by mouth [036]. It may also cause smarting of the skin [371]. Skin sensitization may occur [401].

Polychlorinated biphenyls

CAS NUMBER: 1336-36-3

SYNONYMS: aroclors

NOTE: This is a mixture of chlorinated biphenyl isomers with commercial mixtures named according to the approximate percentage of chlorine in the mixture. The commercial mixtures are known as Aroclor 12XX where the last two digits represent the percent chlorine. Thus, Aroclor 1221 has about 21 percent chlorine in the mixture. The common products are:

Aroclor 1216 CAS NUMBER: 12674-11-2

Aroclor 1221 CAS NUMBER: 11104-28-2

Aroclor 1232 CAS NUMBER: 11141-16-5

Aroclor 1242 CAS NUMBER: 53469-21-9

Aroclor 1248 CAS NUMBER: 12672-29-6

Aroclor 1260 CAS NUMBER: 11096-82-5

Aroclor 1262 CAS NUMBER: 37324-23-5

Aroclor 1268 CAS NUMBER: 11100-14-4

CHEMICAL FORMULA: Variable

MOLECULAR WEIGHT: Variable

WLN: Not available

PHYSICAL DESCRIPTION: Clear, viscous liquids ranging in color from none to light yellow.

SPECIFIC GRAVITY: 1.4 (variable)

DENSITY: Variable

MELTING POINT: Not available

BOILING POINT: 275-420 C

SOLUBILITY: Water: Insoluble DMSO: Not available 95% Ethanol: Not available Acetone: Not available

OTHER SOLVENTS: Not available

OTHER PHYSICAL DATA: Not available

NOTE: The following data is for polychlorinated biphenyls in general:

VOLATILITY: Not available

FIRE HAZARD: Flash Point: Not available

Flammability: Non-flammable except at extremely high temperatures [716]

Autoignition temperature: Not available

LEL: Not available UEL: Not available

REACTIVITY: Though chemically stable, PCB's are very reactive when exposed to ultraviolet light. Over the past several decades the photoreductive dehalogenation reactions of PCB's have been examined in organic solvents, neat thin films and solid phase, and as surface deposits on silica. [717]

STABILITY: Polychlorinated biphenyls are noted for their stability. They are resistant to heat, oxidation and attack by strong acids and bases. [717] PCB's decompose to highly toxic substances including dioxin when subjected to high temperatures [716].

USES: Because of certain properties such as non-flammability except at extremely high temperatures, low electrical conductivity, and stability to chemical and biological breakdown, PCB's have been suited for usage in electrical equipment, hydraulic equipment, and heat transfer systems. [716] PCB's have been used as dielectric fluids, fire retardants, heat transfer agents, hydraulic fluids, plasticizers and in other applications [717].

PCB's have also been used in railroad transformers, mining equipment, carbonless copy paper, pigments, electromagnets, as a microscopy mounting medium and immersion oil, as optical liquids, and in compressors and natural gas pipeline liquids [718].

COMMENTS: See specific data for Aroclor 1254 and other individual Aroclors.

Aroclor 1221

CAS NUMBER: 11104-28-2

NIOSH Registry Number: TQ1352000

ACUTE/CHRONIC HAZARDS: Aroclor 1221 is moderately toxic by ingestion and skin contact. It is a suspected human carcinogen. Experimental reproductive effects have been noted. When heated to decomposition it emits toxic fumes of Cl-. [703]

SYMPTOMS OF EXPOSURE:

See symptoms of exposure to polychlorinated biphenyl mixture above.

Aroclor 1232

CAS NUMBER: 11141-16-5

NIOSH Registry Number: TQ1354000

ACUTE/CHRONIC HAZARDS: Aroclor 1232 is moderately toxic by skin contact and mildly toxic by ingestion. It is a suspected human carcinogen. When heated to decomposition it emits toxic fumes of Cl-. [703]

SYMPTOMS: See symptoms of exposure to polychlorinated biphenyl mixture above.

Aroclor 1242

CAS NUMBER: 53469-21-9

NIOSH Registry Number: TQ1356000

ACUTE/CHRONIC HAZARDS: Aroclor 1242 is a poison by subcutaneous route. It is moderately toxic by ingestion. Human systemic effects by inhalation include pulmonary and liver effects. It is a suspected human carcinogen. Experimental reproductive effects and mutagenic data have been noted. When heated to decomposition it emits toxic fumes of Cl-. [703]

SYMPTOMS: See symptoms of exposure to polychlorinated biphenyl mixture above.

Aroclor 1248

CAS NUMBER: 12672-29-6

NIOSH Registry Number: TQ1358000

ACUTE/CHRONIC HAZARDS: Aroclor 1248 is moderately toxic by skin contact. It is a suspected human carcinogen; experimental teratogenic and reproductive effects have been noted. When heated to decomposition it emits toxic fumes of Cl-. [703]

SYMPTOMS: See symptoms of exposure to polychlorinated biphenyl mixture above.

Aroclor 1254

CAS NUMBER: 11097-69-1

VOLATILITY: Not available

FIRE HAZARD: Flash Point: 222 C (432 F) [102, 107, 421]

Flammability: Combustible. Fires involving this material may be controlled with a dry chemical, carbon dioxide, or Halon extinguisher.

LEL: Not available UEL: Not available

REACTIVITY: This chemical is incompatible with strong oxidizers [102, 107, 346]. It will attack some forms of plastics, rubber, and coatings [102].

STABILITY: This chemical is sensitive to heat [102]. Solutions of this chemical in water, DMSO, 95% ethanol, or acetone should be stable for 24 hours under normal laboratory conditions [700].

USES: This chemical is used in electrical capacitors, electrical transformers, vacuum pumps, gas-transmission turbines, high- temperature dielectrics for electric wires and electrical equipment, heat-exchange fluids, coatings, inks, insecticides, fillers, adhesives, paints and in duplicating papers. It is also used as a plasticizer for cellulosics, vinyl resins and chlorinated rubbers. Formerly used as hydraulic fluids, fire retardants, wax extenders, dedusting agents, pesticide extenders, lubricants, cutting oils, sealants and caulking compounds.

COMMENTS: This chemical contains biphenyls with 54% chlorine. It is composed of 11% tetra-, 49% penta-, 34% hexa-, and 6% heptachlorobiphenyls. Polychlorinated biphenyls (mixture)

NIOSH Registry Number: TQ1360000

ACUTE/CHRONIC HAZARDS: Moderately toxic by ingestion; some poisonous by other routes. Suspected human carcinogens; experimental carcinogens and tumorigens; experimental reproductive effects. As with chlorinated naphthalenes, there are effects on skin and toxicity to liver. Hepatotoxicity appears to be increased with concurrent exposure to carbon tetrachloride. Toxicity increases with chlorine content. Oxides are more toxic than unoxidized materials. Severe liver damage may be fatal. [703] Yusho disease: in 1968, an outbreak of poisoning occurred in Yusho, Japan, involving some 1000 people of all ages (15,000 victims referred to by Umeda), who ingested for several months rice bran oil that had been contaminated with PCBs to the extent of 1500 to 2000 ppm. [722] High exposure to PCBs was associated with reduced birth weight. This group also showed shortened gestational age after adjustment for these same variables. PCBs produce liver tumors in rats. Although human studies have not been conclusive, PCB's should be considered potential human carcinogens. [723] When heated to decomposition PCBs emit highly toxic fumes of Cl-. [703] See SYMPTOMS.

SYMPTOMS: In persons who have suffered systemic intoxication, the usual signs and symptoms are nausea, vomiting, loss of weight, jaundice, edema and abdominal pain. Where the liver damage has been severe the patient may pass into coma and die. [703] A 1968 outbreak of poisoning in Yusho, Japan involved some 1000 people of all ages who ingested for several months rice bran oil contaminated with PCBs to the extent of 1500 to 2000 ppm with dibenzofurans also present. After a latent period of 5 to 6 months, nausea, lethargy, chloracne, brown pigmentation of skin areas and nails, subcutaneous edema of the face, distinctive hair follicles, a cheese-like discharge from the eyes, swelling of eyelids and transient visual disturbances, gastroenteric distress, and jaundice were noted. Infants born to poisoned mothers had decreased birth weight and skin discoloration, and two stillbirths were reported. [722] The following symptoms have also been reported: conjunctival hyperemia, visual and hearing disturbances, increases in diastolic and systolic blood pressure, weakness and numbness of extremities, neurobehavioral and psychomotor impairment after occupational and in utero exposure, GI upset and diarrhea, liver damage, clinical hepatitis, and asymptomatic hyperthyroxinemia. [723]

Aroclor 1260

CAS NUMBER: 11096-82-5

NIOSH Registry Number: TQ1362000

ACUTE/CHRONIC HAZARDS: Aroclor 1260 is moderately toxic by ingestion and skin contact. It is a suspected human carcinogen and an experimental carcinogen. Experimental reproductive effects have been noted. Mutagenic data exists. When heated to decomposition it emits highly toxic fumes of Cl-. [703]

SYMPTOMS: See symptoms of exposure to polychlorinated biphenyl mixture above.

Aroclor 1262

CAS NUMBER: 37324-23-5

NIOSH Registry Number: TQ1364000

ACUTE/CHRONIC HAZARDS: Aroclor 1262 is moderately toxic by skin contact. It is a suspected human carcinogen. When heated to decomposition it emits toxic fumes of Cl-. [703]

SYMPTOMS: See symptoms of exposure to polychlorinated biphenyl mixture above.

Aroclor 1268

CAS NUMBER: 11100-14-4

NIOSH Registry Number: TQ1366000

ACUTE/CHRONIC HAZARDS: Aroclor 1268 is moderately toxic by skin contact. It is a suspected human carcinogen. When heated to decomposition it emits toxic fumes of Cl-. [703]

SYMPTOMS: See symptoms of exposure to polychlorinated biphenyl mixture above.

Polycyclic organic matter

NOTE: Polycyclic organic matter is a compound group under Section 112 of the Clean Air Act. The group includes organic compounds with more than one benzene ring, and which have a boiling point greater than or equal to 100 C. [708] The compounds described here are intended as selected examples of polycyclic organic matter and do not comprise a comprehensive list. Other polycyclic organic matter compounds which appear elsewhere in this database include:

2-Acetylaminofluorene 4-Aminobiphenyl Benzidine 3,3'-Dichlorobenzidine 4-Dimethylaminoazobenzene 4-Nitrobiphenyl

HAP WEIGHTING FACTOR: 1 [713]

Anthracene

CAS NUMBER: 120-12-7

SYNONYMS: Paranaphthalene

CHEMICAL FORMULA: C14H10

MOLECULAR WEIGHT: 178.24 [703]

PHYSICAL DESCRIPTION: Colorless crystals, violet fluorescence [703]

DENSITY: 1.24 @ 27C/4C [703]

MELTING POINT: 217 C [703]

BOILING POINT: 339.9 C [703]

SOLUBILITY: Water: Insoluble [703] Alcohol: Soluble in alcohol @ 1.9/100 @ 20 C [703] Ether: Soluble in ether @ 12.2/100 @ 20 C [703]

OTHER PHYSICAL DATA: Vapor Pressure: 1 mm @ 145.0 C (sublimes) [703] Vapor Density: 6.15 [703]

VOLATILITY:

Vapor pressure: 1 mm Hg @ 145 C (sublimes)

Vapor density: 6.15

FIRE HAZARD: The flash point for anthracene is 121 C (250 F). It is combustible. Fires involving this material should be controlled using a dry chemical, carbon dioxide, foam or Halon extinguisher.

The autoignition temperature is 538 C (1004 F).

LEL: 0.6% UEL: Not available

REACTIVITY: Anthracene darkens in sunlight and reacts with oxidizers. It reacts explosively with flame, Ca(OCl)2 and chromic acid.

STABILITY: Anthracene is sensitive to prolonged exposure to air and light.

USES: Anthracene is an important source of dyestuffs and it is used in the manufacture of anthraquinone, alizarin dyes, insecticides and wood preservatives.

COMMENTS: None

NIOSH Registry Number: CA9350000

ACUTE/CHRONIC HAZARDS: Anthracene is an irritant and may be toxic by ingestion.

SYMPTOMS: Exposure may cause irritation of the eyes and respiratory tract and gastrointestinal irritation if swallowed. Long-term contact may result in pigmentation or carcinogenesis on the skin.

Benzo(a)pyrene

CAS NUMBER: 50-32-8

SYNONYMS: 3,4-Benzopyrene 3,4-Benzpyrene 3,4-Benzylpyrene Benzo(D,E,F)chrysene 6,7-Benzopyrene Benz(a)pyrene 3,4-Benz(a)pyrene 3,4-Benzypyrene

CHEMICAL FORMULA: C20H12

MOLECULAR WEIGHT: 252.32 [703]

PHYSICAL DESCRIPTION: Yellow crystals [703]

SPECIFIC GRAVITY: Not available

MELTING POINT: 179 C [703]

BOILING POINT: 312 C @ 10mm [703]

SOLUBILITY: Water: Insoluble [703] Benzene: Soluble [703] Toluene: Soluble [703] Xylene: Soluble [703]

OTHER PHYSICAL DATA: Not available

VOLATILITY: Not available

FIRE HAZARD: Benzo(a)pyrene is combustible. Fires involving it can be controlled with a dry chemical, carbon dioxide or Halon extinguisher. A water spray may also be used [269].

LEL: Not available UEL: Not available

REACTIVITY: Benzo(a)pyrene is incompatible with strong oxidizers [051,102,107,269]. It readily undergoes nitration and halogenation [395]. Ozone, chromic acid and chlorinating agents oxidize this compound. Benzo(a)pyrene may react with organic and inorganic oxidants including various electrophiles, peroxides, nitrogen oxides and sulfur oxides [051,071]. Hydrogenation occurs with platinum oxide [395].

STABILITY: Benzo(a)pyrene undergoes photo-oxidation after irradiation in indoor sunlight or by fluorescent light in organic solvents. Solutions of it in benzene oxidize under the influence of light and air [395]. Solutions of it in water, DMSO, 95% ethanol or acetone should be stable for 24 hours under normal lab conditions [700].

USES: Benzo(a)pyrene is used extensively in cancer research.

COMMENTS: Benzo(a)pyrene is a polynuclear (five-ring) aromatic hydrocarbon [062,346]. It is found in coal tar, in cigarette smoke, in the atmosphere as a product of incomplete combustion, in the exhaust soot and tar from gasoline and diesel engines, and in oil, water and food [051,071]. Manufacturing sources include coal tar processing, petroleum refining, shale refining, coal and coke processing, kerosene processing and heat and power generation sources. Natural sources include synthesis by various bacteria and algae. Man-caused sources include combustion of tobacco, combustion of fuels, runoff containing greases and oils, and potential roadbed and asphalt leachate [055].

NIOSH Registry Number: DJ3675000

ACUTE/CHRONIC HAZARDS: Benzo(a)pyrene may be harmful by ingestion or inhalation [071,269]. It may be an irritant and when heated to decomposition it emits acrid smoke and toxic fumes of carbon monoxide and carbon dioxide [043,269].

SYMPTOMS: Symptoms of exposure may include irritation of mucous membranes, dermatitis, bronchitis, coughing, dyspnea, conjunctivitis, photosensitization, pulmonary edema, reproductive effects and leukemia [107]. Contact with the skin may result in erythema, pigmentation, desquamation, formation of verrucae and infiltration [395]. It may also cause keratoses which are relatively small, heaped up, scaling, brown plaques on the skin, some of which may be fissured and may itch [401]. Exposure also may cause reddening and squamous eczema of the eye lid margins with only small erosion of the corneal epithelium and superficial changes in the stroma which disappear a month following exposure. Chronic exposure to the fumes and dust of this type of compound may cause discoloration of the cornea and epithelioma of the eye lid margin [102]. Repeated exposure may cause sunlight to have more severe effects on a skin including allergic skin rash. Aplastic anemia may also occur.

Chrysene

CAS NUMBER: 218-01-9

SYNONYMS: 1,2-Benzophenanthrene Benzo(a)phenanthrene 1,2,5,6-Dibenzonaphthalene

CHEMICAL FORMULA: C18H12

MOLECULAR WEIGHT: 228.30 [703]

PHYSICAL DESCRIPTION: Orthorhombic bipyramidal plates from benzene. [703]

DENSITY: 1.274 [703]

MELTING POINT: 254 C [703]

BOILING POINT: 448 C [703]

SOLUBILITY: Water: Insoluble Alcohol: Slightly soluble Ether: Slightly soluble Carbon bisulfide: Slightly soluble Glacial acetic acid: Slightly soluble Boiling benzene: Moderately soluble [703]

Chrysene has been described as generally being only slightly soluble in cold organic solvents, but fairly soluble in some of these solvents when they are hot. [703]

OTHER PHYSICAL DATA: Sublimes easily in vacuo [703]

VOLATILITY: Not volatile

FIRE HAZARD:

Chrysene is combustible. Fires involving this compound may be controlled using a dry chemical, carbon dioxide or Halon extinguisher. A water fog may also be used.

LEL: Not available UEL: Not available

REACTIVITY: Not available

STABILITY: This compound is stable under normal laboratory conditions. When heated to decomposition it emits acrid smoke and fumes [703].

USES: Not available

COMMENTS: Chrysene is found in coal tar. It is formed during distillation of coal, and also in very small amounts during distillation or pyrolysis of many fats and oils. [703]

NIOSH Registry Number: GC0700000

ACUTE/CHRONIC HAZARDS: An experimental carcinogen, neoplastigen and tumorigen by skin contact; human mutagenic data exists. When heated to decomposition, emits acrid smoke and fumes. [703] Has been implicated as an etiologic determinant of chemical carcinogenesis. Able to induce aryl hydrocarbon hydroxylase in cultured human lymphocytes. A significant increase in enzyme induction occurred in chrysene-induced cultures compared with controls. No data are available in humans. Limited evidence of carcinogenicity in animals. [723] Chrysene may be a mild irritant.

SYMPTOMS: Not available

Phenanthrene

CAS NUMBER: 85-01-8

CHEMICAL FORMULA: C14H10

MOLECULAR WEIGHT: 178.24 [703]

PHYSICAL DESCRIPTION: Solid or monoclinic crystals [703]

DENSITY: 1.179 @ 25 C [703]

MELTING POINT: 100 C [703]

BOILING POINT: 339 C [703]

SOLUBILITY: Water: Insoluble [703] Hot alcohol: Soluble [703] Ether: Very soluble [703] Carbon Disulfide: Soluble [703] Benzene: Soluble [703]

OTHER PHYSICAL DATA: Vapor Pressure: 1 mm @ 118.3 C [703] Vapor Density: 6.14 [703]

VOLATILITY:

Vapor pressure: 1 mm Hg @ 20 C

Vapor density: 6.14

FIRE HAZARD: Phenanthrene has a flash point of 171 C (340 F). It is combustible. Fires involving this compound can be controlled using a dry chemical carbon dioxide or Halon extinguisher; a water spray can also be used.

LEL: Not available UEL: Not available

REACTIVITY: Phenanthrene may react with oxidizing materials.

STABILITY: Phenanthrene is stable under normal laboratory conditions.

USES: Phenanthrene is used in the synthesis of dyestuffs, explosives, and as an intermediate in many organic synthetic reactions.

COMMENTS: Phenanthrene is an ingredient in coke oven emissions.

NIOSH Registry Number: SF7175000

ACUTE/CHRONIC HAZARDS: Phenanthrene is an irritant and, when heated to decomposition it, emits acrid smoke and fumes.

SYMPTOMS: Symptoms of exposure may include skin sensitization, dermatitis, bronchitis, cough, dyspnea, respiratory neoplasm, kidney neoplasm, skin irritation, and respiratory irritation.

Pyrene

CAS NUMBER: 129-00-0

SYNONYMS: Benzo(def)phenanthrene

CHEMICAL FORMULA: C16H10

MOLECULAR WEIGHT: 202.26 [703]

PHYSICAL DESCRIPTION: Colorless solid, solutions have a slight blue color [703]

DENSITY: 1.271 @ 23 C [703]

MELTING POINT: 156 C [703]

BOILING POINT: 404 C [703]

SOLUBILITY: Water: Insoluble [703] Organic solvents: Fairly soluble [703]

OTHER PHYSICAL DATA: Melting Point: 151.2 C [706] A condensed ring hydrocarbon [703]

VOLATILITY:

Vapor pressure: 2.60 mm Hg @ 200.4 C; 6.90 mm Hg @ 220.8 C [039]

FIRE HAZARD: Flash point data for pyrene are not available. It is probably combustible. Fires involving this material may be controlled with a dry chemical, carbon dioxide or Halon extinguisher.

LEL: Not available UEL: Not available

REACTIVITY: Pyrene reacts with nitrogen oxides to form nitro derivatives. It also reacts with 70% nitric acid [395].

STABILITY: Pyrene is stable under normal laboratory conditions. Solutions of it in water, DMSO, 95% ethanol or acetone should be stable for 24 hours under normal lab conditions [700].

USES: Pyrene is used in biochemical research and as an intermediate in the synthesis of 3,4-benzpyrene and many other polyaromatic hydrocarbons.

COMMENTS: Tetracene impurities give this compound a yellow color. It occurs in coal tar [031].

NIOSH Registry Number: UR2450000

ACUTE/CHRONIC HAZARDS: Pyrene can be absorbed through the skin [062]. When heated to decomposition it emits acrid smoke and fumes [042].

SYMPTOMS: In rats, inhalation has caused hepatic, pulmonary and intragastric pathologic changes. Cutaneous absorption for 10 days has caused hyperemia, weight loss and hematopoietic changes. Applications for 30 days have produced dermatitis.

1,3-Propane sultone

CAS NUMBER: 1120-71-4

SYNONYMS: 1,2-Oxathiolane-2,2-dioxide

CHEMICAL FORMULA: C3H6O3S

MOLECULAR WEIGHT: 122.15 [702]

PHYSICAL DESCRIPTION: Not available

SPECIFIC GRAVITY: Not available

MELTING POINT: 31 C [702]

BOILING POINT: 180 C (at 30 mm) [702]

SOLUBILITY: Water: 100 mg/L [702]

OTHER PHYSICAL DATA: Heat of Combustion: 440.80 kcal/mole [702]

HAP WEIGHTING FACTOR: 1 [713]

VOLATILITY: Not available

FIRE HAZARD: Not available

LEL: Not available UEL: Not available

REACTIVITY: Not available

STABILITY: When heated to decomposition it emits toxic fumes of SOx [703].

UDRI Thermal Stability Class: 5 [702] UDRI Thermal Stability Ranking: 230 [702]

USES: Not available

COMMENTS: None

NIOSH Registry Number: RP5425OOO

ACUTE/CHRONIC HAZARDS: A poison by subcutaneous route; moderately toxic by skin contact and intraperitoneal routes. An experimental carcinogen, neoplastigen, tumorigen and teratogen; produces experimental reproductive effects. Human mutagenic data exists. It is implicated as a human brain carcinogen. A skin irritant. When heated to decomposition, emits toxic fumes of Sox. [703]

SYMPTOMS: Not available

---

beta-Propiolactone

CAS NUMBER: 57-57-8

SYNONYMS: 2-Oxetanone Hydracrylic acid beta-lactone 3-Propiolactone Propanilide 3-Hydroxypropionic acid beta-lactone

CHEMICAL FORMULA: C3H4O2

MOLECULAR WEIGHT: 72.06

WLN: T40VTJ

PHYSICAL DESCRIPTION: Colorless liquid

SPECIFIC GRAVITY: 1.1460 @ 20/4 C [031,205,395]

DENSITY: 1.148 g/mL @ 20 C [371]

MELTING POINT: -33.4 C [017,031,205,395,421]

BOILING POINT: 162 C (decomposes) [017,031,051,395,421]

SOLUBILITY: Water: 10-50 mg/mL @ 19 C [700] DMSO: >=100 mg/mL @ 19 C [700] 95% Ethanol: >=100 mg/mL @ 19 C [700] Acetone: >=100 mg/mL @ 19 C [700]

OTHER SOLVENTS: Benzene: Miscible [205] Ether: Miscible [031,062,205,395] Chloroform: Miscible [031,062,395] Acetic acid: Miscible [205] Most lipids: Miscible [395] Most organic solvents: Soluble [395,421]

OTHER PHYSICAL DATA: Irritating, pungent odor Refractive index: 1.4131 @ 20 C Boiling point: 150 C @ 750 mm Hg (decomposes); 61 C @ 20 mm Hg [031] Boiling point: 51-52 C @ 11 mm Hg [025] 51 C @ 10 mm Hg [017,031,047,395] Specific gravity: 1.1420 @ 25/4 C; 1.1490 @ 20/20 C [031] Gives acid solution Dipole moment: 3.8 Heat of combustion: -4730 calories/gram

HAP WEIGHTING FACTOR: 1 [713]

VOLATILITY:

Vapor pressure: 3.4 mm Hg @ 25 C [395]

Vapor density: 2.5

FIRE HAZARD: beta-Propiolactone has a flash point of 70 C (158 F) [031,205,269,274]. It is combustible. Fires involving this material may be controlled with a dry chemical, carbon dioxide or Halon extinguisher. A water spray may also be used [371].

LEL: 2.9% [051,071,371] UEL: Not available

REACTIVITY: beta-Propiolactone reacts with alcohol [062,205,395]. It also reacts with amino acids and cysteine [395]. It is incompatible with strong oxidizing agents and strong bases [269]. It is a nucleophilic alkylating agent which reacts readily with acetate, halogen, thiocyanate, thiosulphate, hydroxyl and sulphydryl ions [395]. In saline water, it reacts with chloride [071].

STABILITY: beta-Propiolactone is slowly hydrolyzed in water [031,036,205,346,395]. It has a half-life of about 3 hours in water @ 25 C [395]. It decomposes on standing at room temperature [421]. It can polymerize and rupture the container, especially at elevated temperatures. At 22 C, 0.04% polymerizes per day [371]. It is stable in glass containers at refrigerated temperatures [031, 062,421]. NMR stability screening indicates that solutions of it in DMSO are stable for less than 2 hours; and solutions in acetone are stable for at least 24 hours [700].

USES: beta-Propiolactone is used as a bactericide, fungicide, virucide, vapor sterilant and disinfectant. It is a versatile intermediate in organic synthesis, e.g., in place of acrylic acid in the Diels-Alder diene synthesis, as a chemical intermediate in the synthesis of acrylate plastics and the production of acrylic acid and esters. It is also used in intermediate sterilization of human grafts, and used in the sterilization of blood plasma, vaccines, surgical instruments, and enzymes.

COMMENTS: None

NIOSH Registry Number: RQ7350000

ACUTE/CHRONIC HAZARDS: beta-Propiolactone may be toxic by ingestion, inhalation and skin absorption [036,071,269,371]. It is a strong irritant [025,036,371] and it is corrosive [269]. It is a positive animal carcinogen [042,036,269,346,395], and it is also a human carcinogen [015,269,325,346]. When heated to decomposition it emits acrid smoke and toxic fumes of carbon monoxide and carbon dioxide [042,269].

SYMPTOMS: Symptoms of exposure may include severe irritation of the skin, eyes and respiratory tract [036,371]. Other symptoms may include corrosion and destruction of the tissues of the mucous membranes, upper respiratory tract, eyes, and skin. Inhalation may be fatal as a result of spasm, inflammation, and edema of the larynx and bronchi; chemical pneumonitis; and pulmonary edema [269]. Eye contact can cause tearing [371]. It can also cause corneal opacification [346]. Skin contact can lead to blistering [371]. Repeated or prolonged contact can cause erythema and vesication of the skin [346]. Inhalation can cause irritation of the nose and throat. Ingestion can cause burns of the mouth and stomach [371].

In experimental animals, skin contact has led to skin irritation ranging from erythema to hair loss and scarring [346,421]. Eye contact in animals has caused immediate pain, graying of the cornea, miosis, swollen lids, permanent opacification, and edema of the conjunctiva [099]. Other symptoms of exposure in experimental animals may include liver necrosis, renal tubular damage, and death due to rapid development of spasms, dyspnea, convulsions, and collapse [346].

Propionaldehyde

CAS NUMBER: 123-38-6

SYNONYMS: Methylacetaldehyde Propyl aldehyde Propanal Propylic aldehyde Propional Methyl acetaldehyde

CHEMICAL FORMULA: C2H5CHO

MOLECULAR WEIGHT: 58.08

WLN: VH2

PHYSICAL DESCRIPTION: Colorless liquid

SPECIFIC GRAVITY: 0.8058 @ 20/4 C

DENSITY: Not available

MELTING POINT: -81 C

BOILING POINT: 48.8 C

SOLUBILITY: Water: 50-100 mg/mL @ 18 C [700] DMSO: >=100 mg/mL @ 18 C [700] 95% Ethanol: >=100 mg/mL @ 18 C [700] Acetone: >=100 mg/mL @ 18 C [700]

OTHER SOLVENTS: Ether: Very soluble

OTHER PHYSICAL DATA: Refractive Index: 1.3636 @ 20 C; 1.36460 @ 19 C
Suffocating odor, stench Specific gravity: 0.8192 @ 9.7/4 C; 0.7898 @ 33/4 C

HAP WEIGHTING FACTOR: 1 [713]

VOLATILITY:

Vapor pressure: 235 mm Hg @ 20 C; 687 mm Hg @ 45 C [055]

Vapor density: 2.0 [043,055,371,451]

FIRE HAZARD: Propionaldehyde has a flash point of -9 C (15 F) [029,036,066,205] and it is flammable. Fires involving this material may be controlled with a dry chemical, carbon dioxide or Halon extinguisher.

The autoignition temperature of propionaldehyde is 207 C (405 F) [036,043,371,451].

LEL: 2.9% [036,043,066,451] UEL: 17% [036,043,066,451]

REACTIVITY: Propionaldehyde reacts vigorously with oxidizers [043,058,269]. It reacts with water [058]. It is incompatible with strong bases and strong reducing agents [269]. When mixed with methyl methacrylate, a rapid exothermic reaction may occur [043,066]. Polymerization may occur in the presence of acids or caustics [371].

STABILITY: Propionaldehyde may form explosive peroxides [029,036]. Solutions of it in water, DMSO, 95% ethanol or acetone should be stable for 24 hours under normal lab conditions [700].

USES: Propionaldehyde is used in the manufacture of propionic acid, polyvinyl acetals and other plastics, alkyl resins and plasticizers. It is also used as a disinfectant and preservative, and in the synthesis of rubber chemicals.

COMMENTS: None

NIOSH Registry Number: UE0350000

ACUTE/CHRONIC HAZARDS: Propionaldehyde may be toxic and an irritant. When heated to decomposition it emits toxic fumes.

SYMPTOMS: Symptoms of exposure may include eye, skin, mucous membrane and upper respiratory tract irritation; reddening of the skin, coughing, pulmonary edema, narcosis, nausea, vomiting, diarrhea and respiratory failure.

Propoxur

CAS NUMBER: 114-26-1

SYNONYMS: Baygon o-isopropoxyphenyl methylcarbamate

CHEMICAL FORMULA: $C_{11}H_{15}NO_3$

MOLECULAR WEIGHT: 209.27 [703]

PHYSICAL DESCRIPTION: A white to tan, crystalline solid [703]

SPECIFIC GRAVITY: Not available

MELTING POINT: 91.5 C [702]

BOILING POINT: Not available

SOLUBILITY: Water: slightly soluble [703] 0.2% [702] All polar organic solvents: soluble [703]

OTHER PHYSICAL DATA: Not available

HAP WEIGHTING FACTOR: 1 [713]

VOLATILITY:

Vapor Pressure: 9.7×10^{-6} mm Hg @ 20 C [719]

FIRE HAZARD: Not available

LEL: Not available UEL: Not available

REACTIVITY: Not available

STABILITY: When heated to decomposition it emits toxic fumes of NOx [703]. Unstable in alkaline media [702].

USES: Insecticide (trade name "Baygon") [709]

COMMENTS: None

NIOSH Registry Number: FC3150000

EDITOR'S NOTE:

Some of the following data is for the chemical mixture Baygon 1.5 as noted. Baygon 1.5, a carbamate insecticide, is one of at least three commercial forms of Propoxur.

ACUTE/CHRONIC HAZARDS: Propoxur is a poison via ingestion, subcutaneous, intraperitoneal, intravenous, intramuscular and possibly other routes. It is moderately toxic by inhalation and skin contact. Experimental reproductive effects and mutagenic data exist. It is moderately irritating to skin. When heated to decomposition it emits toxic fumes of NOx. [703]

Baygon 1.5: Inhalation, dermal absorption or ingestion of this material may result in systemic intoxication due to inhibition of the enzyme cholinesterase. The sequence of development of systemic effects varies with the route of entry, and the onset of symptoms may be delayed an hour or more (see SYMPTOMS). If exposure is excessively prolonged, breathing vapors may result in unconsciousness, kidney or lung damage, or even death. Animal studies have shown that this product is moderately toxic orally and minimally toxic dermally. It is a severe eye irritant, possibly causing irreversible eye damage. Chronic effects of exposure: High doses of propoxur induced bladder cancers when fed to rats in one study. Cancer was not induced in several other feeding studies on rats and other mammals. The implications of these studies for humans are not known. Repeated skin contact may result in defatting of the skin by solvents in the product which can lead to redness and irritation of the skin. Chronic overexposure to these solvent compounds may cause symptoms such as mucous membrane irritation, nausea, headache, loss of appetite, weakness and alcohol intolerance. [719]

SYMPTOMS: Baygon 1.5: First symptoms of poisoning may be nausea, increased salivation, lacrimation, blurred vision and constricted pupils. Other symptoms of systemic poisoning include vomiting, diarrhea, abdominal cramping, dizziness and sweating. After inhalation, respiratory symptoms like tightness of chest, wheezing, and laryngeal spasms may be pronounced at first. If the poisoning is severe, then symptoms of convulsions, low blood pressure, cardiac irregularities, loss of reflexes and coma may occur. In extreme cases, death may occur due to a combination of factors such as respiratory arrest, paralysis of respiratory muscles or intense bronchoconstrictions. Complete symptomatic recovery from sublethal poisoning usually occurs within 24-48 hours once the source of exposure is completely removed. The solvents in this product can be irritating to the eyes, nose and throat. In high concentration they may cause symptoms such as headaches, nausea, vomiting, lightheadedness and dizziness from overexposure by inhalation. [719]

Propylene dichloride

CAS NUMBER: 78-87-5

SYNONYMS: 1,2-Dichloropropane

CHEMICAL FORMULA: C3H6Cl2

MOLECULAR WEIGHT: 112.99

WLN: GY1&1G

PHYSICAL DESCRIPTION: Clear colorless liquid

SPECIFIC GRAVITY: 1.1560 @ 20/4 C [017,395]

DENSITY: 1.1558 g/mL @ 20 C [205]

MELTING POINT: -100 C [102,107,269,275]

BOILING POINT: 95-96 C [031,269,275,421]

SOLUBILITY: Water: <0.1 mg/mL @ 21.5 C [700] DMSO: >=100 mg/mL @ 19 C [700] 95% Ethanol: >=100 mg/mL @ 19 C [700] Acetone: >=100 mg/mL @ 19 C [700]

OTHER SOLVENTS: Chloroform: Soluble [017,395] Ether: Soluble [017,172,395,430] Benzene: Soluble [017,395] Ethanol: Soluble [172,395] Alcohol: Soluble [017,430] Most organic solvents: Miscible [031,421,062]

OTHER PHYSICAL DATA: Refractive index: 1.4384 @ 20 C [269,275] 1.437 @ 25 C [172,430] Chloroform-like odor [031,062,395,451] Solidifies below -70 C [031,421] Boiling point: -3.7 C @ 10 mm Hg [017] Specific gravity: 1.1593 @ 20/20 C [043,430] 1.159 @ 25/25 C [031] Specific gravity: 1.1481 @ 24/22 C [052] Density: 1.159 g/mL @ 25 C [421] Odor threshold: 50 ppm [102] Vapor pressure: 1 mm Hg @ -38.5 C; 5 mm Hg @ -17.0 C; 10 mm Hg @ -6.1 C [038] Vapor pressure: 20 mm Hg @ 6.0 C; 60 mm Hg @ 28.0 C; 100 mm Hg @ 39.4 C [038] Vapor pressure: 200 mm Hg @ 57.0 C; 400 mm Hg @ 76.0 C [038] Burning rate (estimated): 3.2 mm/min [371] Liquid surface tension: 29 dynes/cm @ 20 C [371] Liquid water interfacial tension: 37.9 dynes/cm [371] Latent heat of vaporization: 67.7 cal/g [371] Heat of combustion (estimated): 4100 cal/g [371] Heat of fusion: 13.53 cal/g [371] Fire point: 38 C [031] Evaporation rate (butyl acetate=1): >1 [102] 100% Volatile by volume [102] UV max: lambda max (in methanol): 268.5 nm, 264 nm (shoulder), 262 nm, 259 nm (shoulder), 256 nm, 249 nm (shoulder), 243 nm (epsilon=0.601, 0.551, 0.713, 0.625, 0.567, 0.424, 0.351) [052]

HAP WEIGHTING FACTOR: 1 [713]

VOLATILITY:

Vapor pressure: 40 mm Hg @ 19.4 C [038,043,395] 42 mm Hg @ 20 C [055]

Vapor density: 3.9 [043,055,395,451]

FIRE HAZARD: Propylene dichloride has a flash point of 15.6 C (60 F) [043,107,371,451] and it is flammable. Fires involving this material may be controlled with a dry chemical, carbon dioxide or Halon extinguisher. A water spray may also be used [058].

The autoignition temperature of propylene dichloride is 557 C (1035 F) [043, 107,371,451].

LEL: 3.4% [043,066,107,451] UEL: 14.5% [043,066,107,451]

REACTIVITY: Propylene dichloride reacts with strong oxidizers and strong acids [058,102, 269,346]. It also reacts with aluminum [043,058,269,451]. When confined, this reaction can lead to an explosion [043]. It is incompatible with bases and aluminum alloys [269]. It will attack some forms of plastics, rubber and coatings [102].

STABILITY: Propylene dichloride is sensitive to heat [102]. Solutions of it in water, DMSO, 95% ethanol or acetone should be stable for 24 hours under normal lab conditions [700].

USES: Propylene dichloride is used as an intermediate for perchloroethylene and carbon tetrachloride. It is also used as a lead scavenger for antiknock fluids, as a solvent for fats, oils, waxes, gums and resins, in solvent mixtures for cellulose esters and ethers, in scouring compounds, in spotting agents, in metal degreasing agents, as a soil fumigant for nematodes, as a grain fumigant and in dry cleaning fluids.

COMMENTS: Propylene dichloride is one of the most toxic chlorinated hydrocarbons [043].

NIOSH Registry Number: TX9625000

ACUTE/CHRONIC HAZARDS: Propylene dichloride is toxic by ingestion, inhalation and skin absorption [036,102,269,451]. It is an irritant [151,165,172,371] and is narcotic in high concentrations [031,058, 102,151]. When heated to decomposition it emits toxic fumes of carbon monoxide, carbon dioxide, chlorine, hydrogen chloride gas and phosgene [043,058,102,269].

SYMPTOMS: Symptoms of exposure may include irritation of the skin, eyes and respiratory tract [151]. It may also cause irritation of the mucous membranes of the nose, throat and lungs [058]. Other symptoms may include dermatitis by defatting of the skin and more severe irritation if confined against the skin [346]. Exposure can cause smarting of the eyes or respiratory system and smarting and reddening of the skin [371]. Exposure may also lead to hemolytic anemia, disseminated intravascular coagulation, necrosis and failure of the kidneys, liver failure and death [395]. Liver and kidney damage may occur [043,058,269,301] and heart damage may also occur [043] cardiac effects have also been reported [301].. Prolonged exposure may cause nausea, headache, vomiting and central nervous system depression [269]. High concentrations may cause narcosis [031,058,102,151]. Drowsiness and lightheadedness have been reported [102]. Eye contact may cause burning, tearing, reddening and swelling of the eye and surrounding tissue. Skin contact can also cause burning and swelling. Inhalation may lead to coughing, burning sensation, runny nose, sore throat and dizziness [058].

Propylene oxide

CAS NUMBER: 75-56-9

SYNONYMS: 1,2-Epoxy propane Methyl ethylene oxide Methyloxirane Propene oxide 1,2-Propylene oxide

CHEMICAL FORMULA: C3H6O

MOLECULAR WEIGHT: 58.1 [704]

PHYSICAL DESCRIPTION: Colorless liquid with a benzene-like odor; (Note: a gas above 94 F) [704]

SPECIFIC GRAVITY: 0.83 @ 20C/4C [704]

MELTING POINT: Not available

BOILING POINT: 94 F [704]

SOLUBILITY: Water: 41% [704]

OTHER PHYSICAL DATA: Vapor Pressure: 445 mm [704] Freezing Point: -170 F [704]

HAP WEIGHTING FACTOR: 1 [713]

VOLATILITY:

Vapor Pressure: 445 mm [704]

FIRE HAZARD: Flash Point: -35 F [704]

Class IA Flammable Liquid [704]. Severe explosion hazard when exposed to flame. [702]

Autoignition temperature: Not available

LEL: 2.3% UEL: 36

%REACTIVITY: Incompatible or reacts with anhydrous metal chlorides; iron; strong acids, caustics and peroxides. Polymerization may occur due to contamination with alkalis, aqueous acids, amines and acidic alcohols. [704]

Epoxy resin: Mixing propylene oxide and epoxy resin in a waste bottle led to an explosion, probably owing to the polymerization of the oxide catalyzed by the amine accelerator in the resin. [710]

Sodium hydroxide: A drum of crude product containing unreacted propylene oxide and sodium hydroxide catalyst exploded and ignited, probably owing to base-catalyzed exothermic polymerization of the oxide. [710]

STABILITY: Polymerization may occur due to high temperatures [704].

USES: Not available

COMMENTS: Its use as a biological sterilant is hazardous because of ready formation of explosive mixtures with air (2.8-37%). Mixtures with CO_2, though non-explosive, may be asphyxiant and vesicant. these may be ineffective, but neat propylene oxide vapor may be used safely if it is removed by evacuation by water-jet pump. The heat of decomposition has been determined as 1.12 kJ/g. [710]

NIOSH Registry Number: TZ2975000

ACUTE/CHRONIC HAZARDS: Propylene oxide is moderately toxic via oral, inhalation, and dermal routes. [702] Targeted organs include the blood, skin, and respiratory system. [704] 1,2-Propylene oxide is an irritant inhalant that may cause conjunctivitis, corneal burns, and irritation of lungs and skin. Injuries to internal organs, other than the lungs, appears to be insignificant. Exposures to vapors in high concentrations may also cause CNS depression, motor weakness, vomiting and diarrhea. Propylene oxide has produced elevated blood alkylhistidine and chromosomal aberrations in humans. [723]

SYMPTOMS: Symptoms of exposure may include irritation of the eyes, upper respiratory system, and lungs; skin irritation, blistering and burns. [704] Propylene oxide is an irritant that may cause cough, dyspnea, pulmonary edema, and lead to pneumonia. Cyanosis has been seen. In high concentrations it may cause headache, motor weakness, incoordination, ataxia, and coma. Vomiting and diarrhea may be seen after both oral or inhalation exposure. Pure propylene oxide may evaporate without causing burns; if confined to the skin by clothing, it may cause irritation or necrosis. Propylene oxide appears to induce cancers at the site of exposure. Sarcomas were seen at injection sites, and nasal and G.I. cancers were seen with chronic exposures to animals. Propylene oxide can cause dermatitis. Persons with existing skin disorders may be more susceptible to the effects of this agent. [723]

1,2-Propylenimine

CAS NUMBER: 75-55-8

SYNONYMS: 2-Methylazacylopropane 2-Methylaziridine Methylethylenimeine 1,2-Propyleneimine Propylene imine Propylenimine

CHEMICAL FORMULA: C_3H_7N

MOLECULAR WEIGHT: 57.10

WLN: T3MTJ B1

PHYSICAL DESCRIPTION: Colorless, oily, fuming liquid

SPECIFIC GRAVITY: 0.802 [371,395]

DENSITY: 1.4094-1.4109 g/mL @ 25 C [062]

MELTING POINT: -65 C [371,430]

BOILING POINT: 66 C [102,205,346,371]

SOLUBILITY: Water: >=100 mg/mL @ 19 C [700] DMSO: >=100 mg/mL @ 19 C [700] 95% Ethanol: >=100 mg/mL @ 19 C [700] Acetone: >=100 mg/mL @ 19 C [700]

OTHER SOLVENTS: Alcohol: Soluble [205] Petroleum ether: Soluble [205] Most organic solvents: Soluble [062]

OTHER PHYSICAL DATA: Strong ammonia-like odor [102,346,371] Refractive index: 1.4084 [205]

HAP WEIGHTING FACTOR: 10 [713]

VOLATILITY:

Vapor pressure: 112 mm Hg @ 20 C [430]

FIRE HAZARD: 1,2-Propylenimine has a flash point of -15 C (5 F) [275] and it is flammable. Fires involving this material may be controlled with a dry chemical, carbon dioxide or Halon extinguisher.

LEL: Not available UEL: Not available

REACTIVITY: 1,2-Propylenimine may polymerize explosively in the presence of acids or acid fumes [066,102]. It is incompatible with strong oxidizers [102,346]. It reacts with carbonyl compounds, quinones and sulphonyl halides [395]. It may attack some forms of plastics, rubber and coatings [102].

STABILITY: 1,2-Propylenimine is heat sensitive [102,346]. It may decompose in air [025]. It polymerizes easily if uninhibited. It hydrolyzes in hydrochloric acid solutions [395,430]. It reacts slowly with water. It is stable if kept in contact with solid caustic soda (sodium hydroxide) [371].

USES: 1,2-Propylenimine is an organic intermediate used in the production of polymers for use in the paper and textile industries as coatings and adhesives. Its derivatives are also used in the rubber and pharmaceutical industries. It is also used in the modification of latex surface coating resins to improve adhesion.

COMMENTS: None

NIOSH Registry Number: CM8050000

ACUTE/CHRONIC HAZARDS: 1,2-Propylenimine may be toxic by ingestion, inhalation and skin absorption [062]. When heated to decomposition it emits irritating vapors and toxic fumes of nitrogen oxides [042,371].

SYMPTOMS: Symptoms of exposure may include eye irritation [042,102, 301,371]. It also causes irritation of the nose and throat. If inhaled, it may cause nausea, vomiting and difficult breathing [371]. Skin and eye burns may occur [346,371]. Exposure has also resulted in eye damage [099,102,301]. Other symptoms include central nervous system effects [301].

Quinoline

CAS NUMBER: 91-25-5

SYNONYMS: 1-Azanaphthalene Benzo(b)pyridine

CHEMICAL FORMULA: C9H7N

MOLECULAR WEIGHT: 129.16

WLN: T66 BNJ

PHYSICAL DESCRIPTION: Colorless to yellow liquid

SPECIFIC GRAVITY: 1.0929 @ 20/4 C [017,047,430]

DENSITY: 1.095 g/mL @ 20 C [371]

MELTING POINT: -15 C [031,038,058]

BOILING POINT: 238 C [017,036,346,451]

SOLUBILITY: Water: <0.1 mg/mL @ 22.5 C [700] DMSO: >=100 mg/mL @ 24 C [700] 95% Ethanol: >=100 mg/mL @ 24 C [700] Acetone: >=100 mg/mL @ 24 C [700]

OTHER SOLVENTS: Alcohol: Miscible [031,043,205] Carbon disulfide: Soluble [043,062,430] Ether: Miscible [031,043,205] Benzene: Soluble [017,430]

OTHER PHYSICAL DATA: Specific gravity: 1.0900 @ 25/4 C [031,043] Boiling point: 163.2 C @ 100 mm Hg; 136.7 C @ 40 mm Hg [031] Boiling point: 119.8 C @ 20 mm Hg [031] 114 C @ 17 mm Hg [017,025,047] Vapor pressure: 10 mm Hg @ 103.8 C; 20 mm Hg @ 119.8 C [038] Vapor pressure: 40 mm Hg @ 136.7 C; 60 mm Hg @ 148.1 C [038] Burning rate: 4.06 mm/min [371] Penetrating amine odor [346] Odor threshold: 71 ppm [371] Refractive index: 1.6268 @ 20 C [017,025,047,430] Critical temperature: 509 C [371] Liquid surface tension: 45.0 dynes/cm [371] Latent

heat of vaporization (estimated): 86 cal/g [371] Heat of combustion: -8710 cal/g [371]
log P octanol: 2.03/2.06 [055] Neutral to phenolphthalein [031]

HAP WEIGHTING FACTOR: 1 [713]

VOLATILITY:

Vapor pressure: 1 mm Hg @ 59.7 C [038,043,055,430] 5 mm Hg @ 89.6 C [038]

Vapor density: 4.45 [043,055,058,430]

FIRE HAZARD: Quinoline has a flash point of 101 C (214 F) [058,205,275,430]. It is combustible. Fires involving this material may be controlled with a dry chemical, carbon dioxide or Halon extinguisher. A water spray may also be used [058,269,371].

The autoignition temperature of quinoline is 480 C (896 F) [043, 062, 371, 451].

LEL: Not available UEL: Not available

REACTIVITY: Quinoline is a weak base [031]. A potentially explosive reaction may occur with hydrogen peroxide [043,066]. It reacts violently with dinitrogen tetraoxide [043,058,066,269]. It also reacts violently with perchromates [043,269,451]. It is incompatible with (linseed oil + thionyl chloride) and maleic anhydride [043,058,066]. It is also incompatible with strong oxidizers and strong acids [058,269]. Quinoline can be unpredictably violent [043]. It dissolves sulfur, phosphorus and arsenic trioxide [031]. It may attack some forms of plastics [371]. It is a preparative hazard [066].

STABILITY: Quinoline is hygroscopic [031,036,269,430]. It absorbs as much as 22% water [031,430]. It is sensitive to light and moisture [031]. It darkens on storage [031,062]. Solutions of it in water, DMSO, 95% ethanol or acetone should be stable for 24 hours under normal lab conditions [700].

USES: Quinoline is used as a solvent for decarboxylations, resins and terpenes, as a dehydrohalogenation reagent, as a preservative for anatomical specimens, as a flavoring and in medicine as an antimalarial agent. It is also used in the manufacture of dyes, niacin, copper-8-quinolinolate, antiseptics, fungicides, pharmaceuticals and 8-hydroxyquinoline sulfate (which is used as an antiseptic, antiperspirant and deodorant). It is used in the manufacture of quinoline derivatives and in synthetic fuel manufacture.

COMMENTS: Quinoline occurs in coal tar and in "stupp" fat [025].

NIOSH Registry Number: VA9275000

ACUTE/CHRONIC HAZARDS: Quinoline may be toxic by most routes [430]. It may be harmful by ingestion, inhalation and skin absorption [058,269] and it is an irritant [036,058,269]. When heated to decomposition it emits toxic fumes of carbon monoxide, carbon dioxide and nitrogen oxides [043,058,269,371].

SYMPTOMS: Symptoms of exposure may include irritation of the skin, eyes, mucous membranes and upper respiratory tract [058]. Retinitis also may occur [043]. Other symptoms of exposure may include methemoglobinemia, corneal damage, liver damage and kidney damage [301]. Exposure can cause nausea, vomiting, gastrointestinal cramps, fever, dizziness, irregular and rapid pulse and collapse [058,430]. It can also cause irritation of the nose and throat, and headache [371]. Central nervous system effects and toxicity of the retina or optic nerve may also occur [058].

Ingestion may cause irritation of the mouth and stomach [371]. It may also cause drowsiness and abdominal pain. Inhalation may cause coughing and shortness of breath. Skin contact may cause burns, redness pain. It can penetrate the skin to produce internal damage. Eye contact may cause burns and blurred vision [058].

Individuals with pre-existing disease in, or a history of ailments involving the skin, mucous membranes, eyes or central nervous system may be at a greater risk of adverse health effects when exposed to this material. The use of alcoholic beverages may enhance its toxic effects [058].

Quinone

CAS NUMBER: 106-51-4

SYNONYMS: p-Benzoquinone 1,4-Cyclohexadienedione 2,5-Cyclohexadiene-1,4-dione 1,4-Dioxybenzene p-Quinone

CHEMICAL FORMULA: $C6H4O2$

MOLECULAR WEIGHT: 108.10

WLN: L6V DVJ

PHYSICAL DESCRIPTION: Yellow powder

SPECIFIC GRAVITY: 1.318 @ 20/4 C

DENSITY: Not available

MELTING POINT: 115.7 C

BOILING POINT: Sublimes

SOLUBILITY: Water: Slightly soluble DMSO: >=10 mg/mL 95% Ethanol: Soluble Acetone: >=10 mg/mL

OTHER SOLVENTS: Petroleum Ether: Soluble Ligroin: Hot solvent Alkalies: Soluble Ether: Soluble

OTHER PHYSICAL DATA: Penetrating odor resembling that of chlorine Dipole moment: 0.67 Odor perceptible @ or just above 0.1 ppm

HAP WEIGHTING FACTOR: 1 [713]

VOLATILITY: Not available

FIRE HAZARD: The flash point of p-quinone is 293 C. It is combustible. Fires involving this compound can be controlled using dry chemicals, carbon dioxide, or Halon extinguishers.

LEL: Not available UEL: Not available

REACTIVITY: p-Quinone acts as an oxidizing agent and is a very reactive compound.

STABILITY: p-Quinone is unstable to light and heat. It is also unstable in the presence of many organic compounds.

USES: p-Quinone is used as an oxidizing agent, in photography, in the manufacture of dyes, in the manufacture of hydroquinone, for tanning hides, for strengthening animal fibers, as a reagent, and as a fungicides.

COMMENTS: None

NIOSH Registry Number: DK2625000

ACUTE/CHRONIC HAZARDS: Quinone may be toxic and a severe irritant. It may also be an animal carcinogen.

SYMPTOMS: Symptoms of exposure may include dermatitis with discoloration, erythema, formation of papules and vesicles. In severe cases there can be necrotic changes in the skin. Vapors acting on eye can cause serious disturbances, including conjunctivitis and even corneal ulceration. Discoloration of conjunctiva and corneal have been reported.

Radionuclides (including Radon)

NOTE: Radionuclides include atoms which spontaneously undergo radioactive decay. [708] Radium and Radon are two examples of this type of atom.

HAP WEIGHTING FACTOR: The June 1991 proposed NESHAP's rule regarding HAP Weighting Factors did not specifically address radionuclides. EPA advises that a method for calculating radionuclide reductions for early reduction demonstrations will be addressed in the final rule.

Radium

CAS NUMBER: 7440-14-4

ATOMIC FORMULA: Ra

ATOMIC WEIGHT: 226.025 [703]

PHYSICAL DESCRIPTION: A radioactive earth metal; brilliant white, tarnishes in air [703]

DENSITY: 5.5 @ 20C/4C [703]

MELTING POINT: 700 C [703]

BOILING POINT: 1737 C [703]

SOLUBILITY: Water: Decomposes [703]

OTHER PHYSICAL DATA: Not available

VOLATILITY: Not available

FIRE HAZARD: Not available

LEL: Not available UEL: Not available

REACTIVITY: Not available

RADIATION HAZARD:

Natural isotope 223Ra, T 1/2 = 11.4 D, decays to radioactive 219Rn by alphas of 5.5-5.7 MeV. Natural isotope 224Ra, T 1/2 = 3.6 D, decays to radioactive 220Rn by alphas of 5.7 MeV. Natural isotope 226Ra, T 1/2 = 1600 Y, decays to radioactive 222Rn by alphas of 4.8 MeV. Natural isotope 228Ra, T 1/2 = 6.7 Y, decays to radioactive 228Ac by betas of 0.05 MeV. [703]

USES: Radium is used in producing self-luminous paints, neutron sources, and in medicine for the treatment of cancer and other diseases. [706]

COMMENTS: Stored radium should be ventilated to prevent build-up of radon. Inhalation, injection, or body exposure to radium can cause cancer and other body disorders. [706]

Radium is very dangerous, and it must be kept heavily shielded and stored away from possible dissemination by explosion, flood, etc. Ingestion of luminous dial paint prepared from radium caused death in many early dial painters. 226Ra is the parent of radon and the precautions described under 222Rn should be followed. 228Ra is a source of thoron and is a member of the thorium series. [703]

NIOSH Registry Number: Not available

ACUTE/CHRONIC HAZARDS: Radium is a highly radiotoxic element by many routes including inhalation, ingestion, or bodily exposure. It can cause lung cancer, bone cancer, osteitis, skin damage, and blood dyscrasias. Radium can become a source of irradiation to the blood-forming organs by replacing calcium. [703] Several deaths due to aplastic anemia have also been described; the effects on the white corpuscles vary but the main findings are leukopenia, leukocytosis, and leukemia. Radium may cause pulmonary fibrosis. [723]

SYMPTOMS: Symptoms of exposure to radium may include nausea, vomiting, diarrhea, intestinal cramps, salivation, and dehydration. Fatigue, weakness, apathy, fever, and hypotension are the result of neurovascular dysfunction. It also produces such abnormalities as pancytopenia with subsequent infections, thrombocytopenia, hemorrhage, and anemia. The order of bone marrow loss is lymphocytes, granulocytes, platelets, and red blood cells. CNS syndrome: Results of experiments with primates suggest that the "brain" form of the acute radiation syndrome in humans should include mental cloudiness, listlessness, drowsiness, and weakness, followed swiftly by severe apathy, prostration, and lethargy. Although psychological and behavioral disturbances can be expected to occur on exposure to even relatively low doses, the early development of frank objective neurologic signs, such as ataxia, nystagmus, and muscle tremor, will require massive doses. Convulsive seizures as a component of the acute radiation syndrome in humans are rare. After total body irradiation, the early signs and symptoms of nonfatal central nervous system (CNS) injury may disappear prior to the appearance of the gastrointestinal syndrome. When an individual has been exposed to a nonuniform dose, with a relatively high dose to the head, transitory CNS symptoms may develop, but survival is possible provided the dose to hemopoietic tissue and intestines is sufficiently low. [723]

Following whole body exposure, acute radiation syndromes may be divided into cerebral, gastrointestinal (GI), and hematopoietic syndromes. The cerebral syndrome is characterized by vomiting and drowsiness, followed by tremors, ataxia, convulsions, and death within 24 to 72 hours. With acute whole-body doses in the range of 5000 rads or more, constellation of prodromal, CNS, and GI symptoms that include hypotension and shock leading to death. The GI syndrome occurs in the range of 600-2000 rads of acute exposure. Radiation causes death of the crypt cells, resulting ultimately in the intestinal villi becoming denuded; ulceration and hemorrhage develop. The hematopoietic syndrome occurs with acute exposure in the range of 200 to 1000 rads. Leukopenia usually occurs within 48 hours as lymphocytes are highly sensitive to radiation. A valuable diagnostic test is a complete blood count 48 hours after exposure. [723]

Symptoms of external exposure may include dark pigmentation and desquamation of the skin on hands and neck. The fingertips may lose their sensitivity and become wasted; nodules and fissuring may then occur followed by the development of malignant growths. Other findings include bone necrosis in the fingers, and with internal contamination, spontaneous fracturing of the bones, jaw necrosis, osteosarcoma, and carcinoma of the paranasal sinuses. [723]

Radon

CAS NUMBER: 14859-67-7

ATOMIC FORMULA: Rn

ATOMIC WEIGHT: 222 [703]

PHYSICAL DESCRIPTION: Colorless, odorless, inert gas; very dense [703]

SPECIFIC GRAVITY: 4.4 @ -62 C (liquid) [706]

DENSITY: (gas @ 1 atm and 0 C): 9.73 g/L [703] (liquid @ boiling point): 4.4 [703]

MELTING POINT: -71 C [706]

BOILING POINT: -62 C [703]

SOLUBILITY: Water: 224 CC/L [702]

OTHER PHYSICAL DATA: Not available

HAP WEIGHTING FACTOR: 1 [713]

VOLATILITY: Not available

FIRE HAZARD: Not available

LEL: Not available UEL: Not available

REACTIVITY: Radon is a noble gas and thus is relatively unreactive [703]. It has been reported that fluorine reacts with radon, forming radon fluoride. [706]

RADIATION HAZARD:

Natural isotope 220Rn (Thoron, Thorium Series), T 1/2 = 55s, decays to radioactive 216Po by alphas of 6.3 MeV. Natural isotope 222Rn (Uranium Series), T 1/2 = 3.8 D, decays to radioactive 218Po by alphas of 5.5 MeV. The permissible levels are given for 222Rn in equilibrium with its daughters. [703]

USES: Radon is still produced for therapeutic use by a few hospitals in minute tubes, called seeds or needles, for application to patients. This practice has now been largely discontinued. [706]

COMMENTS: Where radium is handled or stored, radon and its daughters build up to an equilibrium value in about a month. Therefore, such areas must be well ventilated. Increased incidence of lung cancer has been associated with elevated levels of radon in homes.. [703]

NIOSH Registry Number: Not available

ACUTE/CHRONIC HAZARDS: The chief hazard is inhalation of gaseous radon and its solid daughters. The solid material travels on dust and similar particulates in air where it is is deposited in the lungs. These particulates have been considered to be a major causative agent in the high incidence of lung cancer found in uranium miners. [703] Radon and its decay products are carcinogenic to humans. Most radon-induced lung cancers are bronchogenic. Histologic lung cancer types that have been associated with environmental radon exposure are squamous cell carcinoma and adenocarcinoma. Histologic lung cancer types noted in uranium miners were predominantly small cell carcinomas, although all cell types were represented. There is an additive relationship between radon exposure and smoking for lung cancer risk. [723]

SYMPTOMS: Not available

Selenium compounds

NOTE: For all listings which contain the word "compounds," the following applies: Unless otherwise specified, these listings are defined as including any unique chemical substance that contains the named chemical as part of that chemical's infrastructure. [708] Information for the elemental form of selenium is given below. Other compounds described are intended as selected examples of selenium compounds and do not comprise a comprehensive list.

HAP WEIGHTING FACTOR: 1 [713]

Selenium

CAS NUMBER: 7782-49-2

ATOMIC FORMULA: Se

ATOMIC WEIGHT: 78.96 [703]

PHYSICAL DESCRIPTION: Steel gray, non-metallic element [703] Element: Amorphous or crystalline, red to gray solid [704]

SPECIFIC GRAVITY: 4.28 @ 20C/4C [704]

MELTING POINT: 170-217 C [703]

BOILING POINT: 690 C [703]

SOLUBILITY: Water: Insoluble [703] Alcohol: Insoluble [703] Ether: Very slightly soluble [703]

OTHER PHYSICAL DATA: Density: 4.81-4.26 [703] Vapor Pressure: 1 mm @ 356 C [703]

VOLATILITY:

Vapor Pressure: 1 mm @ 356 C [703]

FIRE HAZARD: Flash Point: Not available

Combustible solid [704].

Autoignition temperature: Not available

LEL: Not available UEL: Not available

REACTIVITY: Forms explosive products with metal amides. Can react violently with barium carbide; bromine pentafluoride; calcium carbide; chlorates; chlorine trifluoride; chromic oxide; fluorine; lithium carbide; lithium silicon; nickel; nitric acid; nitrogen trichloride; oxygen; potassium; potassium bromate; rubidium carbide; silver bromate; sodium; strontium carbide; thorium carbide; uranium; zinc. [703]

STABILITY: When heated to decomposition it emits toxic fumes of Se [703].

USES: Selenium exhibits both photovoltaic action, where light is converted directly into electricity, and photoconductive action, where the electrical resistance decreases with increased illumination. These properties make selenium useful in the production of photocells and exposure meters for photographic use, as well as solar cells. Selenium is also able to convert alternating current. electricity to direct current., and is extensively used in rectifiers Belowi ts melting point selenium is a p-type semiconductor and is finding many uses in electronic and solid-state applications. It is used in Xerography for reproducing and copying documents, letters, etc. Selenium also is used in the glass industry to decolorize glass and to make ruby-colored glasses and enamels. It is also used as a photographic toner, and as an additive to stainless steel. [706]

COMMENTS: None

NIOSH Registry Number: VS7700000

ACUTE/CHRONIC HAZARDS: A poison by inhalation, intravenous and possibly other routes. An experimental tumorigen and teratogen; experimental reproductive effects; liver damage in experimental animals. Consumption of plants grown in seleniferous areas can cause effects in humans and animals (associated with "alkali disease" in cattle at 5-25 mg/kg). Selenosis in humans has occurred from ingestion of 3.2 mg per day. [703]

SYMPTOMS: Symptoms of exposure may include irritation of the eyes, nose and throat; visual disturbance; headache; chills and fever; dyspnea, bronchitis; metallic taste, garlic breath, gastrointestinal disturbances; dermatitis; skin and eye burns; and in animals:

anemia; liver and kidney damage. [704] Occupational exposure has caused pallor, nervousness, depression, garlic odor of breath and sweat, GI disturbances and dermatitis. In Chinese villages chronic ingestion of 5 mg/day resulted in symptoms including brittle hair, brittle nails with spots and streaks, skin lesions, peripheral anesthesia, acroparaesthesia, pain, and hyperflexia. Similar effects noted in populations with blood levels of 800æg/L. [703]

Selenious acid

CAS NUMBER: 7783-00-8

SYNONYMS: Monohydrated selenium dioxide Selenous acid

CHEMICAL FORMULA: H2O3Se

MOLECULAR WEIGHT: 129.88

WLN: H2 SE-O3

PHYSICAL DESCRIPTION: Colorless crystals

SPECIFIC GRAVITY: 3.004 @ 16/4 C

DENSITY: Not available

MELTING POINT: 70 C

BOILING POINT: Not available

SOLUBILITY: Water: Slightly soluble (very hot) DMSO: Not available 95% Ethanol: Very slightly Acetone: Not available

OTHER SOLVENTS: Ammonia: Insoluble

OTHER PHYSICAL DATA: Not available

VOLATILITY:

Vapor Pressure (mm Hg): 2 @ 15 C; 4.5 @ 35 C; 7 @ 40.3 C [715]

FIRE HAZARD: Not available

LEL: Not available UEL: Not available

REACTIVITY: Oxidized to selenic acid by strong oxidizing agents such as ozone, hydrogen dioxide, chlorine. Reduced to selenium by most reducing agents including hydriodic acid, sulfurous acid, sodium hyposulfite, hydroxylamine salts, hydrazine salts, hypophosphorous acid, phosphorous acid. [715]

STABILITY: Gives off water upon heating and selenium oxide sublimes [715].

USES: Selenious acid is used as a reagent for alkaloids and as an oxidizing agent.

COMMENTS: None

NIOSH Registry Number: VS7175000

ACUTE/CHRONIC HAZARDS: Not available

SYMPTOMS: Not available

Selenium disulfide

CAS NUMBER: 7488-56-4

SYNONYMS: Selenium (IV) disulfide (1:2)

CHEMICAL FORMULA: SeS2

MOLECULAR WEIGHT: 143.09

WLN: SE S2

PHYSICAL DESCRIPTION: Bright orange powder

SPECIFIC GRAVITY: Not available

DENSITY: Not available

MELTING POINT: <100 C

BOILING POINT: Decomposes

SOLUBILITY: Water: Not available DMSO: Not available 95% Ethanol: Not available Acetone: Not available

OTHER SOLVENTS: Carbon disulfide: Soluble Ammonium sulfide: Soluble Aqueous reagent nitric acid: Decomposes

OTHER PHYSICAL DATA: Faint odor

VOLATILITY: Not available

FIRE HAZARD: Selenium disulfide is nonflammable.

LEL: Not available UEL: Not available

REACTIVITY: Not available

STABILITY: Selenium disulfide is stable under normal laboratory conditions.

USES: Selenium disulfide is used for the treatment of seborrheic dermatitis, dandruff, eczemas, dermatomycoses and nonspecific dermatoses.

COMMENTS: None

NIOSH Registry Number: VS8925000

ACUTE/CHRONIC HAZARDS: Selenium disulfide may be toxic and an irritant. Slight absorption occurs after local application to normal skin, but it is absorbed more readily on inflamed or damaged skin. When heated to decomposition it emits toxic fumes.

SYMPTOMS: Symptoms of exposure may include dermatitis, nausea, irritation of the skin, eyes and mucous membranes, garlic odor of breath, pallor, nervousness, depression, gastrointestinal disturbance, giddiness and keratitis. Prolonged contact with the skin may result in burns and contact dermatitis.

Selenium sulfide

CAS NUMBER: 7446-34-6

SYNONYMS: Selenium monosulfide

CHEMICAL FORMULA: SeS

MOLECULAR WEIGHT: 111.02

WLN: SE S

PHYSICAL DESCRIPTION: Orange-yellow tablets or powder

SPECIFIC GRAVITY: Not available

DENSITY: 3.056 g/mL @ 0 C [016,043]

MELTING POINT: 111.03 C [043]

BOILING POINT: Decomposes @ 118-119 C [016,043]

SOLUBILITY: Water: <1 mg/mL @ 21 C [700] DMSO: <1 mg/mL @ 21 C [700] 95% Ethanol: <1 mg/mL @ 21 C [700] Acetone: <1 mg/mL @ 21 C [700] Methanol: <1 mg/mL @ 18 C [700] Toluene: <1 mg/mL @ 18 C [700]

OTHER SOLVENTS: Ether: 1 in 1650 [295] Carbon disulfide: Soluble [016] Chloroform: 1 in 160 [295] Most organic solvents: Practically insoluble [295,455]

OTHER PHYSICAL DATA: Faint odor [295,455]

VOLATILITY: Not available

FIRE HAZARD: Flash point data for selenium sulfide are not available. It is probably combustible. Fires involving this material may be controlled with a dry chemical, carbon dioxide or Halon extinguisher.

LEL: Not available UEL: Not available

REACTIVITY: Selenium sulfide ignition takes place when this compound is ground with silver oxide [451].

STABILITY: Selenium sulfide is stable under normal laboratory conditions. Solutions of it in water, DMSO, 95% ethanol or acetone should be stable for 24 hours under normal lab conditions [700].

USES: Selenium sulfide is an antifungal and antiseborrheic used in combination with a detergent as a shampoo for the treatment of dandruff and as an ointment for application to the lid margins in the treatment of seborrheic blepharitis. It is used topically in veterinary medicine to treat eczemas and dermatomycoses.

COMMENTS: None

NIOSH Registry Number: VT0525000

ACUTE/CHRONIC HAZARDS: Selenium sulfide may be toxic by ingestion [036,151,295,455]. It may be toxic by inhalation [036]. With this type of compound there is a danger of cumulative effects [036]. When heated to decomposition it emits toxic fumes of sulfur oxides and selenium [043].

SYMPTOMS: Symptoms of exposure may include vomiting, anorexia, anemia and fatty degeneration of the liver [295,455]. It may also cause a garlic odor of the breath, metallic taste, weakness, tremors and sweating [295]. Sensitization may occur [151,295,301]. Repeated use may cause loss of hair [295,301,455]. Eye effects may include irritation and keratitis [099]. It may cause conjunctivitis [406,455]. It may also cause eye injuries [151]. Local irritation rarely occurs [151]. Other symptoms may include skin burns, dermatitis, increased oiliness or dryness of hair, and orange tinting of gray hair [406].

Styrene

CAS NUMBER: 100-42-5

SYNONYMS: Ethenylbenzene Phenylethylene Vinylbenzene

CHEMICAL FORMULA: C8H8

MOLECULAR WEIGHT: 104.16

WLN: 1U1R

PHYSICAL DESCRIPTION: Colorless to yellowish, oily liquid

SPECIFIC GRAVITY: 0.9060 @ 20/4 C

DENSITY: 0.909 g/cc @ 20 C [700]

MELTING POINT: -31 to -30.6 C

BOILING POINT: 145-146 C

SOLUBILITY: Water: <1 mg/mL @ 19 C [700] DMSO: >=100 mg/mL @ 19 C [700] 95% Ethanol: >=100 mg/mL @ 19 C [700] Acetone: >=100 mg/mL @ 19 C [700]

OTHER SOLVENTS: Methanol: Soluble Carbon disulfide: Soluble Petroleum ether: Soluble Ether: Soluble Benzene: Soluble

OTHER PHYSICAL DATA: Liquid surface tension: 32.14 dynes/cm Liquid-water interfacial tension: 35.48 dynes/cm Critical temperature: 373 C Critical pressure: 39.46 atm Odor threshold: 0.148 ppm Latent heat of vaporization: 156 Btu/lb Heat of polymerization: -277 Btu/lb Penetrating odor

HAP WEIGHTING FACTOR: 1 [713]

VOLATILITY:

Vapor pressure: 4.3 mm Hg @ 15 C; 9.5 mm Hg @ 30 C; 10 mm Hg @ 35 C

Vapor density: 1.1

FIRE HAZARD: The flash point of styrene is 31.1 C (88 F) and it is flammable. Fires involving this chemical should be controlled using a dry chemical, carbon dioxide or Halon extinguisher.

The autoignition temperature of styrene is 490 C (914 F).

LEL: 1.1% UEL: 6.1

%REACTIVITY: Styrene reacts violently with chlorosulfonic acid, oleum, sulfuric acid, and alkali metal graphite. It also reacts vigorously with oxidizing materials. Styrene readily undergoes polymerization when heated or exposed to light or a peroxide catalyst. The polymerization releases heat and it may react violently. It is incompatible with aluminum chloride. It can corrode copper and copper alloys.

STABILITY: Styrene is sensitive to heat, light and may be sensitive to air. UV spectrophotometric stability screening indicates that solutions of it in ethanol are stable for 24 hours [700].

USES: Styrene is used in the manufacture of plastics, synthetic rubber, resins, and in insulators.

COMMENTS: None

NIOSH Registry Number: WL3675000

ACUTE/CHRONIC HAZARDS: Styrene is an irritant and it may be narcotic in high concentrations. When heated to decomposition it emits acrid fumes.

SYMPTOMS: Symptoms of exposure may include weakness, irritation to the eyes and mucous membranes, conjunctivitis, central nervous system depression, loss of coordination, drowsiness, lack of appetite, nausea and vomiting and dermatitis. It is also narcotic in high concentrations.

Styrene oxide

CAS NUMBER: 96-09-3

SYNONYMS: 1,2-Epoxyethylbenzene Epoxystyrene Phenyloxirane Phenethylene oxide

CHEMICAL FORMULA: C8H8O

MOLECULAR WEIGHT: 120.15

WLN: T3OTJ BR

PHYSICAL DESCRIPTION: Colorless to pale straw-colored liquid

SPECIFIC GRAVITY: 1.0523 @ 16/4 C [017,047,205,395]

DENSITY: Not available

MELTING POINT: -37 C [055,205,269,275]

BOILING POINT: 194 C [055,205,269,275]

SOLUBILITY: Water: <1 mg/mL @ 19.5 C [700] DMSO: >=100 mg/mL @ 19.5 C [700] 95% Ethanol: >=100 mg/mL @ 19.5 C [700] Acetone: >=100 mg/mL @ 19.5 C [700] Methanol: Miscible [062,395,430]

OTHER SOLVENTS: Benzene: Miscible [062,395,430] Ether: Miscible [062,395,430] Carbon tetrachloride: Miscible [395,430] Heptane: Soluble [395] Most organic solvents: Miscible [395]

OTHER PHYSICAL DATA: Specific gravity: 1.0469 @ 25/4 C [042,062] 1.0540 @ 20/20 C [395] 1.054 @ 25/25 C [430] Boiling point: 84-85 C @ 25 mm Hg [017,047] 75 C @ 10 mm Hg [025] 65 C @ 5 mm Hg [025] Refractive index: 1.5338 @ 20 C [205,269,275] 1.5328 @ 25 C [062,395] Spectroscopy data: Lambda max: 250 nm (shoulder), 254 nm, 260 nm, 265 nm (shoulder) (epsilon = 13.2, 14.5, 16, 10) [395] Sweet pleasant odor [055] Absolute perception limit: 0.063 ppm [055] Viscosity: 1.99 centipoise @ 20 C [395]

HAP WEIGHTING FACTOR: 1 [713]

VOLATILITY:

Vapor pressure: 0.3 mm Hg @ 20 C [055,395,430]

Vapor density: 4.14 [042,055]

FIRE HAZARD: Styrene oxide has a flash point of 79 C (175 F) [205,269,275,430]. It is combustible. Fires involving this material may be controlled with a dry chemical, carbon dioxide or Halon extinguisher. A water spray may also be used [269].

LEL: Not available UEL: Not available

REACTIVITY: Styrene oxide is incompatible with oxidizers [042,269]. It is also incompatible with acids and bases [269]. It reacts with 4-(4'-nitro- benzyl)pyridine [395]. It polymerizes exothermally and reacts vigorously with compounds possessing a labile hydrogen (e.g. alcohols and amines) in the presence of catalysts such as acids, bases and certain salts [395, 430].

STABILITY: Styrene oxide is moisture sensitive [269]. UV spectrophotometric stability screening indicates that solutions of it in 95% ethanol are stable for at least 24 hours [700].

USES: Styrene oxide is a reactive intermediate, especially to produce styrene glycol and its derivatives and in the epoxy resin industry it is used as a diluent. It may have applications in the preparation of agricultural and biological chemicals, cosmetics and surface coatings, and in the treatment of textiles and fibers. In Japan it is used as an intermediate for the production of phenylethyl alcohol.

COMMENTS: None

NIOSH Registry Number: CZ9625000

ACUTE/CHRONIC HAZARDS: Styrene oxide is corrosive and can cause burns [269,275]. It may be toxic by ingestion, inhalation or skin absorption [062,269]. It is absorbed slowly through the skin [346,395,430]. It is an irritant [269,346,395,430] and, when heated to decomposition, it emits acrid fumes and toxic fumes of carbon monoxide and carbon dioxide [042,269].

SYMPTOMS: Symptoms of exposure may include severe irritation of the skin and eyes, and skin sensitization [395,430]. It can cause corrosion of tissues [269,275]. Other symptoms may include burns, irritation of mucous membranes and upper respiratory tract, nausea, vomiting and headaches [269]. Exposure may also cause central nervous system depression, hepatic lesions and pain to the eyes [430]. Eye contact has caused severe irritation in rabbits [099].

Symptoms of exposure to a related compound may include drying and cracking of the skin on contact, primary irritation to mucosal surfaces, fatigue, weakness, depression, unsteadiness, feeling of drunkenness, abnormal electroencephalograms and one case of toxic retrobulbar neuritis. Chronic exposure to a related compound has caused peripheral neuropathies (distal hypesthesia and decreased nerve conduction velocities). In experimental animals, this related compound has caused cardiac arrhythmias, renal and hepatic damage, and chemical pneumonitis [151].

2,3,7,8-Tetrachlorodibenzo-p-dioxin

CAS NUMBER: 1746-01-6

SYNONYMS: 2,3,7,8-TCDD

CHEMICAL FORMULA: C12H4Cl4O2

MOLECULAR WEIGHT: 321.96

WLN: T C666 BO IOJ EG FG LG MG

PHYSICAL DESCRIPTION: Colorless to white crystals

SPECIFIC GRAVITY: Not available

DENSITY: Not available

MELTING POINT: 295 C [029,031]

BOILING POINT: Decomposes @ 500 C [051,072,346]

SOLUBILITY: Water: <1 mg/mL @ 25 C [700] DMSO: <1 mg/mL @ 25 C [700] 95% Ethanol: <1 mg/mL @ 25 C [700] Acetone: <1 mg/mL @ 25 C [700] Methanol: 0.01 mg/mL @ 25 C [072,395,900] Toluene: <1 mg/mL @ 20 C [700]

OTHER SOLVENTS: Benzene: 0.57 mg/mL [072,395] Chloroform: 0.37 mg/mL [072,395] Perchloroethylene: 0.68 mg/mL @ 25 C [900] Chlorobenzene: 0.72 mg/mL [072,395] o-Dichlorobenzene: 1.4 mg/mL [072,395] n-Octanol: 0.05 mg/mL [051,072,395] Lard oil: 0.04 mg/mL [395] Hexane: 0.28 mg/mL @ 25 C [900]

OTHER PHYSICAL DATA: Crystals from anisole (melting point: 320-325 C) [031,072] Decomposition begins @ 500 C and is virtually complete within 21 seconds @ a temperature of 800 C [051,072,346] Partition coefficient in water (hexane system): 1000 [072] Lambda max (in chloroform): 248 nm, 310 nm (E = 92.2, 173.6) [395]

HAP WEIGHTING FACTOR: 100,000 [713]

VOLATILITY:

Vapor pressure: 0.00000000064 mm Hg @ 20 C; 0.0000000014 mm Hg @ 25 C [901]

FIRE HAZARD: 2,3,7,8-Tetrachlorodibenzo-p-dioxin is nonflammable [051,072].

LEL: Not available UEL: Not available

REACTIVITY: 2,3,7,8-Tetrachlorodibenzo-p-dioxin is in general an unreactive compound. It is changed chemically when exposed as solutions in iso-octane or n-octanol to ultraviolet light [395]. It undergoes catalytic perchlorination [051,072].

STABILITY: 2,3,7,8-Tetrachlorodibenzo-p-dioxin undergoes slow photochemical degradation and slow bacterial degradation. It is extremely stable but is chemically degraded by temperatures in excess of 500 C or by irradiation with ultraviolet light under certain conditions [051,072]. Photodecomposition is negligible in aqueous solutions [055]. Solutions of it in water, DMSO, 95% ethanol or acetone should be stable for 24 hours when protected from light [700].

USES: 2,3,7,8-Tetrachlorodibenzo-p-dioxin is present in certain herbicide and fungicide formulations, such as 2,4,5-T and the pentachlorophenols. It has been tested for use in flameproofing polyesters and against insects and wood-destroying fungi.

COMMENTS: 2,3,7,8-Tetrachlorodibenzo-p-dioxin is a contaminant created in the manufacture of Agent Orange, a widely used defoliant in Vietnam [031]. It is implicated as the causative agent of various symptoms described by veterans exposed in the war [031,051]. It is considered to be the most toxic chemical manufactured [406].

NIOSH Registry Number: HP3500000

ACUTE/CHRONIC HAZARDS: 2,3,7,8-Tetrachlorodibenzo-p-dioxin may be very toxic [026,031], an irritant, and an allergin. It also may cause eye irritation [043,301].

SYMPTOMS: Symptoms of exposure may include eye irritation, allergic dermatitis, wasting, hepatic necrosis, thymic atrophy, hemorrhage, lymphoid depletion and chloracne [043]. It may cause hypercholesterolemia and psychiatric disturbances [406]. It may also cause hyperpigmentation, liver damage, Hodgkin's lymphoma, raised serum hepatic enzyme levels, disorders of fat metabolism, disorders of carbohydrate metabolism, cardiovascular disorders, urinary tract disorders, respiratory disorders, pancreatic disorders, polyneuropathies, lower extremity weakness, sensorial impairments and neurasthenic or depressive syndromes [395]. Exposure may lead to excessive oiliness of the skin, abdominal pains, excessive flatulence, loss of body weight, oppressive headaches, excessive fatigue, unusual loss of vigor, porphyria cutanea tarda, porphyrinuria, uncharacteristic irritability and high blood cholesterol [173]. It may cause hepatotoxicity, thrombocytopenia, suppression of cellular immunity and death [346]. It may also cause a burning sensation to the eyes, nose and throat, headache, dizziness, nausea, vomiting, itching, redness, swelling of the face, nodules on the face, forearms, shoulders, neck and throat which may progress to come domes and cysts, acneform eruptions, aching muscles (mainly thighs and chest), insomnia, hirsutism, loss of libido, pain, hepatic dysfunction, hyperlipidemia and emotional disorders [301]. Other symptoms may include hemorrhagic cystitis, focal pyelonephritis, arthralgias, diabetes mellitus, burn-like sores, spontaneous abortions, liver damage (fatty changes, mild fibrosis, hemofuscin deposition and degeneration) and death [072].

1,1,2,2-Tetrachloroethane

CAS NUMBER: 79-34-5

SYNONYMS: Acetylene tetrachloride 1,1-Dichloro-2,2-dichloroethane sym-Tetrachloroethane

CHEMICAL FORMULA: $C_2H_2Cl_4$

MOLECULAR WEIGHT: 167.86

WLN: GYGYGG

PHYSICAL DESCRIPTION: Heavy, clear, colorless liquid

SPECIFIC GRAVITY: 1.58658 @ 25/4 C [031,051]

DENSITY: 1.586 g/mL [269,275]

MELTING POINT: -36 C [017,025,038,395]

BOILING POINT: 146 C [036,107,301,346]

SOLUBILITY: Water: <0.1 mg/mL @ 22 C [700] DMSO: >=100 mg/mL @ 20 C [700] 95% Ethanol: >=100 mg/mL @ 20 C [700] Acetone: >=100 mg/mL @ 20 C [700] Methanol: Miscible [031,051,421]

OTHER SOLVENTS: Alcohol: Miscible [205,295] Ethanol: Miscible [031,051,421] Carbon tetrachloride: Miscible [031,051,421] Chloroform: Miscible [031,051,205,421] Petroleum ether: Miscible [031,501,205,421] Oils: Miscible [031,051] Dimethylformamide: Miscible [031,051] Ether: Miscible [031,205,295,421] Benzene: Miscible [031,051,205,421] Carbon disulfide: Miscible [031,051]

OTHER PHYSICAL DATA: Sweetish, suffocating, chloroform-like odor [031,051,421] Refractive index: 1.4940 @ 20 C [017,047] 1.4918 @ 25 C [062,395,430] Odor threshold: 0.5 ppm [371] Evaporation rate (butyl acetate=1): 0.65 [058,102] Specific gravity: 1.600 @ 20/4 C [043,055] 1.593 @ 25/25 C [062] UV max (in methanol): 272 nm (epsilon=0.328) [395] Boiling point: 33.9 C @ 10 mm Hg [017,047] 62 C @ 45 mm Hg [025] Vapor pressure: 1 mm Hg @ -3.8 C; 10 mm Hg @ 33.0 C; 20 mm Hg @ 46.2 C [038] Vapor pressure: 40 mm Hg @ 60.8 C; 60 mm Hg @ 70.0 C; 100 mm Hg @ 83.0 C [038] Vapor pressure: 200 mm Hg @ 102.2 C; 400 mm Hg @ 124.0 C [038] Liquid surface tension: 37.85 dynes/cm [371] Latent heat of vaporization: 55.1 cal/g [371]

HAP WEIGHTING FACTOR: 10 [713]

VOLATILITY:

Vapor pressure: 5 mm Hg @ 21 C [395] 6 mm Hg @ 25 C [421,430]

Vapor density: 5.79 [051,055,371,395]

FIRE HAZARD: 1,1,2,2-Tetrachloroethane is nonflammable [031, 062, 275, 395].

LEL: Not available UEL: Not available

REACTIVITY: 1,1,2,2-Tetrachloroethane reacts violently with nitrogen tetroxide and 2,4-dinitro- phenyl disulfide [043,051,451]. It also reacts violently with nitrates [269]. When it is heated with solid sodium hydroxide, ignitable compounds are formed. It forms impact-sensitive mixtures with potassium and sodium [036, 043,066,451]. Mixtures with sodium-potassium alloy can explode on standing and are impact-sensitive [051,451]. 1,1,2,2-Tetrachloroethane is incompatible with chemically active metals and strong caustics [058,102,346]. In the presence of steam, it is incompatible with hot iron, aluminum and zinc [102,346]. It is incompatible with strong oxidizers [269]. It reacts with bromoform [051]. It may also react with alkalies. It is converted by fuming sulfuric acid [395]. It may attack some forms of plastics, rubber and coatings [102].

STABILITY: 1,1,2,2-Tetrachloroethane degrades slowly on exposure to air [395]. It is sensitive to light in the presence of air [058,102]. It is sensitive to heat [102]. Any water can cause appreciable hydrolysis even at room temperature, and hydrolysis and oxidation become comparatively rapid above 110 C [043]. Oxidative decomposition occurs by ultraviolet radiation. Solutions of it in anhydrous DMSO, ethanol or acetone should be stable for 24 hours under normal lab conditions [700].

USES: 1,1,2,2-Tetrachloroethane is used as a solvent for fats, oils, waxes, resins, cellulose acetate, rubber, copal, phosphorous, sulfur, chlorinated rubber and other organic materials. It is also used as a solvent in certain types of Friedel-Crafts reactions and phthalic anhydride condensations. It is used in the manufacture of paint, varnish and rust removers, in soil sterilization, in weed killer and insecticide formulations, in the determination of theobromine in cacao, as an immersion fluid in crystallography, in the biological laboratory to produce pathological changes in the gastrointestinal tract, liver and kidneys, as an intermediate in the manufacture of trichloroethylene and other chlorinated hydrocarbons having 2 carbon atoms, in cleansing and degreasing metals, in lacquers and in photographic film. It is also used as an alcohol denaturant, fumigant, dry cleaning agent and herbicide. It is used in cement and in the manufacture of bleach and artificial silk, leather and pearls.

COMMENTS: 1,1,2,2-Tetrachloroethane has the highest solvent power of the chlorinated hydrocarbons [031].

NIOSH Registry Number: KI8575000

ACUTE/CHRONIC HAZARDS: 1,1,2,2-Tetrachloroethane may be highly toxic by ingestion, inhalation and skin absorption [102,151,295,371]. It may be readily absorbed by all routes [151]. It is corrosive [051,062,395], a lachrymator and an irritant [043,058,269,371]. When heated to decomposition it emits toxic fumes of carbon monoxide, carbon dioxide, hydrogen chloride gas and phosgene [058,102,269,371].

SYMPTOMS: Symptoms of exposure may include irritation of the skin, eyes, nose, throat, mucous membranes and respiratory tract [058]. Lacrimation may occur [043,058,102,371]. Corrosion of exposed tissues may also occur [051, 062,395]. Other symptoms of exposure may include drowsiness, headache, jaundice, abdominal pain or distress, tremor, fatigue, constipation, insomnia, irritability, anorexia, loss of appetite, pulmonary edema, nephritis, albumin and casts in the urine, and kidney damage [102,346]. Giddiness, and unconsciousness have occurred [036]. Central nervous system effects may include general anesthesia, somnolence, hallucinations and distorted perceptions [043]. Other effects may include narcosis, acute yellow atrophy of the liver, liver cirrhosis, fatty degeneration of the kidneys and heart, brain changes, changes in the peripheral nerves, hemolysis, salivation, restlessness, dizziness, nausea, vomiting, coma and death [043,051]. Monocytosis, dermatitis, liver tenderness and damage, delirium and convulsions may occur [043]. Oliguria, cyanosis, central nervous system depression, uremia, peripheral paresthesia and hypesthesia may also occur [301]. Exposure may cause prickling sensation and numbness of limbs, loss of kneejerk, sweating, paralysis of the interossei muscles of the hands and feet, disappearance of ocular and pharyngeal reflexes, peripheral neuritis, liver dysfunction, general malaise, an unpleasant taste in the mouth, mental confusion, stupor, hematemesis, purpuric rashes and blood changes including an increase in mononuclear leukocytes, progressive anemia and slight thrombocytosis [346]. Gastrointestinal disturbances may occur [269,295,421]. It may also cause liver necrosis, nervousness, incoordination, respiratory failure and enlargement and fatty degeneration of the liver [151]. Heart damage has been reported [102,269]. Neurological disturbances have also been reported [371]. It can cause hepatitis, gastric pain, vertigo and leukopenia [421]. It can also cause a deep dusky coloration of the skin, weight loss, pain over the liver, dark urine and bilirubinuria [102]. Pulmonary damage, gastrointestinal irritation, barely perceptible pulse and shallow and rapid respiration may result [430]. Other symptoms which may occur are polyneuritis (inflammation of the nerves), conjunctivitis, extreme exhaustion, hematuria, mental instability, cardiac irregularity and epigastalgia [051].

Skin contact may result in dryness, scaling and inflammation [151]. Severe lesions may occur [371]. Eye contact may result in burning and serious eye damage [102]. Inhalation may cause a burning sensation, wheezing, coughing, laryngitis and shortness of breath [269]. Ingestion may cause diarrhea and severe mucosal injury [371].

Tetrachloroethylene

CAS NUMBER: 127-18-4

SYNONYMS: Perchlorethylene Perchloroethylene Tetrachloroethene

CHEMICAL FORMULA: C2Cl4

MOLECULAR WEIGHT: 165.83

WLN: GYGUYGG

PHYSICAL DESCRIPTION: Clear colorless liquid

SPECIFIC GRAVITY: 1.6227 @ 20/4 C [017,047]

DENSITY: 1.6230 g/mL @ 20 C [031,205,275,421]

MELTING POINT: -19 C [017,025,047]

BOILING POINT: 121 C [017,036,062,269]

SOLUBILITY: Water: <0.1 mg/mL @ 17 C [700] DMSO: >=100 mg/mL @ 22 C [700] 95% Ethanol: >=100 mg/mL @ 22 C [700] Acetone: >=100 mg/mL @ 22 C [700]

OTHER SOLVENTS: Alcohol: Soluble [455] Chloroform: Miscible [031,205,295] Oils: Miscible [062,395,421,455] Ether: Miscible [031,062,205,395] Benzene: Soluble [017,031,047,430] Most organic solvents: Miscible [421]

OTHER PHYSICAL DATA: Specific gravity: 1.6311 @ 15/4 C [031,043] 1.62 @ 25/25 C [058]] Specific gravity: 1.625 @ 20/20 C [036,395] Boiling point: 14 C @ 10 mm Hg [017,047] 33.2 C @ 30 mm Hg [025] Refractive index: 1.5053 @ 20 C [017] Ether-like odor [031,062,295,371] Odor threshold: 50 ppm [102] Saturation concentration: 126 g/m3 @ 20 C; 210 g/m3 @ 30 C [102] Evaporation rate (butyl acetate=1): 2.8 [102] Critical temperature: 347 C (657 F) [371]

HAP WEIGHTING FACTOR: 1 [713]

VOLATILITY:

Vapor pressure: 14 mm Hg @ 20 C [053,055,102] 15.8 mm Hg @ 22 C [043,051]

Vapor density: 5.83 [043,051,055,102]

FIRE HAZARD: Tetrachloroethylene is nonflammable [029, 033, 275, 451].

LEL: Not available UEL: Not available

REACTIVITY: Tetrachloroethylene is oxidized by strong oxidizing agents (such as sulfuric acid, nitric acid and sulfur trioxide) [395]. It is incompatible with strong bases [269]. It reacts violently with barium, beryllium and lithium [043, 051,102, 451]. It also reacts violently with dinitrogen tetraoxide [036, 043, 053, 066]. Violent reactions occur with aluminum [053,066]. The presence of 0.5% trichloroethylene in this compound has caused an explosion during unheated drying over solid sodium hydroxide [036,066]. Mixtures with (aluminum + zinc oxide) produce dense smoke when burned [066]. This chemical is slowly decomposed by various metals in the presence of moisture [053,295]. It decomposes slowly on contact with water [053,395].

It also decomposes at 700 C on contact with active carbon. Reaction with excess hydrogen in the presence of reduced nickel catalyst produces total decomposition [395]. This chemical will attack some forms of plastics, rubber and coatings [053,102].

STABILITY: Tetrachloroethylene may be slowly decomposed by light [053,295,455]. It may be sensitive to prolonged exposure to air [295,455]. Solutions of it in water, DMSO, 95% ethanol or acetone should be stable for 24 hours under normal lab conditions [058].

USES: Tetrachloroethylene is used as a solvent for fats, greases, waxes, rubbers, gums and caffeine from coffee. It is also used in the dry cleaning industry, in degreasing metals, as an anthelmintic against hookworms (Ancylostoma and Necator), intestinal flukes (Heterophyes), nematodes and trematodes; in veterinary medicine as an anthelmintic, as a chemical intermediate in the manufacture of trichloroacetic acid and fluorocarbons; in textile finishing, in cold cleaning of metals, as a fumigant for insects and rodents, as a drying agent for metals and certain other solids, as a vermifuge, as a heat-transfer medium, in copying machines, in the manufacture of paint removers, in printing inks and in removing soot from industrial boilers.

COMMENTS: None

NIOSH Registry Number: KX3850000

ACUTE/CHRONIC HAZARDS: Tetrachloroethylene is an irritant of the skin, eyes, nose and throat [051,102, 430] and a lachrymator [043]. It is also narcotic in high concentrations [430]. When heated to decomposition it emits toxic gases and vapors such as hydrogen chloride, phosgene, carbon monoxide and carbon dioxide [053,102,269].

SYMPTOMS: Symptoms of exposure may include irritation of the skin and eyes [051,151,269,430]. It may also cause irritation of the nose and throat [051,102,430]. Inhalation of the vapors may cause dizziness [053,102, 269,421]. It may also cause nausea and vomiting [053,151,215,455]. High concentrations of the vapor may cause stupor [346,455]. Central nervous system depression has been reported [051,215,346]. This compound can sensitize the heart to arrhythmias produced by catecholamines [215,301,406]. It may be toxic to the liver and kidneys [053,215,269,430]. High concentrations have been associated with trigeminal neuropathy [406]. It may have anesthetic effects [043, 371,430]. It can also cause irritation of the mucous membranes [269,395,421]. Low concentrations may cause headaches and vertigo [102,215,430]. It may also cause incoordination and unconsciousness [053,102]. Light narcosis may occur [102,151,269]. Prolonged exposure can cause increased salivation, metallic taste, increased hand perspiration, tightness of frontal sinuses, congestion of frontal sinuses, thickness of tongue, and difficulty in motor coordination. Exhilaration and inebriation have been observed [421]. It may cause lightheadedness [430]. Severe direct skin contact can cause skin burns and blistering [053,301]. It may also cause erythema [421]. Other symptoms may include irresponsible behavior, loss of inhibition and ventricular premature beats. Excretion of this compound in mothers' milk can cause obstructive jaundice in newborn infants [301]. Peripheral nerve damage may be indicated by tingling, numbness and muscle weakness [102,301]. Other symptoms may include tremors, sleepiness, fatigue and blurred vision [215]. Exposure in combination with alcohol exacerbates toxic symptoms and may cause a red blotchy appearance of skin on the face and upper body [215,269]. Exposure may also result in acute hepatic necrosis, oliguric uremia and upper respiratory tract irritation [269,395]. It may cause giddiness, sinus inflammation and even death by respiratory arrest [151]. Lung edema, sleeplessness, abdominal pain and constipation may result [395]. Repeated skin contact may cause a dry, scaly and fissured dermatitis [051]. Anesthetic death may occur [346,430]. Anorexia may be another symptom [430]. It may cause gastrointestinal irritation [043]. Other symptoms may include neuritis, defects in memory and slurred speech [102].

Titanium tetrachloride

CAS NUMBER: 7550-45-0

CHEMICAL FORMULA: Cl4Ti

MOLECULAR WEIGHT: 189.70 [703]

PHYSICAL DESCRIPTION: Not available

SPECIFIC GRAVITY: Not available

MELTING POINT: -24.1 C [702]

BOILING POINT: 136.4 C [702]

SOLUBILITY: Water: Solubility

OTHER PHYSICAL DATA: Not available

HAP WEIGHTING FACTOR: 1 [713]

VOLATILITY: Not available

FIRE HAZARD: Flash Point: 7 C [702]

Flammability: Not available

Autoignition temperature: Not available

LEL: Not available UEL: Not available

REACTIVITY: Reacts violently with potassium [702]. Addition of titanium tetrafluoride directly to tetrahydrofuran causes a violent exothermic reaction [710]. Incompatible or reacts with hydrogen fluoride, sulfur nitrides, and urea [710].

CORROSIVITY:

Highly corrosive [702, 703].

STABILITY: Absorbs moisture from the air and evolves dense white fumes [715]. When heated to decomposition it emits toxic fumes of Cl- [703].

USES: Formerly used with potassium bitartrate as a mordant in textile industry, and with dyewoods in dyeing leather; also as smoke-producing screen with ammonia; manufacture of iridescent glass and artificial pearls. [715]

COMMENTS: Keep tightly closed [715].

NIOSH Registry Number: XR1925000

ACUTE/CHRONIC HAZARDS: Titanium tetrachloride is a poison by inhalation. It is a corrosive irritant to skin, eyes and mucous membranes. When heated to decomposition it emits toxic fumes of Cl-. [703] Observed damage to corneas in five human beings who had been severely exposed to these fumes, with particularly severe effects on cornea in patients who died from exposure. [723]

SYMPTOMS: Symptoms of exposure to titanium tetrachloride may include irritation of the eyes, mucous membranes, skin and respiratory tract, and surface skin burns. Marked congestion of mucous membrane of pharynx, vocal cords and trachea; stenosis of larynx, trachea and upper bronchi may occur. [723]

Toluene

CAS NUMBER: 108-88-3

SYNONYMS: Methylbenzene Phenylmethane

CHEMICAL FORMULA: C7H8

MOLECULAR WEIGHT: 92.14

WLN: 1R

PHYSICAL DESCRIPTION: Clear, colorless to amber liquid

SPECIFIC GRAVITY: 0.8669 @ 20/4 C [017,047,430]

DENSITY: 0.867 g/mL [269,275]

MELTING POINT: -95 C [205,371,421,430]

BOILING POINT: 110.6 C [025,051,205,371]

SOLUBILITY: Water: <1 mg/mL @ 18 C [700] DMSO: >=100 mg/mL @ 18 C [700] 95% Ethanol: >=100 mg/mL @ 18 C [700] Acetone: >=100 mg/mL @ 18 C [700]

OTHER SOLVENTS: Alcohol: Soluble [017,053,062,430] Most organic solvents: Miscible [295,421] Ligroin: Soluble [017] Chloroform: Miscible [031,043,051,205] Glacial acetic acid: Miscible [031,051,295] Carbon disulfide: Miscible [031,051] Ether: Miscible [031,043,051,205] Benzene: Soluble [017,053,062]

OTHER PHYSICAL DATA: Refractive index: 1.4968 @ 20 C [269,275] Boiling point: 14.5 C @ 14.5 mm Hg [017] Benzene-like odor [031,062,346,430] Critical temperature: 318.6 C [371] Critical pressure: 30400 mm Hg [371] Floats on water [051,071,371,451] Dissolves slowly in water [051,071,371] 100% Volatile by volume [058] Evaporation rate (butyl acetate = 1): >1 [058] Vapors are heavier than air [058,371] Burns with a smoky flame [295] Odor threshold: 0.17 ppm [371] Burning rate: 5.7 mm/min [371] Liquid surface tension: 29 dynes/cm @ 20 C [371] Liquid water interfacial tension: 36.1 dynes/cm [371] Latent heat of vaporization: 86.1 cal/g [371] Heat of combustion: -9686 cal/g [371] Heat of fusion: 17.17 cal/g [371] Lambda max (in methanol): 269 nm, 265 nm (shoulder), 262 nm, 260 nm (shoulder), 256 nm, 254 nm (shoulder), 249 nm (shoulder), 243 nm (shoulder), 238 nm (shoulder) (epsilon = 227.1, 169.5, 254.6, 212.6, 190.2, 178.2, 125.6, 72.63, 38.8) [052]

HAP WEIGHTING FACTOR: 1 [713]

VOLATILITY:

Vapor pressure: 10 mm Hg @ 6.4 C, 20 mm Hg @ 18.4 C, 40 mm Hg @ 31.8 C [038]

Vapor density: 3.14 [043,051,071,102]

FIRE HAZARD: Toluene has a flash point of 4.4 C (40 F) [033,043,421,430] and it is flammable. Fires involving this material may be controlled with a dry chemical, carbon dioxide or Halon extinguisher. A water spray may also be used [051,071]. Vapor trail flash-back can occur [371,451].

The autoignition temperature for toluene is 536 C (997 F) [036, 062, 371].

LEL: 1.27% [043,058,102,371] UEL: 7.1% [051,058,421,451]

REACTIVITY: Toluene reacts violently with 1,3-dichloro-5,5-dimethyl-2,4-imidazolidid- ione, dinitrogen tetraoxide, concentrated nitric acid, (sulfuric acid + nitric acid), silver perchlorate, bromine trifluoride, uranium hexafluoride, tetranitromethane and oxidizing materials [043]. It is incompatible with oxygen [058]. It may attack some forms of plastics, rubber and coatings [102].

STABILITY: Toluene is hygroscopic [269]. Solutions of it in water, DMSO, 95% ethanol or acetone should be stable for 24 hours under normal lab conditions [700].

USES: Toluene is used to manufacture benzoic acid, benzaldehyde, benzene, phenol, caprolactam, linoleum, toluenediisocyanates (polyurethane resins), toluene sulfonates (detergents), artificial leather and fabric and paper coatings. It is also used in explosives, dyes and other organic compounds, as a solvent for paints, lacquers, gums, most oils, rubber, vinyl organosols and resins, in the extraction of various principles of plants, as a gasoline additive, as a diluent and thinner in nitrocellulose lacquers, as an adhesive solvent in plastic toys and model airplanes, in saccharin, perfumes and scintillation counters, in photogravure color-printing process, in the cleaning industry, in the impregnation of cartridge paper, as a solvent in pharmaceutical and plastic industries, as an insecticide, as a urine preservative for chemical examination and as an asphalt and naphtha constituent.

COMMENTS: Toluene is obtained from coal tar and petroleum [055,421].

NIOSH Registry Number: XS5250000

ACUTE/CHRONIC HAZARDS: Toluene may be toxic and an irritant. It may be harmful by inhalation, ingestion or skin absorption [036,051,071,269] and may be readily absorbed through the skin [269]. When heated to decomposition it emits toxic fumes of carbon dioxide and carbon monoxide [102,269].

SYMPTOMS: Symptoms of exposure may include eye irritation, dilation of the pupils, impairment of reaction in association with fatigue, slight pallor of the fundi, inebriation, ocular disturbances and reddening of vision [099]. It also may cause dizziness, headache, nausea, mental confusion, blood disease and dermatitis [036]. It may also cause giddiness [071]. Other symptoms may include hallucinations or distorted perceptions, narcosis in high concentrations, motor activity changes. An occasional report of chronic poisoning describes an anemia and leucopenia. It causes loss of appetite, a bad taste, lassitude, impairment of coordination and reaction time and an enlarged liver [043]. It may cause central nervous system depression, bone marrow depression, petechial hemorrhages, noncoagulated blood, congestion of all organs, severe bone marrow aplasia, necrosis or fatty degeneration of the heart, liver and adrenals, weakness, euphoria, vomiting, tightness in the chest, blurred vision, tremors, shallow and rapid respiration, ventricular irregularities including fibrillation, paralysis, convulsions, violent excitement or delirium, unconsciousness, kidney or liver damage, skin irritation, scaling and cracking, drowsiness, nervousness, pallor, petechiae, abnormal bleeding, irreversible encephalopathy with ataxia, tremulousness, emotional lability, diffuse cerebral atrophy and death [301]. It may also cause abnormal tendon reflexes, reduced grasping power, decreased finger agility, metabolic acidosis, hepatomegaly, exhilaration, lightheadedness, cardiovascular collapse, cardiac arrhythmias, bradycardia, mydriasis, insomnia, restlessness, staggering gait, lack of self-control, stupor, increased irritability (personality changes), exaggerated mood swings, equilibrium disorders and vertigo [051]. Symptoms of exposure may include skin sensation such as a pins and needles feeling or numbness and difficulty in seeing bright light [102]. It may cause high urinary pH, nightmares, vertical nystagmus and status epilepticus [295]. It may also cause loss of memory, anorexia, palpitation and aplastic anemia [421]. Other symptoms may include renal but also neural and especially cerebellar dystrophy, lacrimation, hilarity, nasal mucous secretion and metallic taste. High concentrations may result in paresthesia, collapse, cardiac sensitization and fatal cardiotoxicity [430]. It may cause irritation to the nose and throat, respiratory arrest, coughing, gagging, distress, pulmonary edema, griping and diarrhea [371]. It may also cause chemical pneumonitis, coma and conjunctivitis [058].

2,4-Toluenediamine

NOTE: The form of this name in the Clean Air Act Amendments Section 112(b) list is 2,4-Toluene diamine. The correct notation for the compound is 2,4-toluenediamine.

CAS NUMBER: 95-80-7

SYNONYMS: 3-Amino-p-toluidine 5-Amino-o-toluidine 2,4-Diaminotoluene 4-Methyl-1,3-benzenediamine

CHEMICAL FORMULA: C7H10N2

MOLECULAR WEIGHT: 122.17

WLN: ZR CZ D1

PHYSICAL DESCRIPTION: Light brown solid

SPECIFIC GRAVITY: 1.045 @ 100 C [058]

DENSITY: Not available

MELTING POINT: 99 C [017,029,062,105]

BOILING POINT: 292 C [029,047,395]

SOLUBILITY: Water: 1-5 mg/mL @ 21 C [700] DMSO: 50-100 mg/mL @ 21 C [700] 95% Ethanol: 10-50 mg/mL @ 21 C [700] Acetone: >=100 mg/mL @ 21 C [700]

OTHER SOLVENTS: Ether: Soluble [017,047,062,205] Benzene: Soluble [017,047] Aqueous sodium carbonate: Soluble [105]

OTHER PHYSICAL DATA: Needles from water; prisms from ethanol Boiling point: 29 C @ 2 mm Hg [017] 148-150 C @ 8 mm Hg [017,029,051] Lambda max: 294 nm in methanol Slight odor

HAP WEIGHTING FACTOR: 1 [713]

VOLATILITY:

Vapor pressure: 1 mm Hg @ 106.5 C; 10 mm Hg @ 151.7 C [051,395]

FIRE HAZARD: 2,4-Toluenediamine has a flash point of 149 C (300 F) [058]. It is combustible. Fires involving this chemical may be controlled with a dry chemical, carbon dioxide or Halon extinguisher.

LEL: Not available UEL: Not available

REACTIVITY: 2,4-Toluenediamine may react with acids, acid chlorides, acid anhydrides, strong oxidizing agents and chloroformates [269]. It is also incompatible with isocyanates [058].

STABILITY: 2,4-Toluenediamine is sensitive to air and moisture. It is sensitive to temperatures greater than 104 F (40 C) [058]. It is also sensitive to prolonged exposure to light [051]. 2,4-Toluenediamine readily becomes oxidized in neutral or alkaline solutions to form a dark product [099]. Solutions of it in water, DMSO, 95% ethanol or acetone should be stable for 24 hours under normal lab conditions [700].

USES: 2,4-Toluenediamine is used in polymerizations; in dyes used for textiles, leather, furs and hair dye formulations; as a source of toluene diisocyanate; a developer for direct dyes; and chain extender and crosslinker.

COMMENTS: 2,4-Toluenediamine was forbidden in hair dye formulations in 1971.

NIOSH Registry Number: XS9625000

ACUTE/CHRONIC HAZARDS: 2,4-Toluenediamine is irritating to the eyes and skin. It can be absorbed through the skin [051,269]. It is harmful if swallowed or inhaled [058]. When heated to decomposition it emits toxic fumes of carbon monoxide, carbon dioxide and nitrogen oxides [042,058,269]. It may also emit fumes of ammonia [058]. There is also evidence that this chemical is an animal carcinogen [015].

SYMPTOMS: Symptoms of exposure may include contact sensitization, dermatitis, keratoconjunctivitis, blepharitis and reversible opacities of the cornea [099]. It causes methemoglobinemia, central nervous system depression, degeneration of the liver, eye irritation, skin blistering, nausea, vomiting, jaundice and anemia [346]. It may also cause skin irritation [042,051,269]. In addition, cyanosis, itching, reddening and tearing of the eyes, temporary discoloration of the skin and kidney damage can occur [058].

2,4-Toluene diisocyanate

CAS NUMBER: 584-84-9

SYNONYMS: 2,4-Diisocyanato-1-methylbenzene Isocyanic acid, 4-methyl-m-phenylene ester

CHEMICAL FORMULA: C9H6N2O2

MOLECULAR WEIGHT: 174.16

WLN: OCNR B1 ENCO

PHYSICAL DESCRIPTION: Clear, colorless to light yellow liquid

SPECIFIC GRAVITY: 1.2244 @ 20/4 C [017,031,205,395]

DENSITY: 1.22 g/mL @ 25 C [058,371]

MELTING POINT: 19.5-21.5 C [031,042,395]

BOILING POINT: 251 C [017,031,036,205]

SOLUBILITY: Water: Decomposes [205] DMSO: >=100 mg/mL @ 23 C [700] 95% Ethanol: Decomposes [031,042,205] Acetone: <1 mg/mL @ 23 C [700] Methanol: >=100 mg/mL @ 18 C [700] Toluene: <1 mg/mL @ 18 C [700]

OTHER SOLVENTS: Benzene: Miscible [031,042,205,395] Ether: Miscible [031,042,205,395] Diglycol monomethyl ether: Miscible [031,395] Carbon tetrachloride: Miscible [031,042,395] Chlorobenzene: Miscible [031,042,395] Kerosene: Miscible [031,042,395] Olive oil: Miscible [031,042,395] Most organic solvents: Soluble [062]

OTHER PHYSICAL DATA: Specific gravity: 1.22 @ 25/15.5 C [062,395] Boiling point: 115-120 C @ 10 mm Hg [269,275] 126 C @ 11 mm Hg [017,031] Boiling point: 124-126 C @ 18 mm Hg [029,395] Vapor pressure: 1 mm Hg @ 80 C [055] Sharp, sweet, fruity, pungent odor Odor threshold: 0.4-2.14 ppm Refractive index: 1.5689 @ 20 C Liquid surface tension: 25 dynes/cm Heat of combustion (estimated): -5720 cal/g

HAP WEIGHTING FACTOR: 10 [713]

VOLATILITY:

Vapor pressure: 0.01 mm Hg @ 20 C [055,062,395] 0.025 mm Hg @ 25 C [058]

Vapor density: 6.0

FIRE HAZARD: 2,4-Toluene diisocyanate has a flash point of 132 C (270 F) [031,062,371,395]. It is combustible. Fires involving this material may be controlled with a dry chemical, carbon dioxide or Halon extinguisher.

The autoignition temperature of 2,4-toluene diisocyanate is >149 C (>300 F) [371].

LEL: 0.9% [042,058,371] UEL: 9.5% [042,058,371]

REACTIVITY: 2,4-Toluene diisocyanate reacts with water with evolution of carbon dioxide [031,036, 058,395]. It may polymerize vigorously on contact with bases and acyl chlorides, or slowly with gas evolution by diffusion of moisture into polythene containers [036,066]. It reacts with compounds containing active hydrogens such as acids and alcohols (reactions may be violent). It is incompatible with strong oxidizers [269,395]. It will cause some corrosion to copper alloys and aluminum [058].

STABILITY: 2,4-Toluene diisocyanate is heat sensitive [058,269,395]. It darkens on exposure to light [031,395]. Polymerization may occur on contact with moisture [058]. Thin layer chromatography stability screening indicates that solutions of it in acetone are stable for at least 24 hours [700]. NMR stability screening indicates that solutions of it in DMSO are stable for less than 24 hours [700].

USES: 2,4-Toluene diisocyanate is used in the manufacture of polyurethane foams, other elastomers and coatings. It is also used as a cross-linking agent for nylon 6.

COMMENTS: The technical product usually contains 20-35% of the 2,6-isomer.

NIOSH Registry Number: CZ6300000

ACUTE/CHRONIC HAZARDS: 2,4-Toluene diisocyanate may be highly toxic by ingestion, inhalation and skin absorption [026,062,269]. It is a strong irritant [062,099,151,269] and high concentrations are extremely destructive to tissues [269]. It is also a lachrymator [058,099,269,421] and an allergin [031]. There is sufficient evidence that this chemical is carcinogenic in animals [015,395]. When heated to decomposition it emits toxic fumes of carbon monoxide, carbon dioxide, nitrogen oxides and hydrogen cyanide [042,058,269].

SYMPTOMS: Symptoms of exposure may include irritation of the skin, eyes and respiratory tract; asthma, sensitization, cough, wheezing, decrease in lung function, bronchitis, pulmonary edema, nausea, vomiting and abdominal pain [058,346]. Other symptoms may include lacrimation, shortness of breath, tightness in the chest, bronchial spasm and pneumonitis [058,421]. The pneumonitis may be with flu-like symptoms (e.g. fever and chills). It may cause runny nose, sore throat, chest discomfort and diarrhea [058]. Ingestion may cause irritation of the gastrointestinal tract, dyspnea and eosinophilia [151]. It may also irritating and corrosive action in the mouth, stomach and tissue of the digestive tract [058]. Exposure may lead to respiratory inflammation, chest congestion, headache, insomnia and burns of the eyes and skin [371]. Exposure may also lead to dryness of the throat, poor memory, difficulty in concentrating, confusion, changes in personality, depression and irritability [395]. Other symptoms may include allergic reactions, burning sensation and laryngitis [269]. Allergic eczema has been reported [031]. Skin contact may cause reddening, swelling and blistering [058,346]. It may also cause rashes and scaling [058]. Severe dermatitis has been reported [036,042]. High concentrations are extremely destructive to tissues of the mucous membranes, upper respiratory tract, eyes and skin [269]. Eye contact causes severe irritation, pain, reddening, swelling, conjunctivitis and corneal damage [058]. Smarting, burning and a prickling sensation may occur [421]. Other eye effects may include disturbances of the corneal epithelium, microcystic corneal epithelial edema with subjective impression of foggy or smoky vision, reduced visual acuity, keratitis and, rarely, photophobia, blepharospasm, severe iridocyclitis and secondary glaucoma. Exposure to rabbit eyes has caused immediate pain, lacrimation, swelling of the lids and conjunctival reaction [099]. Repeated, low-level exposure may produce chronic lung disease [371].

---

o-Toluidine

CAS NUMBER: 95-53-4

SYNONYMS: 1-Amino-2-methylbenzene 2-Amino-1-methylbenzene 2-Methylaniline 2-Methylbenzenamine 1-Methyl-2-aminobenzene 2-Methyl-1-aminobenzene

CHEMICAL FORMULA: C_7H_9N

MOLECULAR WEIGHT: 107.16

WLN: ZR B

PHYSICAL DESCRIPTION: Colorless to light yellow liquid

SPECIFIC GRAVITY: 1.004 @ 20/4 C [042,051,055,058]

DENSITY: 1.008 g/mL @ 20 C [421]

MELTING POINT: -16.3 C [042,051,102]

BOILING POINT: 200-202 C [031,042,051,421]

SOLUBILITY: Water: 5-10 mg/mL @ 15 C [700] DMSO: >=100 mg/mL @ 15 C [700] 95% Ethanol: >=100 mg/mL @ 15 C [700] Acetone: >=100 mg/mL @ 15 C [700]

OTHER SOLVENTS: Carbon tetrachloride: Miscible [395,396] Dilute acids: Slightly soluble [031,042,205,421] Ether: Soluble [017,031,042,205]

OTHER PHYSICAL DATA: Refractive index: 1.5725 @ 20 C Boiling point: 80.1 C @ 10 mm Hg [017,047] 92 C @ 18 mm Hg [025] Specific gravity: 1.008 @ 20/20 C [031,051,061] pKa 9.46 Aromatic, aniline-like odor Spectroscopy data: lambda max 232.5 nm, 281.5 nm (E = 758.6, 144.5) pH 8.0 Heat of combustion: -8990 cal/g

HAP WEIGHTING FACTOR: 1 [713]

VOLATILITY:

Vapor pressure: 0.1 mm Hg @ 20 C; 0.3 mm Hg @ 30 C [055]

Vapor density: 3.69 [042,051,058]

FIRE HAZARD: o-Toluidine has a flash point of 85 C (185 F) [031,042,205,421]. It is combustible. Fires involving this material may be controlled with a dry chemical, carbon dioxide or Halon extinguisher. A water spray may also be used [058].

The autoignition temperature of o-toluidine is 482 C (900 F) [042, 058, 102, 371].

LEL: 1.5% [058,102] UEL: Not available

REACTIVITY: o-Toluidine reacts violently with fuming nitric acid [025,036,042,066]. It can react with oxidizing materials [042,058,269,346]. It is incompatible with strong acids [269]. It will attack some forms of plastics, rubber, and coatings [102].

STABILITY: o-Toluidine is sensitive to air and light (it darkens to a reddish brown color) [031, 051, 061, 396]. It is also sensitive to heat [058, 102, 396]. Solutions of it in water, DMSO, 95% ethanol or acetone should be stable for 24 hours under normal lab conditions [700].

USES: o-Toluidine is used in textile printing dyes, as a vulcanization accelerator, in organic synthesis, as a dye intermediate, and as an antioxidant in the manufacture of rubber. o-Toluidine is also an ingredient in a clinical laboratory reagent used to test blood samples for glucose, in making colors fast to acids, as an intermediate in pharmaceutical manufacturing and synthetic chemicals production, and as an intermediate in the manufacture of pesticides.

COMMENTS: None

NIOSH Registry Number: XU2975000

ACUTE/CHRONIC HAZARDS: o-Toluidine may be toxic by ingestion, inhalation or skin contact [025,036,061, 269]. It can be absorbed through the skin [051,061,269,421]. It is an irritant [051,269,371] and is a suspected human carcinogen [015,269,396,413]. When heated to decomposition it emits toxic fumes of carbon monoxide, carbon dioxide or nitrogen oxides [042,058,269,371].

SYMPTOMS: Symptoms of exposure may include methemoglobinemia, cyanosis (bluish discoloration of the lips, skin and fingernails); and irritation of the skin and eyes [051,102,269,371]. Other symptoms may include headaches and weakness [036,042,051,102,346]. It can cause drowsiness, dizziness, anoxia, and hematuria [102,346]. If ingested, it can cause nausea, vomiting, loss of consciousness and coma [102,371]. It can cause mental confusion, and in severe cases, convulsions [036]. It can also cause difficulty in breathing, air hunger, psychic disturbances, and marked irritation of the kidneys and bladder [042,051]. Excessive drying of the skin may result from repeated or prolonged contact [058]. It can also cause dermatitis and eye burns [346]. It can cause corneal damage [301]. Stupor and death may occur [102]. Repeated inhalation of low concentrations may cause pallor, low-grade secondary anemia, fatigability and loss of appetite [371]. Symptoms of exposure may also be delayed [058,269].

Toxaphene

CAS NUMBER: 8001-35-2

SYNONYMS: Chlorinated camphene Chlorocamphene Octachlorocamphene Polychlorocamphene

CHEMICAL FORMULA: $C_{10}H_{10}Cl_8$

MOLECULAR WEIGHT: 413.80

WLN: L55 A CYTJ CU1 D1 D1 XG XG XG

PHYSICAL DESCRIPTION: Yellow or amber waxy solid

SPECIFIC GRAVITY: 1.63 [051,072]

DENSITY: 1.66 g/mL @ 27 C [051,062,072]

MELTING POINT: 65-90 C [031,043,062,395]

BOILING POINT: Decomposes [051,072]

SOLUBILITY: Water: <1 mg/mL @ 19 C [700] DMSO: >=100 mg/mL @ 19 C [700] 95% Ethanol: 5-10 mg/mL @ 19 C [700] Acetone: >=100 mg/mL @ 19 C [700] Toluene: Soluble [430]

OTHER SOLVENTS: Chloroform: Soluble [051,072] Petroleum oils: Soluble [072,173] Aromatic hydrocarbons: Very soluble [031,043,051,072] Most organic solvents: Soluble [051,062,072,395] Hexane: Soluble [051,072,430] Aliphatic hydrocarbons: Soluble [051,072] Xylene: Soluble [430] Deodorized kerosene: >280 [430] Mineral oil: 55-60 [430]

OTHER PHYSICAL DATA: Density: 1.65 g/mL @ 25 C [051,072] The flash point for a solution of this compound in 10% xylene is 28.9 C (84 F) [051,072] The autoignition temperature for a solution of this compound in 10% xylene is 530 C (986 F) [051,072] Slight pine odor [031,043,151,173] Odor threshold (detection): 0.14 mg/kg [055] Odor threshold in water: 0.0052 mg/L [051] Burning rate: 5.8 mm/min [051] Viscosity: 89 centipoise @ 110 C, 57 centipoise @ 120 C [051] Viscosity: 39.1 centipoise @ 130 C [051] Specific heat: 0.258 cal/g @ 41 C [051]

HAP WEIGHTING FACTOR: 1 [713]

VOLATILITY:

Vapor pressure: 0.2-0.4 mm Hg @ 25 C [173]

FIRE HAZARD: Flash point data for toxaphene are not available. It is probably combustible. Fires involving this material may be controlled with a dry chemical, carbon dioxide or Halon extinguisher.

LEL: Not available UEL: Not available

REACTIVITY: Toxaphene is decomposed in the presence of alkali [033,051,395]. It is corrosive to iron [033] and is incompatible with strong oxidizers [346]. It is noncorrosive in the absence of moisture [395].

STABILITY: Toxaphene is decomposed by sunlight and heat [033, 051, 173, 395].

USES: Toxaphene is an insecticide and pesticide. It is used on cotton crops, cattle, swine, soybeans, corn, wheat, peanuts, lettuce, tomatoes, grains, vegetables, fruit and other food crops. It is used in the control of animal ectoparasites, grasshoppers, army-worms, cutworms and all major cotton pests. It controls livestock pests such as flies, lice, ticks, scab mites and mange. It also controls mosquito larvae, leaf miners, bagworms, church bugs, yellow jackets and caterpillars.

COMMENTS: None

NIOSH Registry Number: XW5250000

ACUTE/CHRONIC HAZARDS: Toxaphene may be toxic by ingestion, inhalation and skin absorption [062]. It also is an irritant [031,072] and a lachrymator [173]. When heated to decomposition it emits toxic fumes of chlorides [072].

SYMPTOMS: Symptoms of exposure may include nausea, mental confusion, jerking of the arms and legs, convulsions, cyanosis, lacrimation, eye pain, headache, vertigo, abdominal pain, liquid stools and weakness [173]. Other symptoms may include somnolence, effect on seizure threshold, coma, allergic dermatitis, skin irritation and liver injury [043]. It may cause central nervous system stimulation, tremors and death [031]. It may also cause salivation, leg and back muscle spasms, vomiting, hyper excitability, shivering, tetanic contractions of all skeletal muscles and respiratory failure [072]. Auditory reflex excitability may occur [301]. Exposure may cause agitation, dry and red skin, and unconsciousness [346].

1,2,4-Trichlorobenzene

CAS NUMBER: 120-82-1

SYNONYMS: 1,2,4-Trichlorobenzol

CHEMICAL FORMULA: C6H3Cl3

MOLECULAR WEIGHT: 181.45

WLN: GR BG DG

PHYSICAL DESCRIPTION: Clear colorless liquid

SPECIFIC GRAVITY: 1.454 @ 20/4 [269,275]

DENSITY: 1.4634 g/mL @ 25 C [062,451]

MELTING POINT: 17 C [017,031,036,205]

BOILING POINT: 213 C [031,043,058,451]

SOLUBILITY: Water: <1 mg/mL @ 21 C [700] DMSO: >=100 mg/mL @ 21 C [700] 95% Ethanol: >=100 mg/mL @ 21 C [700] Acetone: >=100 mg/mL @ 21 C [700]

OTHER SOLVENTS: Benzene: Miscible [031] Ether: Miscible [017,031] Oils: Miscible [062] Petroleum ether: Miscible [031] Carbon disulfide: Miscible [031] Most organic solvents: Miscible [062]

OTHER PHYSICAL DATA: Specific gravity: 1.4634 @ 25/25 C [031] 1.574 @ 10/4 C [055] Boiling point: 84.8 C @ 10 mm Hg [017,047] Vapor pressure: 10 mm

Hg @ 81.7 C [038] Flash point also reported as 98.9 C (210 F) [051,062,071] Odor threshold: ~3 ppm [058] Faint aromatic odor [058] Refractive index: 1.5708 @ 20 C [269,275] 1.5524 @ 25 C [031]

HAP WEIGHTING FACTOR: 1 [713]

VOLATILITY:

Vapor pressure: 1 mm Hg @ 38.4 C [038,043,058] 5 mm Hg @ 67.3 C [038]

Vapor density: 6.26 [043]

FIRE HAZARD: 1,2,4-Trichlorobenzene has a flash point of 110 C (230 F) [031, 043, 058, 205]. It is combustible. Fires involving this material may be controlled with a dry chemical, carbon dioxide or Halon extinguisher. A water spray may also be used [043,051].

The autoignition temperature of 1,2,4-trichlorobenzene is 571 C (1060 F) [451].

LEL: 2.5% @ 150 C [058,451] UEL: 6.6% @ 150 C [058,451]

REACTIVITY: 1,2,4-Trichlorobenzene can react vigorously with oxidizing materials [043,269].

STABILITY: 1,2,4-Trichlorobenzene is stable under normal laboratory conditions.

USES: 1,2,4-Trichlorobenzene is used as a solvent in chemical manufacturing, a dielectric fluid, a heat-transfer medium, a dye carrier, an herbicide intermediate and a degreaser. It is also used in dyes, chemical intermediates, synthetic transformer oils, lubricants and insecticides.

COMMENTS: None

NIOSH Registry Number: DC2100000

ACUTE/CHRONIC HAZARDS: 1,2,4-Trichlorobenzene is toxic by ingestion [053, 269]. It is also harmful by inhalation or skin absorption [269] and it is an irritant of the skin, eyes and lungs.

SYMPTOMS: Symptoms of exposure to 1,2,4-trichlorobenzene include irritation of the skin, eyes, mucous membranes and upper respiratory tract; and headache [051, 058, 269]. Other symptoms may include nausea, dizziness, gastrointestinal disturbances and liver damage [269]. Central nervous system stimulation and dermatitis may occur [051]. Exposure may cause skin burns, drowsiness, incoordination and unconsciousness [346]. Skin contact may cause redness. Eye contact may cause burning, itching and redness [058] and also cause severe pain [099]. Other symptoms include redness of the mucous membranes, burning sensation in the nose and throat, sore throat, coughing, dyspnea and central nervous system effects. Prolonged over-exposure may cause kidney effects. Repeated or prolonged eye contact may lead to conjunctivitis. Persons with preexisting eye or skin conditions, or impaired pulmonary, liver or kidney function may be more susceptible to the effects of this chemical [058].

1,1,2-Trichloroethane

CAS NUMBER: 79-00-5

SYNONYMS: Ethane trichloride beta-Trichloroethane Vinyl trichloride

CHEMICAL FORMULA: C2H3Cl3

MOLECULAR WEIGHT: 133.42

WLN: GYG1G

PHYSICAL DESCRIPTION: Clear colorless liquid

SPECIFIC GRAVITY: 1.4416 @ 20/4 C [031,042,205]

DENSITY: 1.4411 g/mL @ 20 C [395]

MELTING POINT: -37 C [102,269,274,395]

BOILING POINT: 113 C [102,346]

SOLUBILITY: Water: 1-5 mg/mL @ 20 C [700] DMSO: >=100 mg/mL @ 20 C [700] 95% Ethanol: >=100 mg/mL @ 20 C [700] Acetone: >=100 mg/mL @ 20 C [700]

OTHER SOLVENTS: Ether: Soluble [018,047] Chloroform: Soluble [018,047] Ketones: Soluble [061] Esters: Soluble [061]

OTHER PHYSICAL DATA: Refractive index: 1.4711 @ 20 C Pleasant, sweet odor Boiling point: 9.5 C @ 10 mm Hg [018,047] 49.97 C @ 75.9 mm Hg [025]

HAP WEIGHTING FACTOR: 1 [713]

VOLATILITY:

Vapor pressure: 16.7 mm Hg @ 20 C [395] 20 mm Hg @ 21.6 C [038]

Vapor density: 4.63

FIRE HAZARD: 1,1,2-Trichloroethane is nonflammable [025, 031, 061, 102].

LEL: 6.0% [102] UEL: 15.5% [102]

REACTIVITY: 1,1,2-Trichloroethane is incompatible with strong oxidizers and strong bases [102, 269, 346]. It reacts violently with sodium, potassium, magnesium, and aluminum [346]. It will attack some forms of plastics, rubber and coatings [102].

STABILITY: 1,1,2-Trichloroethane is light and heat sensitive [269]. Solutions of it in water, DMSO, 95% ethanol or acetone should be stable for 24 hours under normal lab conditions [700].

USES: 1,1,2-Trichloroethane is used as a solvent for fats, waxes, natural resins, alkaloids and various other organic materials. It is also an intermediate in production of vinylidine chloride and Teflon tubing and it is a component of some adhesives.

COMMENTS: None

NIOSH Registry Number: KJ3150000

ACUTE/CHRONIC HAZARDS: 1,1,2-Trichloroethane may be toxic by ingestion or inhalation [269]. It is an irritant [102,269, 274] and it may be absorbed through the skin and in high concentrations, it is narcotic [102]. It is also a positive animal carcinogen [015]. When heated to decomposition it emits toxic fumes of carbon monoxide, carbon dioxide, hydrogen chloride gas, and phosgene gas [102].

SYMPTOMS: Symptoms of exposure may include irritation of the skin, eyes, nose, mucous membranes, and upper respiratory tract; eye damage; skin cracking and erythema; central nervous system depression; liver and kidney damage; narcosis; drowsiness; incoordination; unconsciousness; and death [102,269,301,346,395]. Other symptoms may include headache; tremor; dizziness; peripheral paresthesia; hypesthesia; or anesthesia [355].

Trichloroethylene

CAS NUMBER: 79-01-6

SYNONYMS: Acetylene trichloride Ethinyl trichloride Ethylene trichloride Trichloroethene 1,1,2-Trichloroethylene 1,2,2-Trichloroethylene

CHEMICAL FORMULA: C2HCl3

MOLECULAR WEIGHT: 131.40

WLN: GYGU1G

PHYSICAL DESCRIPTION: Clear, colorless or pale blue, mobile liquid

SPECIFIC GRAVITY: 1.4649 @ 20/4 C [031,043,205]

DENSITY: 1.460-1.466 g/mL @ 20 C [455]

MELTING POINT: -73 C [017,038,043,395]

BOILING POINT: 87 C [017,036,371,451]

SOLUBILITY: Water: <1 mg/mL @ 21 C [700] DMSO: >=100 mg/mL @ 22 C [700] 95% Ethanol: >=100 mg/mL @ 22 C [700] Acetone: >=100 mg/mL @ 22 C [700]

OTHER SOLVENTS: Alcohol: Miscible [031,051,205,455] Chloroform: Miscible [031,205,295,395] Ether: Miscible [031,205,295,395] Ethanol: Miscible [295,395] Vegetable oils: Miscible [395] Lipids: Highly soluble [421] Most organic solvents: Miscible [062,295]

OTHER PHYSICAL DATA: Specific gravity: 1.456-1.462 @ 25/25 C [062] 1.465 @ 20/20 C [058] Specific gravity: 1.45560 @ 25/4 C [430] 1.4904 @ 4/4 C [031] Specific gravity: 1.4695 @ 15/4 C [031] Boiling point: 25 C @ 73 mm Hg [047] 20 C @ 60 mm Hg [025,031] Boiling point: 67 C @ 400 mm Hg; 48 C @ 200 mm Hg; 31.4 C @ 100 mm Hg [031] Boiling point: -1 C @ 20 mm Hg; -12.4 C @ 10 mm Hg; -22.8 C @ 5 mm Hg [031] Vapor pressure: 94 mm Hg @ 30 C [395] 100 mm Hg @ 32 C [043,051] Vapor pressure: 200 mm Hg @ 48 C; 400 mm Hg @ 67 C; 760 mm Hg @ 86.7 C [038] Chloroform-like odor [031,371,451] Refractive index: 1.4735 @ 27 C [062] 1.4773 @ 20 C [017,047,395] Refractive index: 1.47914 @ 17 C; 1.45560 @ 25 C [031] Photo-reactive [062] Sweet, burning taste [295] pH of solutions: 6.7 to 7.5 [058] Odor threshold: 21.4 ppm [151] Evaporation rate (ether=1): 0.28 [058] UV max (in water): 400-350 nm, 325 nm, 300 nm, 278 nm (A = 0.06, 0.08, 0.10, 1.0) [275] Lambda (at vaporization point): <200 nm [395] Liquid surface tension: 29.3 dynes/cm @ 20 C [371] Liquid water interfacial tension: 34.5 dynes/cm @ 24 C [371] Latent heat of vaporization: 57.2 cal/g [371]

HAP WEIGHTING FACTOR: 1 [713]

VOLATILITY:

Vapor pressure: 60 mm Hg @ 20 C [038,055,301] 77 mm Hg @ 25 C [430]

Vapor density: 4.53 [031,043,173,421]

FIRE HAZARD: Trichloroethylene has a flash point of >93.3 C (>200 F) [700]. It is probably nonflammable [025, 031, 062, 173]. Fires involving this material may be controlled with a dry chemical, carbon dioxide or Halon extinguisher. A water spray may also be used [058,371].

The autoignition temperature of trichloroethylene is 410 C (770 F) [051, 371, 430, 451].

LEL: 12.5% [043,051,058,066] UEL: 90% [043,051,058,066]

REACTIVITY: Trichloroethylene decomposes with strong alkalies [036,043,066,395]. It reacts violently with anhydrous perchloric acid and with many metals, such as aluminum, barium, beryllium, lithium, magnesium and titanium [036,066]. It reacts explosively with dinitrogen tetraoxide and with liquid oxygen [036, 043,066,451]. Corrosion products from hydrolysis during large scale distillation have led to exothermic decomposition [036,066]. This compound decomposes with 1-chloro-2,3-epoxypropane, the mono- and di- 2,3-epoxypropyl ethers of 1,4-butanediol, and 2,2-bis(4(2',3'-epoxypropoxy)phenyl) in the presence of catalytic quantities of halide ions [043,066]. It reacts violently with ozone, potassium hydroxide, potassium nitrate, sodium and sodium hydroxide. It reacts with water under heat and pressure [043]. It is incompatible with oxidizers and reducers [269]. It reacts with sulfhydryl groups [395]. It can dissolve most fixed and volatile oils [031,051]. It is also incompatible with caustic soda and caustic potash [058].

STABILITY: Trichloroethylene is sensitive to exposure to light [031,269,295,395]. It is also sensitive to moisture [031,099,173]. It is heat-sensitive [031,043, 295,421]. It is subject to autooxidation [395]. NMR stability screening indicates that solutions of it in DMSO are stable for at least 24 hours [700].

USES: Trichloroethylene is used in dry cleaning operations, in metal degreasing, and as a solvent for fats, greases, waxes, cellulose ester, ethers, dyeing, oils and household cleaners for walls, clothing and rugs. It is also used as a refrigerant, as a heat exchange liquid, in organic synthesis, as a fumigant, as an inhalation analyzer or anesthetic, in cleaning and drying electronic parts, as a diluent in paints and adhesives, in textile processing, in aerospace operations (flushing liquid oxygen), as an industrial solvent in extraction processes, as an analgesic, as a chain terminator for PVC production and as an extractant in food processing (e.g. for decaffeinated coffee). It has been used in the treatment of trigeminal neuralgia, as a disinfectant and detergent for skin, minor wounds and surgical instruments, and as a chemical intermediate in the production of pesticides, gums, resins, tars, paints, varnishes and specific chemicals such as chloroacetic acid.

COMMENTS: Widespread use of trichloroethylene is declining due to its toxicity [025]. Its use as a solvent is not permitted in some states. The FDA has prohibited its used in food, drugs and cosmetics [062].

NIOSH Registry Number: KX4550000

ACUTE/CHRONIC HAZARDS: Trichloroethylene may be toxic by inhalation, ingestion and skin absorption [036, 058,062,269]. It is an irritant [058,151,269] and it can permeate intact human skin [151]. When heated to decomposition it emits irritating and highly toxic fumes of chlorine, carbon monoxide, carbon dioxide, hydrogen chloride gas and phosgene gas [043,269,346,451].

SYMPTOMS: Symptoms of exposure may include headache, dizziness, nausea, irritation of the eyes and unconsciousness [036,058,346]. Other symptoms may include narcosis, eye effects, somnolence, hallucinations, distorted perceptions, gastrointestinal changes, jaundice, anesthesia, drowsiness, damage to the liver and other organs, and death due to ventricular fibrillation resulting in cardiac failure [043]. It also can cause incoordination and impaired judgement [451]. Central nervous system depression and kidney damage may occur [058,151,430]. Prolonged contact may lead to dermatitis [058,269,295,301]. Exposure can cause vomiting, excitement, irregular pulse, ventricular arrhythmias, pulmonary edema, weight loss, anorexia, fatigue, visual impairment, painful joints and wheezing [301]. Exposure also causes skin irritation, mild euphoria, hypertension, intolerance to alcohol, increased cardiac output, hepatorenal failure, respiratory failure and coma [151]. This compound can also cause mental confusion and gastrointestinal disturbances. Prolonged skin exposure can result in defatting, producing a rough, chapped skin which may result in erythema and possibly secondary infection [430]. Eye effects may include trigeminal and oculomotor nerve paralysis, optic or retrobulbar necrosis, and optic atrophy [099]. Other symptoms may include irritation of the lungs and gastrointestinal tract, inebriation, diarrhea, collapse, amnesia, numbness, weakness of the extremities, psychosis, hemiparesis and impairment of neurological and psychological functions [395]. This chemical may cause burning and damage of the eyes, vertigo, tremors, irregular heartbeat, blurred vision and peripheral neuropathy [346]. It may also cause a burning sensation on the skin, tinnitus, ataxia, convulsions, conjunctivitis, irritation of the nose and throat, abdominal pain, loss of memory, intolerance to tobacco and depression [173]. Difficult breathing, attitude of irresponsibility and lacrimation may occur [371]. Tachypnea, skin vesication and paralysis of the fingers may also occur [421]. It can cause loss of equilibrium and discomfort and pain to the eyes [058]. Symptoms may include increased rate and decreased depth of respiration, apnea, slowed heartbeat, cardiotoxicity, blindness, eczema and burns [295]. Other symptoms include central nervous system damage, systemic damage, tachycardia, acute yellow atrophy of the liver, lightheadedness, lethargy and inability to concentrate [051]. If alcohol is consumed shortly before or after exposure, it may cause Degreaser's flush (in which the skin of the face and arms become extremely red [051,346,421,430].

2,4,5-Trichlorophenol

CAS NUMBER: 95-95-4

SYNONYMS: 2,4,5-T

CHEMICAL FORMULA: C6H3Cl3O

MOLECULAR WEIGHT: 197.46

WLN: QR BG DG EG

PHYSICAL DESCRIPTION: Colorless needles or gray flakes

SPECIFIC GRAVITY: 1.678 @ 25/4 C [042,062]

DENSITY: 1.5 g/mL @ 75 C [055]

MELTING POINT: 68 C [025,031,395]

BOILING POINT: 253 C @ 760 mm Hg [031,051,395]

SOLUBILITY: Water: <1 mg/mL @ 21 C [700] DMSO: >=100 mg/mL @ 19 C [700] 95% Ethanol: >=100 mg/mL @ 19 C [700] Acetone: >=100 mg/mL @ 19 C [700] Methanol: 615 g/100 g @ 25 C [031,051,395] Toluene: 122 g/100 g [031,051,395]

OTHER SOLVENTS: Soybean oil: 79 g/100 g @ 25 C [031,051,395] Ligroin: Soluble [017,047] Carbon tetrachloride: 51 g/100 g @ 25 C [031,051,395] Ether: 525 g/100 g @ 25 C [031,051,205,395] Benzene: 163 g/100 g @ 25 C [031,051,205,395] Liquid petrolatum: 56 g/100 g @ 50 C [031,051,395] Denatured alcohol formula 30: 525 g/100 g @ 25 C [031]

OTHER PHYSICAL DATA: Boiling point: 248 C @ 746 mm Hg (sublimes) [031,395] Vapor pressure: 10 mm Hg @ 117.3 C [038] 400 mm Hg @ 225 C [055] Phenolic odor Weak monobasic acid; K @ 25 C = 0.000000043

HAP WEIGHTING FACTOR: 1 [713]

VOLATILITY:

Vapor pressure: 1 mm Hg @ 72 C [038,051,395] 5 mm Hg @ 102.1 C [038]

Vapor density: >1

FIRE HAZARD: 2,4,5-Trichlorophenol is nonflammable [051,371].

LEL: Not available UEL: Not available

REACTIVITY: 2,4,5-Trichlorophenol produces dioxin in alkaline medium at high temperatures [346].

STABILITY: 2,4,5-Trichlorophenol is stable under normal laboratory conditions. Solutions of it in water, DMSO, 95% ethanol, or acetone should be stable for 24 hours under normal lab conditions [700].

USES: 2,4,5-Trichlorophenol is used in the synthesis of various herbicides and used to produce the defoliant 2,4,5-T. It is also used in cooling towers, paper and pulp mill systems, hide and leather processing and disinfection, adhesives, rubber additives and as a wood preservative. It is also used in the textile industry, household sickroom equipment, food processing plants and equipment, food contact surfaces, hospital rooms and bathrooms as a fungicide or bactericide.

COMMENTS: None

NIOSH Registry Number: SN1400000

ACUTE/CHRONIC HAZARDS: 2,4,5-Trichlorophenol may be very toxic, corrosive and a severe irritant. It may be harmful by all routes of exposure and may be absorbed through the skin. When heated to decomposition it emits toxic fumes of carbon monoxide, carbon dioxide and hydrogen chloride gas [043,269]. There is limited evidence that this compound is a human carcinogen [015,396].

SYMPTOMS: Symptoms of exposure may include irritation of the skin, eyes, nose and throat [036,051,371]. The dust may cause swelling and injury to the eyes [371]. Eye contact may also result in conjunctivitis and slight to moderate corneal injuries [051,151]. Other toxic effects may include decrease of activity, motor weakness and convulsive seizures [346]. It also causes lung, kidney and liver damage; an increase and then a decrease in respiratory rate, decrease in urine output, fever, increased bowel action and collapse [036].

Symptoms of exposure to a related compound may include softening and whitening of the skin followed by development of painful burns, headache, dizziness, rapid and difficult breathing and weakness. Ingestion of a related compound causes severe burns and internal damage. Chronic exposure causes digestive disturbances, nervous disorders, skin eruptions, dermatitis, liver damage and kidney damage [036].

2,4,6-Trichlorophenol

CAS NUMBER: 88-06-2

SYNONYMS: Phenachlor

CHEMICAL FORMULA: C6H3Cl3O

MOLECULAR WEIGHT: 197.45

WLN: QR BG DG FG

PHYSICAL DESCRIPTION: Yellow to pinkish-orange needles or solid

SPECIFIC GRAVITY: 1.675 @ 25/4 C [051,062,071]

DENSITY: Not available

MELTING POINT: 69.5 C [017,025,051,071]

BOILING POINT: 246 C [017,205,269,395]

SOLUBILITY: Water: <0.1 mg/mL @ 18 C [700] DMSO: >=100 mg/mL @ 23 C [700] 95% Ethanol: >=100 mg/mL @ 23 C [700] Acetone: >=100 mg/mL @ 23 C [700] Methanol: 525 g/100 g @ 25 C [031,051,071,395] Toluene: 100 g/100 g @ 25 C [031,051,071,395]

OTHER SOLVENTS: Carbon tetrachloride: 37 g/100 g @ 25 C [031,051,071,395] Turpentine: 37 g/100 g @ 25 C [031,395] Stoddard solvent: 16 g/100 g @ 25 C [031,395] Diacetone alcohol: 335 g/100 g @ 25 C [031,051,071,395] Denatured alcohol formula: 400 g/100 g @ 25 C [031,051,071,395] Pine oil: 163 g/100 g @ 25 C [031,395] Ether: 354 g/100 g [031,051,205,395] Benzene: 113 g/100 g [031,051,071,205] Alcohol: Soluble [043,062]

OTHER PHYSICAL DATA: Strong phenolic odor [031,043,051,062] Volatile with steam, but not from alkaline solution [031] pKa: 7.42 @ 25 C [025] Specific gravity:

1.4901 @ 75/4 C [017,047,205] UV max (in methanol): 296 nm (E=129); 289 nm (E=125) [051,071,395] Vapor pressure: 10 mm Hg @ 120.2 C [038]

HAP WEIGHTING FACTOR: 1 [713]

VOLATILITY:

Vapor pressure: 1 mm Hg @ 76.5 C [038,043,395] 5 mm Hg @ 105.9 C [038]

FIRE HAZARD: 2,4,6-Trichlorophenol is nonflammable [051, 062, 205, 269].

LEL: Not available UEL: Not available

REACTIVITY: 2,4,6-Trichlorophenol is incompatible with acid chlorides, acid anhydrides and oxidizing agents [269]. It can be converted to the sodium salt by reaction with sodium carbonate. It forms ethers, esters and salts by reaction with metals and amines. It undergoes substitution reactions such as nitration, alkylation, acetylation and halogenation. It can also be hydrolyzed by reaction with bases at elevated temperatures and pressures [395]. It reacts with alkalies at high temperatures [346].

STABILITY: 2,4,6-Trichlorophenol is stable up to its melting point [071,395]. Solutions of it in water, DMSO, 95% ethanol or acetone should be stable for 24 hours under normal lab conditions [700].

USES: 2,4,6-Trichlorophenol is used as a fungicide, bactericide, germicide, wood preservative, herbicide, anti-mildew agent for textiles, defoliant, glue preservative, insecticide and antiseptics and in pesticide manufacture.

COMMENTS: The technical grade is 97% pure. It is sometimes contaminated with 1,3,6,8-Tetrachlorodibenzo-p-dioxin, 2,3,7-Trichlorodibenzo-p-dioxin, tri-, tetra-, and penta- chlorodimethoxydibenzofurans, and tetra-, hexa-, penta-, and hepta-chlorodibenzofurans [051, 071].

NIOSH Registry Number: SN1575000

ACUTE/CHRONIC HAZARDS: 2,4,6-Trichlorophenol may be very toxic, corrosive and a severe irritant. It may be harmful by all routes of exposure and may be absorbed through the skin. When heated to decomposition it emits toxic fumes of carbon monoxide, carbon dioxide and hydrogen chloride gas [043,269].

SYMPTOMS: Symptoms of exposure may include irritation of the skin, eyes, nose, throat, mucous membranes and upper respiratory tract [036,051,071, 269]. Contact with the skin may result in redness, edema, dermatitis and chemical burns [071,151]. Contact with the eyes may result in corneal injury and iritis [071,151]. It may also cause lacrimation [099]. Other symptoms may include an increase followed by a decrease in respiratory rate and urinary output, fever, increased bowel action, weakness of movement, collapse, convulsions, lung damage, liver damage and kidney damage [036]. Additional symptoms caused by this type of compound may include painless blanching or erythema of the skin, corrosion, profuse sweating, intense thirst, nausea and vomiting, diarrhea, cyanosis from methemoglobinemia, hyperactivity, hyperthermia, skin rashes (sometimes chloracne), neurological and immunological effects, stupor, blood pressure fall, hyperpnea, abdominal pain, hemolysis, coma, and pulmonary edema followed by pneumonia. Rapid death has been reported. If death from respiratory failure is not immediate, jaundice and oliguria or anuria may occur. Skin sensitivity reactions occur occasionally [395]. Prolonged eye contact may cause eye damage. Depending on the intensity and duration of exposure, effects may include severe destruction of tissue [269].

Triethylamine

CAS NUMBER: 121-44-8

SYNONYMS: N,N-Diethylethanamine

CHEMICAL FORMULA: $C_6H_{15}N$

MOLECULAR WEIGHT: 101.22

WLN: 2N2&2

PHYSICAL DESCRIPTION: Colorless liquid

SPECIFIC GRAVITY: 0.7275 @ 20/4 C

DENSITY: Not available

MELTING POINT: -114.7 C

BOILING POINT: 89.3 C

SOLUBILITY: Water: Soluble DMSO: Not available 95% Ethanol: Soluble Acetone: Very Soluble

OTHER SOLVENTS: Chloroform: Very Soluble Ether: Soluble Benzene: Very Soluble

OTHER PHYSICAL DATA: Refractive index is 1.4010 @ 20 C pKa is 11.01 @ 18 C in water

HAP WEIGHTING FACTOR: 1 [713]

VOLATILITY:

Vapor pressure: 53.5 mm Hg

Vapor density: 3.48

FIRE HAZARD: The flash point of triethylamine is -6.67 C (20 F) and it is flammable. Fires involving this compound may be controlled using dry chemicals, carbon dioxide, or Halon extinguishers.

The autoignition temperature of triethylamine is 450 C (842 F).

LEL: Not available UEL: Not available

REACTIVITY: Triethylamine can react with many oxidizers.

STABILITY: Triethylamine is stable under normal laboratory conditions.

USES: Triethylamine is used in the preparation of quarternary ammonium compounds, as an accelerator activator for rubber, for curing and hardening of polymers, as a corrosion inhibitor, and as a propellant.

COMMENTS: None

NIOSH Registry Number: YE0175000

ACUTE/CHRONIC HAZARDS: Triethylamine may be toxic, a severe irritant, and a lachrymator.

SYMPTOMS: Symptoms of exposure may include irritation of eyes and mucous membranes. Eye irritation and corneal edema have been reported from industria exposure. It also causes pulmonary irritation and injury to the heart, liver and kidneys in experimental animals

---

Trifluralin

CAS NUMBER: 1582-09-8

SYNONYMS: 2,6-Dinitro-N,N-dipropyl-4-(trifluoromethyl)benzenamine

CHEMICAL FORMULA: C13H16F3N3O4

MOLECULAR WEIGHT: 335.32

WLN: FXFFR CNW ENW DN3&3

PHYSICAL DESCRIPTION: Yellowish-orange solid

SPECIFIC GRAVITY: Not available

DENSITY: 1.294 g/mL @ 25 C

MELTING POINT: 46-47 C

BOILING POINT: Decomposes

SOLUBILITY: Water: <0.1 mg/mL @ 22.5 C [700] DMSO: >=100 mg/mL @ 18 C [700] 95% Ethanol: 10-50 mg/mL @ 18 C [700] Acetone: >=100 mg/mL @ 18 C [700]

OTHER SOLVENTS: Xylene: Very soluble Stoddard solvent: Very soluble

OTHER PHYSICAL DATA: Heat of combustion: -5020 cal/g Boiling point: 139-140 C @ 4.2 mm Hg

HAP WEIGHTING FACTOR: 1 [713]

VOLATILITY:

Vapor pressure: 0.000199 mm Hg @ 29.5 C [055]

FIRE HAZARD: Trifluralin has a flash point of >85 C (>185 F) [355,371]. It is combustible. Fires involving this material may be controlled with a dry chemical, carbon dioxide or Halon extinguisher. A water spray may also be used [371].

LEL: Not available UEL: Not available

REACTIVITY: Not available

STABILITY: Trifluralin is sensitive to exposure to light [168]. Solutions of it in water, DMSO, 95% ethanol or acetone should be stable for 24 hours when protected from light [700].

USES: Trifluralin is used as a pre-emergence herbicide, especially for cotton plants.

COMMENTS: None

NIOSH Registry Number: XU9275000

ACUTE/CHRONIC HAZARDS: Trifluralin may be toxic and an irritant. It is also an experimental carcinogen. When heated to decomposition it emits toxic fumes of hydrogen fluoride and nitrogen oxides. It can also be irritating to the skin and eyes.

SYMPTOMS: Symptoms of exposure may include irritation of the skin, eyes, gastrointestinal tract, and respiratory tract, convulsions, and coma.

2,2,4-Trimethylpentane

CAS NUMBER: 540-84-1

SYNONYMS: Isobutyltrimethylethane Isooctane (DOT) [703]

CHEMICAL FORMULA: C_8H_{18}

MOLECULAR WEIGHT: 114.26 [703]

PHYSICAL DESCRIPTION: Clear liquid; odor of gasoline [703]

DENSITY: 0.692 @ 20C/4C [703]

MELTING POINT: -107 C [702]

BOILING POINT: 99.2 C [703]

SOLUBILITY: Water: Insoluble [702]

OTHER PHYSICAL DATA: Freezing Point: -116 C [703] Vapor Pressure: 40.6 mm @ 21 C [703] Vapor Density: 3.93 [703]

HAP WEIGHTING FACTOR: 1 [713]

VOLATILITY:

Vapor Pressure: 40.6 mm @ 21 C [703]

Vapor Density: 3.93 [703]

FIRE HAZARD: Flash Point: 10 F [703]

Highly flammable; a very dangerous fire hazard when exposed to heat, flame, oxidizers [702, 703]. To fight fire, use CO2, dry chemical. [703]

Autoignition temperature: 779 F [703]

LEL: 1.1% UEL: 6.0

%REACTIVITY: Can react vigorously with reducing materials [702, 703].

STABILITY: Explosive in the form of vapor when exposed to heat or flame. When heated to decomposition it emits acrid smoke and irritating fumes. [703]

USES: In determining octane numbers of fuels; in spectrophotometric analysis; as solvent and thinner [715].

COMMENTS: None

NIOSH Registry Number: SA3320000

ACUTE/CHRONIC HAZARDS: High concentrations of 2,2,4-Trimethylpentane can cause narcosis. When heated to decomposition it emits acrid smoke and irritating fumes. [703]

SYMPTOMS: Not available

Vinyl acetate

CAS NUMBER: 108-05-4

SYNONYMS: Acetic acid, vinyl ester

CHEMICAL FORMULA: C4H6O2

MOLECULAR WEIGHT: 86.09

WLN: 1VO1U1

PHYSICAL DESCRIPTION: Colorless mobile liquid

SPECIFIC GRAVITY: 0.9317 @ 20/4

DENSITY: Not available

MELTING POINT: -93.2 C

BOILING POINT: 72.2 -2.3 C @ 760 mm/hg

SOLUBILITY: Water: Insoluble DMSO: Not available 95% Ethanol: Not available Acetone: soluble

OTHER SOLVENTS: Alcohol: Miscible Chloroform: Soluble Most organic solvents: Soluble Carbon tetrachloride: Soluble Ether: Soluble Benzene: Soluble

OTHER PHYSICAL DATA: Refractive index: 1.3959 @ 20 C

HAP WEIGHTING FACTOR: 1 [713]

VOLATILITY:

Vapor pressure: 100 mm/hg at 21.5 C

Vapor density: 3

FIRE HAZARD: Fires involving vinyl acetate should be extinguished with foam, CO2 and/or dry chemicals.

The autoignition temperature of vinyl acetate is 426 C (799 F).

LEL: 2.5% UEL: 13.4

%REACTIVITY: Vinyl acetate can react with oxidizing materials and can also react violently with 2-aminoethanol, chlorosulfonic acid, ethylene diamine, ethylene imine, hydrochloric acid, oleum, peroxides, nitric acid, sulfuric acid and HF.

STABILITY: Vinyl acetate polymerizes in light to a colorless, transparent solid.

USES: Vinyl acetate is used in polymerized form for plastic masses, films, lacquers, latex, paints, paper coatings, adhesives and textile finishing.

COMMENTS: None

NIOSH Registry Number: AK0875000

ACUTE/CHRONIC HAZARDS: Vinyl acetate may be toxic and the vapors may be narcotic in high concentrations.

SYMPTOMS: Not available

Vinyl bromide

CAS NUMBER: 593-60-2

SYNONYMS: Bromoethene Bromoethylene

CHEMICAL FORMULA: C2H3Br

MOLECULAR WEIGHT: 106.96

WLN: E1U1

PHYSICAL DESCRIPTION: Colorless gas

SPECIFIC GRAVITY: 1.49 @ 20/4 C [025}

DENSITY: 1.4933 g/mL @ 20 C [047,395,421]

MELTING POINT: -139.5 C [016,047,205,395]

BOILING POINT: 15.8 C [016,025,205,395]

SOLUBILITY: Water: Insoluble [036,043,205,395] DMSO: Not available 95% Ethanol: Soluble [051,395,430] Acetone: Soluble [016,395,421,430]

OTHER SOLVENTS: Alcohol: Soluble [016,051,421] Benzene: Soluble [016,051,395,421] Ether: Soluble [016,051,395,421] Chloroform: Soluble [016,051,395,421]

OTHER PHYSICAL DATA: Boiling point: 16 C @ 175 mm Hg [275] Refractive index: 1.4410 @ 20 C [016,047,430] Pleasant odor [058] Specific volume: 0.218 m3/kg @ 1 atm and 21.1 C [058] Vapor pressure: 1 mm Hg @ -95.4 C, 5mm Hg @ -77.8 C, 10 mm Hg @ -68.8 C, 20 mm Hg @ -58.8 C, 40 mm Hg @ -48.1 C, 60 mm

Hg @ -41.2 C, 100 mm Hg @ -31.9 C, 200 mm Hg @ -17.2 C, 400 mm Hg @ -1.1 C [038]

HAP WEIGHTING FACTOR: 1 [713]

VOLATILITY:

Vapor pressure: 895 mm Hg @ 20 C [395]

Vapor density: 3.7 [395,451]

FIRE HAZARD: Vinyl bromide has a flash point of <-8 C (<18 F) [025,036,066] and it is flammable. Fires involving this material may be controlled with a dry chemical, carbon dioxide or Halon extinguisher. A water spray may also be used [036,043,051].

The autoignition temperature of vinyl bromide is 472 C (882 F) [051].

LEL: 9% [058,430,451] UEL: 15% [430,451]

REACTIVITY: Vinyl bromide reacts violently with oxidants [043,051,066,395]. It is incompatible with active metals and organometallics. Air, oxygen and peroxides should be avoided [058]. Vinyl bromide can react with copper and its alloys to form unstable chemical by-products [052]. It is an explosion hazard at moderate concentrations [051].

STABILITY: Vinyl bromide polymerizes rapidly in sunlight [025,051,066,395]. It is heat sensitive [058].

USES: Vinyl bromide is used as a flame-retarding agent for acrylic fibers and plastics and as an alkylating agent in organic synthesis. It is used in pharmaceuticals, fumigants, rubber substitutes, leather and fabricated metal products. It is also used in the preparation of plastics by polymerization and copolymerization, as an intermediate, for preparing films and for laminating fibers.

COMMENTS: None

NIOSH Registry Number: KU8400000

ACUTE/CHRONIC HAZARDS: Vinyl bromide may be toxic and an irritant. When heated to decomposition it emits toxic fumes of bromide [043].

SYMPTOMS: Symptoms of exposure may include dizziness and narcosis. The liquid irritates the eyes and may irritate the skin by its defatting action [036]. Other symptoms may include disorientation, skin burns and skin irritation [058]. Mucous membrane irritation, lung irritation, liver and kidney damage may also occur [301]. Exposure to this compound may cause sleepiness [346]. Other symptoms may include anesthesia and death [430].

Vinyl chloride

CAS NUMBER: 75-01-4

SYNONYMS: Chloroethylene Chloroethene

CHEMICAL FORMULA: C2H3Cl

MOLECULAR WEIGHT: 62.5

WLN: G1U1

PHYSICAL DESCRIPTION: Colorless gas

SPECIFIC GRAVITY: 0.9106 @ 20/4 C

DENSITY: Not available

MELTING POINT: -154 C

BOILING POINT: -13.9 C

SOLUBILITY: Water: Slightly soluble DMSO: Not available 95% Ethanol: Soluble Acetone: Not available

OTHER SOLVENTS: Ether: Very soluble

OTHER PHYSICAL DATA: Not available

HAP WEIGHTING FACTOR: 10 [713]

VOLATILITY: Not available

FIRE HAZARD: Vinyl chloride has a flash point of -110 F (oc). Fires involving vinyl chloride should be extinguished with carbon dioxide, foam and/or dry chemical.

LEL: Not available UEL: Not available

REACTIVITY: Vinyl chloride can react with oxidizing agents and many other organic compounds.

STABILITY: Vinyl chloride should be protected from light and heat.

USES: Vinyl chloride is used in the plastics industry, as refrigerant, and as an intermediate in organic synthesis.

COMMENTS: Vinyl chloride cam present a severe explosion risk at 30,000 ppm. Use en aerosol sprays prohibited. Large fires of this material are practically inextinguishable.

NIOSH Registry Number: KU9625000

ACUTE/CHRONIC HAZARDS: Vinyl chloride may be toxic and an irritant.

SYMPTOMS: Symptoms of exposure may include systemic disorder; inhalation of high concentration of gas may cause narcosis (paralysis) and may be fatal. Other symptoms may include conjunctivitis and corneal burns, dermatitis, headache, and dizziness.

Vinylidene chloride

CAS NUMBER: 75-35-4

SYNONYMS: 1,1-Dichloroethylene

CHEMICAL FORMULA: C2H2Cl2

MOLECULAR WEIGHT: 96.94

WLN: GYGU1

PHYSICAL DESCRIPTION: Clear colorless liquid

SPECIFIC GRAVITY: 1.218 @ 20/4 C [055,430]

DENSITY: Not available

MELTING POINT: -122.5 C [031,055,421,430]

BOILING POINT: 31.7 C @ 760 mm Hg [346,421,430]

SOLUBILITY: Water: 5-10 mg/mL @ 21 C [700] DMSO: >=100 mg/mL @ 21 C [700] 95% Ethanol: >=100 mg/mL @ 21 C [700] Acetone: >=100 mg/mL @ 21 C [700]

OTHER SOLVENTS: Chloroform: Soluble [017,205] Ether: Soluble [017,047,205] Benzene: Soluble [017,047,205,395] Most organic solvents: Soluble [031,430]

OTHER PHYSICAL DATA: This compound is inhibited Mild, sweet odor resembling chloroform [031,346,395,430] Refractive index: 1.4254 @ 20 C [275] Heat of polymerization: -185 cal/g [371] Reid vapor pressure: 18.3 psia [371] Odor detectable @ 500-1000 ppm [430]

HAP WEIGHTING FACTOR: 10 [713]

VOLATILITY:

Vapor pressure: 500 mm Hg @ 20 C [055,421] 591 mm Hg @ 25 C [395]

Vapor density: 3.25 [055,395]

FIRE HAZARD: Vinylidene chloride has a flash point of -10 C (14 F) [062,421] and it is flammable. Fires involving this compound may be controlled using a dry chemical, carbon dioxide or Halon extinguisher. The vapor is heavier than air and may travel a considerable distance to a source of ignition and flash back [371,451].

The autoignition temperature for vinylidene chloride is 570 C (1058 F) [043,451].

LEL: 7.3% [025,043,371,451] UEL: 16.0% [043,371,430,451]

REACTIVITY: Vinylidene chloride reacts with alcohols and halides [025] and it rapidly absorbs oxygen forming violently explosive peroxides [025,066,395]. It may react violently with chlorotrifluoroethylene [036]. It may also react with chlorosulfonic acid, oleum and nitric acid [451].

STABILITY: When stored between -40 C and 25 C in the absence of inhibitor and in the presence of air, it rapidly absorbs oxygen with formation of highly unstable compounds [066,430]. Light and water tend to promote self-polymerization [066]. Sunlight, air and heat can cause polymerization [371]. NMR stability screening indicates that solutions of it in DMSO are stable for at least 24 hours [700].

USES: Vinylidene chloride is used as an intermediate in the production of plastics, as a comonomer with vinyl chloride, acrylonitrile, acrylates, etc., to form various kinds of Saran, in adhesives, in lacquer resins, for concrete and mortar strengthening, in latexes used as barrier coatings and in reinforced polyesters, printing inks and composites for use in furniture and marble.

COMMENTS: Vinylidene chloride is often inhibited with 0.02% monomethyl ether of hydroquinone.

NIOSH Registry Number: KV9275000

ACUTE/CHRONIC HAZARDS: Vinylidine chloride is toxic, an irritant and a lachrymator, and it is narcotic in high concentrations. When heated to decomposition it emits toxic fumes of chlorides [043].

SYMPTOMS: This compound is irritating to the skin and mucous membranes [031,301, 371]. It is narcotic in high concentrations [031,301,371]. Central nervous system depression may also occur [430]. Symptoms of exposure may also include dizziness, difficult breathing and burning of the skin and eyes [371] and lacrimation [275].

Xylenes (isomers and mixture)

NOTE: This CAS number and the following data refer to a mixture of xylenes.

CAS NUMBER: 1330-20-7

SYNONYMS: Component 1 (83%): Xylenes Dimethylbenzene Xylol
Component 2 (17%): Ethyl benzene

CHEMICAL FORMULA: C8H10

MOLECULAR WEIGHT: 106.17

WLN: Variable

PHYSICAL DESCRIPTION: Colorless liquid

SPECIFIC GRAVITY: 0.860 @ 20/4 C [269,274]

DENSITY: Not available

MELTING POINT: Not available

BOILING POINT: 137-144 C [053,269,274]

SOLUBILITY: Water: <1 mg/mL @ 22 C [700] DMSO: >=100 mg/mL @ 22 C [700] 95% Ethanol: >=100 mg/mL @ 22 C [700] Acetone: >=100 mg/mL @ 22 C [700]

OTHER SOLVENTS: Absolute alcohol: Miscible [031] Ether: Soluble [062] Most organic solvents: Miscible [421]

OTHER PHYSICAL DATA: Refractive index: 1.4970 @ 20 C Aromatic odor

HAP WEIGHTING FACTOR: 1 [713]

VOLATILITY:

Vapor pressure: 6.72 mm Hg @ 21 C [042] 10 mm Hg @ 28 C [053]

Vapor density: 3.7

FIRE HAZARD: Xylene has a flash point of 25.6 C (78 F) [700] and it is flammable. Fires involving this material may be controlled with a dry chemical, carbon dioxide or Halon extinguisher.

The autoignition temperature for xylene is 466 C (870 F) [058].

LEL: 1.0 % [036,053] UEL: 7.0% [036,053]

REACTIVITY: Xylenes react with oxidizing agents [042,053,269]. The mixture will attack some forms of plastics, rubber and coatings. It readily dissolves fats, oils and waxes [430].

STABILITY: Xylene is stable under normal laboratory conditions. Solutions of it in water, DMSO, 95% ethanol or acetone should be stable for 24 hours under normal lab conditions [700].

USES: Xylenes is used as an aviation gasoline, in protective coatings, as a solvent for alkyd resins, lacquers, enamels, and rubber cements, in the synthesis of organic chemicals, in the manufacturing dyes, and as a sterilizing catgut. Xylenes also may be used with Canada balsam as oil-immersion in microscopy and cleaning agent in microscopic technique.

COMMENTS: None

NIOSH Registry Number: ZE2100000

ACUTE/CHRONIC HAZARDS: Xylenes are an irritant and may be toxic if ingested. When heated to decomposition, this mixture emits toxic fumes of carbon monoxide and carbon dioxide [269], and it can be narcotic in high concentrations [031,051,053,269].

SYMPTOMS: Symptoms of exposure to xylenes may include irritation of the eyes, nose and throat; drying and defatting of the skin which may lead to dermatitis; chemical pneumonitis, pulmonary edema, hemorrhage, central nervous system depression, dizziness, staggering, drowsiness, unconsciousness, anorexia, nausea, vomiting and abdominal pain [346]. It may be narcotic in high concentrations [031,051,053,269]. It may also cause headache, fatigue, lassitude, irritability and gastrointestinal disturbances [421]. It can also cause reversible eye damage, a burning sensation in the mucous membranes, salivation, bloody vomit, impaired motor coordination, slurred speech, ataxia, stupor, coma, tremors, shallow respiration, ventricular irregularities, paralysis and convulsions [301]. It causes eye irritation and foggy vision [099].

---

o-Xylene

CAS NUMBER: 95-47-6

SYNONYMS: o-Xylol 1,2-Dimethylbenzene

CHEMICAL FORMULA: C8H10

MOLECULAR WEIGHT: 106.17

WLN: 1R B1

PHYSICAL DESCRIPTION: Colorless liquid

SPECIFIC GRAVITY: 0.897

DENSITY: 0.880 g/mL @ 20 C

MELTING POINT: -25 to -23 C

BOILING POINT: 143-145 C

SOLUBILITY: Water: Insoluble (see data for Xylenes mixture) DMSO: Very soluble (see data for Xylenes mixture) 95% Ethanol: Very soluble (see data for Xylenes mixture) Acetone: Very soluble (see data for Xylenes mixture)

OTHER SOLVENTS: Petroleum ether: Very soluble Carbon tetrachloride: Very soluble Ether: Very soluble Benzene: Very soluble

OTHER PHYSICAL DATA: Refractive index: 1.5055 @ 20 C

HAP WEIGHTING FACTOR: 1 [713]

VOLATILITY:

Vapor pressure: 10 mm Hg @ 32.1 C

Vapor density: 3.66

FIRE HAZARD: The flash point of o-xylene is 17.1 C (63 F) and it is flammable. Fires involving this material should be controlled using a dry chemical, carbon dioxide or Halon extinguisher.

The autoignition temperature of o-xylene is 464 C (867 F).

LEL: Not available UEL: Not available

REACTIVITY: o-Xylene may react with oxidizing materials.

STABILITY: o-Xylene is stable under normal laboratory conditions.

USES: o-Xylene is used as a solvent, in the manufacture of dyes and other organics, in microscopy oil-immersions, as a microscope cleaning agent, a vitamin, in insecticide formulations, and in motor fuels. It is also used in pharmaceutical synthesis.

COMMENTS: None

NIOSH Registry Number: ZE2450000

ACUTE/CHRONIC HAZARDS: o-Xylene is an irritant and may be toxic if ingested. It may also be narcotic in high concentrations.

SYMPTOMS: Symptoms of exposure may include headache, dizziness, skin and eye irritation, severe coughing, respiratory distress, pulmonary edema, nausea, vomiting, cramps, coma and kidney and liver damage.

m-Xylene

CAS NUMBER: 108-38-3

SYNONYMS: 1,3-Dimethylbenzene m-Xylol

CHEMICAL FORMULA: C8H10

MOLECULAR WEIGHT: 106.18

WLN: 1R C

PHYSICAL DESCRIPTION: Colorless liquid

SPECIFIC GRAVITY: 0.8642 @ 20/4 C

DENSITY: 0.8684 g/mL @ 15 C

MELTING POINT: -47.87 C

BOILING POINT: 139.1 C

SOLUBILITY: Water: Insoluble (see data for Xylenes mixture) DMSO: Very soluble (see data for Xylenes mixture) 95% Ethanol: Very soluble (see data for Xylenes mixture) Acetone: Very soluble (see data for Xylenes mixture)

OTHER SOLVENTS: Petroleum ether: Very soluble Ether: Very soluble Benzene: Very soluble

OTHER PHYSICAL DATA: Refractive index: 1.4972 @ 10 C Boiling point: 28.1 C @ 10 mm Hg

HAP WEIGHTING FACTOR: 1 [713]

VOLATILITY:

Vapor pressure: 10 mm Hg @ 28.3 C

Vapor density: 3.66

FIRE HAZARD: The flash point of m-xylene is 29.4 C (85 F) and it is flammable. Fires involving this material should be controlled using a dry chemical, carbon dioxide or Halon extinguisher.

The autoignition temperature of m-xylene 528 C (982 F).

LEL: Not available UEL: Not available

REACTIVITY: m-Xylene may react with oxidizing materials.

STABILITY: m-Xylene is stable under normal laboratory conditions.

USES: m-Xylene is used as a solvent, in the manufacture of dyes and other organics, in microscopy oil-immersions, as a microscope cleaning agent, a vitamin, in insecticide formulations, and in motor fuels. It is also used in pharmaceutical synthesis.

COMMENTS: None

NIOSH Registry Number: ZE2275000

ACUTE/CHRONIC HAZARDS: m-Xylene is an irritant and may be toxic if ingested.

SYMPTOMS: Symptoms of exposure may include headache, dizziness, skin and eye irritation, severe coughing, respiratory distress, pulmonary edema, nausea, vomiting, cramps, coma and kidney and liver damage.

---

p-Xylene

CAS NUMBER: 106-42-3

SYNONYMS: 1,4-Dimethyl benzene

CHEMICAL FORMULA: C8H10

MOLECULAR WEIGHT: 106.17

WLN: IRD

PHYSICAL DESCRIPTION: Clear colorless liquid

SPECIFIC GRAVITY: 0.8611 @ 20/4 C

DENSITY: Not available

MELTING POINT: 13.3 C

BOILING POINT: 138.3 C

SOLUBILITY: Water: Insoluble (see data for Xylenes mixture) DMSO: Very soluble (see data for Xylenes mixture) 95% Ethanol: Very soluble (see data for Xylenes mixture) Acetone: Very soluble (see data for Xylenes mixture)

OTHER SOLVENTS: Ether: Very soluble Benzene: Very soluble

OTHER PHYSICAL DATA: Boiling point: 27.2 C @ 10 mm Hg Refractive index: 1.4958 @ 20 C

HAP WEIGHTING FACTOR: 1 [713]

VOLATILITY:

Vapor pressure is 10 mm Hg at 27.3 C

Vapor density is 3.66

FIRE HAZARD: The flash point of p-xylene is 27.2 C and it is flammable. Fires involving this compound may be controlled using dry chemical, carbon dioxide or Halon extinguishers.

The autoignition temperature of p-xylene is 525 C (986 F).

LEL: Not available UEL: Not available

REACTIVITY: p-Xylene may react with oxidizing materials.

STABILITY: p-Xylene is stable under normal laboratory conditions.

USES: p-Xylene is used as a solvent, in the manufacture of dyes and other organics, in microscopy oil-immersions, as a microscope cleaning agent, a vitamin, in insecticide formulations, and in motor fuels. It is also used in pharmaceutical synthesis.

COMMENTS: None

NIOSH Registry Number: ZE2625000

ACUTE/CHRONIC HAZARDS: p-Xylene is an irritant and may be toxic if ingested. It may also be narcotic in high concentrations.

SYMPTOMS: Symptoms of exposure may include headache, dizziness, skin and eye irritation, severe coughing, respiratory distress, pulmonary edema, nausea, vomiting, cramps, coma and kidney and liver damage.

Index

A

B

C

D

E

F

L

M

N

O

P

R

S

T

W